AF892945

Lecture Notes in Operations Research

Lecture Notes in Operations Research is an interdisciplinary book series which provides a platform for the cutting-edge research and developments in both operations research and operations management field. The purview of this series is global, encompassing all nations and areas of the world.

It comprises for instance, mathematical optimization, mathematical modeling, statistical analysis, queueing theory and other stochastic-process models, Markov decision processes, econometric methods, data envelopment analysis, decision analysis, supply chain management, transportation logistics, process design, operations strategy, facilities planning, production planning and inventory control.

LNOR publishes edited conference proceedings, contributed volumes that present firsthand information on the latest research results and pioneering innovations as well as new perspectives on classical fields. The target audience of LNOR consists of students, researchers as well as industry professionals.

More information about this series at https://link.springer.com/bookseries/16741

Norbert Trautmann · Mario Gnägi
Editors

Operations Research Proceedings 2021

Selected Papers of the International Conference of the Swiss, German and Austrian Operations Research Societies (SVOR/ASRO, GOR e.V., ÖGOR), University of Bern, Switzerland, August 31 – September 3, 2021

Springer

Editors
Norbert Trautmann
Department of Business Administration
University of Bern
Bern, Switzerland

Mario Gnägi
Department of Business Administration
University of Bern
Bern, Switzerland

ISSN 2731-040X ISSN 2731-0418 (electronic)
Lecture Notes in Operations Research
ISBN 978-3-031-08622-9 ISBN 978-3-031-08623-6 (eBook)
https://doi.org/10.1007/978-3-031-08623-6

This Springer imprint is published by the registered company Springer Nature Switzerland AG
The registered company address is: Gewerbestrasse 11, 6330 Cham, Switzerland

Preface

This book contains a selection of refereed short papers presented at the *International Conference on Operations Research (OR 2021)*, jointly organized by the Operations Research Societies of Switzerland (SVOR/ASRO), Germany (GOR e.V.), and Austria (ÖGOR). Amid the global coronavirus pandemic, the OR 2021 was hosted online by the University of Bern from August 31 to September 3, 2021. It was the first-ever virtual event in the OR conference series established in 1971.

Designed around its special theme *Business Analytics for Data-Driven Decision Making*, the OR 2021 offered an excellent platform for academics and practitioners to present their most recent research and discuss trends in the industry. We welcomed more than 500 participants from operations research, management science, mathematics, data science, and analytics. The scientific program included 360 live presentations via Zoom, 12 plenary and semi-plenary talks, and interactive sessions with editors from leading journals and business partners.

This volume presents a selection of 63 papers–including contributions from the winners of the GOR PhD Thesis Award and the GOR Master Thesis Award–that have been carefully reviewed and accepted for publication by the stream chairs and selected referees (acceptance rate of 80%). The topics of the chosen papers reflect the broad spectrum of the OR discipline, ranging from new mathematical algorithms to applications in areas such as energy and environment, healthcare management, finance, or mobility and traffic.

We would like to thank everyone who contributed to the success of the virtual OR 2021 conference: the members of the Program and Organizing Committee, the stream chairs, our incredible plenary and semi-plenary speakers, the generous sponsors and partners, the event staff from the University of Bern and, last but not least, the participants and presenters who joined us from all over the world. You all made this event extraordinary.

April 2022

Norbert Trautmann
Mario Gnägi

Organization

The Chair of Quantitative Methods in Business Administration of the University of Bern organized the OR 2021 conference in cooperation with the Operations Research Societies of Switzerland (SVOR/ASRO), Germany (GOR e.V.), and Austria (ÖGOR).

Organizing Committee

Norbert Trautmann (Chair, SVOR/ASRO)	University of Bern
Nina Ackermann	University of Bern
Mario Gnägi (SVOR/ASRO)	University of Bern
Nicklas Klein (SVOR/ASRO)	University of Bern
Nadine Saner	University of Bern

Program Committee

Norbert Trautmann (Chair, SVOR/ASRO)	University of Bern
Philipp Baumann (SVOR/ASRO)	University of Bern
Richard Hartl (ÖGOR)	University of Vienna
Natalia Kliewer (GOR e.V.)	Free University of Berlin
Raimund Kovacevic (ÖGOR)	TU Wien
Peter Letmathe (GOR e.V.)	RWTH Aachen University
Steffen Rebennack (GOR e.V.)	Karlsruhe Institute of Technology
Bernard Ries (SVOR/ASRO)	University of Fribourg
Angelika Wiegele (ÖGOR)	University of Klagenfurt

Stream Chairs

Klaus Altendorfer
Christiane Barz
Philipp Baumann
Michael Becker-Peth
Michael Breitner
Dirk Briskorn
Reinhard Bürgy
Karl Doerner
Jan Fabian Ehmke
Olivier Gallay
Margaretha Gansterer
Andrea Gaunersdorfer
Jutta Geldermann
Jochen Gönsch
Ralph Grothmann
Martin Grunow
Patrick Hirsch
Stefan Irnich
Paul Kaufmann
Hans Kellerer
Robert Klein
Natalia Kliewer
Max Klimm
Andreas Klinkert
Ulf Lorenz
Elmar Lukas
Virginie Lurkin
Russell McKenna
Maximilian Moll
Roberto Montemanni
Dominik Möst
Sven Müller
Sophie Parragh
Ulrich Pferschy
Marion Rauner
Steffen Rebennack
Bernard Ries
Andrea Emilio Rizzoli
Matteo Salani
Katja Schimmelpfeng
Michael Schneider
Jens Schulz
Thomas Setzer

Claudius Steinhardt
Christian Stummer
Norbert Trautmann
Tina Wakolbinger

Sponsoring Institutions

FICO
Gurobi
INFORM
Springer

Supporting Institutions

Association of European Operational Research Societies (EURO)
Bern Welcome
Swiss Academy of Engineering Sciences (SATW)
Swiss National Science Foundation (SNSF)
Zoom Video Communications

Contents

Analytics

Decision Analysis and Support

Discrete and Combinatorial Optimization

Supply Chain and Production Management

Systems Modeling and Simulation

Award Winners

Operations Management in the Sharing Economy: An Integrated Perspective of Item-Sharing and Crowdshipping

Moritz Behrend(✉)

Kiel University, 24098 Kiel, Germany
moritz.behrend@bwl.uni-kiel.de
https://www.scm.bwl.uni-kiel.de

Abstract. The uprising of the sharing economy yields a variety of new concepts on how members of a community can provide services to each other. Two of these concepts are item-sharing and crowdshipping. They have been implemented in practice in isolation so far but seem to complement each other well. We therefore propose the conceptual idea of an integrated platform on which user requests of both concepts are handled jointly. This paper revisits several studies that address the operations management of such a platform, including optimization models and methods for the corresponding decision making, and managerial insights derived from computational experiments. The results confirm that the two concepts are mutually beneficial and that a well-designed operations management allows to effectively serve demands announced on such a platform.

Keywords: Sharing economy · Item-sharing · Crowdshipping · Operations management

1 Motivation

The sharing economy subsumes business models that provide services to a community as a whole rather than to an individual [1]. These services can differ substantially in the needs they serve and their approach to do this, but it all comes down to the coordination of resources. While some concepts such as shared spaces (e.g. Airbnb) and shared mobility (e.g. Uber or ShareNow) are very successful and/or popular already nowadays, others are developing rather slowly. Two of these concepts are item-sharing and crowdshipping. In item-sharing, members grant others a temporary access to the items they own, such as tools or leisure equipment. In crowdshipping, private drivers offer to execute delivery jobs for other people on trips they would make anyway. These two concepts are so far dealt with in practice on separate platforms that operate in complete isolation. However, they could complement each other well when the available transportation capacity in crowdshipping is utilized to support the

N. Trautmann and M. Gnägi (Eds.): OR 2021, LNOR, pp. 3–8, 2022.
https://doi.org/10.1007/978-3-031-08623-6_1

cumbersome peer-to-peer item exchanges in item-sharing. After all, the need-based access to items in item-sharing necessitates frequent transports of small shipments between requesting consumers.

This paper sheds light on possible benefits of integrating item-sharing and crowdshipping on a single platform. To this end, we discuss the operations management of such a platform and we derive managerial insights from numerical experiments. The following sections briefly summarize the main contributions of the related studies [2–5] that I conducted in this context as part of my PhD thesis.

2 Integrating Item-Sharing and Crowdshipping

The potential of item-sharing and the idea of a conceptual integration with crowdshipping are formally analyzed for a first time in [2]. This paper employs mathematical optimization to support the operator of an integrated platform in assigning supplies to requests for item-sharing and, if applicable, also in assigning a crowdshipper the task to conduct the delivery. To this end, an assignment problem is formulated and two heuristics are proposed, one of which is based on a problem decomposition and the other is based on finding a feasible set of combinations within an incompatibility graph.

Computational experiments reveal that item-sharing exhibits strong economies of density. The integration with crowdshipping increases both profitability and service level of a platform. To achieve high service quality, two alternative ways are considered of how crowdshipping can facilitate item exchanges between consumers. The inclusion of *home deliveries*, where a crowdshipper delivers a package directly to a consumer's home, has a stronger effect on profitability, particularly because of the extra-service consumers may be willing to pay for. In *neighborhood deliveries*, crowdshippers and consumers collaboratively conduct the transport, which allows for further item exchanges. A sensitivity analysis shows that home deliveries become less attractive if higher compensations must be paid to crowdshippers, see Fig. 1. Furthermore, a platform's operational performance substantially depends on the willingness of consumers and crowdshippers to invest time in item transportation.

3 How to Harness Crowdshippers' Full Transportation Capacity

The benefit of an integrated platform was investigated in [2] based on scenarios where crowdshippers accept to conduct at most one delivery per trip. However, crowdshippers may also accept to conduct multiple deliveries if they want to get more involved in crowdshipping. The goal of [3] is to quantify the potential of capacitated crowdshipping and to derive managerial insights on how a platform shall incentivize crowdshippers to improve operational performance.

Crowdshipping with more than one delivery per trip requires a routing of crowdshippers on their way from their origin to their destination via intermediate

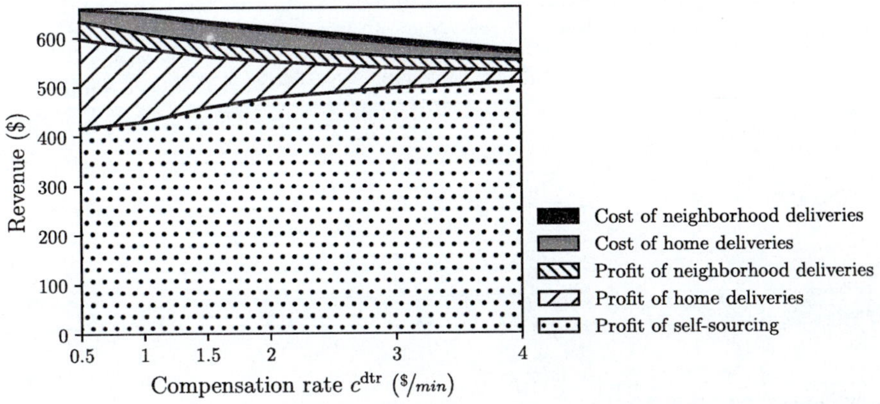

Fig. 1. The integration with crowdshipping allows for more profitable matchings [2].

pick-up and drop-off locations. Such a *detour routing* determines the travel time increase due to crowdshipping, which allows to respect crowdshippers' preferences when assigning delivery jobs to trips. This is not possible through simply generalizing the solution methods of [2] regarding the number of jobs that can be assigned to a trip but rather requires dedicated solution methods. Therefore, [3] proposes a new exact solution method in which first a label setting algorithm determines feasible crowdshipping routes and then a set packing problem selects a combination of those routes that maximizes profit. An excerpt of the algorithm's mechanism is depicted in Fig. 2. The final leg to the destination $d(k_1)$ is always implied. Items that are taken home ($\vartheta_l > 0$) can be assigned to consumers j_1 or j_2 through a neighborhood delivery. For more details we refer to [3].

The proposed method is much faster for the case of single-capacitated crowdshipping than the assignment problem formulations of [2]. Furthermore, if crowdshippers accept to conduct multiple deliveries, higher profits are possible. However, the magnitude of this increase is strongly linked to crowdshipper's willingness to extent their trips since the execution of more deliveries also requires more detouring. As crowdshippers accept longer detours and offer to conduct multiple deliveries, the planning complexity increases substantially. [3] therefore also shows how to turn the label setting procedure into a heuristic.

4 Conceptual Ideas for More Efficient Sharing Practices

Item-sharing is an inherently multi-periodic process as the same items are successively used by multiple consumers in the course of time. [4] therefore introduces the multi-period variant of the integrated item-sharing and crowdshipping problem. In this context, it proposes the concept of *request chaining* in which items are directly forwarded from one consumer to the next without returning the items to intermediate storage locations in-between two requests. This interpretation of item-sharing results in interdependencies between periods and it raises

	Label propagation	New label	Route	ϑ_l	Satisfied requests per column(s)
a)	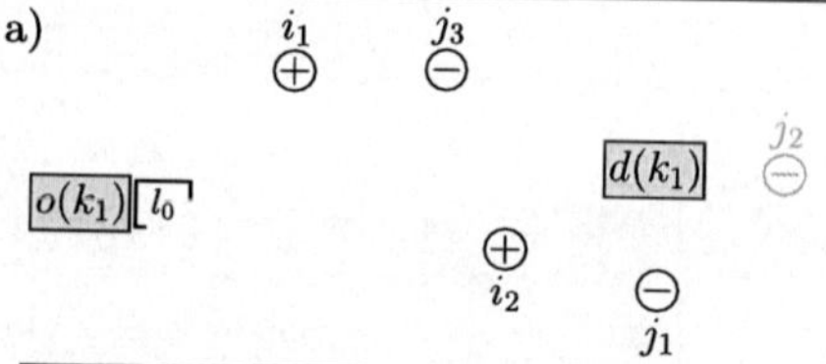	l_0	$o(k_1)$-$d(k_1)$	0	$\emptyset$
b)	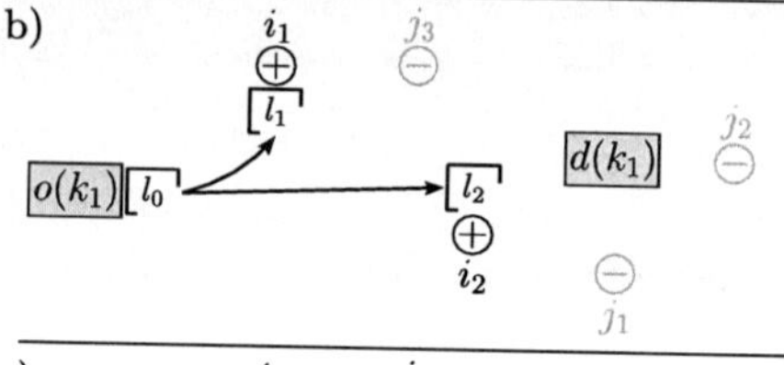	l_1 l_2	$o(k_1)$-i_1-$d(k_1)$ $o(k_1)$-i_2-$d(k_1)$	1 1	$\{j_1\}$, $\{j_2\}$ $\{j_1\}$, $\{j_2\}$
c)	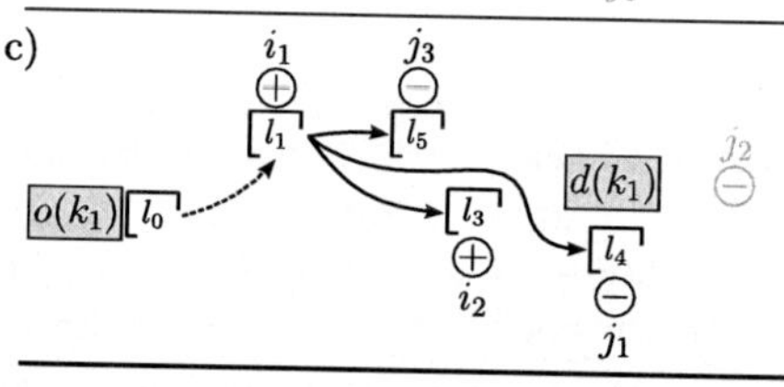	l_3 l_4 l_5	$o(k_1)$-i_1-i_2-$d(k_1)$ $o(k_1)$-i_1-j_1-$d(k_1)$ $o(k_1)$-i_1-j_3-$d(k_1)$	2 0 0	$\{j_1, j_2\}$ $\{j_1\}$ $\{j_3\}$

Fig. 2. An example for propagating labels [3].

the question to what extent a more foresighted planning outperforms previously developed solution methods for a period-to-period decision making.

For this purpose, a binary program is proposed that supports the operational planning of a platform operator in 'routing' items through a network of consumer locations over time. It decides for every leg of an item's route which request to serve next and to whom to assign the responsibility of transportation. As the considered time horizon increases in size, not all requests and crowdshipping trips may be available at each planning point in time. To deal with these dynamics, the binary program is also embedded into a rolling horizon framework.

Computational experiments show that a platform with request chaining can substantially increase profits if the operational planning for the upcoming period also considers information of subsequent periods. This is because a more foresighted long-term planning makes it less likely that items are assigned to remote request locations where they get stuck. Assignments like these occur frequently in a period-to-period planning for scenarios with only a few consumers and limited options to exchange items between them so that a multi-period planning is particularly beneficial in such cases. The planning complexity increases quickly as the considered time horizon is extended. In a dynamic setting with incomplete information, critical parameters are the lead-time of request and trip announcements prior to their realization and the number of periods consumers and crowdshippers shall be notified in advance.

5 A Sustainability Comparison Between Buying and Renting

Item-sharing may turn out as an environmental friendly form of consumption as only a few, highly utilized items are required to serve overall customer demand. However, a shared use of items also necessitates frequent transports between consumers, which may counterbalance the positive environmental benefit of producing fewer items. Against this background, [5] compares the environmental impact of sales-based and sharing-based business models. The sustainability comparison is based on carbon dioxide equivalents (CO_2e) that are released during the production of items and their repeated transportation between consumers.

The resulting eco-assessment of item production and transportation is depicted in Fig. 3. The compared variants are the traditional selling of items to consumers from stores (*buy*); station-based sharing (*sb*) and station-based sharing with request pooling (*sbp*), where the latter allows for a better resource management in times of high demand; and the free-floating sharing concept of [4], once with crowdshipping (ffc) and once without (ff). The set $\mathcal{W}$ contains the fixed locations for the variants *buy*, *sb*, and *sbp* at which the items are supplied, e.g. stores or sharing stations. The results clearly show that total emissions in ownership are much more sensitive to the life cycle emissions of items (e^{item}) than those in the sharing-based business models, due to the need for an additional $n = 1000 - 75 = 925$ items. At the same time, travel-related emissions are higher when items are shared (see total emissions for the variants at $e^{\text{item}} = 0$ kg CO_2e). The intersection A marks the point where the additional

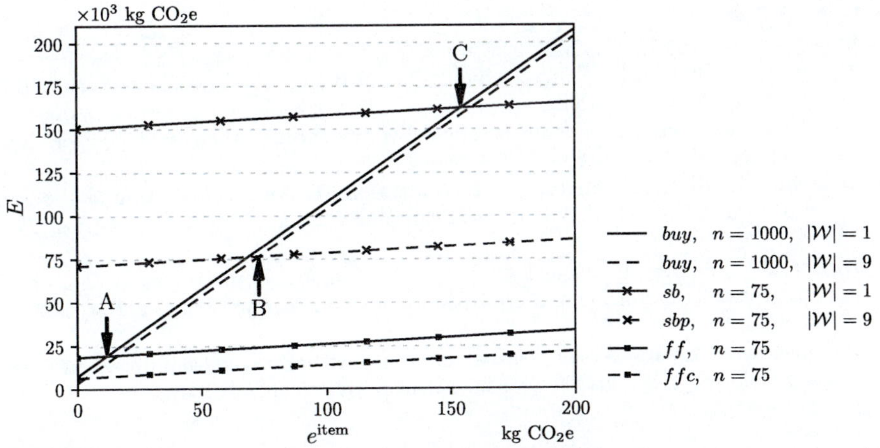

Fig. 3. Total emissions per business model subject to the carbon footprint of items [5].

traveling of 72,700 − 29,600 = 43,100 km in free-floating sharing (ff) is equivalent to producing and selling 925 additional items with a carbon footprint of $e^{item} = 11.83$ kg CO_2e.

The combination of efficient direct item exchanges and the demand for only a few items to serve requests in the free-floating sharing variants results in an overall low environmental impact. They are therefore considered a very promising approach to reduce consumption-based emissions for various item types with a broad spectrum of carbon footprints.

6 Conclusion and Outlook

A major implication of this work for future research is that the operations management of item-sharing platforms constitutes a research domain on its own, next to the well-studied operations management of other sharing economy concepts such as car sharing, bike sharing, or ride-sharing. This work lays the foundation for further analyses by providing benchmark results in the dimensions profit, service level, and environmental impact. The results show that an effective operations management of item-sharing platforms shows great potential for serving consumer requests with shared items, particularly in densely populated areas. The results further demonstrate that crowdshipping is a promising logistics concept to facilitate item exchanges.

Future research can particularly extend the free-floating concept as this one raises many further open questions, both from an operations management perspective and from other disciplines such as behavioral research. Eventually, an upscaling of item-sharing leads to a broad and interesting research area in which not only platform operators but also other parties such as manufacturers of items are affected.

References

1. Botsman, R., Rogers, R.: What's Mine Is Yours. Harper Business, New York (2011)
2. Behrend, M., Meisel, F.: The integration of item-sharing and crowdshipping: can collaborative consumption be pushed by delivering through the crowd? Transp. Res. Part B **111**, 227–243 (2018)
3. Behrend, M., Meisel, F., Fagerholt, K., Andersson, H.: An exact solution method for the capacitated item-sharing and crowdshipping problem. Eur. J. Oper. Res. **279**, 589–604 (2019)
4. Behrend, M., Meisel, F., Fagerholt, K., Andersson, H.: A multi-period analysis of the integrated item-sharing and crowdshipping problem. Eur. J. Oper. Res. **292**(2), 483–499 (2021)
5. Behrend, M.: Buying versus renting: on the environmental friendliness of item-sharing. Transp. Res. Part D: Transp. Environ. **87**, 102407 (2020)

Scheduling and Packing Under Uncertainty

Franziska Eberle(✉)

London School of Economics and Political Science, London, UK
franziska.eberle@posteo.de

Abstract. Uncertainty in the input parameters is a major hurdle when trying to directly apply classical results from combinatorial optimization to real-word challenges. Hence, designing algorithms that handle incomplete knowledge provably well becomes a necessity. In view of the above, the author's thesis [5] focuses on scheduling and packing problems under three models of uncertainty: stochastic, online, and dynamic. For this report, we highlight the results in online throughput maximization as well as dynamic multiple knapsack.

Keywords: Scheduling · Packing · Approximation algorithms · Online algorithms · Dynamic algorithms · Uncertainty

1 Motivation

Incomplete information, e.g., frequently changing or a priori unknown problem parameters, is a major challenge when translating combinatorial optimization results to recommendations for real-world applications since. A particular solution may perform well on some specific input data or estimation thereof, but once the data is slightly perturbed or new tasks need to be performed, the solution may become arbitrarily bad or even infeasible. Thus, either solving the problem under uncertainty or efficiently updating the solution becomes a necessity. The author's thesis explores several models for uncertainty in two fundamental fields of combinatorial optimization: scheduling and packing.

Scheduling problems arise whenever scarce resources have to complete a set of tasks while optimizing some objective. Possible applications range from the industrial sector with production planning via the service sector with delivery tasks to the information sector with data processing and cloud computing. Given the undeniable effects of climate change we are facing today, efficient solutions to any of these problems are paramount. Efficiency might refer to the minimal necessary duration for which the system is running to process all tasks or the maximal achievable throughput in a given time interval.

Packing problems typically appear whenever items have to be assigned to resources with capacities. The most obvious applications are transportation processes, e.g., loading of trucks, that take place at every stage of the manufacturing process until the delivery to the client. Further, packing problems can also be

N. Trautmann and M. Gnägi (Eds.): OR 2021, LNOR, pp. 9–14, 2022.
https://doi.org/10.1007/978-3-031-08623-6_2

found in computing clusters and financial applications: Capacities have to be obeyed and costs have to be minimized when assigning virtual machines to real servers in computing clusters. Or, when making new investment decisions, certain aspects, such as risk or overall volume, have to be bounded while maximizing profit. These examples are all packing problems as they require compliance with capacity constraints while maximizing the total value of packed items.

Incomplete information may be caused by various reasons, such as the unpredictable arrival of new tasks or having only estimates of input parameters at hand. In any of these cases, we are interested in finding provably good solutions in reasonable time, i.e., we would like to design algorithms that deal with incomplete information while performing sufficiently well. This article focuses on two models of uncertainty in the input: *online*, where part of the input is revealed only incrementally, and *dynamic*, where the input is subject to small changes.

Our Contribution. In Sect. 2, we investigate the problem of online throughput maximization with and without commitment requirements. This chapter is based on two publications, [4] and [6]. In Sect. 3, we investigate dynamic multiple knapsack problems; this is based on an unpublished article [2].

2 Online Throughput Maximization

We investigate a scheduling problem where jobs with processing times p_j arrive online over time at their release dates r_j and have to be processed by m identical parallel machines before their respective deadlines d_j. The goal is to preemptively maximize the throughput, i.e., the number of jobs that complete before their deadlines. Since the algorithm has to deal with incomplete information, we use standard competitive analysis to evaluate our algorithms, where we bound the ratio between the solution found by an online algorithm and the offline optimum.

Hard instances for online algorithms involve "tight" jobs, i.e., jobs that need to be scheduled immediately and without interruption in order to complete on time [1]. To circumvent these difficulties, we enforce some relative slack for each job in the interval defined by its release date and its deadline. We assume that each job's interval has length at least $(1+\varepsilon)p_j$ for some slackness parameter $\varepsilon > 0$, which is given to the online algorithm. By setting $\varepsilon = 1$, we obtain constant competitive ratios, and thus, we additionally assume $\varepsilon \leq 1$.

2.1 Non-committed Scheduling

To gain some intuition for our algorithm, we describe the underlying design principles: The threshold algorithm never migrates jobs between machines. Moreover, a running job j can only be preempted by significantly smaller jobs $j^\star$, i.e., $p_{j^\star} < \frac{\varepsilon}{4} p_j$, and j cannot start when its remaining slack is less than $\frac{\varepsilon}{2} p_j$.

The intelligence of the threshold algorithm lies in how it admits jobs. The actual scheduling decision then is simple and independent of the admission of jobs: at any point in time and on each machine, schedule the shortest job that has been

Algorithm 1.1. Threshold algorithm

Scheduling routine: At any time τ and on any machine i, run the job with shortest processing time that has been admitted to i and has not yet completed.

Event: Upon release of a new job at time τ:
  Call threshold preemption routine.

Event: Upon completion of a job j at time τ:
  Call threshold preemption routine.

```
Threshold preemption routine:
j* ← a shortest available job at τ, i.e.,
    j* ∈ arg min{p_j | j ∈ J, r_j ≤ τ and d_j − τ ≥ (1 + ε/2)p_j}
i ← 1
while j* is not admitted and i ≤ m do
    j ← job processed on machine i at time τ
    if j = ∅ do
        admit job j* to machine i
        call threshold preemption routine
    else-if p_{j*} < (ε/4)p_j do
        admit job j* to machine i
        call threshold preemption routine
    else
        i ← i + 1
```

admitted to this machine and has not completed its processing time. A formal description is given in Algorithm 1.1. In [4], we show the following theorem.

Theorem 1. *Let* $0 < \varepsilon \leq 1$. *The threshold algorithm is* $\Theta(\frac{1}{\varepsilon})$*-competitive for online throughput maximization.*

This result is in a sense best possible as we give a matching lower bound [4].

Theorem 2. *Every deterministic online algorithm has a competitive ratio* $\Omega(\frac{1}{\varepsilon})$.

2.2 Scheduling with Commitment

We now focus on the question how to handle *commitment* in online throughput maximization. Modeling commitment addresses the issue that a high-throughput schedule may abort jobs close to their deadlines in favor of many shorter and more urgent tasks, which may not be acceptable for the job owner. We distinguish three different models for scheduling with commitment: (i) *commitment upon job arrival*, (ii) *commitment upon job admission*, and (iii) δ*-commitment.* In the first, most restrictive model, an algorithm must decide immediately at a job's release date if the job will be completed or not. In the second model, an algorithm may discard a job any time before its start, its admission. In the third model, an online algorithm must commit to complete a job when its slack has reduced from the original slack requirement of at least εp_j to δp_j for $0 < \delta < \varepsilon$, i.e., at the latest at $d_j - (1 + \delta)p_j$.

For commitment upon arrival, we give the following strong lower bound, which even holds for randomized algorithms [4].

Theorem 3. *No randomized online algorithm has a bounded competitive ratio for commitment upon arrival.*

For the remaining two commitment models, we design an online algorithm that matches the lower bound of Theorem 2 and show the following theorem. The details of the algorithm and its analysis can be found in [6].

Theorem 4. *Consider throughput maximization on parallel identical machines with or without migration. There is an $\mathcal{O}(\frac{1}{\varepsilon-\delta'})$-competitive online algorithm with commitment, where $\delta' = \frac{\varepsilon}{2}$ in the commitment-upon-admission model and $\delta' = \max\{\delta, \frac{\varepsilon}{2}\}$ in the δ-commitment model.*

The challenge in online scheduling with commitment is that, once we commit to complete a job, the remaining slack of this job has to be spent very carefully. The high-level objectives of our admission scheme are:

(i) Never start a job for the first time if its remaining slack is too small,
(ii) during the processing of a job, admit only significantly shorter jobs, and
(iii) for each admitted shorter job, block some time period during which no other jobs of similar size are accepted.

We have used the first two quite natural goals in Sect. 2.1, while the third goal is crucial for our tight result when scheduling with commitment. The intuition is the following: suppose we committed to complete a job with processing time 1 and have only a slack of $\mathcal{O}(\varepsilon)$ left before the deadline of this job. Suppose that c substantially smaller jobs of size $\frac{1}{c}$ arrive, where c is the competitive ratio we aim for. Without accepting any of them, we do not achieve c-competitiveness. Conversely, accepting too many fills up the slack and, thus, leaves no room for admitting even smaller jobs. The idea is to keep the flexibility for future small jobs by only accepting a $\frac{1}{c}$-fraction of jobs of similar size (within a factor 2).

More precisely, we distinguish two time periods with different regimes for accepting jobs. During the *scheduling interval* of job j, a more restrictive acceptance scheme ensures the completion of j, whereas in the *blocking period*, we guarantee the completion of previously accepted jobs. In contrast to the threshold algorithm, where the processing time of the currently scheduled job provides a uniform acceptance threshold, this distinction enables us to ensure the completion of every admitted job without being too conservative in accepting jobs.

3 Dynamic Knapsack

In the most basic variant, the Knapsack problem, we are given a knapsack with capacity $S \in \mathbb{N}$ and a set $\mathcal{J}$ of n items, where $\mathcal{J} = [n]$, and each item j has a size $s_j \in \mathbb{N}$ and a value $v_j \in \mathbb{N}$. The goal is to find a subset of items, $P \subseteq [n]$, with maximal total value $v(P) = \sum_{j \in P} v_j$, and with total size $s(P) = \sum_{j \in P} s_j \leq$

S. In the more general Multi-Knapsack problem, we are given m knapsacks with capacities S_i for $i \in [m]$. Here, the task is to select m disjoint subsets $P_1, \ldots, P_m \subseteq \mathcal{J}$ such that P_i satisfies $s(P_i) \le S_i$ and the total value $\sum_{i=1}^{m} v(P_i)$ is maximized.

The Knapsack problem is NP-hard [9] while Multi-Knapsack is strongly NP-hard, even for identical knapsack capacities [7]. As a consequence, knapsack variants have been extensively studied through the lens of approximation algorithms. Of particular interest are *approximation schemes*, families of polynomial-time algorithms that compute for any constant $\varepsilon > 0$ a $(1+\varepsilon)$-approximate solution. For Knapsack, there are approximation schemes with running time polynomial in n and $\frac{1}{\varepsilon}$ [8]. For Multi-Knapsack, only an exponential dependency on $\frac{1}{\varepsilon}$ is possible, unless $\mathrm{P} = \mathrm{NP}$ [3]. All these algorithms are *static*, i.e., they have access to the entire instance, and the instance is not subject to changes.

Given the dynamics of real-world instances, it is natural to ask for *dynamic algorithms* that adapt to small changes in the instance while spending only little computation time: During the execution of the algorithm, items and knapsacks arrive and depart and the algorithm needs to maintain an approximate knapsack solution with an *update time* polylogarithmic in the number of items in each step. A dynamic algorithm can be seen as a data structure that implements these updates efficiently and supports relevant query operations. We are the first to analyze knapsack problems in the context of dynamic algorithms.

In [2], we present dynamic algorithms for maintaining approximate knapsack solutions. Our algorithms are *fully dynamic* which means that in an update operation they can handle both, the arrival or departure of an item and the arrival or departure of a knapsack. Further, we consider the *query* model, in which an algorithm is not required to store the solution explicitly in memory such that the solution can be read in linear time. Instead, the algorithm may maintain the solution implicitly with the guarantee that a query about the packing of an item can be answered in poly-logarithmic time. Our main result is the following theorem.

Theorem 5. *For each $\varepsilon > 0$, there is a dynamic $(1+\varepsilon)$-approximate algorithm for Multi-Knapsack with update time $2^{f(1/\varepsilon)}\left(\frac{\log n}{\varepsilon}\right)^{\mathcal{O}(1/\varepsilon)}(\log m)^{\mathcal{O}(1)}$, where $f\left(\frac{1}{\varepsilon}\right)$ is quasi-linear. Item queries can be answered in time $\mathcal{O}(\log m)\left(\frac{\log n}{\varepsilon}\right)^{\mathcal{O}(1)}$ and the solution P can be output in time $\mathcal{O}(|P| + \log m)\left(\frac{\log n}{\varepsilon}\right)^{\mathcal{O}(1)}$.*

The high-level idea for Multi-Knapsack is to partition the given knapsacks based on their capacity, creating two subproblems of Multi-Knapsack. This separation allows us to design algorithms that exploit the structural properties specific to each subproblem. One subproblem consists of relatively few (though not constantly many) knapsacks, but they are the largest of the instance. While the small number of these *special* knapsacks offers more algorithmic freedom, this freedom is necessary since great care has to be taken when computing a solution: There may be items of high value that only fit into special knapsacks.

The second subproblem contains almost all remaining smaller knapsacks. Here, the handling of the sheer number of these *ordinary* knapsacks is the major challenge. On the upside, mistakes are forgiven more easily, allowing us to even discard a small fraction of knapsacks entirely.

Additionally, we create a third partition of knapsacks that lies in-between the two subproblems (w.r.t. knapsack capacity). It consists of knapsacks that contribute negligible value to an optimal solution. This property induces the precise partitioning and allows us to consider the knapsacks as empty *extra* knapsacks, which we use to place leftover items not packed in the subproblems.

A big challenge with this divide-and-conquer approach is to decide which item is assigned to which of the two subproblems. Clearly, for some—*special*—items this question is answered by their size as they only fit into special knapsacks, unlike the remaining—*ordinary*—items. In fact, for them the allocation is so problematic that we downright put a number of high-value ordinary items into extra knapsacks. To handle the rest, we guess the total size of ordinary items put in special knapsacks by an optimal solution, add a virtual knapsack with capacity equal to this guess to the ordinary subproblem, and solve it with the not yet packed ordinary items as input. The input for the special subproblem then consists of all special items together with bundles of the ordinary items packed in the virtual knapsack.

Acknowledgments. I would like to thank my supervisor Nicole Megow for her continuous and excellent support and my colleagues and co-authors for many fruitful and inspiring discussions.

References

1. Baruah, S.K., Haritsa, J.R., Sharma, N.: On-line scheduling to maximize task completions. In: RTSS, pp. 228–236. IEEE Computer Society (1994). https://doi.org/10.1109/REAL.1994.342713
2. Böhm, M., et al.: Fully dynamic algorithms for knapsack problems with polylogarithmic update time. CoRR, abs/2007.08415 (2020). https://arxiv.org/abs/2007.08415
3. Chekuri, C., Khanna, S.: A polynomial time approximation scheme for the multiple knapsack problem. SIAM J. Comput. **35**(3), 713–728 (2005). https://doi.org/10.1137/S0097539700382820
4. Chen, L., Eberle, F., Megow, N., Schewior, K., Stein, C.: A general framework for handling commitment in online throughput maximization. Math. Prog. **183**, 215–247 (2020). https://doi.org/10.1007/s10107-020-01469-2
5. Eberle, F.: Scheduling and packing under uncertainty. Ph.D. thesis, University of Bremen (2020). https://doi.org/10.26092/elib/436
6. Eberle, F., Megow, N., Schewior, K.: Optimally handling commitment issues in online throughput maximization. In: Proceedings of ESA 2020, pp. 41:1–41:15 (2020). https://doi.org/10.4230/LIPIcs.ESA.2020.41
7. Garey, M.R., Johnson, D.S.: Computers and Intractability: A Guide to the Theory of NP-Completeness. W. H. Freeman, New York (1979)
8. Ibarra, O.H., Kim, C.E.: Fast approximation algorithms for the knapsack and sum of subset problems. J. ACM **22**(4), 463–468 (1975). https://doi.org/10.1145/321906.321909
9. Karp, R.M.: Reducibility among combinatorial problems. In: Complexity of Computer Computations, The IBM Research Symposia Series, pp. 85–103. Plenum Press, New York (1972). https://doi.org/10.1007/978-1-4684-2001-2_9

Integrated Service- and Charge-Scheduling for Electric Commercial Vehicles

Patrick Sean Klein(✉)

TUM School of Management, Technical University of Munich,
80333 Munich, Germany
patrick.sean.klein@tum.de
https://osm.wi.tum.de

Abstract. Driven by climate change, rising environmental awareness, and financial incentives, more and more logistics providers integrate electric commercial vehicles (ECVs) into their fleets. Here, challenges such as limited availability of charging infrastructure, variable energy prices, and battery degradation lead to an integrated planning problem dealing with scheduling and charging decisions. We study this joint scheduling and charging problem and propose an exact algorithm based on Branch-and-Price. A comprehensive numerical study benchmarks our approach against an equivalent flow-based mixed-integer formulation, showing that the developed algorithm outperforms commercial solvers, and assesses the algorithm's scalability on larger instances. We show that problems instances that reflect problem sizes encountered in practice can be solved with computational times below an hour.

Keywords: Electric vehicle scheduling · Charge scheduling · Branch and price

1 Introduction

Increasing societal and political environmental awareness, resulting from climate change, as well as local and global emission problems, call for a paradigm change towards sustainable transportation systems. Herein, ECVs are seen as a viable option that promise up to 20% reduction in life-cycle greenhouse-gas emissions compared to internal combustion engine vehicles (ICEVs) while also lowering operational costs. Launching and operating ECV fleets however, bears a new and unresolved planning problem. Long recharging times and limited availability of public recharging infrastructure often require investments into private, on-premise recharging stations. As these remain costly, the number of chargers available generally remains limited such that charging bottlenecks arise and day-ahead planning of charging operations is necessary. Moreover, accelerated battery degradation, caused by, high-voltage (fast) charging, may offset the otherwise lower operational costs of ECVs such that sub-optimal charging patterns need to be avoided when scheduling charging operations. Many operators are billed according to time-of-use (TOU) energy tariffs, which charge varying electricity rates depending on the

N. Trautmann and M. Gnägi (Eds.): OR 2021, LNOR, pp. 15–20, 2022.
https://doi.org/10.1007/978-3-031-08623-6_3

time of consumption. In such scenarios, it becomes necessary to schedule charging operations accordingly as this leads to a non-trivial trade-off between incurring accelerated battery degradation and profiting from periods with low energy prices. Clearly, the impact of charge-scheduling is limited by the service schedules, such that operators who wish to remain cost-competitive should design vehicle schedules with charge schedules in mind and vice versa. Addressing this joint planning problem requires an integrated planning approach combining vehicle scheduling problems (VSPs) with charge-scheduling. However, publications dealing with VSPs have either focused on conventional vehicles and did not account for charging-related concerns, such as battery health, station capacity and variable energy prices (Adler and Mirchandani 2017; Yao et al. 2020), or remain computationally intractable for problem sizes relevant in practice (van Kooten Niekerk et al. 2017). Publications focusing on (depot) charge-scheduling problems have been limited in a similar fashion, either assuming a simplified model of the charging process (Sassi and Oulamara 2014, 2017), remaining heuristic, or proposing algorithms that do not scale to problem sizes encountered in practice (Pelletier et al. 2018). This work aims to close this gap by developing an exact branch and price (B&P) algorithm for an integrated charge- and service operation scheduling problem, accounting for battery degradation, TOU energy prices, limited availability of charging infrastructure, and non-linear battery behavior. For this purpose, Sect. 2 outlines the problem setting and develops a set-covering based formulation. Section 3 then summarizes our approach to solve this problem, followed by an evaluation of the algorithm's computational performance in Sect. 4. Finally, Sect. 5 concludes this article.

2 Problem Setting

We consider a (depot) charge scheduling problem faced by logistics service providers (LSPs) in the context of electric vehicle routing. Here, an electrified fleet of vehicles $\mathcal{K}$ needs to service a set of customers over a planning horizon of several days. We assume that routes have been pre-computed for each vehicle such that tour assignment, consumption, and duration are known in advance. Departure times are subject to scheduling, i.e., may be shifted in time, but must be completed within a certain, tour-specific, time window. Vehicles may be recharged at the depot using a set of (heterogeneous) charging stations. For this purpose, each station $f \in \mathcal{F}$ is equipped with C_f chargers for parallel charging. Charging incurs costs according to the battery degradation caused and the energy price at the time of charging. In this problem setting, the LSP's aim is to find a cost-minimal schedule of charging- and service-operations for each vehicle $k \in \mathcal{K}$ such that i) battery capacity constraints are respected, ii) each tour is serviced by its assigned vehicle, iii) departure time windows are met, and iv) charger capacity is not violated. We model this optimization problem using a set-covering formulation over the set of vehicle schedules. For this purpose, we discretize the planning horizon into a set of equidistant periods $p \in \mathcal{P}$. To ensure the feasibility of the original problem, we pursue a conservative discretization approach for tour departure time windows and charging operations. More precisely, tour departure and arrival are shifted to the beginning and end of their respective periods, and vehicles occupy chargers for entire periods at a time. To maintain accuracy, we still allow time-continuous charging such that vehicles may spend only a fraction of a period charging. We denote the set of all feasible vehicle schedules for vehicle $k \in \mathcal{K}$ as $\mathcal{A}_k$ and let $\mathcal{A} = \cup_{k \in \mathcal{K}} \mathcal{A}_k$ be the set of all schedules. This distinction is necessary as charging schedules may not be compatible with every vehicle due to different tour assignments. We further use binary parameters $y_{\omega,p,f}$ to indicate if $\omega \in \mathcal{A}$ schedules a charging operation at $f \in \mathcal{F}$ in period $p \in \mathcal{P}$. Finally, binary variables x_ω^k assign schedules to vehicles. We obtain the following optimization problem:

$$z_{MP} = \min \sum_{k \in \mathcal{K}} \sum_{\omega \in \mathcal{A}_k} x_\omega^k c(\omega) \tag{1a}$$

$$\sum_{k \in \mathcal{K}} \sum_{\omega \in \mathcal{A}_k} x_\omega^k \cdot y_{\omega,p,f} \leq C_f \qquad f \in \mathcal{F}, p \in \mathcal{P} \tag{1b}$$

$$\sum_{\omega \in \mathcal{A}_k} x_\omega^k = 1 \qquad k \in \mathcal{K} \tag{1c}$$

$$x_\omega^k \in \{0,1\} \qquad \omega \in \mathcal{A}_k \tag{1d}$$

MIP 1 assembles a fleet schedule of minimal cost from the set of feasible vehicle schedules $\mathcal{A}_k$ according to two sets of constraints. First, capacity constraints (1b) enforce charger capacity limitations. Second, covering constraints (1c) ensure that exactly one schedule is assigned to each vehicle. Constraints (1d) state the decision variables' domain.

3 Methodology

As $\mathcal{A}$ is huge for instances of reasonable size, solving MIP 1 using conventional, i.e., LP-based, branch and bound (B&B) remains computationally intractable. As a mitigation, we propose a B&P-based approach. Our algorithm essentially solves a LP relaxation of MIP 1, the so-called *master problem*, using a column-generation procedure at each node of the B&B tree. This procedure is outlined in Sect. 3.1. We propose a problem-specific branching rule to ensure a balanced B&B tree, detailed in Sect. 3.2. Finally, we utilize a primal heuristic, summarized in Sect. 3.3, to find good upper bounds early during the search.

3.1 Column Generation

The fundamental idea of column generation is to approach MIP 1 with only a small subset of schedules $\mathcal{A}$, generated iteratively. For this purpose, each iteration first solves a linear relaxation of MIP 1 to then generate a new schedule $\omega \in \mathcal{A}_k$ based on the dual variables of Constraints (1b) and (1c), denoted $\pi_{p,f}^{(1)}$ and $\pi_k^{(2)}$, respectively. This corresponds to solving the so-called *pricing problem* (Eq. 2), which seeks to generate schedules with negative reduced costs, i.e., schedules which would enter the basis of MIP 1 in the next iteration.

$$\min_{k\in\mathcal{K},\omega\in\mathcal{A}_k} c(\omega) - \pi_k^{(2)} - \sum_{p\in\mathcal{P}}\sum_{f\in\mathcal{F}} y_{\omega,p,f}\cdot\pi_{p,f}^{(1)}. \tag{2}$$

The procedure terminates when (2) is positive, i.e., when the basis of MIP 1 is optimal. We express our pricing problem as a shortest path problem with resource constraints (SPPRC) over a time-expanded network, which we solve by using a problem-specific labeling algorithm. Here, we address non-linear charging, battery degradation, and variable energy prices with a novel label representation: instead of considering every reachable state of charge (SoC) when charging at some charger $f\in\mathcal{F}$, our labels capture charging trade-offs in so-called cost profiles. These state the maximum SoC reachable at the end of a (partial) path ρ when spending a total of c along ρ, potentially replenishing additional energy at f. This allows to delay the decision on how much to charge at f until another station or the network sink is reached. Here, the trade-off becomes explicit and a finite subset of non-dominated charging decisions at f can be identified.

3.2 Branching

Column generation deals with the linear relaxation of MIP 1 and thus may terminate with a fractional solution. We mitigate this issue by embedding our column generation procedure into a B&B algorithm. Here, our branching rule bases on the following observation: In every fractional solution of MIP 1 it holds that $\exists(p,f)\in\mathcal{P}\times\mathcal{F}$ such that $\sum_{\omega\in\mathcal{A}} x_\omega y_{\omega,p,f} = C_f$. We resolve such conflicts by creating *up* and *down* branches based on the charger allocation $y_{\omega,p,f}$. For this purpose, we identify the vehicle k which relies most on charger f in period p, formally, $k = \arg\max_{k'\in\mathcal{K}} \sum_{\omega\in\mathcal{A}_{k'}} x_\omega\cdot y_{\omega,p,f}$, and create *up* and *down*-branches that cut off all solutions where vehicle k uses/does not use charger f in period p, respectively.

3.3 Primal Heuristic

We propose a primal heuristic based on *Sub-MIPing* that works exclusively with the variables of the master problem. Its fundamental idea is to remove a subset of columns present in the current node's (N) reduced column pool $\tilde{\mathcal{A}}^N$ and resolve the resulting problem using IP. We select schedules to remove from $\tilde{\mathcal{A}}^N$ based on quality (i.e., cost) and diversity. Here, diversity is measured through the number of shared charger allocations. We execute this procedure before branching on non-integral nodes unless an upper bound has already been found.

4 Numerical Experiments

We benchmark the performance of our B&P algorithm on two sets of randomly generated instances. The first set comprises small instances (2 vehicles, 1 charger, 1 day) with varying numbers of tours (1, 2) and departure time window lengths

(0%, 25%, 50%, given as a percentage of tour duration). Other parameters, such as tour timing, duration, and charging speed are drawn randomly. We use these to benchmark our algorithm against an equivalent IP. The second set contains instances of varying sizes to demonstrate the scalability of our algorithm. The reference setting spans a planning horizon of 3 days and comprises a fleet of 6 vehicles, each operating 2 tours per day with 30% flexibility. The depot is equipped with $\max(1, \lfloor \frac{|\mathcal{K}|}{2} \rfloor)$ fast and $|\mathcal{K}|$ slow chargers. We generate instances with varying fleet sizes and planning horizon lengths based on this setting. We used `CPLEX 20.1` limited to a single thread and 3600 s of runtime on a `Intel(R) i9-9900, 3.1` GHz CPU with `16 GB` of `RAM` for our computational study. Table 1 shows the results of our benchmark on the small instances.

Table 1. Results of our experiments on small instances.

Parameters		MIP				Branch & Price				#total
#tours	flexibility	runtime	gap	#infeas.	#timeout	runtime	gap	#infeas.	#timeout	
1	Static	2917.92	72.08	2	8	22.79	0.00	6	0	10
	25%	3600.00	100.00	0	10	67.14	0.00	0	0	10
	50%	3600.00	100.00	0	10	117.03	0.00	0	0	10
2	Static	1453.33	15.81	6	4	0.73	0.00	6	0	10
	25%	2523.36	57.35	3	7	3.59	0.00	4	0	10
	50%	3600.00	90.33	0	10	8.25	0.00	2	0	10

The first two columns state the parameter values used during instance generation. The total number of instances generated for each setting is reported in the last column. Columns runtime, gap, #infeas. and #timeout state averages for the runtime [s], the relative gap between best upper and lower bound [%], the number of infeasible instances, and the number of instances not solved within the time limit.

As can be seen, our algorithm outperforms the mixed integer program (MIP) formulation on all instance types, solving each generated instance in less than 500 s. Figures 1a and 1b show the scalability of our algorithm w.r.t. fleet size and planning horizon length. Each dot corresponds to an individual instance.

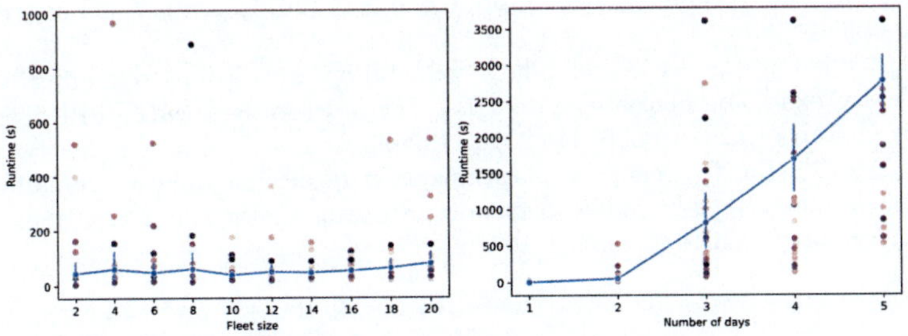

(a) Varying fleet size. (b) Varying planning horizon length.

Fig. 1. Runtimes on large instances

The blue lines give the average runtime over all instances. As is illustrated, our algorithm scales to a fleet size of 20 vehicles and manages to solve instances spanning up to 5 days.

5 Conclusion

We presented a novel charge- and service operation scheduling problem where a fleet of electric vehicles fulfills a set of service operations under the assumption of limited charging station capacity, variable energy prices, battery degradation, and non-linear charging behavior. We developed an exact algorithm based on B&P to solve the proposed problem. A novel labeling algorithm, a state-of-the-art primal heuristic, and problem-specific branching rules establish the efficiency of our algorithm, which is demonstrated in an accompanying numerical study. This study asserts the competitiveness of our algorithm through a benchmark against an equivalent mixed-integer formulation, showing that our algorithm significantly outperforms commercial solvers and scales to instances of larger size, optimally solving instances with planning horizons of up to 5 days or 20 vehicles within one hour, allowing an application in practice.

References

Adler, J., Mirchandani, P.B.: The vehicle scheduling problem for fleets with alternative-fuel vehicles. Transp. Sci. **51**(2), 441–456 (2017). https://doi.org/10.1287/trsc.2015.0615

van Kooten Niekerk, M.E., van den Akker, J.M., Hoogeveen, J.A.: Scheduling electric vehicles. Public Transp. **9**(1), 155–176 (2017). https://doi.org/10.1007/s12469-017-0164-0

Pelletier, S., Jabali, O., Laporte, G.: Charge scheduling for electric freight vehicles. Transp. Res. Part B Methodol. **115**(115), 246–269 (2018). https://doi.org/10.1016/j.trb.2018.07.010

Sassi, O., Oulamara, A.: Simultaneous electric vehicles scheduling and optimal charging in the business context: case study. In: IET Conference Proceedings, Institution of Engineering and Technology, vol. 5, pp. 6.3–6.3(1) (2014). https://doi.org/10.1049/cp.2014.0954

Sassi, O., Oulamara, A.: Electric vehicle scheduling and optimal charging problem: complexity, exact and heuristic approaches. Int. J. Prod. Res. **55**(2), 519–535 (2017). https://doi.org/10.1080/00207543.2016.1192695

Yao, E., Liu, T., Lu, T., Yang, Y.: Optimization of electric vehicle scheduling with multiple vehicle types in public transport. Sustain. Cities Soc. **52**, 101862 (2020). https://doi.org/10.1016/j.scs.2019.101862

Approximation Schemes for Machine Scheduling

Marten Maack([✉])

Heinz Nixdorf Institute and Department of Computer Science, Paderborn University, Paderborn, Germany
marten.maack@hni.uni-paderborn.de

Abstract. Makespan minimization on identical parallel machines, or machine scheduling for short, is a fundamental problem in combinatorial optimization. In this problem, a set of jobs with processing times has to be assigned to a set of machines with the goal of minimizing the latest finishing time of the jobs, i.e., the makespan. While machine scheduling in NP-hard and therefore does not admit a polynomial time algorithm guaranteed to find an optimal solution (unless P=NP), there is a well-known polynomial time approximation scheme (PTAS) for this problem, i.e., a family of $(1+\varepsilon)$-approximations for each $\varepsilon > 0$. The question of whether there is a PTAS for a given problem is considered fundamental in approximation theory. The author's dissertation considers this question for several variants of machine scheduling, and the present work summarizes the dissertation.

Keywords: Scheduling · Parallel machines · Makespan · Approximation · PTAS

1 Introduction

Consider the problem of makespan minimization on identical parallel machines, or *machine scheduling* for short. The input of this problem is a set $\mathcal{M}$ of m machines and a set $\mathcal{J}$ of n jobs each with a processing time or size p_j, and the goal is to find a schedule $\sigma : \mathcal{J} \to \mathcal{M}$ minimizing the makespan $C_{\max}(\sigma) = \max_{i \in \mathcal{M}} \sum_{j \in \sigma^1(i)} p_j$. Already in the 1960s Graham [4,5] proposed the simple list scheduling algorithm for machine scheduling and showed that it is a 2-approximation, i.e., for any input instance the objective value of the schedule produced by the algorithm is at most twice as big as the optimal value. In this algorithm, the jobs are simply considered one after another in any order and assigned to a least loaded machine. Graham also showed that the algorithm is a $\frac{4}{3}$-approximation if the jobs are sorted by non-increasing processing time. These results are considered to be among the first approximation algorithms and have been achieved even before the term has been coined by Johnson [12] in 1974. Since machine scheduling is known to be NP-hard, a polynomial time algorithm guaranteed to solve any instance optimally cannot be

N. Trautmann and M. Gnägi (Eds.): OR 2021, LNOR, pp. 21–26, 2022.
https://doi.org/10.1007/978-3-031-08623-6_4

hoped for. However, in 1987 Hochbaum and Shmoys [6] presented the first polynomial time approximation scheme (PTAS) for this problem and thereby showed that any approximation ratio strictly greater than 1 can be achieved. Formally, a PTAS is a family of approximation algorithms containing a $(1+\varepsilon)$-approximation for each $\varepsilon > 0$. Since then, there have been a plethora of results concerning approximation schemes for machine scheduling and variants thereof.

The author's dissertation [14] seeks to deepen the understanding in this area of research. For this purpose, several approximation schemes have been designed, with new techniques introduced and working for a wide spectrum of variants of machine scheduling. Furthermore, inapproximability results were designed that rule out the existence of approximation schemes for certain cases under the assumption that P≠NP. The corresponding results have been presented in the papers [7–11,15], and the present work briefly summarizes these results.

Before the results are discussed in the following sections, we introduce some concepts and notation regarding approximation schemes and variants of machine scheduling. First, a PTAS is called *efficient* (EPTAS) if it has a running time of the form $f(1/\varepsilon)|I|^{\mathcal{O}(1)}$, where f is some computable function and $|I|$ the input length. Furthermore, an EPTAS is called *fully polynomial* (FPTAS) if the function f is a polynomial.

The first problem variant is called *unrelated scheduling*. In this problem the processing time of a job depends on the machine it is processed on, that is, the input includes a processing time p_{ij} for each machine i and job j. Moreover, *uniform scheduling* can be seen as a special case of unrelated scheduling where each job j has a size p_j, each machine i a speed s_i, and the processing time of job j on machine i is given by $p_{ij} = p_j/s_i$. Finally, *restricted assignment* is another special case of unrelated scheduling with $p_{ij} \in \{p_j, \infty\}$, i.e., each job j has either a fixed processing time p_j or cannot be processed by a given machine at all. It is easy to see, that machine scheduling is a special case of both restricted assignment and uniform scheduling.

2 Unrelated Scheduling with Few Types

In [7], the variant of unrelated scheduling with a constant number K of machine types is considered. Two machines i and i' have the same type if $p_{ij} = p_{i'j}$ holds for each job j. In many application scenarios this setting is plausible, e.g., when considering computers which typically only have a very limited number of different types of processing units. Note that for $K = 1$, we have the problem of machine scheduling. For unrelated scheduling, there is a 2-approximation due to Lenstra, Shmoys and Tardos [13] who also showed that there is no better than 1.5-approximation for restricted assignment, unless P=NP. For machine scheduling, on the other hand, an EPTAS is known [1]. The first PTAS for the case with a constant number of types was presented by Bonifaci and Wiese [2].

In [7] the first EPTAS for unrelated scheduling with a constant number of machine types is presented. For this result, the dual approximation approach by Hochbaum and Shmoys [6] is employed to get a guess T of the optimal makespan. Then, the problem is further simplified via geometric rounding of the processing times. Next, a mixed integer linear program (MILP) is formulated based on the classical configuration ILP with a constant number of integral variables that encodes a relaxed version of the problem. The fractional variables of the MILP have to be rounded and this is achieved with a properly designed flow network utilizing flow integrality and causing only a small error. With an additional error, the obtained solution can be used to construct a schedule with makespan $(1+\mathcal{O}(\varepsilon))T$ which in turn suffices to get an EPTAS. The running time of the EPTAS can be further improved using techniques for configuration ILPs. Furthermore, three other problem variants are studied and it is shown that the result can be expanded to them as well.

3 Interval and Resource Restrictions

While we cannot hope for a PTAS for the restricted assignment problem, there is a line of research focusing on types of restrictions that change this situation. One example is the hierarchical case in which the machines are totally ordered and each job is eligible on a set of consecutive machines including the first machine. A PTAS for this case was presented by Ou, Leung and Li [16].

The paper [15] focuses on two arguably rather natural variants of restrictions, namely interval and resource restrictions. In the case of interval restrictions the machines can be totally ordered such that jobs are eligible on consecutive machines. The main result is that there is no PTAS for the interval case unless P=NP. This result is achieved using a reduction from a hand-crafted satisfiability problem with a particularly neat and symmetric structure. There are several gadgets in the reduction, but at its core there is a truth assignment and a clause gadget. Roughly speaking, in the former, decisions regarding the truth assignments of the variables are made, and in the latter, these decisions are evaluated with respect to the clauses of the satisfiability instance. The main challenge is connecting these two gadgets in a way that does not require encoding too much information into the processing times of the jobs. In particular, an instance of the restricted assignment problem with interval restrictions is constructed in which all processing times are integral and upper bounded by some constant T such that the overall processing time of the jobs equals $|\mathcal{M}|T$. Then it is shown that there exists a schedule with makespan T for the instance, if and only if the given instance of the satisfiability problem is a yes-instance. This rules out the existence of an approximation algorithm with rate smaller than $(T+1)/T$ and a PTAS in particular.

The variant with resource restrictions can be defined as follows: There is a fixed number of (renewable) resources, each machine has a capacity, and each job a demand for each resource. A job is eligible on a machine if its demand is at most as big as the capacity of the machine for each resource. For one resource, this is equivalent to the mentioned hierarchical case and it can be shown that the variant with two resources contains the interval case. It is shown that there is no polynomial time approximation algorithm with a rate smaller than $48/47 \approx 1.02$ or 1.5 for scheduling with resource restrictions with 2 or 4 resources, respectively, unless P=NP.

4 Structural Parameter Restrictions

The paper [11] also considers special types of assignment restrictions. In particular, a graph framework based on the restrictions is introduced. In the primal graph, the jobs are the nodes and are adjacent if they share an eligible machine. In the dual graph, on the other hand, the machines are the nodes and two machines are adjacent if there is a job that is eligible on both of them. Lastly, the incidence graph is a bipartite graph that includes both jobs and machines as nodes, and a job node is adjacent to a machine node if the job is eligible on the machine (see Fig. 1). Cases in which these graphs have certain structural properties are studied.

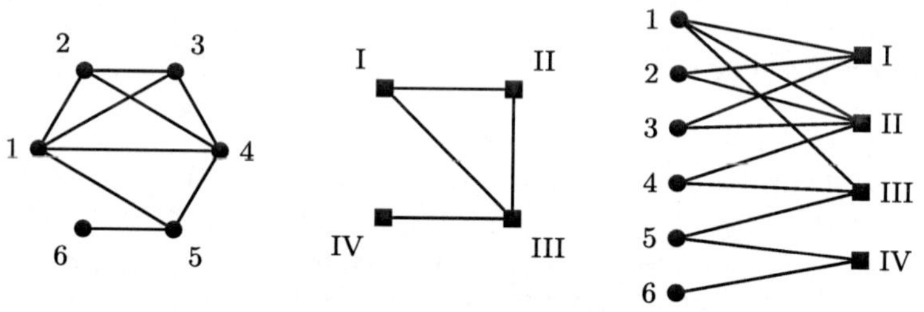

Fig. 1. Primal, dual and incidence graph for an instance with 6 jobs and 4 machines.

The main result of this paper is a PTAS for the case that the incidence graph has a constant rank- or cliquewidth. Intuitively, the cliquewidth measures how far a graph deviates from the simple structure of a complete graph. There are certain graph decompositions associated with theses width parameters that enable the design of a dynamic program. The dynamic program then can be combined with rounding techniques to yield the PTAS result. Moreover, if the incidence graph is a so-called bi-cograph, then the graph decomposition is particularly simple and the bi-cograph case already generalizes most of the previously known PTAS results. Furthermore, the more famous graph parameter treewidth is considered with respect to the primal, dual, and incidence graph and used in the design of further approximation schemes and exact algorithms.

5 Machine Scheduling with Setup Times

Integer linear programs of configurations, or configuration IPs, are a classical tool in the design of algorithms for scheduling and packing problems where a set of items has to be placed in multiple target locations. Herein, a configuration describes a possible placement on one of the target locations, and the IP is used to choose suitable configurations covering the items. In the paper [8], an augmented IP formulation is presented, which is called the module configuration IP. It can be described in the framework of n-fold integer programming, which has recently received a lot of attention with many new algorithmic results (see [3] for an overview). As an application, variants of machine scheduling are considered in which jobs can be split and scheduled on multiple machines. However, before a part of a job can be processed, an uninterrupted setup time depending on the job has to be paid. For both of the variants that jobs can be executed in parallel or not, EPTAS results are obtained. Previously, only constant factor approximations of $5/3$ and $4/3 + \varepsilon$, respectively, were known. Furthermore, an EPTAS is presented for a problem where classes of (non-splittable) jobs are given, and a setup has to be paid for each batch of jobs of each class of jobs being executed on one machine. The results are based on rounding techniques, a careful analysis of near-optimal solutions, and the mentioned new algorithms for n-fold integer programming. Intuitively, the main idea is to perform two steps simultaneously using the module configuration IP. The first step is to split the jobs or form batches of jobs, respectively, and to combine them with the respective setup times, and then covering the resulting so-called modules with configurations in the second step. In a follow-up work [9] the non-splittable problem on uniform machines is considered and a PTAS is achieved.

6 Machine Scheduling with a Shared Resource

In the last work [10], two related variants of machine scheduling are considered: single resource constrained scheduling on identical parallel machines and a generalization with resource dependent processing times. In both problems, jobs require a certain amount of an additional resource and have to be scheduled on machines minimizing the makespan, while at every point in time a given resource capacity is not exceeded. In the first variant of the problem, the processing times and resource amounts are fixed, while, in the second, the former depends on the latter. Both problems contain the problem of bin packing with cardinality constraints as a special case, and, therefore, they are strongly NP-complete even for a constant number of machines larger than three, which can be proven by a reduction from 3-Partition. Furthermore, if the number of machines is part of the input, we cannot hope for an approximation algorithm with absolute ratio smaller than $3/2$. These problems arise naturally in different contexts, e.g., in highly parallelized computing where simultaneously active jobs share common memory, or in production logistics where additional personnel may speed up certain tasks.

For both problems, asymptotic fully polynomial time approximation schemes (AFPTAS) are presented: For any $\varepsilon > 0$ a schedule of length at most $(1 + \varepsilon)$ times the optimum plus an additive term of $\mathcal{O}(p_{\max} \log(1/\varepsilon)/\varepsilon)$ is provided, and the running time is polynomially bounded in $1/\varepsilon$ and the input length. Up to now, only approximation algorithms with absolute approximation ratios were known. The results are achieved via intricate LP and rounding techniques and the result for the second problem uses the first as a subroutine.

References

1. Alon, N., Azar, Y., Woeginger, G.J., Yadid, T.: Approximation schemes for scheduling on parallel machines. J. Scheduling **1**(1), 55–66 (1998)
2. Bonifaci, V., Wiese, A.: Scheduling unrelated machines of few different types. CoRR abs/1205.0974 (2012)
3. Eisenbrand, F., Hunkenschröder, C., Klein, K., Koutecký, M., Levin, A., Onn, S.: An algorithmic theory of integer programming. CoRR abs/1904.01361 (2019)
4. Graham, R.L.: Bounds for certain multiprocessing anomalies. Bell Syst. Tech. J. **45**(9), 1563–1581 (1966)
5. Graham, R.L.: Bounds on multiprocessing timing anomalies. J. SIAM Appl. Math. **17**(2), 416–429 (1969)
6. Hochbaum, D.S., Shmoys, D.B.: Using dual approximation algorithms for scheduling problems theoretical and practical results. J. ACM **34**(1), 144–162 (1987)
7. Jansen, K., Maack, M.: An EPTAS for scheduling on unrelated machines of few different types. Algorithmica **81**(10), 4134–4164 (2019)
8. Jansen, K., Klein, K., Maack, M., Rau, M.: Empowering the configuration-IP - new PTAS results for scheduling with setups times. In: 10th Innovations in Theoretical Computer Science Conference, ITCS 2019, San Diego, California, USA, 10–12 January 2019, pp. 44:1–44:19 (2019)
9. Jansen, K., Maack, M., Mäcker, A.: Scheduling on (un-)related machines with setup times. In: 2019 IEEE International Parallel and Distributed Processing Symposium, IPDPS 2019, Rio de Janeiro, Brazil, 20–24 May 2019, pp. 145–154 (2019)
10. Jansen, K., Maack, M., Rau, M.: Approximation schemes for machine scheduling with resource (in-)dependent processing times. ACM Trans. Algorithms **15**(3), 31:1–31:28 (2019)
11. Jansen, K., Maack, M., Solis-Oba, R.: Structural parameters for scheduling with assignment restrictions. Theor. Comput. Sci. **844**, 154–170 (2020)
12. Johnson, D.S.: Approximation algorithms for combinatorial problems. J. Comput. Syst. Sci. **9**(3), 256–278 (1974)
13. Lenstra, J.K., Shmoys, D.B., Tardos, É.: Approximation algorithms for scheduling unrelated parallel machines. Math. Program. **46**, 259–271 (1990)
14. Maack, M.: Approximation schemes for machine scheduling. Ph.D. thesis, University of Kiel, Germany (2019)
15. Maack, M., Jansen, K.: Inapproximability results for scheduling with interval and resource restrictions. In: 37th International Symposium on Theoretical Aspects of Computer Science, STACS 2020, Montpellier, France, 10–13 March 2020, vol. 154, pp. 5:1–5:18 (2020)
16. Ou, J., Leung, J.Y.T., Li, C.L.: Scheduling parallel machines with inclusive processing set restrictions. Nav. Res. Logist. (NRL) **55**(4), 328–338 (2008)

Prescriptive Analytics for Commodity Procurement Applications

Christian Mandl(✉)

School of Management, Deggendorf Institute of Technology, Deggendorf, Germany
christian.mandl@th-deg.de

Abstract. In this work, we investigate the implications of commodity price uncertainty for optimal procurement and inventory control decisions. While the existing literature typically relies on the *full information paradigm*, i.e., optimizing procurement and inventory decisions under full information of the underlying stochastic price process, we develop and test different data-driven approaches that optimize decisions under very limited statistical model assumptions. Our results are data-driven policies and decision rules that can support commodity procurement managers, inventory managers as well as commodity merchants. We furthermore test all optimization models based on real data from different commodity classes (i.e., metals, energy and agricultural).

This paper is a summary of the author's dissertation (Mandl C. (2019). Optimal Procurement and Inventory Control in Volatile Commodity Markets - Advances in Stochastic and Data-Driven Optimization, [1]).

Keywords: Price uncertainty · Procurement · Inventory control · Data-driven optimization · Machine learning

1 Motivation

During the COVID-19 pandemic, commodity prices increased significantly. A prominent example was timber whose prices increased by more than 200%. But also other commodities that are relevant for many industries such as aluminum and copper were hit by price increases of more than 50%. Similar price explosions were observable for agricultural commodities such as for example corn and wheat (see [2]).

In practice, almost every company is affected by fluctuating commodity prices: firms that purchase raw material for manufacturing as well as companies demanding for packaging material or energy to run their production facilities and transports. Depending on the industry, commodity purchasing cost make up a proportion of 17 to 44% of the total cost of a company (see [3]).

Consequently, commodity price volatility constitutes a significant exogenous risk factor for both commodity-processing and commodity-trading firms and ask for appropriate decision-making in terms of procurement quantities, contracting volumes and inventory levels.

N. Trautmann and M. Gnägi (Eds.): OR 2021, LNOR, pp. 27–32, 2022.
https://doi.org/10.1007/978-3-031-08623-6_5

While commodity-processing firms try to hedge price risk through anticipative inventories (operational hedging) or contracting (financial hedging), commodity merchants try to take advance of price fluctuations through purchasing at a rather low prices, storing over a certain period of time and selling at rather high prices.

In any case, physical commodity procurement states a complex decision problem under uncertainty that is typically solved through stochastic dynamic programming (SDP) in literature and practice (see, e.g., [7,8]). This requires to select a prediction model upfront in order to model uncertainty in commodity prices. The prediction model then serves as an input for the optimization model (i.e., SDP) to derive optimal operating policies (e.g., procurement policies or inventory control policies) under consideration of further operational constraints (e.g., service level objectives or capacity restrictions).

However, the obtained operating policies are optimal only under the assumption of *full information* about the prediction model and its parameters. In practice, however, prediction models are typically misspecified and can therefore lead to adverse procurement and inventory control decisions.

In this work, we *(i)* quantify the effect of price model misspecification for commodity operations and *(ii)* aim at relaxing the *full information paradigm* through novel machine-learning-inspired data-driven optimization approaches that either optimize commodity operations decisions based on raw data itself with very limited statistical model assumptions (see [4,5] for newsvendor applications) or allow for switches between different prediction model specifications (see [6] for theoretical foundation of Markov regime switching models).

Our developed models contribute to the emerging literature on data-driven optimization and machine learning being capable to leverage internal and external feature data that is increasingly available at companies - however focusing on optimal operational decisions rather than predictions.

2 Problem Statement

In the light of the *full information paradigm* introduced in Sect. 1, we study three fundamental optimization problems in physical commodity procurement and merchant operations:

1. Optimal forward contracting of commodities from a procurement perspective.
2. Optimal inventory control under purchase price and demand uncertainty from a procurement perspective.
3. Optimal storage operations from a commodity merchant's perspective.

2.1 Problem Setting 1: Forward Contracting

The first setting [9] addresses the multi-period stochastic procurement problem of optimizing the firm's position in commodity forward markets. Forward (or futures) contracts allow to fix a commodity price p_t^τ at time t for delivery at a future date $\tau > t$ (e.g. front-month, two-months-ahead, etc.). The forward curve $\vec{F}_t = (p_t^\tau, \tau > t)$ evolves stochastically over time. Therefore, even when relying on the Rational Expectations Hypothesis that states that $p_t^\tau = E_t[p_\tau^0]$ with $E[\cdot]$ as the expected value, the decision maker cannot be certain about the market's expectation in period $t+1$. Nonetheless, expectations in period $t+1$ affect decisions in period $t+1$, which again can affect the optimal here-and-now decision in period t. Existing approaches solve the problem under the *full information paradigm* using a pre-specified stochastic forward curve model as an input for a stochastic dynamic program. These approaches suffer from two major problems: (*i*) The well-known curse of dimensionality in dynamic programming and *(ii)* price model and generalization errors. Therefore, we address the following research questions:

- How can firms efficiently operationalize feature data (i.e., potential price drivers) for optimal commodity forward contracting decisions without full information assumptions on the price models characterization?
- What is the economic value of feature data and prescriptive analytics for commodity-purchasing firms?

2.2 Problem Setting 2: Storage (Procurement Perspective)

The second setting [10] addresses the multi-period stochastic inventory control problem under purchase price and demand uncertainty. Commodity storage is an appropriate price risk mitigation strategy in particular for storable commodities without liquid forward markets. In order to derive optimal inventory control policies, the existing literature assumes a stochastic price model that is well-defined in class and parameters.

However, the price model and its parameters can change over time - e.g., due to changes in price regimes after economic crashs or natural disasters. Ignoring changes in the underlying price dynamics (i.e., regime switches) may lead to significantly suboptimal procurement and inventory control decisions. We therefore formulate the following research questions:

- How do price regime switches affect the structure of the optimal inventory control policy?
- What are the cost implications of misspecifying the underlying price model?
- Can Bayesian learning based on information updates improve inventory control decisions for commodities?

2.3 Problem Setting 3: Storage (Merchant Perspective)

The third setting [11] addresses the multi-period inventory control problem from a merchant's perspective with purchase, storage and selling decisions aiming at maximizing profits under random purchase and sales prices (purchase low, store, sell high). Inventory trading plays a fundamental role in commodity industry and asks for stochastic optimization approaches to optimally control inventory levels of for instance gas storage caverns, grain silos or metal warehouses. In practice, storage decisions are typically optimized based on reoptimization heuristics that use the forward price curve as deterministic best-estimates of future spot prices. Reoptimization heuristics are computationally efficient and have been shown to perform near-optimal under the *full information paradigm.* Motivated by the fact that decision-makers in practice do not decide under full price model information, we formulate the following research questions:

- How does the state-of-the-art reoptimization heuristic perform in backtesting settings on real commodity price data?
- How to effectively solve the merchant's storage problem in a data-driven and learning-enabled way and what is the value of data-driven storage policies?

3 Results

3.1 Problem Setting 1: Forward Contracting

We show that the optimal procurement policy of problem setting 1 is characterized by a simple threshold structure that characterizes the optimal procurement quantity y_t:

$$y_t = \begin{cases} [d_{t+\tau} - I_t^\tau]^+ & \text{if } p_t^\tau \leq P_t^\tau \\ 0 & \text{if } p_t^\tau > P_t^\tau. \end{cases} \tag{1}$$

If the current forward (or futures) price p_t^τ is less than or equal to a threshold P_t^τ, then one should hedge the corresponding future period's demand $d_{t+\tau}$ minus the already hedged quantity I_t^τ, otherwise do nothing.

Rather than deriving price threshold P_t^τ via dynamic programming under price model assumptions, we train P_t^τ based on price and feature data X_{it} of features $i = 1, ..., N$ over a historical data sample $t = 1, ..., T$:

$$P_t^\tau(X) = \beta_0^\tau + \sum_{i=1}^{N} \beta_i^\tau X_{it} \tag{2}$$

Feature coefficients β_i^τ (including intercept β_0^τ) are trained with mixed-integer linear programming (MILP) models under the objective function of purchase cost minimization rather than the minimization of price prediction errors. The MILP models are extended by standard regularization functions from machine learning in order to avoid overfitting. For further technical details, we refer to [9].

Besides our numerical tests in a controlled simulation environment, we additionally back-tested the algorithms in a case study for the procurement of natural gas at the European TTF market under consideration of a large feature set (e.g., related commodity prices, macroeconomic data and weather data). For the period 2007–2017, we find that the data-driven procurement policy outperforms several benchmarks in terms of out-of-sample purchase cost achieving a cost performance of on average 5.68% above the perfect foresight cost, i.e., minimal procurement cost if commodity price evolution would have been known prior to decision-making.

3.2 Problem Setting 2: Storage (Procurement Perspective)

By means of stochastic dynamic programming, we show that under switches in the underlying price process, a regime belief-dependent base-stock policy fully characterizes the optimal procurement quantity y_t given the current inventory level I_t:

$$y_t = \begin{cases} [S_t(p_t, \vec{\pi_t}) - I_t & \text{if } I_t < S_t \\ 0 & \text{if } I_t \geq S_t. \end{cases} \tag{3}$$

The base-stock level S_t is a function of the current spot price p_t and a regime belief $\vec{\pi_t}$. We introduce a Markov regime switching (MRS) approach that learns switches in the underlying commodity price process parameters based on the latest price observations. The approach is able to adjust inventory policy parameters accordingly based on a Bayesian learning scheme that updates the regime belief $\vec{\pi_t}$. For further technical details, we refer to [10].

Based on both simulation experiments and experiments on real commodity price data (i.e., for corn and zinc), we show that ignoring regime switches can lead to substantially suboptimal procurement and storage decisions with cost deviations of up to 13%. The results show that a data-driven approach that utilizes Bayesian statistics can significantly improve inventory control decisions compared to traditional stochastic inventory optimization. We find that the value of Bayesian learning is particularly high when unit inventory holding cost and demand uncertainty are low. However, the computation of the optimal base-stock level $S_t(p_t, \vec{\pi_t})$ suffers from the curse of dimensionality. Therefore, we test several suboptimal control policies showing that certainty-equivalent control (CEC) without permanent information updates can be a good approximation for optimal control.

3.3 Problem Setting 3: Storage (Merchant Perspective)

Based on six major exchange-traded commodities (i.e., copper, gold, crude oil, natural gas, corn and soybean), we run extensive performance backtests. We find that the popular reoptimization heuristic (RH) that has been shown to perform near-optimal relative to the full information SDP, can perform significantly suboptimal in backtests based on real data and can even lead to unprofitable storage operations.

Motivated by this observation, we formulate MILP models to train inventory policy parameters, i.e., price thresholds P_t and base-stock levels for injection S_t^i and base-stock levels for withdrawal S_t^o in a data-driven way as a function of features (i.e., futures prices, spot price history and analyst forecasts). Our results show that data-driven storage policies achieve a profit of 26.7% of the perfect foresight solution, while RH achieves only 12.0%. However, we also show that the performance of data-driven policies strongly depends on feature selection and further input parameters (e.g., training set length) that need to be selected carefully in a-priori backtest experiments.

Acknowledgments. While working on this thesis, the author was doctoral student at the chair of logistics and supply chain management (Prof. Dr. Stefan Minner) in the School of Management at Technische Universität München.

References

1. Mandl, C.: Optimal Procurement and Inventory Control in Volatile Commodity Markets - Advances in Stochastic and Data-Driven Optimization. Dissertation, TUM School of Management, Technische Universität München (2019)
2. Refinitiv Eikon. https://eikon.thomsonreuters.com/index.html
3. Beschaffungsmanagement: Rohstoffkosten-Management. Einkaufsexperten fordern jetzt aktives Management von Rohstoffkosten für schlechtere Zeiten. Beschaffungsmanagement 04/2009, pp. 12–13 (2009)
4. Beutel, A.-L., Minner, S.: Safety stock planning under causal demand forecasting. Int. J. Prod. Econ. **140**(2), 637–645 (2012)
5. Ban, G.-Y., Rudin, C.: The big data newsvendor: practical insights from machine learning. Oper. Res. **67**(1), 90–108 (2019)
6. Hamilton, J.D.: A new approach to the economic analysis of nonstationary time series and the business cycle. Econometrica **57**(2), 357–384 (1989)
7. Haköz, C., Seshadri, S.: Supply chain operations in the presence of a spot market: a review with discussion. J. Oper. Res. Soc. **58**(11), 1412–1429 (2007)
8. Secomandi, N., Seppi, D.J.: Real options and merchant operations of energy and other commodities. Found. Trends Technol. Inf. Oper. Manag. **6**(3–4), 161–331 (2012)
9. Mandl, C., Minner, S.: Data-driven optimization for commodity procurement under price uncertainty. Manuf. Serv. Oper. Manag. (2021). https://doi.org/10.1287/msom.2020.0890
10. Mandl, C., Minner, S.: When Do Commodity Spot Price Regimes Matter for Inventory Managers? Working Paper (2021). https://ssrn.com/abstract=3011340
11. Mandl, C., Nadarajah, S., Minner, S., Gavirneni, N.: Data-driven storage operations: Cross-commodity backtest and structured policies. Prod. Oper. Manage. (2022). https://onlinelibrary.wiley.com/doi/full/10.1111/poms.13683

Rebalancing in Shared Mobility Systems – Competition, Feature-Based Mode Selection and Technology Choice

Layla Martin(✉)

Operations, Planning, Accounting, and Control Group,
Eindhoven University of Technology, Eindhoven, The Netherlands
l.martin@tue.nl

Introduction

Shared mobility systems or vehicle sharing schemes, such as carsharing and ridesharing, gain global importance, as they help alleviating two major challenges of today's society: congestion and emissions. However, to be able to tackle these challenges, customers must be able to find vehicles nearby within a short period of time. Vehicles may be unavailable because they are currently rented out, broken, or because of *spatio-temporal demand imbalances*, i.e., vehicles agglomerating in a part of the operating area while no vehicles are available elsewhere. In the case of one-way and free-floating systems, users can rent vehicles at one location, and return them elsewhere. For example, party-goers might use carsharing to get to a bar, but public transit on their return trip, or customers may use public transportation to get to a shopping mall, but vehicles with sufficient trunk-space for their return trip. Temporally, demand imbalances can for example arise because of customers commuting to work. Multiple data-driven studies show such demand imbalances throughout different cities and services, two examples are [1,2]. Ignoring these demand imbalances would result in lower profits, but also highly varying service levels [3]. Thus, operators *rebalance* their fleet, i.e., they move vehicles from locations with an expected excess in supply to locations with an expected excess in demand (*operator-based* rebalancing), or use incentives and discounts such that users adapt their behavior (*user-based* rebalancing). While user-based rebalancing has the potential to reduce the necessity for operator-based rebalancing, operator-based rebalancing remains the method of choice for very high and random demand imbalances (see, for example, [4,5]). For a literature overview on the rebalancing problem, we refer to the recent literature review of [6], as well as the broader reviews in [7,8] which summarize the operations management and transportation science literature on vehicle sharing. The author's thesis tackles three extensions of the operator-based rebalancing problem, namely competition, feature-based mode selection, and technology choice [9]. In the following, we devote one section to each of the extensions.

N. Trautmann and M. Gnägi (Eds.): OR 2021, LNOR, pp. 33–38, 2022.
https://doi.org/10.1007/978-3-031-08623-6_6

1 Competition in Rebalancing Shared Mobility Systems

In recent years, competition has between carsharing operators has been increasing worldwide [10,11]. [12] investigates how much operational profit operators can gain by considering the fact that they are competing in a city when rebalancing, and how much operational profit they loose due to competing. We also study drivers of profit gains and losses. We model the rebalancing problem under competition as a variant of a vehicle routing problem with profits, called the Competitive Pickup and Delivery Orienteering Problem (C-PDOP). If multiple operators place a vehicle at the same location, they share the profits. To solve the C-PDOP for its pure-strategy Nash equilibria, we present two algorithms, namely Iterated Best Response (IBR) and Potential Function Optimization (PFO). The latter is only applicable in a restricted version of the C-PDOP, namely if all operators receive the same payoff from a customer (homogeneous payoffs), customers do not have preferences for either operator (indifferent customer choice), and if profits of placing vehicles at nearby locations are not inter-dependent (unit-demand stations). The former algorithm can find pure-strategy Nash equilibria even in the more general case if they exist. Using the proposed model and algorithms, and a large set of artificial instances as well as a case study, we answer three research questions:

How much can operators gain from considering the presence of competition in their rebalancing operations with regards to gross profits? Put differently, what is the price of ignoring the presence of competition?

In experiments using artificial data, we observe profit gains of up to several orders of magnitude. In a Munich, Germany, case study with two operators, profit gains of 35% on average over both operators are achievable, if operators ignore competition otherwise. Assuming that operators expect their competitors to place vehicles at all locations, operators gain 12% of their profits on average by considering competition.

How much is lost by competing in comparison to jointly optimizing fleet rebalancing with regard to gross profits, and how do alternative business models under competition compare to each other?

Welfare-maximization, i.e., the joint optimization of the routing, but keeping the routes separate, does not substantially increase profits, compared to the Nash equilibrium solution. Commonly, the small profit gains do not justify the additional coordination effort for the operators. Merging both fleets, or outsourcing all rebalancing activities to a common third-party provider, on the other hand, has the potential for substantial profit improvements compared to the Nash equilibrium solution. We prove that under two minor restrictions, merging the fleets, or outsourcing rebalancing operations to the same service increases profits. In the Munich case study, operators lose approx. 10% by competing rather than merging or outsourcing their rebalancing activities.

Which features drive the gains from considering competition, and the losses due to the presence of competition?

If many customers have multiple memberships, and if a larger number of operators offers their service in a city, i.e., if competition is fierce, profit gains due to considering competition increases while the profit loss due to considering competition decreases. If the imbalance is larger, i.e., if the operators must rebalance more vehicles, the percentage profit loss due to considering competition decreases, while the absolute profit loss increases. Assuming marginally decreasing payoffs of additional vehicles at a station, we show that larger stations decrease the potential profit gain. Also, profit gains due to considering competition increase if operators receive different payoffs for serving the same customer (under full competition), customers have preferences (only for the less preferred operator).

2 Feature-Based Mode Selection in Rebalancing Shared Mobility Systems

Most literature on rebalancing vehicle sharing systems assumes that vehicles can be loaded onto a truck, or that workers of the operator drive the vehicles themselves. If workers drive vehicles, they usually combine multiple rebalancing operations, and thus have to continue to the next vehicle. They can use (foldable) bikes or public transit, or hitch rides with colleagues. In [13] we study the mode selection problem of carsharing operators. To understand which modes to use in which city, and why, we first solve a large number of instances for their optimal mode, or their optimal modal combination, and use this data to build classifiers that predict the mode based on a large set of features such as velocity, cost, and distances. These classifiers are linear regression for rebalancing costs per mode, permitting to select the cost-minimal mode, logistic regression for the probability a mode is optimal, and three decision trees using different subsets of features to predict the best mode directly. The structure of the classifiers, i.e., the weights in the linear and logistic regression classifiers, and the decision rules in the decision trees indicate the importance of the features in the modal choice. This permits us two answer the following two research questions:

Can a good mode be selected a-priori based upon features of the fleet and city?

Linear regression, logistic regression, and one of the three suggested decision trees determines the best mode with an accuracy of more than 90%. At the same time, the excess cost of misclassification, i.e., the cost difference between the best mode and the mode suggested by a classifier, is low with less than 10% of the total cost on average over all misclassified instances. This is because misclassification commonly occurs if multiple modes are good choices. We show that in most misclassified instances, multiple instances using the same features result in different optimal modes, and in most instances, using multiple modes simultaneously, i.e., hybridization, further reduces costs.

Which features drive the choice of the optimal rebalancing mode?

Most commonly, the optimal mode is bike or truck. Bike is preferred over truck, if worker wages are not excessively high, and costs for vehicles (fuel, wear, tear, deprecation) are not very low. Other modes, i.e., public transit and car, only become the best mode, if both bike and truck restricted, e.g., by very low accessibility by truck and very low velocities for the mode bike.

3 Technology Choice in Rebalancing Shared Mobility Systems

In addition to the revolutionary nature of the "sharing economy", the mobility market also faces the advent of driverless vehicle technology [14]. [15] focuses on the strategic technology choice problem of shared mobility operators arising when driverless vehicles become available. We address the question of how many driverless, and how many human-driven vehicles a shared mobility system operator should procure, i.e., a fleet size and mix problem. Compared to human-driven vehicles, driverless vehicles are more expensive to procure due to the additional technology. On the other hand, they are more profitable operationally, due to lower rebalancing costs, and possibly higher contribution margins if human-driven vehicles require a driver. We introduce a bound-and-enumerate algorithm for the technology choice problem, and a semi-Markov decision process as well as a fluidic approximation for the rebalancing problem, both of which can be solved using linear programming. The fluidic approximation extends [16] to the case with multiple vehicle types. Using the developed algorithms on both artificial instances as well as two case studies using data of the New York taxi commission and of Chinese ride-hailing provider DiDi, we answer four research questions:

Should shared mobility operators use fleets with only one vehicle type, or mix among driverless and human-driven vehicles?

In many instances, the possibility to introduce driverless vehicles in the fleet mix increases profits. Particularly, operators benefit from driverless vehicles if contribution margins differ between vehicle types; then, operators exclusively use driverless vehicles. On small instances, mixed fleets are often not optimal, since operators loose pooling benefits. For larger instances, however, mixed fleets are commonly profitable if contribution margins are equal for both vehicle types, including the case study using DiDi data.

In case mixed fleets are optimal, how does the optimal fleet structure look like?

If mixed fleets are profitable, the optimal fleet usually consists of more human-driven than driverless vehicles. In artificial instances with mixed fleets, the fraction of driverless vehicles is around 20%, and the fraction of driverless vehicles increases if the instance has a high imbalance, e.g., due to different arrival rates at different stations. The absolute number of driverless vehicles is not too low in most instances with mixed fleets, since the availability of driverless vehicles would be very low otherwise, reducing profits.

Under which circumstances can shared mobility operators benefit from introducing driverless vehicles in their fleet?

We observe that operators benefit from introducing driverless vehicles in their fleet mix, unless operational profits, i.e., contribution margins minus costs, as well as rebalancing costs are almost equal for driverless and human-driven vehicles, and investment costs differ substantially. In a case study using NYC taxi data, the imbalance is even so large, that the operator optimally only procures driverless vehicles, while the lower imbalance and higher demand in the DiDi case study results in the optimality of a mixed fleet.

How much can operators gain with respect to total profits from using mixed fleets containing driverless vehicles?

Operators should consider driverless vehicles even if their service was not profitable with only human-driven vehicles, since the higher operational profits can make the service profitable. If contribution margins differ and instances are highly imbalanced, profit gains can be several orders of magnitude. The profit gain due to mixed fleets compared to purely driverless fleets is lower, on average around 2%. This is because mixed fleets are only optimal if contribution margins are equal, i.e., profit differences are solely due to differences in investment and rebalancing costs.

Acknowledgements. While working on this thesis, the author was a doctoral student in the School of Management at the Technical University of Munich under the supervision of Prof. Stefan Minner who co-authored all papers. In addition to her supervisor, she would also like to thank her further co-authors Diogo Poças and Andreas S. Schulz (first paper), and Marco Pavone and Maximilian Schiffer (third paper). The author gratefully acknowledges the support of Deutsche Forschungsgemeinschaft as part of the research training group 2201 (Advanced Optimization in a Networked Economy).

References

1. Wagner, S., Brandt, T., Neumann, D.: In free float: developing business analytics support for carsharing providers. Omega **59**, 4–14 (2016)
2. Huang, K., de Almeida Correia, G., An, K.: Solving the station-based one-way carsharing network planning problem with relocations and non-linear demand. Transp. Res. Part C Emerg. Technol. **90**, 1–17 (2018)
3. Hao, W., Martin, L.: Prohibiting cherry-picking: Regulating vehicle sharing services who determine fleet and service structure. Transp. Res. Part E: Logistics Transp. Rev. **161** (2022). https://www.sciencedirect.com/science/article/pii/S1366554522000849
4. Chen, L., Mislove, A., Wilson, C.: Peeking beneath the hood of Uber. In: Proceedings of the 2015 Internet Measurement Conference, pp. 495–508 (2015)
5. Guda, H., Subramanian, U.: Your Uber is arriving: managing on-demand workers through surge pricing, forecast communication, and worker incentives. Manag. Sci. **65**(5), 1995–2014 (2019)
6. Illgen, S., Höck, M.: Literature review of the vehicle relocation problem in one-way car sharing networks. Transp. Res. Part B Methodol. **120**, 193–204 (2019)
7. Laporte, G., Meunier, F., Wolfler Calvo, R.: Shared mobility systems: an updated survey. Ann. Oper. Res. **271**(1), 105–126 (2018). https://doi.org/10.1007/s10479-018-3076-8
8. He, L., Mak, H.-Y., Rong, Y.: Operations management of vehicle sharing systems. In: Hu, M. (ed.) Sharing Economy. SSSCM, vol. 6, pp. 461–484. Springer, Cham (2019). https://doi.org/10.1007/978-3-030-01863-4_19
9. Martin, L.: Rebalancing in shared mobility systems – competition, feature-based mode selection and technology choice. Doctoral dissertation, Technische Universität München (2020)
10. Kortum, K., Schönduwe, R., Stolte, B., Bock, B.: Free-floating carsharing: city-specific growth rates and success factors. Transp. Res. Procedia **19**, 328–340 (2016)

11. Perboli, G., Ferrero, F., Musso, S., Vesco, A.: Business models and tariff simulation in car-sharing services. Transp. Res. Part A Policy Pract. **115**, 32–48 (2018)
12. Martin, L., Minner, S., Poças, D., Schulz, A.S.: The competitive pickup and delivery orienteering problem for balancing carsharing systems. Transp. Sci. **55**(6), 1232–1259 (2021)
13. Martin, L., Minner, S.: Feature-based selection of carsharing relocation modes. Transp. Res. Part E Logist. Transp. Rev. **149**, 102270 (2021)
14. Qi, W., Shen, Z.: A smart-city scope of operations management. Prod. Oper. Manag. **28**(2), 393–406 (2019)
15. Martin, L., Minner, S., Pavone, M., Schiffer, M.: It's all in the mix: technology choice between driverless and human-driven vehicles in sharing systems. Working Paper (2021)
16. Braverman, A., Dai, J., Liu, X., Ying, L.: Empty-car routing in ridesharing systems. Oper. Res. **67**(5), 1437–1452 (2019)

Optimal Line Planning in the Parametric City

Berenike Masing(✉)

Zuse Institute Berlin, Berlin, Germany
masing@zib.de

Abstract. We formulate the line planning problem in public transport as a mixed integer linear program ($MILP$), which selects both passenger and vehicle routes, such that travel demands are met with respect to minimized travel times for both operators and users. We apply $MILP$ to the *Parametric City*, a generic city model developed by Fielbaum et al. [2]. While the infrastructure graph and demand are entirely rotation symmetric, asymmetric optimal line plans can occur. Using group theory, we analyze the properties of symmetric solutions and introduce a *symmetry gap* to measure their deviation of the optimum. We also develop a $1 + \frac{1+\sqrt{2}}{g}$-approximation algorithm, depending only on the cost related parameter g. Supported by computational experiments, we conclude that in practice symmetric line plans provide good solutions for the line planning problem in the Parametric City.

Keywords: Line planning · City modelling · Symmetry · Mixed integer programming · Approximation algorithm

1 Introduction

The goal of line planning is to determine the most efficient routes, as well as frequencies of service in order to satisfy travel demands in a city. We do so with the help of a mixed integer linear programming problem ($MILP$) formulation, whose objective is to minimize both operator as well as passenger travel times combined by a scalarization parameter. It considers all circuits in a graph as potential lines and all simple paths as passenger routes. Good line plans must generally be computed—at great expense due to the model size—for each city individually, their solutions are difficult to compare and cannot be applied to other cities. Our approach is therefore use the *Parametric City*, a generic model developed by Fielbaum et al. [2] for the purpose of designing transportation services. It can be adjusted to represent the most characteristic aspects of the city, such as its geography, as well as the degree of mono-, polycentricity and dispersion. The Parametric City is entirely rotation symmetric—it is therefore natural to assume that this symmetry is reflected in the optimal line plans. However, there are cases, in which the optimal line plans are asymmetric. Our main attention is on this influence of symmetry: On the optimal solutions and how much a symmetric solution deviates from its optimum. We examine in which

N. Trautmann and M. Gnägi (Eds.): OR 2021, LNOR, pp. 39–44, 2022.
https://doi.org/10.1007/978-3-031-08623-6_7

cases optimal solutions must be symmetric, when they can be utilized as good approximations and in which cases it is detrimental to assume symmetry in the line plans.

2 The Parametric City

2.1 The Model

We choose the Parametric City [2] as city representative, since this model balances generality and simplicity – it can represent any city and its prominent features, while remaining simple enough to be analyzable. It is comprised of a infrastructure graph $\mathcal{G} = (V, A)$ (see Fig. 1) with $2n + 1$ vertices and a demand $d_{s,t}, (s, t) \in V \times V$ (see Table 1). Table 2 gives an overview of the parameters.

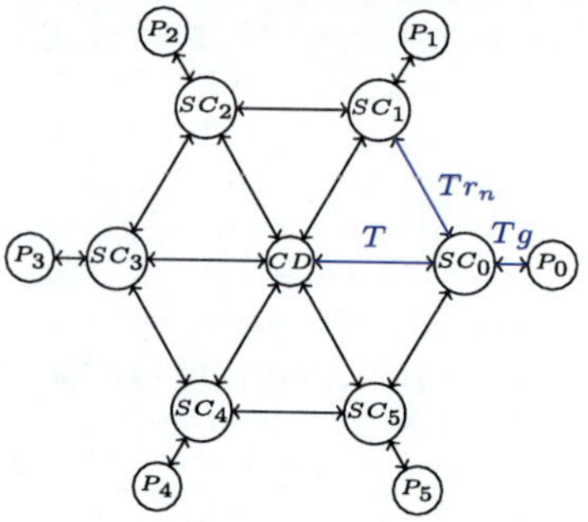

Fig. 1. Graph $\mathcal{G}$ with $n = 6$

Table 1. Demand $d_{s,t}$ (not listed vertex-pairs correspond to $d_{s,t} = 0$)

s,t	SC_i	$SC_j, j \neq i$	CD
P_i	$\frac{aY}{n}\beta$	$\frac{aY}{n(n-1)}\gamma$	$\frac{aY}{n}\alpha$
SC_i	0	$\frac{(1-a)Y}{n(n-1)}\tilde{\gamma}$	$\frac{(1-a)Y}{n}\tilde{\alpha}$

Table 2. Parameters in the Parametric City, f.o.t. = fraction of travelers $\in [0, 1]$

n	no. of subcenters/peripheries
T	arc length (SC_i, CD)
g, r_n	factors for arc length $(SC_i, P_i), (SC_i, SC_{i\pm1})$
Y	total patronage
a	f.o.t. from P_i
$\alpha\,(\tilde{\alpha})$	f.o.t. from $P_i\,(SC_i)$ to CD
β	f.o.t. from P_i to SC_i
$\gamma\,(\tilde{\gamma})$	f.o.t. from $P_i\,(SC_i)$ to SC_j, $i \neq j$
	$\alpha + \beta + \gamma = 1$, $\tilde{\alpha} + \tilde{\gamma} = 1$, $\alpha/\gamma = \tilde{\alpha}/\tilde{\gamma}$

2.2 Rotation Symmetry

The graph $\mathcal{G}$ is evidently rotation symmetric. While the demand matrix is not symmetric in the usual sense, the demand itself is rotation symmetric as well: E.g., the demand from a periphery P_i to the central business district CD is the same as that of any other periphery P_j to CD. This notion of symmetry can be precisely defined with the help of group actions:

The group $G = \mathbb{Z}/n\mathbb{Z}$ acts on the vertices of $\mathcal{G}$ by rotation around CD. This action extends to any tuple of vertices, in particular arcs, path, and lines. We denote by $G \cdot x$ the group orbit of a vertex (tuple) x. E.g., $G \cdot SC_0$ corresponds to the set of all subcenters, and $G \cdot CD = \{CD\}$. With this group action, we describe the rotation symmetry of the demand by the property of $d_{s,t} = d_{s',t'}$ for all $(s', t') \in G \cdot (s, t)$ for any vertex tuple $(s, t) \in V \times V$.

3 The Line Planning Problem

We formulate the line planning problem ($MILP$) as a mixed integer program using two types of variables: $y_p \in \mathbb{R}$ for the passenger flow on path $p \in P$, and $f_l \in \mathbb{N}$ for the frequency of line $l \in L$. P is the set of all simple paths, while L is the line pool consisting of all simple directed cycles in $\mathcal{G}$. We denote the sets P_a and L_a as the set of paths and lines which use arc $a \in A$. Further, $P_{s \to t}$ is the set of all s-t-paths. This follows [1], with a few minor changes: We restrict to only one mode of transport, do not include line-activation costs, and expand the line pool to include all circular, not only bidirectional lines.

$(MILP)$	$(MILP_A)$	(1)
$\min \mu \sum_{l \in L} c_l f_l + (1-\mu) \sum_{p \in P} c_p y_p$	$\min \mu \sum_{a \in A} c_a F_a + (1-\mu) \sum_{p \in P} c_p y_p$	(2)
$\text{s.t.} \sum_{p \in P_{s \to t}} y_p = d_{s,t}$	$\text{s.t.} \sum_{p \in P_{s \to t}} y_p = d_{s,t} \quad \forall (s,t) \in V \times V$	(3)
$\sum_{p \in P_a} y_p - \sum_{l \in L_a} f_l K \leq 0$	$\sum_{p \in P_a} y_p - F_a K \leq 0 \quad \forall a \in A$	(4)
$\sum_{l \in L_a} f_l \leq \Lambda$	$F_a \leq \Lambda \quad \forall a \in A$	(5)
	$\sum_{a \in \delta_v^+} F_a - \sum_{a \in \delta_v^-} F_a = 0 \quad \forall v \in V$	(6)
$f_l \in \mathbb{N} \quad \forall l \in L$	$F_a \in \mathbb{N} \quad \forall a \in A$	(7)
$y_p \geq 0$	$y_p \geq 0 \quad \forall p \in P$	(8)

For a solution (f, y) of $MILP$, $f = (f_l)_{l \in L}$ is called the line plan and $y = (y_p)_{p \in P}$ the passenger flow. A line l is part of the line plan if and only if $f_l > 0$, analogously for the passenger flow.

We refer to [5] for an explanation of the constraints. The objective is a combination of operator and user costs respectively and are scalarized by parameter $\mu \in [0, 1]$. The larger μ, the more focus lies on the minimization of operator costs, while a small μ aims at user-friendly line plans. We consider the running and travel times as the total length of a line or path, i.e., $c_l = \sum_{a \in l} c_a$ and $c_p = \sum_{a \in p} c_a$, where c_a is the length of arc $a \in A$ as defined in the Parametric City, cf. Fig. 1.

As costs depend on the arc-lengths of the routes only, we can reformulate and hence reduce the model size significantly: Instead of considering the large line pool as variables, one can consider the frequencies of all aggregated lines traversing an arc, i.e., by considering $F_a := \sum_{l \in L_a} f_l$. To model the circulations of the lines, we can impose standard flow conservation constraints (6), where δ_v^+ and δ_v^- denote the set of out- and incoming arcs at node v respectively. The entire arc-based mixed integer linear programming problem $MILP_A$ can be found on the right of Definition 3. In fact, the following holds:

Lemma 1. *$MILP$ and $MILP_A$ are equivalent, in the sense that for a feasible solution (F, y) to $MILP_A$ there exists a feasible solution (f, y) to $MILP$ with $cost_A(F, y) = cost(f, y)$ and vice versa.*

4 Symmetry

Definition 1 (Symmetric Solution). *Consider a solution (f, y) to $MILP$ and the equivalent solution (F, y) to $MILP_A$. Then*

1. *(f, y) is line-symmetric if $f_l = f_{l'}$ for all $l' \in G \cdot l, l \in L$,*
2. *(f, y) or (F, y) is path-symmetric if $y_p = y_{p'}$ for all $p' \in G \cdot p, p \in P$,*
3. *(f, y) is arc-symmetric if $\sum_{l \in L_a} f_l = \sum_{l \in L_{a'}} f_l$ for all $a' \in G \cdot a, a \in A$,*
4. *(F, y) is arc-symmetric if $F_a = F_{a'}$ for all $a' \in G \cdot a, a \in A$.*

The solution (f, y) is symmetric if it is line- and path-symmetric, while (F, y) is symmetric if it is arc- and path-symmetric.

Proposition 1 (Sufficient condition for symmetry). *A line-symmetric, arc-symmetric or path-symmetric optimal solution is sufficient for the existence of an entirely symmetric optimal solution.*

Thus, to determine an optimal value of symmetric solutions, it is enough to impose symmetric arc-frequency conditions (Definition 1, 4) on $MILP_A$. We denote the model as $MILP_{sym}$. As we have only six orbits on the set of arcs in the Parametric City, $MILP_{sym}$ reduces to a problem with a fixed number of variables. Due to further geometric properties, this number gets reduced even further to only three variables. This has significant consequences: As was proven by Lenstra [4], a mixed integer programming problem with a fixed number of variables can be solved in polynomial time. Hence:

Proposition 2. *The symmetric line planning problem $MILP_{sym}$ is solvable in polynomial time.*

Given a feasible general solution (F, y), it is always possible to construct a feasible symmetric solution (F^s, y^s) by taking (rounded) averages per orbit. This allows for an estimate of how much the optimal solution deviates from a symmetric one at most:

$$cost(F^s, y^s) - cost(F, y) \leq \mu 2T(1 + r_n)(n - 1). \tag{9}$$

Consequently, if we optimize for user comfort only, i.e., if $\mu = 0$, the existence of a symmetric optimal line plan is guaranteed. However, for other values of μ we introduce the *symmetry gap* $\Gamma := \frac{OptVal(MILP_{sym})}{OptVal(MILP_A)}$. This gives us the means to measure the quality of a symmetric solution compared to an asymmetric one.

Proposition 3. *The relative symmetry gap Γ is bound by:*

$$\Gamma \leq C_n(\alpha, \gamma) \leq C_n \leq 1 + \frac{(1+\sqrt{2})}{g}.$$

The values of $C_n(\alpha, \gamma)$ and C_n can be computed efficiently by using the bound (9) and the optimal value of the LP-relaxation, which can be computed analytically; a detailed description can be found in [5]. Due to $MILP_{sym}$ being solvable in polynomial time and in combination with Proposition 3, we can formulate the following proposition:

Proposition 4. *The algorithm that arises from restricting the line planning problem to symmetric solutions is a $1 + \frac{(1+\sqrt{2})}{g}$-approximation algorithm for the Parametric City if the cost related parameter g is fixed.*

To ascertain how the gap behaves when g goes to 0, consider the following instance: Set $g = 1/n, \mu = 1$ and choose a very large capacity, e.g., set $K = Y$. In other words, the entire patronage can fit into a single vehicle. Regardless of any other parameter choice, the optimal frequency plans of $MILP_A$ and $MILP_{sym}$ are the ones displayed in Fig. 2 on the left and right respectively. The corresponding gap becomes arbitrarily large for $n \to \infty$. However, this extreme example is constructed and does not reflect realistic input data. To assess whether and how often in practice purely asymmetric solutions occur, and how large the gap becomes, we solved multiple large batches of Parametric City instances for realistic parameter choices.

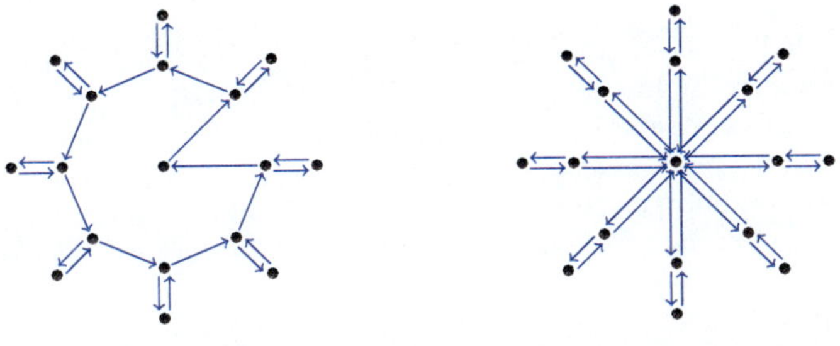

Fig. 2. Optimal frequency plans: general vs. symmetric model

5 Computational Results

We performed multiple computations, choosing various geometry-related parameters in the Parametric City, as well as different values of μ. For each choice, we computed the optimal solution for all demand parameters $\alpha, \gamma \in [0.025, 0.95]$ and a step size of 0.025. Each problem was solved with Gurobi 9 [3] to optimality (with a tolerance of 10^{-4} in the relative MIP gap)in three variations: the standard $MILP_A$, its symmetric version $MILP_{sym}$, as well as the restriction to both rotation and reflection symmetric solutions by imposing additional reflection symmetry constraints. See below a representative example for the choices of $n = 8, g = 1/3, Y = 24000, K = 100, \mu = 1, a = 0.8$. Surprisingly, even with realistic results, purely asymmetric solutions can be found, as becomes evident on the left of Fig. 3. When comparing different values of μ, one can also observe that their number increases with μ. For a more in-depth comparison see [5]. Also noticeable is the fact that reflection symmetry occurs in the rotation symmetric solutions roughly as often as not. This observation is unexpected, since the demand is also reflection symmetric. When looking at the symmetry gap Γ however, it becomes evident that the difference in costs is extremely small. For this instance, the actually computed deviation of symmetric solutions from the optimal is less than 2%, as becomes evident on the right of Fig. 3 and is significantly smaller than the theoretical upper bounds also depicted (cf. Proposition 3).

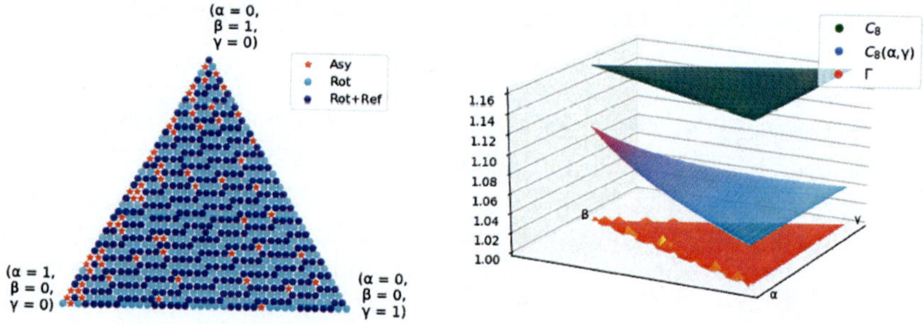

Fig. 3. Realistic input data

We conclude that in practice, city planners are justified in assuming symmetric solutions: The line plans can either be considered as optimal, due to the somewhat idealized underlying city model, or one can use them as good and easy to compute approximations. With the exception of when operator costs are ignored, the possibility of asymmetric solutions should be kept in mind however, as unfortunate parameter choices can lead to a large deviation in costs.

References

1. Borndörfer, R., Grötschel, M., Pfetsch, M.E.: A column-generation approach to line planning in public transport. Transp. Sci. **41**(1), 123–132 (2007). https://doi.org/10.1287/trsc.1060.0161
2. Fielbaum, A., Jara-Diaz, S., Gschwender, A.: A parametric description of cities for the normative analysis of transport systems. Netw. Spat. Econ. **17**(2), 343–365 (2016). https://doi.org/10.1007/s11067-016-9329-7
3. Gurobi Optimization, LLC: Gurobi optimizer reference manual (2020). http://www.gurobi.com
4. Lenstra, H.W.: Integer programming with a fixed number of variables. Math. Oper. Res. **8**(4), 538–548 (1983). http://www.jstor.org/stable/3689168
5. Masing, B.: Optimal line planning in the parametric city. Master's thesis, Technische Universität Berlin (2020)

A Semidefinite Approach for the Single Row Facility Layout Problem

Jan Schwiddessen[1,2(✉)]

[1] Alpen-Adria-Universität Klagenfurt, Klagenfurt, Austria
Jan.Schwiddessen@aau.at
[2] Technische Universität Dortmund, Dortmund, Germany
Jan.Schwiddessen@tu-dortmund.de

Abstract. This paper summarizes several results from the author's Master's thesis. We propose a new semidefinite approach for the $\mathcal{NP}$-hard *single row facility layout problem* (SRFLP) which is the problem of arranging n facilities of given lengths on a straight line, while minimizing a weighted sum of distances between all facility pairs. We tighten existing semidefinite relaxations with further inequalities and use an algorithmic method not yet considered for the SRFLP. Our new approach outperforms all other approaches in the literature and significantly reduces the best known duality gaps for all unsolved benchmark instances with $n \leq 100$. Many instances with up to $n = 81$ are solved for the first time.

Keywords: Single row facility layout · Semidefinite optimization

1 The Single Row Facility Layout Problem

In the *single row facility layout problem* (SRFLP), n one-dimensional facilities with positive integer lengths ℓ_i for each facility $i \in [n] := \{1, \dots, n\}$ and pairwise integer weights c_{ij}, $i, j \in [n]$, $i < j$, are given. The task is to find a layout or permutation $\pi \in \Pi_n$ of the facilities that minimizes the total weighted sum of center-to-center distances between all facility pairs:

$$\min_{\pi \in \Pi_n} \sum_{i,j \in [n],\, i<j} c_{ij} d_{ij}^{\pi}, \qquad \text{(SRFLP)}$$

where Π_n denotes the set of all layouts of n facilities and d_{ij}^{π} denotes the distance between facilities i and j with respect to their centroids in the layout π.

If all facilities have unit length and all weights c_{ij} have values in $\{0, 1\}$, the SRFLP reduces to the *minimum linear arrangement problem* that is already $\mathcal{NP}$-hard in the strong sense [4]. An important practical application of this problem is given by the optimal ordering of machines in flexible manufacturing systems [6]. The two best approaches in the literature use linear [1] and semidefinite [7,8] programming relaxations. Instances with $n \leq 42$ were solved in [7].

In Sect. 2, we outline the approaches in [1] and [8]. Based on these, we improve existing semidefinite relaxations in Sect. 3 and propose to use an algorithmic

N. Trautmann and M. Gnägi (Eds.): OR 2021, LNOR, pp. 45–51, 2022.
https://doi.org/10.1007/978-3-031-08623-6_8

method not yet considered for the SRFLP. Finally, we summarize our numerical results in Sect. 4.

2 The Most Successful Approaches for the SRFLP

An integer programming formulation (BTW) of the SRFLP is given in [1]:

$$\min \sum_{i,j\in[n],\, i<j} c_{ij} \sum_{k\in[n]\setminus\{i,j\}} \ell_k \quad b_{ikj} + \sum_{i,j\in[n],\, i<j} c_{ij} \frac{\ell_i+\ell_j}{2} \tag{1}$$

$$\begin{aligned}
\text{s.t.}\quad & b_{ijk}+b_{ikj}+b_{jik}=1, && i,j,k\in[n],\ i<j<k, && (2)\\
& b_{ihj}+b_{ihk}+b_{jhk}\le 2, && i,j,k,h\in[n],\ |\{i,j,k,h\}|=4,\ i<j<k, && (3)\\
& -b_{ihj}+b_{ihk}+b_{jhk}\ge 0, && i,j,k,h\in[n],\ |\{i,j,k,h\}|=4,\ i<j<k, && (4)\\
& +b_{ihj}-b_{ihk}+b_{jhk}\ge 0, && i,j,k,h\in[n],\ |\{i,j,k,h\}|=4,\ i<j<k, && (5)\\
& +b_{ihj}+b_{ihk}-b_{jhk}\ge 0, && i,j,k,h\in[n],\ |\{i,j,k,h\}|=4,\ i<j<k, && (6)\\
& b_{ijk}\in\{0,1\}, && i,j,k\in[n],\ |\{i,j,k\}|=3,\ i<k, && (7)
\end{aligned}$$

where the so-called *betweenness variables* $b_{ijk}\in\{0,1\}$ have the meaning

$$b_{ijk}=\begin{cases}1, & \text{if facility } j \text{ lies between } i \text{ and } k,\\ 0, & \text{otherwise.}\end{cases} \tag{8}$$

As shown in [1], the linear relaxation of (BTW) can be strengthened by a family of inequalities containing inequalities (4)–(6): let $\beta\le n$ be an even integer and let $R\subseteq[n]$ be such that $|R|=\beta$. Then for each $r\in R$, and for any partition (S,T) of $R\setminus\{r\}$ such that $|S|=\beta/2$, the inequality

$$\sum_{p,q\in S,\, p<q} b_{prq} + \sum_{p,q\in T,\, p<q} b_{prq} \le \sum_{p\in S,\, q\in T,\, p<q} b_{prq} \tag{9}$$

is valid [1] and facet-defining [11] for the *betweenness polytope* $\mathcal{P}_{BTW}$. We obtain inequalities (4)–(6) for $\beta=4$. Thus, inequalities (4)–(6) are also facet-defining, whereas inequalities (3) do not define facets in general [11].

Lower bounds for the SRFLP were computed in [1] by solving the linear relaxation of (BTW) enhanced with inequalities (9) for $\beta=6$ as cutting planes. These lower bounds turned out to be the optimal values of all instances considered in [1]. Instances with up to $n=35$ were solved in several hours.

Despite these strong results, it is easy to find instances with $n=6$ for which the lower bounds do not coincide with the optimal values. One may wonder, why no branch-and-cut approach was proposed in [1]. The reason is that the (dual) simplex method shows a very poor performance when applied to (BTW). Thus, the occurring linear programs in the cutting plane approach were solved with interior-point methods. Indeed, our tests showed that interior-points methods would outperform the dual simplex method in a branch-and-cut approach, although all linear programs have to be solved from scratch. However, the dual bounds increase very slowly.

The best approach for the SRFLP in the literature goes back to [8] and uses semidefinite relaxations to compute tight lower bounds. To state these relaxations, we introduce bivalent *ordering variables* y_{ij}, $i < j \in [n]$, $i < j$, with the meaning

$$y_{ij} = \begin{cases} +1, & \text{if facility } i \text{ lies to the left of facility } j, \\ -1, & \text{otherwise.} \end{cases} \tag{10}$$

We collect all ordering variables (10) in a vector y and consider the matrix

$$\overline{Y} := \begin{pmatrix} 1 \\ y \end{pmatrix} \begin{pmatrix} 1 \\ y \end{pmatrix}^{\top} = \begin{pmatrix} 1 & y^{\top} \\ y & yy^{\top} \end{pmatrix} = \begin{pmatrix} 1 & y^{\top} \\ y & Y \end{pmatrix},$$

where $Y_{ij,kl} = y_{ij}y_{kl}$. The semidefinite relaxation in [8] then reads

$$\begin{aligned} \min \quad & \langle C, Y \rangle + K \\ \text{s.t.} \quad & Y_{ij,jk} - Y_{ij,ik} - Y_{ik,jk} = -1, \qquad i,j,k \in [n],\ i < j < k, & (11) \\ & \operatorname{diag}(\overline{Y}) = e, \quad \overline{Y} \succeq 0, \\ & \overline{Y} \in \mathcal{M}, \quad \overline{Y} \in \mathcal{LS}. & (12) \end{aligned}$$

Here, the cost matrix C and the constant K have to be chosen appropriately, and e denotes the vector of all ones. Moreover, $\mathcal{M}$ denotes the *metric polytope* that is defined by the set of all *triangle inequalities*, known to be facet-defining for the cut polytope [3]. For a subset $\{p_1, p_2, p_3\}$ of three rows (or columns) of a symmetric matrix $X \in \mathbb{R}^{p \times p}$, the triangle inequalities can be written as

$$\sum_{1 \le i < j \le 3} \delta_i \delta_j X_{p_i, p_j} \ge -1, \tag{13}$$

where $\delta_k \in \{-1, 1\}$, $k = 1, 2, 3$. Thus, there are $\mathcal{O}(n^6)$ triangle inequalities in our case. The set $\mathcal{LS}$ contains $\mathcal{O}(n^5)$ 'matrix cuts' that can be derived from equations (11). The latter are also called *3-cycle-equations*.

In [8], a partial Lagrangian approach is used to dualize the 3-cycle equations (11), triangle inequalities (13), and the matrix cuts. This results in a convex but non-smooth dual function that is optimized using a bundle method. The value of the dual function and a subgradient are obtained by solving a max-cut problem using interior-point methods. Using this approach, instances with $n \le 42$ were solved in less than two hours in [7]. Tight lower bounds for instances with up to $n = 100$ were obtained in several days of computing time.

3 A New Semidefinite Approach

Strengthened Relaxation. We exclude the matrix cuts $\overline{Y} \in \mathcal{LS}$ (12) from our relaxations, since they barely improve the lower bounds. Instead we strengthen the semidefinite relaxations with the so-called *pentagonal*, *hexagonal*, and *heptagonal* inequalities that are also facet-defining for the cut polytope [3].

For a subset $\{p_1, \ldots, p_5\}$ of row indices of a symmetric matrix $X \in \mathbb{R}^{p \times p}$, all choices $\delta_k \in \{-1, 1\}$, $k = 1, \ldots, 5$, the pentagonal inequalities can be written as

$$\sum_{1 \leq i < j \leq 5} \delta_i \delta_j X_{p_i, p_j} \geq -2.$$

There are $\mathcal{O}(n^{10})$ pentagonal inequalities for an SRFLP instance with n facilities. Hence, exact separation by enumeration is not computationally feasible. However, there is one important subset of only $\mathcal{O}(n^6)$ pentagonal inequalities that greatly improves the semidefinite relaxations for large-scale instances.

Reconsider the inequalities (9) for $\beta = 6$: for any $R = \{i, j, k, l, m, r\} \subseteq [n]$ and any partition (S, T) of $R \setminus \{r\}$ with $|S| = 3$, the resulting inequality is a pentagonal inequality on the facility pairs (i, r), (j, r), (k, r), (l, r), (m, r) if the betweenness variables (8) are written as quadratic expressions (see [7]) of the ordering variables (10). Interpreting the (unordered) pairs as the edge set of an undirected graph yields a star with the six vertices $\{i, j, k, l, m, r\}$ and the center vertex r. Thus, we call all pentagonal inequalities with the above structure *starlike pentagonal inequalities* and denote the set of all starlike pentagonal inequalities by $\mathcal{P}^*$. We propose the following semidefinite relaxation for the SRFLP:

$$\begin{aligned}
\min\ & \langle C, Y \rangle + K \\
\text{s.t. } & Y_{ij,jk} - Y_{ij,ik} - Y_{ik,jk} = -1, \quad i, j, k \in [n],\ i < j < k, \\
& \operatorname{diag}(\overline{Y}) = e, \quad \overline{Y} \succeq 0, \\
& \overline{Y} \in \mathcal{M}, \quad Y \in \mathcal{P}^*.
\end{aligned} \qquad (\mathrm{SDP}_{\mathcal{P}^*})$$

Proposition 1. *The optimal value of ($SDP_{\mathcal{P}^*}$) is greater than or equal to the optimal value of the linear relaxation of model (BTW) enhanced by the cutting planes (9) for $\beta = 6$.*

Additionally, we also use generic pentagonal, hexagonal, and heptagonal inequalities that are separated heuristically. For each type, we use two simple separation routines that try to find highly violated inequalities. Both routines are called several times until a prescribed time limit is reached. The first routine chooses a small random subset of row indices and enumerates all respective inequalities on these row indices. The second routine implements a 1-opt local search heuristic. It starts with a random inequality and swaps a single row index until a local maximum of violation is found.

Primal Heuristics. In order to find good primal feasible solutions, i.e., a permutation of the facilities, we first apply the famous Goemans-Williamson hyperplane rounding heuristic [5] to the (approximate) solution of our semidefinite relaxation. This gives us a vector $\overline{y}$ of ordering variables that might not represent a permutation, i.e., $\overline{y}$ might violate the 3-cycle equations (11). Similar to [8], we use two different strategies to make such a vector feasible and then apply a 2-opt local search heuristic by swapping two facilities in each step.

The first strategy borrows an idea from [2]. For each facility we assign to it the number of other facilities that *should* lie on the left according to the definition of the ordering variables (10). We then compute a feasible solution by sorting the facilities in an ascending order, where ties can be broken arbitrarily.

The second strategy directly works on the 3-cycle equations (11). The idea is that we use $\overline{y}$ to construct a new feasible vector y step-by-step. In each iteration we choose an undetermined entry y_{ij} of y and assign the value $\overline{y}_{ij}$ to it. We then check all 3-cycle-equations (11) and fix all entries in y that can only attain one specific value, i.e., that are implicitly fixed by prior assignments. We iterate with this procedure until all entries of y are fixed.

Algorithmic Approach. We compute tight lower bounds for the SRFLP by approximately solving our semidefinite relaxations in a cutting plane approach. For almost all benchmark instances in the literature, these lower bounds are sufficient to prove the optimality of a feasible layout found by our primal heuristics.

Our semidefinite relaxations can be written in the more compact form:

$$\min\left\{\langle C, X\rangle : \mathcal{A}(X) \leq a,\ \mathcal{B}(X) = b,\ X \succeq 0\right\}, \tag{SDP}$$

where all inequalities are symbolically collected as $\mathcal{A}(X) \leq a$ and all equations as $\mathcal{B}(X) = b$. We then use the family of semidefinite bounds presented in [10] and adapt the algorithmic approach in [9] to solve (SDP) approximately, i.e., for some decreasing penalty parameter $\alpha > 0$ we approximately solve the regularized dual problem

$$\begin{aligned} \sup &\left\{-a^\top\lambda - b^\top\mu \;-\; \tfrac{\alpha}{2}N^2 - \tfrac{1}{2\alpha}\left\|\left[C + \mathcal{A}^\top(\lambda) + \mathcal{B}^\top(\mu)\right]_-\right\|_F^2\right\} \\ \text{s.t. } &\lambda \geq 0,\ \mu \text{ free,} \end{aligned} \tag{DSDP$_\alpha$}$$

where N is the order of the matrix variable X, $\|\cdot\|_F$ denotes the Frobenius norm, and $[\cdot]_-$ is the projection onto the cone of negative semidefinite matrices.

(DSDP$_\alpha$) is a convex optimization problem with a differentiable objective function [10] and is solved using a quasi-Newton method. For any triple (α, λ, μ) with $\alpha > 0$ and $\lambda \geq 0$, the objective function of (DSDP$_\alpha$) yields a lower bound on the optimal value of the SRFLP. Since strong duality holds in our SRFLP setting [8], we can get arbitrarily close to the usual semidefinite bound of (SDP) by approximately solving (DSDP$_\alpha$) for $\alpha > 0$ small enough [10].

In each iteration of the cutting plane algorithm, (DSDP$_\alpha$) is optimized for a fixed $\alpha > 0$ until the primal variable $X = -\frac{1}{\alpha}\left[C + \mathcal{A}^\top(\lambda) + \mathcal{B}^\top(\mu)\right]_-$ almost satisfies all linear constraints, see also [9,10]. We then use the approximate solution to compute feasible layouts and to add a few thousand highly violated cuts. For each value of the penalty parameter α, we first only separate triangle inequalities. Later we also separate starlike pentagonal inequalities and finally also general pentagonal, hexagonal, and heptagonal inequalities. We drop inequalities whose Lagrange multipliers are close to zero.

4 Results and Conclusion

The new semidefinite approach proposed in Sect. 3 has been tested on many benchmark instances from the literature with up to $n = 100$. For the first time, instances with $n = 81$ have been solved. While instances with $n \leq 70$ were solved in a couple of hours, computing the proof of optimality for even larger instances can take more than two days. For each unsolved instance, the best known duality gap in the literature is reduced by a factor of 10 to 1000.

A second version that only solves ($\text{SDP}_{\mathcal{P}*}$) without further heuristically separated pentagonal, hexagonal, and heptagonal inequalities was also tested. It uses slightly different settings of numerical parameters and was designed to be more time efficient. It is much faster than any other approach in the literature, but sometimes fails on solving instances with $n \geq 56$. However, the duality gaps are always close to zero.

We observed that adding the starlike pentagonal inequalities leads to a significant improvement of the semidefinite relaxations. General pentagonal and heptagonal inequalities also improve the lower bounds for large-scale instances, but the inclusion of hexagonal inequalities does not pay off.

To conclude, semidefinite relaxations for the SRFLP were improved and a suitable algorithmic approach was used to solve the resulting large-scale semidefinite programs. The algorithmic approach allows a good trade-off between tight bounds and fast computation times. Using our semidefinite bounds in a branch-and-bound approach could lead to further improvements in the future.

References

1. Amaral, A.R.S.: A new lower bound for the single row facility layout problem. Discret. Appl. Math. **157**(1), 183–190 (2009)
2. Anjos, M.F., Kennings, A., Vannelli, A.: A semidefinite optimization approach for the single-row layout problem with unequal dimensions. Discret. Optim. **2**(2), 113–122 (2005)
3. Deza, M.M., Laurent, M.: Geometry of Cuts and Metrics, vol. 15. Springer, Heidelberg (2009). https://doi.org/10.1007/978-3-642-04295-9
4. Garey, M.R., Johnson, D.S., Stockmeyer, L.: Some simplified NP-complete problems. In: Proceedings of the Sixth Annual ACM Symposium on Theory of Computing, pp. 47–63 (1974)
5. Goemans, M.X., Williamson, D.P.: .878-approximation algorithms for MAX CUT and MAX 2SAT. In: Proceedings of the Twenty-Sixth Annual ACM Symposium on Theory of Computing, pp. 422–431 (1994)
6. Heragu, S.S., Kusiak, A.: Machine layout problem in flexible manufacturing systems. Oper. Res. **36**(2), 258–268 (1988)
7. Hungerländer, P., Rendl, F.: A computational study and survey of methods for the single-row facility layout problem. Comput. Optim. Appl. **55**(1), 1–20 (2012)
8. Hungerländer, P., Rendl, F.: Semidefinite relaxations of ordering problems. Math. Program. **140**(1), 77–97 (2013)
9. Krislock, N., Malick, J., Roupin, F.: BiqCrunch: a semidefinite branch-and-bound method for solving binary quadratic problems. ACM Trans. Math. Softw. (TOMS) **43**(4), 1–23 (2017)

10. Malick, J., Roupin, F.: On the bridge between combinatorial optimization and nonlinear optimization: a family of semidefinite bounds for 0–1 quadratic problems leading to quasi-newton methods. Math. Program. **140**(1), 99–124 (2013)
11. Sanjeevi, S., Kianfar, K.: A polyhedral study of triplet formulation for single row facility layout problem. Discret. Appl. Math. **158**(16), 1861–1867 (2010)

Novel Planning Problems in E-commerce Warehousing

Felix Weidinger(✉)

Technical University of Darmstadt, Hochschulstraße 1, 64289 Darmstadt, Germany
felix.weidinger@tu-darmstadt.de
https://www.or.wi.tu-darmstadt.de

Abstract. E-commerce led to drastic changes in end consumer behavior over the last years. In consequence, today's retailers often need to provide online channels to stay attractive and competitive. The logistic processes behind this online front-end, needed to smoothly ship small orders, picked from a large assortment in little time, however, are complex and differ from classical warehousing processes to a large extent. As a result, novel warehousing systems have evolved. These can either be adaptions of classical warehousing systems or novel warehousing systems, often based on new technological developments. This paper summarizes today's challenges and describes some new planning problems in modern e-commerce warehouses.

Keywords: Warehousing · E-commerce · Logistics

1 Impact of E-commerce on Logistic Processes

Over the past years, online shopping has enjoyed growing popularity among consumers worldwide. With the changing consumer habits, retailers are forced to develop their supply chains in order to meet the new demand structures. This change can be observed particularly impressively in the warehouses of retailers.

Modern e-commerce warehousing systems are designed to quickly assemble small orders, often including only one or two items, from a large assortment. In times of normal demand, this requirement must be met cost-effectively, whereas, in times of high demand, e.g., in the pre-Christmas season or during Singles Day, a high level of scalability of the systems is necessary to be able to offer the customer a constant quality of service [1,8]. For traditional warehousing systems, these requirements can only be met with massive use of resources - if at all.

The gap between traditional systems and increasing requirements is becoming ever larger as e-commerce retailers are trying to improve their market position in the competitive field of e-commerce retailing with ever new service offerings. For example, with premium shipping programs, which guarantee delivery within the next or even the same day. These programs massively increase the time pressure on order processing. Exceptionally high time restrictions can often be observed for grocery products, for which online retailers compete directly with a dense

N. Trautmann and M. Gnägi (Eds.): OR 2021, LNOR, pp. 52–56, 2022.
https://doi.org/10.1007/978-3-031-08623-6_9

network of branches of stationary retailers. The unicorn start-up Gorillas, for example, promise delivery within 10 min after ordering [3].

Modern warehousing systems designed for picking e-commerce orders adapt traditional systems by reorganizing critical processes or use hardware-based innovations to eliminate bottlenecks. This new generation of warehousing systems includes a variety of different solutions, each of which has new system-specific planning problems. If processes are restructured, fundamental assumptions can change, so that established solution procedures are no longer applicable. Furthermore, new planning problems can arise whenever new technical solutions need to be controlled optimally. The paper gives examples of already established e-commerce warehousing systems and illustrates altered and novel planning problems occurring in such systems.

2 Mixed Shelves Storage Systems

Mixed shelves storage warehouses are adaptions of classical picker-to-parts warehouses, where pickers, often equipped with some kind of picking cart, walk through the warehouse collecting ordered products from person-high racks. Other than in the traditional case, however, goods arriving at a mixed shelves storage warehouse are depalletized and items of the same product, often referred to as stock-keeping unit (SKU), are stowed in different positions of the warehouse. This way, items of a SKU to be picked can be found close by, irrespective of the current position of the picker, and, therefore, unproductive walking times of the pickers can be reduced when processing orders shaped like typical e-commerce orders [6].

The organizational adaptions in such systems, compared to a classical warehouse where items of the same SKU are not actively scattered, result in novel variations of well-known planning problems. For example, new variations of the storage assignment and picker routing problem.

2.1 Storage Assignment in Mixed Shelves Storage Warehouses

Classical storage assignment strategies (see, e.g., [2]) are most often based on the concept of a central depot, where each picking tour starts and ends. In mixed shelves warehouses, however, pickers constantly walk through the storage area while picking orders in dynamic overlapping batches and continually handing off single orders to a distributed network of depots. This results in picking tours not being closed anymore. As, in consequence, a single starting point of each tour, e.g., a central depot in classical warehouses, does not exist anymore, classical storage assignment strategies are not applicable in such systems and a new problem variation of the storage assignment problem occurs [6].

In [6], the authors propose an approach that aims at maximizing the level of scatter within the warehouse. This way, the main advantage of mixed shelves storage, e.g., short travel distances of the pickers towards an item of a demanded SKU irrespective of their current position, is exploited as far as possible.

With this aim in mind, an optimization problem, minimizing the sum of maximum walking distances to reach the closest item of a SKU starting from a set of so-called measuring points, is defined. Hereby, the measuring points can be seen analogously to the depot in classical storage assignment. The difference, however, is that there are multiple measuring points and, further, the optimization problem needs to assign each SKU to multiple storage positions.

2.2 Picker Routing in Mixed Shelves Storage Warehouses

Similar to the storage assignment problem, the picker routing problem in mixed shelves storage warehouses is a new variation of a classical planning problem in warehousing. Although the objective aimed at, e.g., finding a shortest feasible picking tour, remains unchanged, organizational adaptions to the needs of e-commerce warehousing alter the problem significantly for mixed shelves storage systems.

In the vast majority of cases, the shelves within a warehouse are arranged in a rectangular manner. The structured layout, hereby, leads to a likewise structured distance matrix for the picker routing problem. In classical warehouses, with only one storage position per SKU, the strong structure of the distance matrix can be exploited to solve the picker routing problem in polynomial time depending on the number of aisles [4].

The picker routing problem in mixed shelves storage warehouses with a rectangular layout, however, belongs to the class of NP-hard optimization problems, although distance matrices for this setting do not differ from the classical case. The increased problem complexity is caused by an additional selection component, as in mixed shelves storage warehouses each SKU demanded by an order is provided by several storage positions and not all of them need to be visited by the picker during the tour [5]. The picker routing problem, therefore, is another example of a changed problem structure caused by adapting a classical picker-to-parts warehousing strategy to the needs of modern e-commerce warehousing.

3 Robotic Mobile Fulfillment Systems

Another way of avoiding unproductive walking times of pickers is realized by so-called robotic mobile fulfillment systems. Here, the pickers stay within their picking stations, and SKUs, stored in portable racks, are brought to them by autonomously driving robots (see Fig. 1a). By utilizing a fleet of autonomously driving robots in combination with basic warehousing equipment, a parts-to-picker system is realized that overcomes major drawbacks of classical parts-to-picker systems in e-commerce warehousing, as, for example, high investment costs and inflexible hardware. As a result, these systems can be found in a wide variety of e-commerce warehouses, manufactured by many different system providers.

Novel systems, as robotic mobile fulfillment systems, of course, come with new interesting planning problems. Details on the planning hierarchy in these

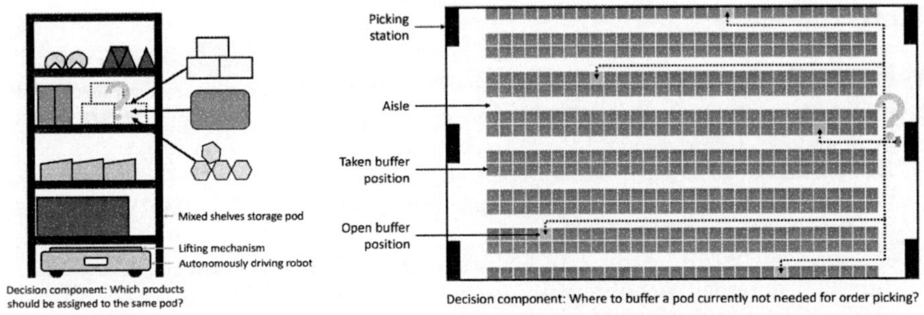

(a) pod and robot (b) warehouse from an aerial view

Fig. 1. Sketches of two described decision problems

systems can be found in the appendix of [7]. A vivid example of new planning problems within robotic mobile fulfillment systems is the storage assignment decision. The SKUs are, comparable to picker-to-parts mixed shelves storage, scattered over multiple portable racks, often denoted as pods. In consequence, one aspect to consider during storage assignment is the number of pods a SKU should be spread over. A further element of the storage assignment decision is deciding which SKUs should share the same pod (at least once, see Fig. 1a), because they might often be ordered together.

As pods are movable racks, however, the pods themselves need to be temporarily assigned to buffer positions within the warehouse whenever they are currently not needed at a picking station (see Fig. 1b), such that an additional component of the problem arises. In conclusion, a multi-level storage assignment problem with many components can be observed, substantially different from classical storage assignment because of many novel aspects.

4 Conclusion

The examples in this paper show how online channels, meant to sell products via the internet, have also had an impact on logistics processes, and in particular on warehousing. New warehousing systems have emerged and are still emerging, coming with many interesting planning problems that are either variations of already known problems or completely novel. Although many efforts have already been made to make those warehousing systems as performant as possible, taking into account the needs associated with e-commerce warehousing, they are continuously challenged by new trends. Important trends that are likely to lead to interesting new approaches in the near future are extremely short delivery times, the shipment of fresh food, and the simultaneous distribution of goods across many sales channels, known as omnichannel strategies. Facing these challenges, the field of warehousing will stay a vivid field of research.

References

1. Boysen, N., De Koster, R., Weidinger, F.: Warehousing in the e-commerce era: a survey. Eur. J. Oper. Res. **277**(2), 396–411 (2019). https://doi.org/10.1016/j.ejor.2018.08.023
2. De Koster, R., Le-Duc, T., Roodbergen, K.J.: Design and control of warehouse order picking: a literature review. Eur. J. Oper. Res. **182**(2), 481–501 (2007). https://doi.org/10.1016/j.ejor.2006.07.009
3. O'Hear, S. (2020). https://techcrunch.com/2020/12/11/gorillas/. Accessed 12 July 2021. Internet Archive: https://web.archive.org/web/20210712141757/https://techcrunch.com/2020/12/11/gorillas/
4. Ratliff, H.D., Rosenthal, A.S.: Order-picking in a rectangular warehouse: a solvable case of the traveling salesman problem. Oper. Res. **31**(3), 507–521 (1983). https://doi.org/10.1287/opre.31.3.507
5. Weidinger, F.: Picker routing in rectangular mixed shelves warehouses. Comput. Oper. Res. **95**, 139–150 (2018)
6. Weidinger, F., Boysen, N.: Scattered storage: how to distribute stock keeping units all around a mixed-shelves warehouse. Transp. Sci. **52**(6), 1412–1427 (2018). https://doi.org/10.1287/trsc.2017.0779
7. Weidinger, F., Boysen, N., Briskorn, D.: Storage assignment with rack-moving mobile robots in KIVA warehouses. Transp. Sci. **52**(6), 1479–1495 (2018). https://doi.org/10.1287/trsc.2018.0826
8. Weidinger, F., Boysen, N., Schneider, M.: Picker routing in the mixed-shelves warehouses of e-commerce retailers. Eur. J. Oper. Res. **274**(2), 501–515 (2019). https://doi.org/10.1016/j.ejor.2018.10.021

Analytics

Machine Learning Constructives and Local Searches for the Travelling Salesman Problem

Tommaso Vitali[1], Umberto Junior Mele[1,2(✉)], Luca Maria Gambardella[1,2], and Roberto Montemanni[3]

[1] Università della Svizzera Italiana, 6900 Lugano, Switzerland
[2] Dalle Molle Institute for Artificial Intelligence, IDSIA-SUPSI, 6900 Lugano, Switzerland
umbertojunior.mele@idsia.ch
[3] Department of Sciences and Methods for Engineering, University of Modena and Reggio Emilia, 42122 Reggio Emilia, Italy

Abstract. The *ML-Constructive* heuristic is a recently presented method and the first hybrid method capable of scaling up to real scale traveling salesman problems. It combines machine learning techniques and classic optimization techniques. In this paper we present improvements to the computational weight of the original deep learning model. In addition, as simpler models reduce the execution time, the possibility of adding a local-search phase is explored to further improve performance. Experimental results corroborate the quality of the proposed improvements.

Keywords: Travelling Salesman Problem · Machine learning · Hybrid heuristic · Combinatorial optimization · Artificial intelligence

1 Introduction

The Travelling Salesman Problem (TSP) is one of the most investigated problems in the Combinatorial Optimization (CO) field. This is partly due to the fact that it belongs to the set of NP-Hard problems, which makes it particularly challenging. Moreover, the many practical problems that can be reduced to this – such as in Kumar *et al.* [10] where models of the TSP are presented to be used in the manufacture of microchips – make it even more attractive. At the same time, the full potentials of Machine Learning (ML) and Deep Learning (DL) techniques are becoming increasingly recognized in the CO field [2].

T. Vitali and U. J. Mele—Equal contribution.

Umberto Junior Mele was supported by the Swiss National Science Foundation through grants 200020-182360: "Machine learning and sampling-based metaheuristics for stochastic vehicle routing problems".

All the code for the experiments, along with an *online* Compendium can be found in the GitHub repository at https://github.com/tommivitali/ML-Constructive_LS.

N. Trautmann and M. Gnägi (Eds.): OR 2021, LNOR, pp. 59–65, 2022.
https://doi.org/10.1007/978-3-031-08623-6_10

Mele et al. [17] recently introduced *ML-Constructive*, a promising constructive approach that computes fast solutions in two separate phases. The first phase uses ML to create a sub-solution with the most reliable edges. The second phase employs a classic heuristic to complete the tour. Here we introduce an extension to the original idea to enhance the performance of the *ML-Constructive* algorithm.

In Sect. 2, we formally state the Travelling Salesman Problem and present a brief literature review. A high-level description of the plain method is presented in Sect. 3. In Sect. 4, we present all the changes we made to improve the algorithm, and this section contains the novelties of the present paper. Finally, in Sect. 5, the results of the new approach we propose are shown and discussed.

2 Problem Description and Literature Review

2.1 The Travelling Salesman Problem

Let consider the complete graph $G = (V, E)$, where $V = \{1, ..., n\}$ is a set of $|V| = n$ nodes, and $E = \{e_{ij} : i, j \in V \text{ with } i \neq j\}$ is a set of edges connecting nodes to each other. Also, let c_{ij} be the cost for edge e_{ij} connecting node i to node j. The objective of the Travelling Salesman Problem is to find the shortest possible tour that visits each node exactly once, and then gets back to the first node [1]. The NP-hard nature of the problem makes it fundamental the development of algorithms that compute approximate solutions with a good confidence even on large instances.

An effective way to heuristically reduce the complexity of a TSP is to consider only subsets of edges when building a feasible tour. A Candidate List CL_i for node i is defined as the set of edges that contains the most likely edges to be part of the optimal tour. There exist different methods to create candidate lists, the simplest one of which is to consider only the edges connecting the k closest nodes to each node i.

2.2 A Brief Literature Review

It is well known that an efficient way to solve large Combinatorial Optimization problems is to employ the *Divide-and-Conquer* paradigm. Such a paradigm is promising for addressing CO problems with Machine Learning as well. Since ML models suffer from a intrinsic generalization problem trying to scale up to large instances [6] due to well-known ML limits (e.g. imbalanced training) [13].

Valuable surveys describing recent approaches using ML to generate solutions for CO problems are in [2,16]. Different approaches suggesting DL networks to solve the TSP with end-to-end methodology have been presented among which the studies carried out by Miki *et al.* [18], Kool *et al.* [9] and Mele *et al.* [15].

The best proposal at the moment is the *ML-Constructive* heuristic [17], which focuses on the development of an efficient interaction between Machine Learning and Combinatorial Optimization techniques. It uses candidate lists (CLs) as input to the ML model, and is able to scale up with satisfactory results. Other approaches attempting to solve the scalability issue were introduce by Fu *et al.* [6] and by Fitzpatrick *et al.* [5], where Machine Learning is used to construct CLs and then classical heuristics for the tour construction are applied.

3 The Original ML-Constructive Heuristic

The *ML-Constructive* heuristic is a constructive hybrid algorithm composed by two phases. The first phase exploits Machine Learning's ability in detecting specific patterns to create an initial partial solution. This solution comprises the edges most likely to be part of an optimal tour according to the ML learnt patterns. The second phase instead uses a well-known heuristic to complete the solution. In fact, some difficulties may arise with ML where data is not adequate, further details can be found in Mele *et al.* [17].

In order to initialize the problem, reduce the search space and create valid inputs for the Machine Learning model, *ML-Constructive* initially computes a candidate list for each node. Then, a list L_P of promising edges is created, such that it contains all the edges connecting the closest two vertices for each CL. The Machine Learning is in charge of checking when an edge $l \in L_P$ has to be used or not for the solution. To do so, it learns the probability that these edges have of being optimal by considering just the CL of the considered l edge as input. Initially the insertion feasibility of edge l is checked considering the current partial solution, then the ML predicts the probability of l being an optimal edge. If this probability is greater than a certain threshold, the edge will be inserted in the current partial solution.

The order in which these requests are tackled is a fundamental choice. In *ML-Constructive*, the list L_P is sorted according to the positions in the CL and the non-decreasing cost values. The edges connecting the nearest node in the CL are placed before, then those which are second closest follow.

To complete the tour obtained during the Machine Learning phase, the Clarke-Wright (CW) heuristic [4] was used. Note that no change is made to the edges inserted during the first phase.

4 Improvements to ML-Constructive

The original algorithm uses a ResNet architecture [7] to confirm the addition of an edge in the solution. Such an architecture carries a high computational cost we would like to avoid. Our first contribution is to attempt to reduce it by replacing the ResNet with a different ML model. Several alternative ML models were examined. In addition, since *ML-Constructive* has some shortcomings in the heuristic part too, a different CL constructor and a third phase are introduced as well. The new CL constructor exploits the Delaunay triangolarization [11] to speed up the creation of the lists. The third phase instead increases the quality of the complete solution by introducing a local search on the most uncertain edges since the CW solution can be largely improved. We point out that the second phase is kept unchanged here from the original algorithm. As shown

Table 1. Performance of ML models.

Dataset	Model	Accuracy	Balanced accuracy	Precision	TPR	FPR
1st edge	Baseline [17]	0,782	0,499	0,867	0,885	0,886
	Linear [21]	0,436	0,650	0,975	0,359	0,059
	ResNet [17]	0,715	0,694	0,915	0,762	0,339
	SVM [8]	0,500	0,664	0,959	0,427	0,100
	Ensemble [3,20]	0,525	0,679	0,962	0,456	0,099
2nd edge	Baseline [17]	0,501	0,500	0,511	0,512	0,512
	Linear [21]	0,560	0,620	0,839	0,341	0,101
	ResNet [17]	0,504	0,538	0,816	0,104	0,028
	SVM [8]	0,458	0,547	0,763	0,198	0,104
	Ensemble [3,20]	0,411	0,514	0,722	0,075	0,047

by Mele *et al.* [17], even when the *ML-Constructive* is able to predict all the optimal edges in L_P, sometimes it does not reach the complete optimal solution in the end. The methods are described in the remainder of this section, while the implementation details can be found in the *online compendium.*

4.1 First Phase: Machine Learning Models

To find a ML model that works accurately and in a short time, several ML models were tried out and tested. Their performances in terms of predictions quality and tour construction are shown in Table 1 and 2, respectively. Five-thousand instances were randomly generated to train these models, with $100 \leq n \leq 1000$. The points were sampled in the unit side square, and the optimal solutions were computed with the Concorde solver [1]. The cost between each node in the CL_i of node i were employed as input of the ML, where the cost is the euclidean distance between vertices. In addition, it is also provided a vector indicating whether that edge is in the current partial solution or not. Given such vector of dimension $(k+1) \cdot k$, the ML model is asked to predict if the first or second neighbor in the CL_i of node i is optimal.

With the aim of preserving a consistent balance between training and testing, each CL_i in the training set were filtered according to the partial optimal solution found using *ML-Constructive* constraints and iterations [17].

To accomplish the task several approaches were engaged: the baseline predictor which randomly predicts using the empirical probabilities of the CL positions [17], the same ResNet architecture introduced by Mele *et al.* [17], a linear classifier [21], a linear SVM [8], and finally an Ensemble [20] voting classifier including also an XGBoost [3]. The latter shows the best performance on the test set. Since the first edge occurrence is quite over-represented we applied an under-sampling technique as well [12]. More details on the training settings can be found in the *online* compendium.

Table 2. Statistics computed on the results of the modified ML-Constructive heuristic executed on 54 TSPLIB instances. Each column refers to a different ML model.

	B [17]	NN [17]	Lin [21]	SVM [8]	ENS [20]	ML-C [17]	SVM+LS	OPT [17]	OPT+LS
avg	12,66	8,82	9,46	7,98	9,20	8,03	5,56	4,47	2,96
std	2,99	1,95	2,75	1,84	2,57	1,87	1,61	2,45	2,35
best	0/54	9/54	6/54	10/54	4/54	25/54	48/54	49/54	47/54
time	0,932	1,909	3,286	2,605	5,559	9,822	631,665	0,483	36,632

Several classic metrics are shown in Table 1, an higher True Positive Rate (TPR) and a lower False Positive Rate (FPR) are preferable [17]. The *ML-Constructive* was tested on 54 TSPLIB instances [19].

4.2 Third Phase: Local Search

The *ML-Constructive* provides good approximated tours, which can however still be improved. These tours have some flaws, since it is possible to get some crossing edges in them. To obtain better solutions we extend the heuristic with a further step, which employs a 2-opt local search [14]. Generally, such local search compares every possible couple of edges. However, since we assume the choices made with the help of the network in the first phase are correct, here we try to improve only the edges obtained during the second phase. The edges that have been inserted by the ML models (first phase) will not be modified.

5 Results

To compare the results obtained by the original version of *ML-Constructive* from [17] and what it is proposed in this work, preliminary experiments were carried out on the same 54 instances selected by Mele *et al.* [17] from the TSPLIB library [19], and reported in this paper. More instances should be considered to draw more solid conclusions. The size of the instances considered varies between 100 and 1748. A brief recap of the results – with the heuristic executed using several ML models in the first phase – is shown in Table 2. A more detailed version of this table can be found in the *online* compendium, where the results of each instance are shown and discussed.

The first column *B* is the baseline, while *NN* confirms an edge if it connects the nearest node in the CL. The other columns show the performance using some ML models; the column ML-C shows the results of the execution of the original *ML-Constructive* algorithm [17]. On the two columns indicated by "LS" is performed also the 2-opt local search as a third phase of the algorithm. Clearly, this leads overall to better performance: 2-opt moves are applied only if they bring a better tour length.

The introduction of new ML models has brought an improvement in terms of computational burden for the first phase. In terms of quality, the use of SVM also brought an improvement (not significant) compared to ML-C from [17]. The result is attractive as it also leads to an improvement in speed of about 4x. More work must be carried out to improve the accuracy of the Machine Learning decision-taker, and we noticed that keeping low FPR is preferable to having high TPR.

The local research introduced shows an improvement in terms of solution quality as well, although more effort is required to bring the gap of the tour established after the local search to zero. Overall, the changes we made led to better performance with respect to the original *ML-Constructive* from [17], apart from a few particular instances. The promising results obtained by the "optimal" ML policy (OPT) suggest that there's room for improvement along this direction. Recall that the OPT policy is derived on the assumption that the ML decision-taker can correctly predict all the optimal edges in L_P without making any mistakes.

References

1. Applegate, D.L., Bixby, R.E., Chvátal, V., Cook, W.J.: The Traveling Salesman Problem: A Computational Study. Princeton University Press, Princeton (2006)
2. Bengio, Y., Lodi, A., Prouvost, A.: Machine learning for combinatorial optimization: a methodological tour d'horizon. Eur. J. Oper. Res. **290**(2), 405–421 (2020)
3. Chen, T., Guestrin, C.: XGBoost: a scalable tree boosting system. In: Proceedings of the 22nd ACM SIGKDD International Conference on Knowledge Discovery and Data Mining, pp. 785–794 (2016)
4. Clarke, G., Wright, J.W.: Scheduling of vehicles from a central depot to a number of delivery points. Oper. Res. **12**(4), 568–581 (1964)
5. Fitzpatrick, J., Ajwani, D., Carroll, P.: Learning to sparsify travelling salesman problem instances. In: Stuckey, P.J. (ed.) CPAIOR 2021. LNCS, vol. 12735, pp. 410–426. Springer, Cham (2021). https://doi.org/10.1007/978-3-030-78230-6_26
6. Fu, Z.H., Qiu, K.B., Zha, H.: Generalize a small pre-trained model to arbitrarily large TSP instances. arXiv preprint: arXiv:2012.10658 (2020)
7. He, K., Zhang, X., Ren, S., Sun, J.: Deep residual learning for image recognition. In: Proceedings of the IEEE Conference on Computer Vision and Pattern Recognition, pp. 770–778 (2016)
8. Hearst, M.A., Dumais, S.T., Osuna, E., Platt, J., Scholkopf, B.: Support vector machines. IEEE Intell. Syst. Appl. **13**(4), 18–28 (1998)
9. Kool, W., Van Hoof, H., Welling, M.: Attention, learn to solve routing problems! arXiv preprint: arXiv:1803.08475 (2018)
10. Kumar, R., Luo, Z.: Optimizing the operation sequence of a chip placement machine using TSP model. IEEE Trans. Electron. Packag. Manuf. **26**(1), 14–21 (2003)
11. Lee, D.T., Schachter, B.J.: Two algorithms for constructing a Delaunay triangulation. Int. J. Comput. Inf. Sci. **9**, 219–242 (1980)
12. Liu, X.Y., Wu, J., Zhou, Z.H.: Exploratory undersampling for class-imbalance learning. IEEE Trans. Syst. Man Cybern. **39**, 539–550 (2008)
13. Marcus, G.: Deep learning: a critical appraisal. arXiv preprint: arXiv:1801.00631 (2018)

14. Martin, O., Otto, S.W., Felten, E.W.: Large-step Markov chains for the TSP incorporating local search heuristics. Oper. Res. Lett. **11**, 219–224 (1992)
15. Mele, U.J., Chou, X., Gambardella, L.M., Montemanni, R.: Reinforcement learning and additional rewards for the traveling salesman problem. In: Proceedings of the 8th International Conference on Industrial Engineering and Applications, pp. 198–204 (2021)
16. Mele, U.J., Gambardella, L.M., Montemanni, R.: Machine learning approaches for the traveling salesman problem: a survey. In: Proceedings of the 8th International Conference on Industrial Engineering and Applications, pp. 182–186 (2021)
17. Mele, U.J., Gambardella, L.M., Montemanni, R.: A new constructive heuristic driven by machine learning for the traveling salesman problem. Algorithms **14**(9), 267 (2021)
18. Miki, S., Yamamoto, D., Ebara, H.: Applying deep learning and reinforcement learning to traveling salesman problem. In: International Conference on Computing, Electronics and Communications Engineering, pp. 65–70. IEEE (2018)
19. Reinelt, G.: TSPLIB: a traveling salesman problem library. ORSA J. Comput. **3**(4), 376–384 (1991)
20. Wolpert, D.H.: Stacked generalization. Neural Netw. **5**, 241–259 (1992)
21. Yu, H.F., Huang, F.L., Lin, C.J.: Dual coordinate descent methods for logistic regression and maximum entropy models. Mach. Learn. **85**, 41–75 (2011)

An a-priori Parameter Selection Approach to Enhance the Performance of Genetic Algorithms Solving Pickup and Delivery Problems

Cornelius Rüther(✉), Shabanaz Chamurally, and Julia Rieck

Institute for Business Administration and Information Systems, Operations Research Group, University of Hildesheim, Universitätsplatz 1, 31141 Hildesheim, Germany
{ruether,chamurally,rieck}@bwl.uni-hildesheim.de

Abstract. Solving a pickup and delivery problem with, e. g., multiple depots, time windows, and heterogeneous vehicles is a challenging routing task. Due to the complexity, a meta-heuristic approach (e. g., a genetic algorithm) with sufficiently good solution quality is recommended. Genetic algorithms contain multiple operators such as the crossover and mutation operators that are called with certain probabilities. However, selecting appropriate probability values (parameters) for these operators strongly depend on the data structure of the given instances. For each new instance, the best parameter configuration must be found to enhance the overall solution quality. In this paper, an a-priori parameter selection approach based on classifying new instances to clusters is presented. Beforehand, a bayesian optimization approach with gaussian processes is used to find the best parameters for each cluster. The a-priori parameter selection is evaluated on four well-known pickup and delivery problem data sets, each with 60 instances and different number of depots.

Keywords: Parameter selection · Grouping genetic algorithm · Bayesian optimization · Pickup and delivery problem

1 Problem Identification

In the context of pickup and delivery problems, transportation requests which consist of shipping a quantity of goods from an origin (pickup location/customer) to a destination (delivery location/customer) have to be executed. Goods of different requests can be carried together as long as the vehicles' capacities are not exceeded. The vehicles (e. g., with heterogeneous capacities) typically start empty at a depot in the beginning of the time horizon, visit customers during the day and return to the depot empty again, and in many problems several depots are considered. At each location, the corresponding goods have to be loaded/unloaded within a time window specified by the customer. Hence, the resulting problem is a multi-depot pickup and delivery problem with time windows and heterogeneous vehicle fleets (MDPDPTWHV). Due to the problem's

N. Trautmann and M. Gnägi (Eds.): OR 2021, LNOR, pp. 66–72, 2022.
https://doi.org/10.1007/978-3-031-08623-6_11

complexity, an extensive solution approach (in particular a meta-heuristic) must be applied to achieve a good solution quality (cf. [5,6]). In the following, the promising grouping genetic algorithm (GGA) presented in [6] will be used. Contrary to a classical GA, the GGA is characterized by the fact that the representation within the genotype is group-oriented such that a gene contains all requests served by the corresponding vehicle. However, it should be noted that the solution quality of the approach strongly depends on the chosen probabilities (parameters) for the individual operators within the GGA. Moreover, it is important to also consider the structure of the instance to be solved when choosing a suitable parameter configuration [6]. In order to find appropriate parameters automatically, the so-called *parameter tuning problem* (PTP) has to be solved, which is done effectively in [6] by a *bayesian optimization* (BO) approach . However, solving the PTP is a very time-consuming task and thus it cannot be done for each new problem instance individually. Thus, an a-priori parameter selection approach has to be developed in order to find a parameter configuration for the GGA for a new instance efficiently beforehand.

The paper is structured as follows: In Sect. 2, a short overview of contributions using machine learning techniques to improve meta-heuristics is given. The a-priori selection approach is presented in Sect. 3 followed by the results in Sect. 4. Finally, the paper closes with a short discussion in Sect. 5.

2 Related Work

During the last years, there has been a rise in interest for meta-heuristics in the field of optimization and machine learning. Although it is evident that the performance of a meta-heuristic depends on its configured parameters values, the academic community has not formally addressed the PTP until the end of the last century. For a detailed description of different parameter tuning approaches for meta-heuristics, the interested reader is referred to [4].

In order to address the energy minimization vehicle routing problem, Cooray and Rupasinghe [1] developed a genetic algorithm which was enhanced through k-means clustering. The authors identified the mutation rate as particularly relevant for parameter tuning. Using k-means clustering, they first clustered the instances into three clusters. Different mutation rates were applied on each cluster to calculate energy minimization percentages. The results showed that certain mutation rates work better on specific clusters. Based on this observation, in this contribution, an a-priori parameter selection for new instances is developed, where in contrast to [1] the parameters of a given cluster are chosen purposefully through a BO approach (see [6] for a detailed description of the BO approach).

In the context of solving vehicle routing problems, Gutierrez-Rodríguez et al. [3] studied the selection of proper meta-heuristics via meta-learning. The paper defines two sets of meta-features that can be considered to characterize different routing problem instances. In order to explore desirable structural characteristics of good solutions, Arnold and Kenneth [2] argue that knowledge about a problem is highly valuable when designing efficient heuristics. The authors showed how this knowledge can be generated based on data mining by defining several metrics to measure an instance and a solution. They defined a set of metrics that can be used to capture the characteristics of an instance.

Considering the MDPDPTWHV, Rüther and Rieck [6] developed an effective BO approach with gaussian processes to improve GGA's parameter configurations for 12 pre-defined problem classes. By optimizing the probability parameters of six mutation and five repair operator variants, it is shown that the BO is able to enhance the solution quality of the GGA in each problem class and its computational time. In addition, the authors argue that different data structures require different parameter configurations to obtain a sufficient solution quality.

Although finding appropriate parameter configurations for improving meta-heuristics is a well-known problem, an a-priori approach learning from known data in order to select good parameters for new instances with machine learning techniques has not been presented in the literature yet. In this paper, different meta-features proposed in the literature are used to find the best parameter settings for mutation and repair operators of a GGA solving the MDPDPTWHV.

3 Selecting Parameter Configuration for the Grouping Genetic Algorithm

In order to select appropriate parameters a-priori, features describing the structure of MDPDPTWHV instances have to be developed (see Subsect. 3.1). Then, these features are used to identify clusters of similar instances by using k-means (see Subsect. 3.2) in order to learn good parameter configuration for the GGA with which all instances of a cluster is to be solved. As the BO proposed in [6] is a reasonable approach for tuning the parameters of the GGA, the procedure is also used in this contribution in order to find parameter configurations (see Subsect. 3.3). Finally, a classification approach is used to find a-priori the best parameter configuration for new instances (see Subsect. 3.4).

3.1 Feature Selection

Most meta-heuristics expose operators that have strong impact on the performance of the algorithm and on the quality of the solutions obtained. Often, the meta-heuristic adaptation requires an instance-based calibration of the parameters of its operators, as a good initial parameter setting can vary considerably from problem to problem and between problem instances. To discover characteristics that are typical for an instance, the structure of an instance has to be transformed into some quantitative metrics (features). Thus, a set of features relevant to the MDPDPTWHV is extracted. Table 1 describes the 16 features that are used in the approach at hand. In the following, the features based on time windows are described.

Each customer i is associated with a time window (TW) $[e_i, l_i]$ in which the service of loading/unloading goods has to be started. The feature *average TW* is computed through the mean value of all TW lengths, where n is the respective number of customers:

Table 1. Feature definition

Feature	Definition
Vehicles capacity/Cost	Sum of all vehicle capacities/costs using all vehicles
Vehicle/Request count	Number of vehicles available/requests
Service duration	Sum of all service times
Total demand	Sum of all customer demands
Total profit	Sum of all gross profits
Average TW	Avg. time window size
Average TW overlap	Avg. of time window overlaps between customers
Max pickup-delivery distance	Max. dist. between a pickup and it's delivery node
Min pickup-delivery distance	Min. dist. between a pickup and it's delivery node
Avg pickup-delivery distance	Avg. distance between a pickup node and it's corresponding delivery node
Variance pickup-delivery distance	Var. of the dist. between each pickup node and it's corresponding delivery node
Degree of capacity utilization	Deg. of cap. util. based on number of vehicles used
Avg. customer-depot distance	Avg. dist. between the customers nodes and depots
Variance customer-depot distance	Var. in the dist. between customer nodes and depots

$$\frac{1}{n}\sum_{i=1}^{n}(l_i - e_i)\,. \tag{1}$$

The average TW overlap is a measure of the relationships between the TWs of all customers in an asymmetric instance:

$$\frac{1}{(n-1)\cdot n}\sum_{i=1}^{n}\sum_{j=1,j\neq i}^{n}\max(l_i - e_j, 0)\,. \tag{2}$$

3.2 *k*-means Clustering

As an intuitive clustering approach, the k-means clustering is applied in order to cluster the n_I instances in each data set into k separate disjoint clusters $C_1, \ldots, C_k$ based on the n_f features defined in Sect. 3.1. Typically, k-means clustering is carried out such that the *squared euclidean distance* between each instance x_i and the center m_j of the cluster C_j to which x_i belongs is minimized, i.e.

$$\sum_{i=1}^{n_I}\min_{j=1,\ldots,k}\sum_{f=1}^{n_f}\left(x_i^f - m_j^f\right)^2, \tag{3}$$

where x_i^f and m_j^f are the evaluation of feature f in instance x_i and in the center of cluster C_j, respectively.

3.3 Bayesian Optimization

Optimizing the parameters of meta-heuristics is a *black-box* problem, since the relationship between the parameter configurations of the algorithm and its performance cannot be measured metrically. Thus, BO which finds a global minimum $x^* = \arg\min_{x \in X} \varphi(x)$ of an unknown *black-box function* φ is a reasonable approach. The method iteratively learns a probabilistic model of *gaussian processes* that estimates φ by known function values $\varphi(x)$ and a surrogate, called *acquisition function* [6]. Here, function φ describes the GGA's solution quality depending on the parameters x of six mutation and five repair operator variants. In order to determine the parameter configuration for a cluster, the BO approach is applied on a random sample of instances of the cluster considered.

3.4 An a-priori Parameter Selection

The clusters found from the k-means clustering are used to classify existing instances into k disjoint clusters upon which the BO finds a good parameter configuration for each cluster. To find a good parameter configuration for a new instance x_i, first the set of features given in Subsect. 3.1 are extracted for x_i. Then, the corresponding cluster C_k for instance x_i is specified by using the nearest neighbor approach, i. e., the cluster C_k that is the closest to the feature vector of x_i regarding the center m_k is chosen. Finally, the parameter configuration found by the BO for cluster C_k is applied on x_i.

4 Evaluation

In order to build the a-priori parameter selection approach, an extended version of the data sets proposed by Rüther and Rieck [5] (*training data*), which are created by an instance generator, is used. There are data sets with 4, 6, 8, and 9 depots (i. e., 4D, 6D, 8D, 9D instances) considered each consisting of 1,200 instances. Hence, the training data set contains 4,800 instances in total. The k-means method is applied on each depot-wise training data set with respect to the derived features. To find the optimal number of clusters k, the elbow method is applied. The BO approach is used to find an appropriate parameter configuration for all instances in each of the found clusters. A practical data set (*evaluation data*) proposed by Rüther and Rieck [7] containing 60 instances per data set is taken into account to evaluate the a-priori approach. For the instances coming from the evaluation data, the corresponding clusters are determined and then the GGA with the parameter configuration found by the BO for the respective cluster is applied (*optimized GGA*). Since the initial parameter configuration of the GGA in [6] showed promising results, this GGA configuration is used as baseline method on the instances of the evaluation data (*initial GGA*).

Table 2 presents the results of this study. Here, μ_ϵ is the mean value and σ_ϵ is the standard deviation of the relative error which is determined through comparing the initial and optimized GGA instance-wise to the best solution as

Table 2. Mean value μ_ϵ and std. deviation σ_ϵ of relative error for GGA's configurations

	Approach	μ_ϵ	σ_ϵ	μ_{CPU}	#best
4D	Optimized GGA	0.35%	0.67%	693.78 s	40
	Initial GGA	1.41%	1.50%	757.66 s	20
6D	Optimized GGA	0.20%	0.48%	1968.05 s	45
	Initial GGA	2.20%	2.56%	2150.22 s	15
8D	Optimized GGA	0.30%	0.77%	3585.35 s	44
	Initial GGA	1.81%	2.00%	3908.35 s	16
9D	Optimized GGA	0.20%	0.47%	3357.35 s	44
	Initial GGA	1.58%	1.48%	3580.19 s	16

in [6]. The results in Table 2 show that the optimized GGA provides better results than the initial GGA. In particular, the mean relative error in each data set is smaller and the optimized GGA is more stable, since the standard deviations are smaller (i. e., if the optimized GGA does not find the best solution for an instance, the found solution is only slightly worse than the one by the initial GGA). In addition, the number of best solutions found #best and computational time on average μ_{CPU} have been improved through a-priori parameter selection.

5 Discussion

This paper presents an a-priori parameter selection approach based on k-means clustering and bayesian optimization for selecting parameter configurations to solve new problem instances. To do so, instances are represented into a vector of meaningful features upon which a machine learning method is applied to improve parameter configurations. Using an evaluation over 240 multi-depot instances, it is shown that appropriate parameter configurations can be found. It is worth mentioning that the idea presented can be used with any machine learning method to learn parameter configurations of a meta-heuristic based on instance features, which is therefore supposed to be considered for future research. Moreover, it should be examined whether the best parameter configuration of a cluster is also the best or under the top configurations for each instance of the cluster.

References

1. Cooray, P., Rupasinghe, T.: Machine learning-based parameter tuned genetic algorithm for energy minimizing vehicle routing problem. J. Ind. Eng. **2017**, 1–13, 100087 (2017). https://doi.org/10.1155/2017/3019523
2. Florian, A., Sörensen, K.: What makes a VRP solution good? The generation of problem-specific knowledge for heuristics. Comput Oper Res **106**, 280–288 (2019)

3. Gutiérrez-Rodríguez, A.E., Conant-Pablos, S.E., Ortiz-Bayliss, J.C., Terashima-Marín, H.: Selecting meta-heuristics for solving vehicle routing problems with time windows via meta-learning. Expert Syst. Appl. **118**, 470–481 (2019)
4. Huang, C., Li, Y., Yao, X.: A survey of automatic parameter tuning methods for metaheuristics. IEEE Trans. Evol Comput. **24**(2), 201–216 (2019)
5. Rüther, C., Rieck, J.: A grouping genetic algorithm for multi depot pickup and delivery problems with time windows and heterogeneous vehicle fleets. In: Paquete, L., Zarges, C. (eds.) EvoCOP 2020. LNCS, vol. 12102, pp. 148–163. Springer, Cham (2020). https://doi.org/10.1007/978-3-030-43680-3_10
6. Rüther, C., Rieck, J.: A bayesian optimization approach for tuning a genetic algorithm solving practical-oriented pickup and delivery problems, 1–11 (2022). https://www.uni-hildesheim.de/fb4/institute/bwl/betriebswirtschaft-und-operations-research/forschungprojekte/working-papers/
7. Rüther, C., Rieck, J.: Bundle selection approaches for collaborative practical-oriented Pickup and Delivery Problems. EURO J. Transp. Logist. **11**, 100087 (2022). https://doi.org/10.1016/j.ejtl.2022.100087

Decision Analysis and Support

A Decision Support System to Optimize Production Quantities with Respect to Product Phase-Out Costs

Fabian Leuthold[1](✉), Katrin Hügel[1], Oliver Mörl[2], and Harold Tiemessen[1]

[1] Institute of Modeling and Simulation, Eastern Switzerland University of Applied Sciences, 9000 St. Gallen, Switzerland
{fabian.leuthold,katrin.huegel,harold.tiemessen}@ost.ch
[2] Leica Geosystems AG, 9435 Heerbrugg, Switzerland
oliver.moerl@leica-geosystems.com

Abstract. The increasing pressure to innovate requires that products are replaced by successor products at ever shorter intervals. Manufacturing companies are therefore increasingly faced with the planning of product phase-outs. A central question in the planning of phase-outs is how many units of the discontinued products should still be manufactured before their fabrication is ceased. At Leica Geosystems, a production company in the surveying industry, this decision-making process takes place in dedicated interdepartmental planning meetings. In this paper, we describe a Decision Support System that enables the management to develop different phase-out scenarios, compare them with each other, and thus make informed planning decisions.

Keywords: Decision support systems · Production and inventory systems · Mixed-integer programming

1 Introduction

Leica Geosystems is a leading provider in the high-end segment of the surveying industry. In addition to the claim to meet the highest quality requirements, the company is also particularly characterized by a high degree of innovation in product development. For supply chain management, this results in the challenge of carefully planning the replacement of product families. The first task here is to determine the quantities of products still to be manufactured in such a way that the residual value of the components and semi-finished products remaining at the end is minimized [1]. For complex high-tech products, such as complete solutions in the field of surveying, hundreds of components can be affected when a product family is replaced. The bill of material (BOM) defines how many components of each type are used in a product. A BOM-based comparison of discontinued and successor products provides information about discontinued components (see Fig. 1). Each discontinued component has a certain value and a current inventory. The companies' phase-out strategies are limited by a wide

N. Trautmann and M. Gnägi (Eds.): OR 2021, LNOR, pp. 75–80, 2022.
https://doi.org/10.1007/978-3-031-08623-6_12

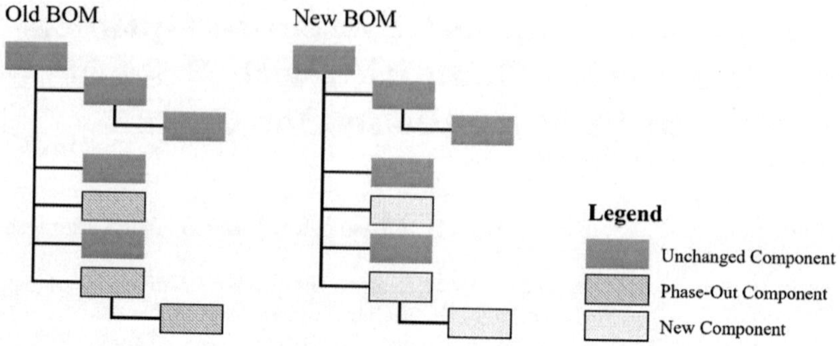

Fig. 1. Comparison of discontinued and successor product at BOM-level (own illustration)

variety of purchase obligations such as lot sizes, minimum purchase quantities, and quantity contracts, which should be considered in the analysis. Moreover, estimates from the sales department on the opportunities to sell discontinued products should be taken into account. Interesting in this context is a recent study showing that domain experts make better phase-out demand forecasts than the best performing forecast models [2].

Since today's Enterprise Resource Planning (ERP) Systems do not offer adequate support in these matters, manufacturing companies typically use rather naive spreadsheet calculations. To address this need, we propose a Decision Support System (DSS) that combines mathematical optimization functionality with scenario analysis to deal with sales scenarios to support decision-making in the process of product phase-out control. The DSS will be used during live planning meetings. In order to determine the optimal production quantities, we have developed a Mixed Integer Program (MIP).

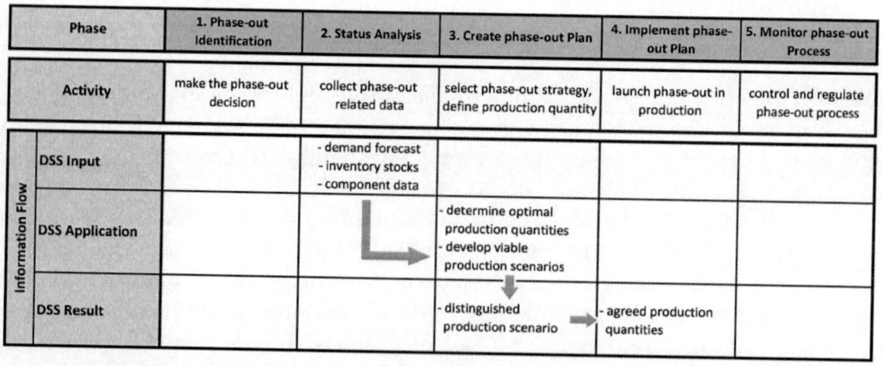

Fig. 2. DSS support and information flow (own illustration based on Wagner [3])

Figure 2 visualizes the preparations and the use of the DSS in the various phase-out phases according to Wagner [3].

The problem discussed in this paper has some similarities to the well-known Last Time Buy (LTB) problem. In both settings we have one last opportunity to increase stock of components (and products) and the goal is to minimize total cost while anticipating on future demand/sales. LTB models are in particular relevant to spare parts logistics. In contrast to our proposed phase-out model, LTB models usually consider single parts (thus no bill of materials), assume known demand distributions and do not impose constraints on lot sizes. Some more advanced LTB models consider alternative sourcing options, such as repair of returned broken items. For a detailed discussion of LTB models, we refer to Behfard et al. [4].

Our main contribution consists of two parts:

- We develop a MIP model for minimizing scrapping costs during product phase-out for products with single-level BOM structures, subject to minimum and maximum production quantities and a wide range of real-world technical and operational constraints. We use an ILP solver to solve realistic problem instances and show that the MIP approach is responsive and outperforms solutions constructed by human planners by up to 30%.
- We develop a DSS that contains a MIP model and a MIP solver for minimizing scrapping costs for user-specified minimum and maximum production quantities. In order to take into account difficult-to-quantify and case-specific expert knowledge, the DSS also allows to calculate manually adjusted scenarios, to visualize the impacts on key performance indicators and to compare the results with those from MIP-optimized solutions. The tool thus supports interactive decision making and empowers the purchasing department, production department and marketing & sales department to determine an implementable and coordinated phase-out strategy of high acceptance.

2 Problem Description and Model Formulation

The MIP presented in the following provides the optimization model of the DSS and allows to determine the production quantities as well as the related order quantities of the involved components to minimize the scrapping costs.

Sets and Indices

i: Product indices for identification of all products $\in I = \{1, ..., n\}$

j: Component indices for identification of all components $\in J = \{1, ..., m\}$

J^K: Subset of all components with quantity contract

$J^{\overline{K}}$: Subset of all components without quantity contract

Parameters

$A_i \in \mathbb{N}_0$: Lower limit for production quantity of the product of type i

$B_i \in \mathbb{N}_0$: Upper limit for production quantity of the product of type i

$K_j \in \mathbb{N}_0$: Number of components in stock of type j
$S_{ij} \in \mathbb{N}_0$: Number of components of type j in product of type i
$C_j \in \mathbb{R}$: Price of a component of type j
$R_j \in \mathbb{N}_0$: Service requirement for components of type j
$L_j \in \mathbb{N}$: Lot size for orders of components of type j
$M_j \in \mathbb{N}_0$: Minimum purchase quantity for orders of components of type j
$N_j \in \mathbb{N}_0$: Quantity from quantity contract for components of type j
$Q \in \mathbb{R}$: A large number (exceeding largest order quantity)

Decision Variables

$X_i \in \mathbb{N}_0$: Production quantity of products of type i
$G_j \in \mathbb{N}_0$: Order quantity of components of type j
$U_j \in \mathbb{Z}$: Number of lots of components of type j
$w_j \in \{0, 1\}$: 1, if $G_j \neq 0$, 0 else
$z_j \in \{0, 1\}$: 1, if $G_j \leq M_j$, 0 else

$$min \quad \sum_{j \in J} G_j \cdot C_j + \sum_{j \in J} K_j \cdot C_j - \sum_{i \in I, j \in J} X_i \cdot S_{ij} \cdot C_j - \sum_{j \in J} R_j \cdot C_j \tag{1}$$

$$\forall i \in I \quad A_i \leq X_i \tag{2}$$

$$\forall i \in I \quad B_i \geq X_i \tag{3}$$

$$\forall j \in J \quad G_j \geq R_j - K_j + \sum_{i \in I} X_i \cdot S_{ij} \tag{4}$$

$$\forall j \in J \quad U_j \geq \frac{G_j}{L_j} \tag{5}$$

$$\forall j \in J^{\overline{K}} \quad G_j \leq w_j \cdot Q \tag{6}$$

$$\forall j \in J^{\overline{K}} \quad G_j \geq -w_j \cdot Q \tag{7}$$

$$\forall j \in J^{\overline{K}} \quad G_j \geq M_j - (1 - w_j) \cdot Q \tag{8}$$

$$\forall j \in J^{\overline{K}} \quad G_j \leq M_j + (1 - z_j) \cdot Q \tag{9}$$

$$\forall j \in J^{\overline{K}} \quad G_j \geq U_j \cdot L_j - z_j \cdot Q \tag{10}$$

$$\forall j \in J^{\overline{K}} \quad G_j \leq U_j \cdot L_j + z_j \cdot Q \tag{11}$$

$$\forall j \in J^K \quad G_j = N_j \tag{12}$$

The objective function (1) minimizes the overall scrapping costs: The value of ordered components and components at stock is reduced by the components consumed by production and the ones reserved for service work (spare parts). Constraints (2) and (3) ensure that the production quantities are within the permissible ranges. Constraint (4) guarantees that the components' order quantities together with their stocks cover the required component numbers claimed by the production quantities as well as the service needs, whereas constraint (5) makes sure that the lot sizes are respected. The constraints (6) to (11) have the effect that if components without a quantity contract are ordered, either the minimum purchase quantity or a multiple of their lot size can be ordered. Finally, constraint (12) guarantees that order quantity contracts are respected.

3 DSS-Based Phase-Out Negotiation

In order to determine a suitable phase-out scenario using the DSS, first, the parameter values described above are collected from the ERP system. Then, during an interdepartmental meeting, the experts determine the optimal production quantities for each product as well as the minimal scrapping cost related to that scenario. Thereafter, expert knowledge is taken into account from the purchasing department, production department and marketing/sales department. In this step, alternative viable phase-out scenarios are developed by modifying the production quantities of the products within the allowed ranges. Based on these negotiations, a common consensus is formed and a commitment to the most appropriate and implementable scenario is created.

4 Numerical Experiments

The DSS has been tested on the basis of historical phase-outs. These cases involved product families with two products and a few dozen components. It could be shown that the scrapping costs could be reduced by 20–30% with the help of the DSS presented here. For all historical cases, an optimal phase-out configuration could be determined within less than 30 s, which is perfectly fine for the intended application[1].

5 Conclusions and Future Work

The DSS presented in this paper provides the company management with a tool to interactively determine the production quantities and the related order quantities of the involved components for the product phase-out during planning meetings. This way, expert knowledge from purchasing, production, and marketing & sales can be taken into account in order to find the best applicable phase-out scenario. A DSS based on the work presented here is currently in use at Leica Geosystems AG, in Heerbrugg, Switzerland, where it supports the planning of upcoming product phase-outs.

In the future we plan to expand the optimization model for handling hierarchical BOMs and to support more elaborate cost models. From a usability point of view, an integration into an existing ERP system and the implementation of an automated data acquisition would make the solution more efficient and user-friendly.

[1] Measured under Microsoft® Windows® 10 using the Matlab® 2020a MILP solver on Intel® Core™ i7 CPU@1.9 GHz.

References

1. Schinzinger, R.: A procedure for the optimal phase-out of product lines. IEEE Trans. Eng. Manag. **EM-18**(4), 139–146 (1971)
2. Ahmed, S.: Phase-out demand forecasting: predictive modeling on forecasting product life cycle. Master's thesis (2020)
3. Wagner, R.: Production Phase-Out: Process Modelling Including Adaptions for Production Planning and Control. BoD–Books on Demand (2017)
4. Behfard, S., van der Heijden, M.C., Al Hanbali, A., Zijm, W.H.: Last time buy and repair decisions for spare parts. Eur. J. Oper. Rese. **244**(2), 498–510 (2015)

A Facility Location Problem with Minimum Workload and Penalties for Waste Collection

Meritxell Pacheco Paneque(✉), Vera Fischer, and Reinhard Bürgy

University of Fribourg, Fribourg, Switzerland
meritxell.pacheco@unifr.ch

Abstract. A facility location problem to place waste collection facilities is introduced. We present two different strategies to deal with residents whose access to the opened collection facilities is restricted (e.g., they are too far away). First, we consider a penalty to be borne by the municipality for each resident bringing its waste to a location that is considered to be unacceptable. Second, we provide a collection service performed by a vehicle such that each resident is either served by a collection facility or a vehicle collecting its waste. We compare both formulations with a simple example that considers a small neighborhood of a Swiss municipality.

Keywords: Facility location · Mixed-integer linear programming · Waste collection · Outliers

1 Introduction

In most Swiss municipalities, a curbside system consisting of heavy trucks that stop at almost each household is used for non-recoverable waste collection. Due to the many stops of the trucks, this strategy causes high fuel consumption, emissions and noise. These effects can be alleviated by requesting residents to bring their waste to collection facilities comprising large containers that are distributed throughout the municipality. When a container of a collection facility is full, a truck transports an empty container from the disposal facility (depot) to the collection facility and replaces it. The truck then transports the full container back to the disposal facility and discharges it.

We formulate this optimization problem as a facility location problem (FLP, e.g., [1,2]) with two particular features. First, for a facility to be placed at a candidate location, its collected waste must be greater or equal than a minimum workload that justifies its opening. Minimum required workloads are typically used in the location of preventive health care facilities (e.g., [3]). In this application, the workload at a facility needs to be greater than a certain threshold to ensure the quality of services (accreditation) and to justify the allocation of public funding.

Second, we take into account residents' preferences when allocating them to collection sites. This is modeled with a given list called preference list that

N. Trautmann and M. Gnägi (Eds.): OR 2021, LNOR, pp. 81–86, 2022.
https://doi.org/10.1007/978-3-031-08623-6_13

for each resident ranks the candidate locations according to some convenience measure (e.g., walking distance, proximity to interesting points). The preference list is divided in two parts based on a given threshold on the level of acceptability of candidate locations (e.g., maximum walking distance). Residential buildings might bring their waste to unacceptable locations. This is typically the case for distant residential buildings, since allocating them to acceptable locations would involve the placement of multiple collection sites, exerting a disproportionately strong influence over the final solution. These residential buildings are known as outliers, and the municipality needs to put in place a strategy to handle them.

Two variations of the FLP are formulated in [4] to handle outliers in a meaningful way: the robust facility location problem (R-FLP), which consists of placing facilities such that the service cost to any subset of at least p facilities is minimized, and the facility location with penalties problem (FLP-P), which decides for each resident to either service them and pay the service cost to its nearest facility, or to pay the penalty. The FLP-P has been studied earlier for the prize collecting traveling salesman problem (e.g., [5]) and will be considered for our problem.

Another possibility is to provide an alternative service to outliers and visit them with a vehicle, which has been recently referred to as location-or-routing problem (LoRP, e.g., [6]). In this setting, a resident can be *covered* (served) either by a collection facility or by a vehicle departing from an open facility subject to maximum route length and vehicle capacity constraints. The objective is to minimize the total weighted cost of opening facilities, vehicle routing and residents coverage by open facilities. The LoRP is closely related to the location-routing problem (LRP, e.g., [7]), which enables to model locational problems while paying attention to the underlying issues of vehicle routing.

In this paper, we consider two strategies to deal with outliers in our problem and propose a mixed-integer linear programming (MILP) formulation for each of them. In the first strategy, we incorporate outliers in the formulation via penalties. If a resident has to use a collection facility that is placed in an unacceptable location, the municipality has to pay a penalty to compensate it. In contrast to the FLP-P, our penalty costs depend on the convenience measure for residential buildings to unacceptable locations, and service costs to acceptable locations are not considered. We call this problem FLP with penalties and minimum workload (FLP-PW). In the second strategy, we formulate our problem as a LoRP. A vehicle located at the disposal facility performs routes to visit outliers and collect their waste. As opposed to the classical LoRP, we do not start the routes at one of the open facilities and do not consider maximum route length. For the sake of simplicity, we assume that the collection facilities can gather as much waste as needed and we ignore coverage cost in the objective. We call this problem LoRP with minimum workload (LoRP-W).

This paper is organized as follows. In Sect. 2 we formally define the problem. In Sects. 3 and 4 we present a MILP formulation for each of the two strategies, respectively. In Sect. 5 we test both formulations and enumerate some avenues for future research.

2 Problem Description

Let F be a set of candidate locations for the collection facilities and V the set of residential nodes. For each location $j \in F$, we need to decide whether or not to place a collection facility there. Each node $i \in V$ might refer to one residential building or might aggregate several ones. We denote its waste by w_i and its non-empty preference list by $V_i^{\text{pref}} \subseteq F$. We assume that V_i^{pref} is totally ordered, with $\text{pref}(i, j)$ denoting the index of candidate location j in V_i^{pref}. Hence, $\text{pref}(i, j) < \text{pref}(i, j')$ indicates that location j is preferred over location j' by residential node i.

A collection facility can only be placed at a candidate location if it gathers a minimum waste (minimum workload) $W^{\min}$. As pointed out in Sect. 1, for each node $i \in V$, we split its preference list V_i^{pref} into an acceptable V_i^{ac} and an unacceptable V_i^{unac} sub-list of candidate locations, i.e., $V_i^{\text{pref}} = (V_i^{\text{ac}}, V_i^{\text{unac}})$. This division meets an assumed threshold on the acceptability of locations. Thus, it might be that one of the two sub-lists is empty, but not both. We consider a fixed cost c_j to place a collection facility at location j. The goal of both the FLP-PW and LoRP-W is to find a subset $S \subseteq F$ with minimum total cost as defined in Sects. 3 and 4, respectively.

3 FLP-PW

In the FLP-PW, we assume a penalty cost r_{ij} associated with residential node i bringing the waste to location $j \in V_i^{\text{unac}}$ to be borne by the municipality. Let y_j be a binary variable that is equal to 1 if a collection facility is placed at location $j \in F$ and x_{ij} be a binary variable that is equal to 1 if the waste w_i is allocated to the collection facility located at j. The FLP-PW is formulated in (1). The objective function (1a) defines the goal of minimizing the total cost, which is calculated as the sum of the opening costs of collection facilities and the penalty costs associated with unacceptable locations. Constraints (1b) ensure that the waste of each residential node is assigned to exactly one facility in its preference list. Constraints (1c) impose that the residential node is allocated to the first collection facility $j \in V_i^{\text{pref}}$ that is opened. Constraints (1d) enforce the minimum workload $W^{\min}$ on open collection facilities. Constraints (1e) ensure that a residential node can be allocated to a collection facility only if it is opened. Constraints (1f) define the domain of the decision variables.

$$\min \sum_{j \in F} c_j y_j + \sum_{i \in V} \sum_{j \in V_i^{\text{unac}}} r_{ij} x_{ij} \tag{1a}$$

$$\text{s.t.} \sum_{j \in V_i^{\text{pref}}} x_{ij} = 1 \qquad \forall i \in V \tag{1b}$$

$$\sum_{j' \in V_i^{\text{pref}} : \text{pref}(i,j') > \text{pref}(i,j)} x_{ij'} \leq 1 - y_j \qquad \forall i \in V, j \in V_i^{\text{pref}} \tag{1c}$$

$$\sum_{i \in V: j \in V_i^{\text{pref}}} w_i x_{ij} \geq W^{\min} y_j \qquad \forall j \in F \tag{1d}$$

$$x_{ij} \leq y_j \qquad \forall i \in V, j \in F \tag{1e}$$

$$x_{ij}, y_j \in \{0,1\} \qquad \forall i \in V, j \in F \tag{1f}$$

4 LoRP-W

Let $G = (V, A)$ be a directed graph with node set V and arc set A. Each arc $(i, i') \in A$ represents the shortest path between nodes $i \in V$ and $i' \in V$ and has length $\ell_{ii'}$. We consider a single vehicle with capacity Q to collect the waste. It can perform as many routes as needed. The vehicle is located at a disposal facility σ, where it departs and dumps the collected waste. Note that, for the sake of simplicity, we do not allow for splits, i.e., the waste at each residential node cannot be split up between multiple routes. To this end, we assume $w_i \leq Q, \forall i \in V$.

Let y_j and x_{ij} have the same meaning as in the FLP-PW. Furthermore, we denote by $u_{ii'}$ the binary variable that is equal to 1 if the vehicle goes from residential node i to i'. To prevent subtours, we associate a non-negative continuous variable $f_{ii'}$ with each arc (i, i') that indicates the quantity of waste that traverses such arc. The LoRP-W is formulated in (2). The total cost in the objective function (2a) is calculated as the sum of the opening costs of collection facilities and the driving costs associated with the performed routes. Constraints (2b) ensure that a residential node is either served by a collection facility or visited in a route. Constraints (2c), (2d) and (2e) have the same meaning as constraints (1c), (1d) and (1e) in the FLP-PW, respectively. Constraints (2f) define the degree constraints. Constraints (2g) ensure that the net flow out of any visited residential node must be the waste associated with that node. Constraints (2h) to (2j) define the domain of the decision variables. Note that constraints (2j) link the variables f with the variables u. If $u_{ii'} = 1$, i.e., the vehicle visits both residential nodes i and i', then the flow passing through the arc connecting them cannot exceed the vehicle's capacity Q.

$$\min \sum_{j \in F} c_j y_j + \sum_{i \in V \cup \{\sigma\}} \sum_{i' \in V \cup \{\sigma\}} \ell_{ii'} u_{ii'} \tag{2a}$$

$$\text{s.t.} \sum_{j \in V_i^{\text{pref}}} x_{ij} + \sum_{i' \in V \cup \{\sigma\}} u_{ii'} = 1 \qquad \forall i \in V \tag{2b}$$

$$\sum_{j' \in V_i^{\text{pref}}: \text{pref}(i,j') > \text{pref}(i,j)} x_{ij'} \leq 1 - y_j \qquad \forall i \in V, j \in V_i^{\text{pref}} \tag{2c}$$

$$\sum_{i \in V: j \in V_i^{\text{pref}}} w_i x_{ij} \geq W^{\min} y_j \qquad \forall j \in F \tag{2d}$$

$$x_{ij} \leq y_j \qquad \forall i \in V, j \in F \tag{2e}$$

$$\sum_{i' \in V \cup \{\sigma\}} u_{ii'} - \sum_{i' \in V \cup \{\sigma\}} u_{i'i} = 0 \qquad \forall i \in V \tag{2f}$$

$$\sum_{i' \in V \cup \{\sigma\}} f_{ii'} - \sum_{i' \in V \cup \{\sigma\}} f_{i'i} = w_i u_{ii'} \qquad \forall i \in V \qquad (2g)$$

$$x_{ij}, y_j \in \{0,1\} \qquad \forall i \in V, j \in F \qquad (2h)$$

$$u_{ii'} \in \{0,1\} \qquad \forall i, i' \in V \cup \{\sigma\} \qquad (2i)$$

$$0 \leq f_{ii'} \leq Q u_{ii'} \qquad \forall i, i' \in V \cup \{\sigma\} \qquad (2j)$$

5 Preliminary Results

We test both formulations on a dataset that represents a small neighborhood of a Swiss municipality with 57 residential buildings, 411 inhabitants, an area of 0.13 km^2 and a total waste production of 727 waste units. Graph G contains 97 nodes, out of which 33 are residential nodes, and 307 arcs, out of which 172 are incident to the disposal facility.

In both formulations, we assume for the first part of the objective $c_j = 500$ as the fixed cost for placing a collection facility at location j. To be able to compare the two formulations, we define the variable costs for the second part of the objective as follows. In FLP-PW, we assume that a resident i brings its waste to an unacceptable location $j \in V_i^{\text{unac}}$ by a private car driving the following route $i - j - i$. The penalty costs are then defined as $r_{ij} = d_{ij} + d_{ji}$, where d_{ij} is the shortest path distance from residential node i to collection facility j calculated from the underlying graph. In a similar way, we define the length of an arc $(i, i') \in A$ in LoRP-W as $\ell_{ii'} = d_{ii'}$, where $d_{ii'}$ is the shortest path distance from residential node i to i'. Hence, the second objective is to minimize the total distance driven by the private cars of the residents in FLP-PW or the collection vehicle in LoRP-W. Given the total waste (727 units), we consider a minimum workload $W^{\min} = 100$ and a vehicle capacity $Q = 200$. Finally, to generate the preference lists, we test various values for the maximum walking distance and sort the candidate locations for each residential node in increasing order with respect to the walking distance. Thus, the unacceptable locations are those placed at a walking distance larger than the maximum walking distance.

Table 1 includes the obtained results for the maximum walking distance values of $\{50, 100, 150, 200\}$. Note that the outlier cost refers to the total distance driven by the vehicles (collection vehicle or private cars of the residents). Together with the facility cost it adds up to the total cost which is represented in the objective function value. Both formulations found optimal solutions for all instances. As expected, the number of outliers substantially decreases as the maximum walking distance increases. The same can be observed for the number of open facilities. The FLP-PW is more expensive with respect to the total cost and tends to open more facilities than the LoRP-W. Thus, the FLP-PW covers more residents by facilities than its counterpart. The LoRP-W on the other hand accepts more outliers. Overall, the results show that both strategies (formulations) can be used for the presented problem and provide meaningful solutions to the instances generated for this small example.

Note that the proposed formulations can be computationally too expensive. For this small example, the results are obtained within seconds. For instances

Table 1. Results of both formulations for maximum walking distance values of $\{50, 100, 150, 200\}$.

Max. walk. dist.	Formulation	Obj. fun.	Facilities		Outliers	
			# Open fac.	Facility cost	# Outliers	Outlier cost
50	FLP-PW	9647	6	3000	14	6647
50	LoRP-W	9088	3	1500	23	7588
100	FLP-PW	8349	6	3000	8	5349
100	LoRP-W	5917	3	1500	11	4417
150	FLP-PW	6521	3	1500	7	5021
150	LoRP-W	5013	3	1500	6	3513
200	FLP-PW	4726	2	1000	3	3726
200	LoRP-W	4024	2	1000	3	3024

with a larger number of nodes, additional methodologies that speed up the solution approach, such as decomposition techniques or other heuristics, might need to be applied to efficiently produce good quality solutions. Furthermore, an exhaustive calibration of the parameters of the model needs to be performed to generate more realistic results and compare both strategies in greater detail.

References

1. Cornuéjols, G., Nemhauser, G., Wolsey, L.: The uncapicitated facility location problem. Cornell University Operations Research and Industrial Engineering (1983)
2. Ghiani, G., Laganà, D., Manni, E., Triki, C.: Capacitated location of collection sites in an urban waste management system. Waste Manag. **32**(7), 1291–1296 (2012)
3. Dogan, K., Karatas, M., Yakici, E.: A model for locating preventive health care facilities. CEJOR **28**(3), 1091–1121 (2019). https://doi.org/10.1007/s10100-019-00621-4
4. Charikar, M., Khuller, S., Mount, D.M., Narasimhan, G.: Algorithms for facility location problems with outliers. In: Proceedings of the 12th ACM-SIAM Symposium on Discrete Algorithms, vol. 1, pp. 642–651 (2001)
5. Bienstock, D., Goemans, M.X., Simchi-Levi, D., Williamson, D.: A note on the prize collecting traveling salesman problem. Math. Program. **59**(1), 413–420 (1993). https://doi.org/10.1007/BF01581256
6. Arslan, O.: The location-or-routing problem. Transp. Res. Part B Methodol. **147**, 1–21 (2021)
7. Prodhon, C., Prins, C.: A survey of recent research on location-routing problems. Eur. J. Oper. Res. **238**(1), 1–17 (2014)

Determining and Resolving Conflicts in the Configuration of High-Performance Pumps

Tobias Fischer[1], Christopher Hamkins[2], Sebastian Velten[1(✉)], and Pascal Wortel[1]

[1] Fraunhofer Institute for Industrial Mathematics ITWM, Kaiserslautern, Germany
{tobias.fischer,sebastian.velten,pascal.wortel}@itwm.fraunhofer.de
[2] KSB SE & Co. KGaA, Frankenthal, Germany
christopher.paul.hamkins@ksb.com

Abstract. We consider a configuration problem for high-performance pumps, where the configurations that are technically possible are specified in restriction tables. For this problem we propose a *Constraint Programming* based feasibility model and present an algorithm for determining and resolving conflicts. Computational results show that the approach is very well suited for various use cases in practice.

Keywords: Constraint programming · Product model · Minimal conflicts · Conflict resolution

1 Introduction

High-performance pumps and valves are often sold in markets with extremely diverse customer requirements. Although they are assembled from a standardized set of parts kept in stock, hardly any two orders are exactly the same. To be able to fulfill the customer's specification but ensure that the product can be manufactured and that all the technical, pricing, lead time and business limits are respected, a product configuration software is used.

The software is used by sales people to configure and quote the product and to later automatically generate the parts lists and work plans needed in the factory. Sometimes the configuration software will report that there is a conflict with the choices made to satisfy the customer requirements. Due to the complexity of the product model, it is practically impossible for the user to resolve these manually. Automated assistance for conflict resolution is needed.

The project "Development of new methods, algorithms and software for configuration and selection of complex product systems" is supported by a grant from the European Union European Fund for Regional Development.

N. Trautmann and M. Gnägi (Eds.): OR 2021, LNOR, pp. 87–92, 2022.
https://doi.org/10.1007/978-3-031-08623-6_14

2 The Configuration Problem

The product is described by a set of features $\mathcal{F}$ with finite domains. The goal is to have a valid configuration with an assignment of a specific value to each feature such that all the constraints are fulfilled. The constraints fall into two categories, a consistent product model background that remains the same, and a set of user decisions and domain restrictions that vary for each solution attempt.

The product structure is a tree-like maximal bill of materials (BOM). The tree nodes can themselves also have sub-nodes, etc., but loops are not allowed and the depth is of course finite. The nodes in the BOM are optional and may or may not exist in a particular configuration depending on feature evaluations.

The main constraints are formulated using restriction tables and conditional assignments. A restriction table lists allowed combinations of values for a set of features. A combination of values not corresponding to a line in the table is not allowed. The restriction tables are assigned to nodes and are only active if their node exists. This can lead to nodes being rejected in a solution if the tables they contain cannot be fulfilled due to other constraints. A conditional assignment is a requirement that a feature take on a certain value if a condition holds.

A typical model has around 500 features with 4 values on average, 300 relevant nodes, 150 conditional assignments and about 600 restriction tables with 1 to 10 columns each and containing 140,000 lines in total for all tables.

3 The CP Model

We model the configuration problem using a *Constraint Programming* (CP) approach. The CP formulation is the basis for finding feasible configurations (using a backtracking search based on feature and evaluation priorities) and for determining conflicts and resolutions. For the description of the model we use decision variables and constraints provided by Google's OR-Tools CP library (see [2]).

For each feature $f \in \mathcal{F}$ we define an integer decision variable x_f with domain $\{-1, 0, \ldots, |f| - 1\}$, where $|f|$ is the number of possible evaluations for f. The value -1 is used to represent intentional nonevaluation of a feature. Furthermore, for each node n in the set of all nodes $\mathcal{N}$ we define a binary decision variable x_n which is equal to 1 if n exists and 0 otherwise.

The most important constraints are the table restrictions. For each node $n \in \mathcal{N}$ let $\mathcal{R}_n$ be the set of its restriction tables. The set of all restriction tables is denoted by $\mathcal{R} = \bigcup_{n \in \mathcal{N}} \mathcal{R}_n$. A restriction table $r \in \mathcal{R}$ constrains the possible combinations of values for p_r features by a set $\mathcal{A}_r$ of q_r vectors of length p_r.

As the nodes are optional and for $n \in \mathcal{N}$ the restriction tables in $\mathcal{R}_n$ need only to be satisfied if n exists, we introduce a *local feature variable* x_f^n for each feature f referenced in a restriction table in $\mathcal{R}_n$. Each of these local variables x_f^n has the same domain as x_f and we add the constraint

$$x_n \leq \left(x_f = x_f^n\right)\texttt{.Var()} \tag{1}$$

to the CP formulation. In this constraint, `(...).Var()` is a binary decision variable which is equal to 1 if the inner constraint is satisfied and 0 otherwise (see [2]). Thus, (1) enforces that x_f and x_f^n are equal if node n exists.

Given the local feature variables, the restriction tables are then represented in the CP formulation by the Google OR-Tools constraint

$$\texttt{AllowedAssignment}\left(\left(x^n_{f_1}, \ldots, x^n_{f_{p_r}}\right), \mathcal{A}_r\right), \tag{2}$$

which ensures the vector of values assigned to $\left(x^n_{f_1}, \ldots, x^n_{f_{p_r}}\right)$ is an element of $\mathcal{A}_r$. In combination with (1) these constraints make sure that the restriction tables are respected if node n exists. Note that the usage of the local feature variables is necessary as `(...).Var()` is not available for `AllowedAssignment(...)` (see [2]).

Apart from the table restrictions, the nodal existences each depend on the values of certain features and some other feature's values depend on the existence of certain nodes. In the CP model this is reflected by conjunctions and disjunctions of value assignments to feature and node existence variables. Furthermore, some features $f \in \mathcal{F}$ must not take a certain set of values $\mathcal{V}_f^t$ if the feature `OPRTN_PLNTS`, which selects the production plant, is set to a certain value t:

$$\max_{v \in \mathcal{V}_f^t} (x_f = v)\,\texttt{.Var()} \leq 1 - (x_{\texttt{OPRTN\_PLNTS}} = t)\,\texttt{.Var()}. \tag{3}$$

In addition to these constraints, which describe the feasible configurations of the product in general, the CP model can be extended by explicit user requests. The requests either lead to instantiations of decision variables (assignment of a certain value to a feature or determination of node existence) or further domain restrictions for feature variables.

4 Conflict Determination and Resolution

Especially in early stages of product modeling or when users specify many feature assignments, the CP model may be infeasible. In these cases, the determination of the *minimal conflicts* is crucial. They describe the core of the infeasibility and thus form the basis for resolutions that fulfill the specifications as far as possible.

In the following we present a bottom-up approach combined with a binary search for finding all minimal conflicts of a set of constraints. We use this approach since the minimal conflicts are usually rather small. Other procedures for determining all minimal conflicts can for example be found in [3] (top-down approach) and [1] (simultaneous construction of maximal satisfiable and minimal conflicting constraint sets).

Let $\mathcal{C}$ be a set of constraints. $\mathcal{C}$ is a conflict if not all constraints in $\mathcal{C}$ can be fulfilled at the same time. $\mathcal{C}$ is minimal if no proper subset of $\mathcal{C}$ is a conflict. Moreover, let $\texttt{S}$ be a set of sets. Then $\mathcal{H}$ is a hitting set of $\texttt{S}$ if $\mathcal{S}' \cap \mathcal{H} \neq \emptyset$ for all $\mathcal{S}' \in \texttt{S}$. $\mathcal{H}$ is minimal if no proper subset of $\mathcal{H}$ is a hitting set.

Let `MinC` be the set of minimal conflicts found so far and `H` be the set of minimal hitting sets of `MinC`. Then the bottom-up approach for finding all minimal conflicts of a conflict $\mathcal{C}$ has the following steps:

Algorithm 1: Determination of one minimal conflict.

Input: Conflict $\mathcal{C}'$.
Result: Minimal conflict $\mathcal{C}^*$.
if $|\mathcal{C}'| = 1$ **then**
  Return $\mathcal{C}'$.
Set $L := c_1, \ldots, c_N$ (ordered constraints of $\mathcal{C}'$), $n_{\min} = 0$, $n_{\max} = N$, $\mathcal{C}^* = \emptyset$.
while *true* **do**
  while *true* **do**
    Set $n^* = \lceil (n_{\max} - n_{\min})/2 \rceil$ and $\tilde{\mathcal{C}} = \mathcal{C}^* \cup \{c_1, \ldots, c_{n^*}\}$.
    if $\tilde{\mathcal{C}}$ *is a conflict* **then**
      if $n^* = n_{\max}$ **then**
        $\mathcal{C}^* = \mathcal{C}^* \cup \{c_{n*}\}$ and **Break.**
      else
        Set $n_{\max} = n^*$.
    else
      Set $n_{\min} = n^*$.
  if $\mathcal{C}^*$ *is a conflict* **then**
    Break.
  else
    Set $n_{\min} = 0$, $n_{\max} = n^* - 1$.
Return $\mathcal{C}^*$.

1. Set $\mathcal{C}' = \mathcal{C}$.
2. Find a minimal conflict $\mathcal{C}^* \subseteq \mathcal{C}'$ by Algorithm 1. $\texttt{MinC} := \texttt{MinC} \cup \{\mathcal{C}^*\}$.
3. Update H with respect to $\mathcal{C}^*$.
4. For all $\mathcal{H} \in \texttt{H}$:
   - If $\mathcal{C} \backslash \mathcal{H}$ is a conflict: Set $\mathcal{C}' = \mathcal{C} \backslash \mathcal{H}$, go to Step 2.
   - Otherwise: Remove $\mathcal{H}$ from H.
5. Return MinC.

The update of H with respect to $\mathcal{C}^*$ in Step 3 is done as follows:

- Extend each element of H by each element of $\mathcal{C}^* : \texttt{H} := \{\mathcal{H} \cup c : \mathcal{H} \in \texttt{H}, c \in \mathcal{C}^*\}$.
- Remove all duplicate and non minimal hitting sets from H.

In addition, note that $\mathcal{C} \backslash \mathcal{H}$ resolves all minimal conflicts of the current set MinC. Therefore, if $\mathcal{C} \backslash \mathcal{H}$ is a conflict, this set contains a new minimal conflict and no duplicates are generated. Finally, the approach guarantees that MinC is complete (i.e. all minimal conflicts have been found) as in Step 5 the complements $\mathcal{C} \backslash \mathcal{H}$ for all hitting sets $\mathcal{H}$ of MinC are feasible.

Given MinC, resolutions are determined by relaxing minimal hitting sets of this set and applying the prioritized backtracking search.

5 Use Cases and Computational Results

5.1 Conflict Resolution

In this use case the end user would like to be informed which of the user decisions cause conflicts with the product model and be given suggestions for changing

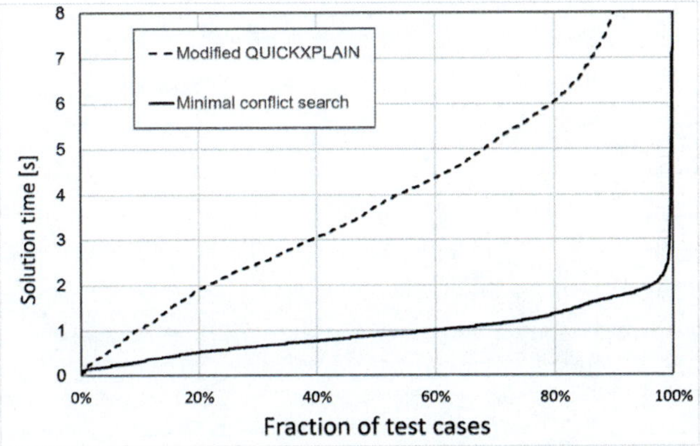

Fig. 1. Solution time for conflict resolution use case

them to get a valid configuration. During product configuration a user proceeds by selecting values for some of the features and the system is to return a valid assignment for the rest of them or to determine that no valid solution exists and return a number of suggestions on how to resolve the problem by changing some of the decisions. Normally the user proceeds feature by feature with the system responding after each choice, so a fast response is very important.

Since this is the same as the use case described in [4] the following variant of this approach was implemented as a baseline. The user decisions were prioritized based on the order in which they were made. Because of the step-by-step decision procedure the QUICKXPLAIN algorithm in [4] simply reports "relax your last decision" as resolution, rather unsatisfying for the end user. As the product model is soluble without user decision constraints, and each feature value can occur in at least one solution, there are always at least two possible conflict resolutions, so the algorithm was extended to also find all relaxations of single user decisions that resolve the conflict and if there are less than two, also a general relaxation different than "relax your last decision".

Figure 1 shows a comparison of the solution times for the algorithm developed here (Sect. 4) with the modified QUICKXPLAIN for about 6000 test cases taken from actual and conflicting end user interactions. It can be seen that the new algorithm usually completes in 25% of the time that the baseline required. Further, in general it reports more possible conflict resolutions (potentially one for each minimal hitting set). At the far right of the figure we see that there are a few cases in which both algorithms required substantially more time or didn't complete, reminding us that the problem at hand remains NP-hard.

5.2 Conflict Analysis

In this use case the product modeler assigns values to some features and finds that no solution is possible, although the feature assignment is intended to be

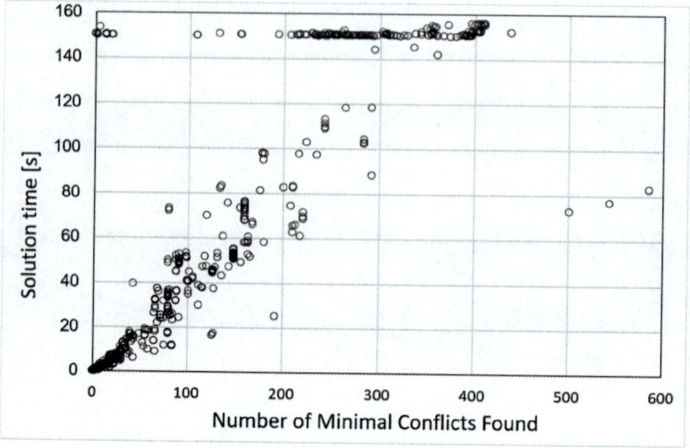

Fig. 2. Running time for conflict analysis use case

a valid configuration. In this case, the product modeler needs to know which of the constraints in the model prevent the assignments to check the formulation of those constraints for errors. The response time is not so important, but the completeness of the answer is. Manual analysis of the conflict usually takes days, so if the system can automatically deliver a complete analysis in less time it is a great benefit for the modeler looking for problems in the model.

The performance of the algorithms for conflict analysis was demonstrated on 3700 test cases with conflicts, also taken from real end user interactions. Figure 2 shows the processing time required vs. the number of minimal conflicts discovered. The solution time of up to a few minutes is perfectly acceptable for practical use. The large number of minimal conflicts appears surprising at first but can be explained due to (necessary) redundancies in the restriction tables.

For the tests, a time limit of roughly 150 s was enforced. When the time limit was reached, the algorithm had always reported some minimal conflicts. Since every single minimal conflict needs to be addressed to resolve the overall conflict, reporting even one minimal conflict is a help for the product modelers.

References

1. Bailey, J., Stuckey, P.J.: Discovery of minimal unsatisfiable subsets of constraints using hitting set dualization. In: Hermenegildo, M.V., Cabeza, D. (eds.) PADL 2005. LNCS, vol. 3350, pp. 174–186. Springer, Heidelberg (2005). https://doi.org/10.1007/978-3-540-30557-6_14
2. Google: Google's OR-Tools (2020). https://developers.google.com/optimization/
3. Han, B., Lee, S.-J.: Deriving minimal conflict sets by CS-trees with mark set in diagnosis from first principles. IEEE Trans. Syst. Man Cybern. B **29**(2), 281–286 (1999)
4. Junker, U.: QUICKXPLAIN: preferred explanations and relaxations for over-constrained problems. In: Proceedings of the 19th National Conference on Artificial Intelligence, 16th Conference on Innovative Applications of Artificial Intelligence, San Jose, California, USA, 25–29 July 2004, pp. 167–172 (2004)

Discrete and Combinatorial Optimization

The Knapsack Problem with Special Neighbor Constraints on Directed Co-graphs

Steffen Goebbels[1](✉), Frank Gurski[2], and Dominique Komander[2]

[1] Faculty of Electrical Engineering and Computer Science, iPattern Institute, Niederrhein University of Applied Sciences, 47805 Krefeld, Germany
steffen.goebbels@hsnr.de

[2] Institute of Computer Science, Algorithmics for Hard Problems Group, University of Düsseldorf, 40225 Düsseldorf, Germany
{frank.gurski,dominique.komander}@hhu.de

Abstract. The knapsack problem is one of the best known and most fundamental NP-hard problems in combinatorial optimization. We consider two knapsack problems which contain additional constraints in the form of directed graphs whose vertex set corresponds to the item set. In the 1-neighbor knapsack problem, an item can be chosen only if at least one of its successors is chosen. In the all-neighbors knapsack problem, an item can be chosen only if all of its successors are chosen. For both problems, we consider uniform and general profits and weights. Since all these problems generalize the knapsack problem, they are NP-hard. This motivates us to consider the problem on special graph classes. Therefore, we restrict these problems to directed co-graphs, i.e., directed complement reducible graphs, that are precisely those digraphs which can be defined from the single vertex graph by applying the disjoint union, order and series composition. We show polynomial time solutions for the uniform problems on directed co-graphs and pseudo-polynomial time solutions for the general problems on directed co-graphs. These results improve known worst-case runtimes in comparison to constraints given by unrestricted digraphs.

Keywords: Knapsack problem · Neighbor constraints · Directed co-graphs

1 Introduction

In the last years, the interest in knapsack problems related to graphs has grown strongly. These include, for example, the knapsack problem with conflict or forcing graphs, the subset sum problem with digraph restrictions, and the partially ordered knapsack problem. A *directed graph* or *digraph* is a pair $G = (A, E)$, where A is a finite set of *vertices* (here the item set of a knapsack instance) and $E \subseteq \{(u, v) \mid u, v \in A,\ u \neq v\}$ is a finite set of ordered pairs of distinct vertices, the *directed edges* or *arcs*. For a vertex $u \in A$, the set

N. Trautmann and M. Gnägi (Eds.): OR 2021, LNOR, pp. 95–100, 2022.
https://doi.org/10.1007/978-3-031-08623-6_15

$N_G^+(u) = \{v \in A \mid (u,v) \in E\}$ is called the *set of all successors.* We analyze the knapsack problem with additional constraints based on a digraph $G = (A, E)$ from [1,2]. The *all-neighbors constraint* prescribes that an item $u \in A$ can be chosen into a feasible knapsack solution $A' \subseteq A$ only if all its successors in $N_G^+(u)$ are also chosen, i.e.,

$$u \in A' \Rightarrow N_G^+(u) \subseteq A'. \tag{1}$$

The *1-neighbor constraint* prescribes that an item $u \in A$ can be chosen into $A' \subseteq A$ only if it does not have a successor or if at least one of its successors in $N_G^+(u)$ is chosen, i.e.,

$$\left(u \in A' \wedge N_G^+(u) \neq \emptyset\right) \Rightarrow N_G^+(u) \cap A' \neq \emptyset. \tag{2}$$

This allows us to state the following optimization problems given in [2].

Name: Knapsack with all-neighbor (1-neighbor) constraint, KPaN (KP1N)
Instance: A set $A = \{a_1, \dots, a_n\}$ of $n \geq 1$ items, a capacity $c \in \mathbb{N}_0 := \{0, 1, 2, \dots\}$, and a digraph $G = (A, E)$. Every item a_j has a size $s_j \in \mathbb{N}_0$, $s_j \leq c$, and a profit $p_j \in \mathbb{N}_0$.
Task: Find a subset A' of A that maximizes $p(A') := \sum_{a_j \in A'} p_j$ subject to

$$s(A') := \sum_{a_j \in A'} s_j \leq c, \tag{3}$$

and (1) (or (2), respectively).

The restriction of KPaN and KP1N to uniform sizes and profits ($s_j = p_j = 1$ for $j = 1, \dots, n$) is denoted by uniform KPaN and uniform KP1N, respectively.

Following [2], we consider the four problems $\{\text{1-neighbor}, \text{all-neighbor}\} \times \{\text{general}, \text{uniform}\}$. For every instance of KPaN and KP1N, $A' = \emptyset$ and, if $s(A) \leq c$, $A' = A$ are feasible solutions.

The class of *directed co-graphs* (*di-co-graphs*) is recursively defined as follows.

(i) Every digraph on a single vertex $(\{u\}, \emptyset)$, is a di-co-graph.
(ii) If $G_1 = (V_1, E_1)$ and $G_2 = (V_2, E_2)$ are vertex-disjoint di-co-graphs, then following graphs are also di-co-graphs:
  (a) the *disjoint union* $G_1 \oplus G_2$, which is defined as the digraph with vertex set $V_1 \cup V_2$ and edge set $E_1 \cup E_2$,
  (b) the *order composition* $G_1 \oslash G_2$, defined by their disjoint union plus all possible edges directed from V_1 to V_2, and
  (c) the *series composition* $G_1 \otimes G_2$ defined by their disjoint union plus all possible edges between V_1 and V_2 in both directions.

Every expression X using these operations is called a *di-co-expression* and digraph(X) is the defined digraph with $|X|$ vertices. A tree structure for every di-co-graph, denoted as *di-co-tree*, can be derived in linear time, see [4]. The leaves of the di-co-tree represent the vertices of the digraph, and the inner nodes of the di-co-tree correspond to the operations applied on the sub-expressions defined by the subtrees.

Table 1. Time complexity of knapsack problems with neighbor constraints on a digraph

	All-neighbor constraint		1-neighbor constraint	
	Digraph	Di-co-graph	Digraph	Di-co-graph
Uniform	Strongly NP-hard, [2]	$\mathcal{O}(n^3)$	Strongly NP-hard, [2]	$\mathcal{O}(n^3)$
General	Strongly NP-hard, [2]	$\mathcal{O}(n(P+1) \cdot \max\{n, P+1\})$	Strongly NP-hard, [2]	$\mathcal{O}(nP^2 + n^2)$

Table 1 summarizes complexity results for unrestricted digraphs as well as for di-co-graphs that are shown in subsequent sections. Some solutions are pseudo-polynomial. Their running times depend on the profit sum $P = \sum_{j=1}^{n} p_j \leq n \cdot \max_{1 \leq j \leq n} p_j$ which is an upper bound for the profit of an optimal solution. The runtime bounds $\mathcal{O}(n^3)$ for uniform KPaN and KP1N on a di-co-graph follow directly from the bounds of the general problems by setting $p_j = s_j = 1$ and $P = n$.

2 General Problems on Directed Co-graphs

We consider an instance I of general KPaN on a di-co-graph G with n vertices and m arcs. A binary di-co-expression X, that represents G, can be computed in $\mathcal{O}(n+m) \subseteq \mathcal{O}(n^2)$ time, see [4]. For a subexpression X' of X let $F(X', p)$ be the minimum size of a feasible solution of KPaN on digraph(X') where we replace c by ∞ in Eq. (3) while the profit is exactly p. We set $F(X', p)$ to ∞, whenever there is no such feasible solution. Algorithm 1 shows how to use dynamic programming to compute all values $F(X', p')$, $p' \in \{0, \ldots, P\}$, in $\mathcal{O}((P+1)^2 n)$ time when traversing the di-co-tree of X from the leaves to the root to find all sub-expressions X' so that previously computed values of F can be used.

If one also considers the effort bounded by $\mathcal{O}(n^2)$ for computing the di-co-tree and $\mathcal{O}(n)$ to compute P, we find the optimum $\text{OPT}(I) = \max\{p \mid F(X, p) \leq c\}$ in $\mathcal{O}(n(P+1)\max\{n, P+1\})$ time.

A subset sum problem with digraph constraint (SSG) is defined in [6]. This is a general KPaN problem on a digraph for which item sizes and profits are equal. The bound for SSG on directed co-graphs given in [7] can be verified with Algorithm 1. In fact, the dynamic program is structured similarly to the proof techniques in [7].

A di-co-graph can possess cycles and does not necessarily contain sinks (vertices without successors). Therefore, the induced digraph of a feasible, non-empty KPaN or KP1N solution does not necessarily have sinks. Let L and R be di-co-expressions. A feasible KP1N solution on digraph(L) without sinks of digraph(L) is also a feasible solution with respect to digraph$(L \oslash R)$ because every solution vertex already has a successor in digraph(L) which belongs to the feasible solution. But this is not true for feasible KP1N solutions on digraph(L) that contain at least one sink of digraph(L). To get useful information about the sinks within a solution, we use an extended data structure. For some subexpression X' of X let

Algorithm 1. General KPaN on di-co-graphs: Determine minimal solution size $F(X,p)$ for a given profit p from previously computed values for sub-expressions of X.

```
F(X,p) := ∞, (V,E) := digraph(X)
if p = 0 then F(X,p) := 0
else if |X| = 1 then let a_i ∈ V be the vertex of digraph(X).
    if p = p_i then F(X,p) := s_i
else if X = L ⊕ R then          ▷ no edges between vertices of digraph(L) and digraph(R)
    for p' := 0; p' ≤ p; p' := p' + 1 do
        S := F(L,p') + F(R,p − p')
        if F(X,p) > S then F(X,p) := S
else
    (V_L, E_L) := digraph(L), (V_R, E_R) := digraph(R)
    P_L := Σ_{a_i∈V_L} p_i, P_R := Σ_{a_i∈V_R} p_i, S_L := Σ_{a_i∈V_L} s_i, S_R := Σ_{a_i∈V_R} s_i
    if X = L ⊘ R then     ▷ all vertices in digraph(R) are successors of vertices in digraph(L)
        if p ≤ P_R then F(X,p) := F(R,p)
        else F(X,p) := S_R + F(L, p − P_R)
    if X = L ⊗ R then                  ▷ no vertex or all vertices in a solution
        if p = P_L + P_R then F(X,p) = S_L + S_R
if F(X,p) > c then F(X,p) := ∞
```

Algorithm 2. This dynamic knapsack program computes the minimal size $\hat{F}(X,p)$ of a non-empty set of vertices (with sizes and profits) from the vertices of digraph(X) such that their sum of profits equals $p \in \{0,\ldots,P\}$.

```
if |X| = 1 then let a_i ∈ V be the vertex of digraph(X).
    if p = p_i then F̂(X,p) := s_i
    else F̂(X,p) := ∞
else                                   ▷ X = L ⊕ R or X = L ⊘ R or X = L ⊗ R
        ▷ we consider solutions that do not contain vertices of either digraph(L) or digraph(R):
    F̂(X,p) := min{F̂(L,p), F̂(R,p)}
                ▷ other solutions consist of vertices of digraph(L) and digraph(R):
    for p' = 1; p' < p; p' := p' + 1 do
        if F̂(X,p) > F̂(L,p') + F̂(R,p − p') then F̂(X,p) := F̂(L,p') + F̂(R,p − p')
```

$F(X',p,k)$ be the minimum size of a feasible solution of KP1N on digraph(X') where c is replaced by ∞ in Eq. (3) and the profit is exactly p. Additionally, the feasible solution must contain at least one sink of digraph(X') if $k=1$ and does not include a sink of digraph(X') if $k=0$. We set $F(X',p,k)$ to ∞ whenever there is no such feasible solution.

Profits (and sizes) of vertices a_i are allowed to be zero. Thus, a zero profit can be realized with either an empty solution or with non-empty solutions that only contain vertices a_i with zero profit $p_i = 0$. In contrast to empty solutions, vertices with zero profit can be used to fulfill neighbor constraints (see Algorithm 3, case $X = L \oslash R$ for $p'' = 0$, and case $X = L \otimes R$ for $p' = 0$ or $p' = p$). To differentiate between empty and non-empty zero profit solutions, we introduce function $\tilde{F}$ that is defined as F with the difference that now instead of feasible solutions only non-empty feasible solutions are considered. If there are only empty feasible solutions then $\tilde{F}(X',p,k) = \infty$. Thus, we have

$$F(X,p,k) = \begin{cases} \tilde{F}(X,p,k), & \text{if } p > 0 \vee k = 1, \\ 0, & \text{else.} \end{cases}$$

Algorithm 3. General KP1N on di-co-graphs: Determine minimal size $\tilde{F}(X, p, k)$ of a non-empty solution for a given profit p from previously computed values for sub-expressions of X.

$\tilde{F}(X,p,k) := \infty$
if $|X| = 1$ **then** let $a_i \in V$ be the only vertex.
  if $k = 1 \wedge p = p_i$ **then** $\tilde{F}(X,p,k) := s_i$
else if $X = L \oplus R$ **then** ▷ no edges between vertices of digraph(L) and digraph(R)
  $\tilde{F}(X,p,k) := \tilde{F}(L,p,k)$ ▷ solution restricted to vertices of digraph(R) has zero profit
  if $\tilde{F}(X,p,k) > \tilde{F}(R,p,k)$ **then** $\tilde{F}(X,p,k) := \tilde{F}(R,p,k)$ ▷ zero profit if restricted to L
  for $p' = 1$; $p' < p$; $p' := p' + 1$ **do** ▷ positive profit with respect to both L and R
    if $k = 0$ **then** ▷ there must be no sink
      $S := \tilde{F}(L,p',0) + \tilde{F}(R,p-p',0)$
      if $\tilde{F}(X,p,k) > S$ **then** $\tilde{F}(X,p,k) := S$
    else ▷ there has to be at least one sink
      $S_1 := \tilde{F}(L,p',1) + \tilde{F}(R,p-p',0)$, $S_2 := \tilde{F}(L,p',0) + \tilde{F}(R,p-p',1)$
      $S_3 := \tilde{F}(L,p',1) + \tilde{F}(R,p-p',1)$
      for $i = 1$; $i \leq 3$; $i := i + 1$ **do**
        if $\tilde{F}(X,p,k) > S_i$ **then** $\tilde{F}(X,p,k) := S_i$
else if $X = L \oslash R$ **then** ▷ all vertices in digraph(R) are successors of vertices in digraph(L)
  if $k = 0$ **then** $\tilde{F}(X,p,k) := \tilde{F}(L,p,k)$
  if $\tilde{F}(X,p,k) > \tilde{F}(R,p,k)$ **then** $\tilde{F}(X,p,k) := \tilde{F}(R,p,k)$
  for $p'' = 0$; $p'' < p$; $p'' := p'' + 1$ **do**
    $S_R := \tilde{F}(R,p'',k)$
    if $S_R < \infty$ **then**
      $S_L := \hat{F}(L,p-p'')$, see Algorithm 2
      if $\tilde{F}(X,p,k) > S_L + S_R$ **then** $\tilde{F}(X,p,k) := S_L + S_R$
else ▷ $X = L \otimes R$ such that digraph(X) has no sinks
  if $k = 0$ **then**
    $\tilde{F}(X,p,k) := \tilde{F}(L,p,0)$, $S_R := \tilde{F}(R,p,0)$
    if $S_R < \tilde{F}(X,p,k)$ **then** $\tilde{F}(X,p,k) := S_R$
      ▷ combine solutions with at least one vertex from both digraphs such
      ▷ that the neighbor constraint is fulfilled due to the series composition:
    for $p' = 0$; $p' \leq p$; $p' := p' + 1$ **do**
      $S_L := \hat{F}(L,p')$, see Algorithm 2
      $S_R := \hat{F}(R,p-p')$, see Algorithm 2
      if $S_L + S_R < \tilde{F}(X,p,k)$ **then** $\tilde{F}(X,p,k) := S_L + S_R$

While traversing vertices u of di-co-tree T with root r of di-co-graph G in a bottom-up order, we compute $\tilde{F}(X(u), p, k)$ for expression $X(u)$ representing a subtree of T with root u and integers $0 \leq p \leq P$, $k \in \{0, 1\}$, with Algorithm 3. Within this algorithm, we solve a knapsack problem without graph restrictions. This is done via Algorithm 2 which computes $\hat{F}(X, p)$ as the smallest size of a non-empty knapsack solution on vertices of digraph(X) without any graph restrictions and without a capacity bound. This solution obtains exactly profit p. We first compute all values of $\hat{F}$ by also traversing di-co-tree T in bottom-up order in $\mathcal{O}((P + 1)^2 n)$ time. Then, we apply Algorithm 3 in the same order to also compute all values of $\tilde{F}(X, p, k)$ in $\mathcal{O}((P+1)^2 n)$ time. By considering $\mathcal{O}(n^2)$ time to get the di-co-tree and $\mathcal{O}(n)$ to compute P, the optimization problem

$$\mathrm{OPT}(I) = \max\{p \in \{0, \ldots, P\} \mid p = 0 \vee \exists_{k \in \{0,1\}} \tilde{F}(X(r), p, k) \leq c\}$$

and thus, the general KP1N problem can be solved in $\mathcal{O}(P^2 n + n^2)$ time.

3 Outlook

An extended version [5] of this paper, dealing also with other graph types, was accepted during publishing time. *Minimal series-parallel digraphs* (*msp-digraphs*) can be represented by a tree structure similar to di-co-graphs, and the tree structure can be constructed in linear time, see [8]. This allows us to obtain the same bounds as for di-co-graphs. The only difference is that we have to track the existence of sources (vertices without predecessors) instead of dealing with sinks.

A *directed tree* is a directed graph for which the underlying undirected graph is a tree. In directed trees, we allow opposite edges. An *out-rooted tree* (*in-rooted tree*) is an orientation of a tree with a distinguished root such that all arcs are directed away from (to) the root. Whereas out- and in-rooted trees are special cases of msp-digraphs, KPaN and KP1N problems on arbitrary directed trees with n vertices can also be solved with dynamic programming to obtain polynomial and pseudo-polynomial bounds, respectively. Uniform KPaN and KP1N can be treated in $\mathcal{O}(n^3)$ time, the same bound as for di-co-graphs and msp-digraphs. The runtime of general KPaN on directed trees is bounded by $\mathcal{O}(n(P+1)(P+n))$, and general KP1N can be solved in $\mathcal{O}(nP^2+n)$ time. These bounds are obtained by transforming the tree to a binary tree before applying dynamic programming. Since directed trees, di-co-graphs and msp-digraphs have bounded directed clique-width [3], it remains open whether the results can be extended to constraints given by classes of digraphs of bounded directed clique-width.

References

1. Borradaile, G., Heeringa, B., Wilfong, G.: The 1-neighbour knapsack problem. In: Iliopoulos, C.S., Smyth, W.F. (eds.) IWOCA 2011. LNCS, vol. 7056, pp. 71–84. Springer, Heidelberg (2011). https://doi.org/10.1007/978-3-642-25011-8_6
2. Borradaile, G., Heeringa, B., Wilfong, G.: The knapsack problem with neighbour constraints. J. Discrete Algorithms **16**, 224–235 (2012)
3. Courcelle, B., Olariu, S.: Upper bounds to the clique width of graphs. Discrete Appl. Math. **101**(1–3), 77–114 (2000)
4. Crespelle, C., Paul, C.: Fully dynamic recognition algorithm and certificate for directed cographs. Discrete Appl. Math. **154**(12), 1722–1741 (2006)
5. Goebbels, S., Gurski, F., Komander, D.: The knapsack problem with special neighbor constraints. Math. Meth. Oper. Res. **95**(1), 1–34 (2021). https://doi.org/10.1007/s00186-021-00767-5
6. Gourvès, L., Monnot, J., Tlilane, L.: Subset sum problems with digraph constraints. J. Comb. Optim. **36**(3), 937–964 (2018). https://doi.org/10.1007/s10878-018-0262-1
7. Gurski, F., Komander, D., Rehs, C.: Solutions for subset sum problems with special digraph constraints. Math. Meth. Oper. Res. **92**(2), 401–433 (2020). https://doi.org/10.1007/s00186-020-00718-6
8. Valdes, J., Tarjan, R., Lawler, E.: The recognition of series-parallel digraphs. SIAM J. Comput. **11**, 298–313 (1982)

Oriented Vertex and Arc Coloring of Edge Series-Parallel Digraphs

Frank Gurski(✉), Dominique Komander, and Marvin Lindemann

Institute of Computer Science, Algorithmics for Hard Problems Group, University of Düsseldorf, 40225 Düsseldorf, Germany
{frank.gurski,dominique.komander,marvin.lindemann}@hhu.de

Abstract. We study the oriented vertex and arc coloring problem on edge series-parallel digraphs which are related to the well known series-parallel graphs. The oriented class of edge series-parallel digraphs is recursively defined from pairs of vertices connected by a single arc and applying the parallel and series composition, which leads to specific orientations of undirected series-parallel graphs. We re-prove the known bound of 7 for the oriented chromatic number and the oriented chromatic index of series-parallel digraphs and we show that these bounds are tight even for edge series-parallel digraphs. Further, we give linear time solutions for computing the oriented chromatic number and the oriented chromatic index of minimal series-parallel digraphs and edge series-parallel digraphs.

Keywords: Series-parallel digraphs · Oriented vertex-coloring · Oriented arc-coloring · Linear time solutions

1 Introduction and Preliminaries

In this paper we consider graph colorings for *oriented graphs*, i.e. digraphs with no loops and no opposite arcs.

Courcelle introduced oriented vertex colorings [2], in particular an *oriented r-vertex-coloring* of an oriented graph $G = (V, E)$ is a mapping $c : V \to \{1, \dots, r\}$ such that:

- $c(u) \neq c(v)$ for every $(u, v) \in E$,
- $c(u) \neq c(y)$ for every $(u, v) \in E$ and $(x, y) \in E$ with $c(v) = c(x)$.

The *oriented chromatic number* of G, denoted by $\chi_o(G)$, is the smallest r such that there exists an oriented r-vertex-coloring for G.

An oriented r-vertex-coloring of an oriented graph G corresponds to an oriented graph H on r vertices, such that there exists a homomorphism from G to H. We then call H the *color graph* of G.

In the oriented chromatic number problem (OCN) there is given an oriented graph G and an integer r and one has to decide whether there is an oriented r-vertex-coloring for G. If r is not part of the input, we call the related problem

N. Trautmann and M. Gnägi (Eds.): OR 2021, LNOR, pp. 101–106, 2022.
https://doi.org/10.1007/978-3-031-08623-6_16

the r-Oriented Chromatic Number (OCN_r). If $r \leq 3$, we can decide OCN_r in polynomial time, while OCN_4 is still NP-complete [5]. Further, for every transitive acyclic digraph and every minimal series-parallel digraph OCN can be solved in linear time [3].

An *oriented r-arc-coloring* of an oriented graph $G = (V, E)$ is a mapping $c : E \to \{1, \ldots, r\}$ such that:

- $c((u,v)) \neq c((v,w))$ for every two arcs $(u,v) \in E$ and $(v,w) \in E$
- $c((u,v)) \neq c((y,z))$ for every four arcs $(u,v) \in E$, $(v,w) \in E$, $(x,y) \in E$, and $(y,z) \in E$, with $c((v,w)) = c((x,y))$.

Moreover, the *oriented chromatic index* of G, denoted by $\chi'_o(G)$, is the smallest r such that G has an oriented r-arc-coloring.

In the oriented chromatic index problem (OCI) we have an oriented graph G and an integer r and we need to decide whether there is an oriented r-arc-coloring for G. If r is not part of the input, we call the related problem the r-Oriented Chromatic Index (OCI_r). If $r \leq 3$, then we can decide OCI_r in polynomial time, while OCI_4 is NP-complete [6].

The *line digraph* $LD(G)$ of digraph G has a vertex for every arc in G and an arc from u to v if and only if $u = (x,y)$ and $v = (y,z)$ for vertices x, y, z from G [4]. We call digraph G the *root digraph* of $LD(G)$.

Observation 1 **([6]).** *Let G be an oriented graph. Then, it holds that*

1. $\chi'_o(G) = \chi_o(LD(G))$ *and*
2. $\chi'_o(G) \leq \chi_o(G)$

The definition of oriented vertex-coloring and oriented arc-coloring was often used for undirected graphs, where the maximum value $\chi_o(G')$ or $\chi'_o(G')$ of all possible orientations G' of a graph G is considered. In this sense it has been shown in [8] that every series-parallel graph has oriented chromatic number at most 7 and that this bound is tight. Further, due [7] every series-parallel graph has oriented chromatic index at most 7, this bound is also tight. These results lead to (not necessarily tight) upper bounds for the oriented chromatic number and the oriented chromatic index of edge series-parallel digraphs.

In this paper we re-prove the bound of 7 for the oriented chromatic index (Corollary 3) and the oriented chromatic number (Theorem 2) of edge series-parallel digraphs and we show that these bounds are tight even for edge series-parallel digraphs (Expressions X_3 and X_4). Furthermore, we give linear time solutions for computing the oriented chromatic index (Theorem 1) and the oriented chromatic number (Theorem 3) of edge series-parallel digraphs.

2 Minimal Vertex Series-Parallel Digraphs

We recall the definition of minimal vertex series-parallel digraphs.

Definition 1 (Minimal Vertex Series-Parallel Digraphs [9]). *The class of* minimal vertex series-parallel digraphs, msp-digraphs *for short, is recursively defined as follows.*

(i) Every digraph on a single vertex $(\{v\}, \emptyset)$, *denoted by* v, *is a* minimal vertex series-parallel digraph.

(ii) If $G_1 = (V_1, E_1)$ *and* $G_2 = (V_2, E_2)$ *are vertex-disjoint minimal vertex series-parallel digraphs and* O_1 *is the set of vertex of outdegree* 0 *(set of sinks) in* G_1 *and* I_2 *is the set of vertices of indegree* 0 *(set of sources) in* G_2, *then*
   (a) the parallel composition $G_1 \cup G_2 = (V_1 \cup V_2, E_1 \cup E_2)$ *is a* minimal vertex series-parallel digraph *and*
   (b) the series composition $G_1 \times G_2 = (V_1 \cup V_2, E_1 \cup E_2 \cup (O_1 \times I_2))$ *is a* minimal vertex series-parallel digraph.

An expression X using the operations of Definition 1 is called an *msp-expression* while digraph(X) is the digraph defined in X.

Using the recursive structure of msp-digraphs in [3] we show the following bound.

Proposition 1 ([3]). *Let* G *be an msp-digraph. Then, it holds that* $\chi_o(G) \leq 7$.

We can also bound the oriented chromatic number of an msp-digraph G using the corresponding root digraph G' which is an esp-digraph (see Sect. 3). By Lemma 1, Observation 1(1.), and Corollary 3 it holds that $\chi_o(G) = \chi_o(LD(G')) = \chi'_o(G') \leq 7$.

For the optimality of the shown bound, we recall from [3] the msp-expression

$$\begin{aligned} X_1 = v_1 \times (v_2 \cup v_3 \times (v_4 \cup v_5 \times v_6)) \times (v_7 \cup (v_8 \cup v_9 \times v_{10}) \\ \times (v_{11} \cup v_{12} \times v_{13})) \times (v_{14} \cup (v_{15} \cup (v_{16} \cup v_{17} \times v_{18}) \\ \times (v_{19} \cup v_{20} \times v_{21})) \times (v_{22} \cup (v_{23} \cup v_{24} \times v_{25}) \times v_{26})) \times v_{27}. \end{aligned}$$

Since $\chi_o(\text{digraph}(X_1)) = 7$ the bound of Proposition 1 is best possible.

Proposition 2 ([3]). *Let* G *be an msp-digraph. Then, the oriented chromatic number of* G *can be computed in linear time.*

By Proposition 1 and Observation 1(2.) we know the following bound.

Corollary 1. *Let* G *be an msp-digraph. Then, it holds that* $\chi'_o(G) \leq 7$.

For the optimality of the shown bound we recursively define msp-expressions Y_i. Y_0 defines a single vertex graph and for $i \geq 1$ we define $Y_i = (Y_0 \cup Y_{i-1} \times Y_{i-1})$ in order to define $X_2 = Y_0 \times Y_0 \times Y_6 \times Y_0 \times Y_0$. Since $\chi'_o(\text{digraph}(X_2)) = 7$, which was computed by a computer program, we know that that the bound of Corollary 1 is best possible.

3 Edge Series-Parallel Digraphs

Undirected series-parallel graphs are graphs with two distinguished vertices called terminals, formed recursively by parallel and series composition [1, Section 11.2].

Proposition 3 ([8]). *Let G' be an orientation of a series-parallel graph G. Then, it holds that $\chi_o(G') \leq 7$.*

In [8] it was also shown that the bound is tight. For the chromatic index of orientations of undirected series-parallel graphs Observation 1(2.) and Proposition 3 lead to the following bound.

Corollary 2. *Let G' be some orientation of a series-parallel graph G. Then, it holds that $\chi'_o(G') \leq 7$.*

In [7] it was shown that the bound is tight.

We recall the definition of edge series-parallel digraphs, originally defined as edge series-parallel multidigraphs.

Definition 2 (Edge Series-Parallel Multidigraphs [9]). *The class of* edge series-parallel multidigraphs, esp-digraphs *for short, is recursively defined as follows.*

(i) Every digraph of two distinct vertices joined by a single arc $(\{u, v\}, \{(u, v)\})$, denoted by (u, v), is an edge series-parallel multidigraph.
(ii) If $G_1 = (V_1, A_1)$ and $G_2 = (V_2, A_2)$ are vertex-disjoint minimal edge series-parallel multidigraphs, then
  (a) the parallel composition *$G_1 \cup G_2$, which identifies the source of G_1 with the source of G_2 and the sink of G_1 with the sink of G_2, is an* edge series-parallel multidigraph *and*
  (b) the series composition *$G_1 \times G_2$, which identifies the sink of G_1 with the source of G_2, is an* edge series-parallel multidigraph.

An expression X using the operations of Definition 2 is called an *esp-expression* and digraph(X) the defined digraph.

Lemma 1 ([9]). *An acyclic multidigraph G with a single source and a single sink is an esp-digraph if and only if $LD(G)$ is an msp-digraph.*

Oriented Arc-Colorings. Since every esp-digraph is an orientation of a series-parallel graph by Corollary 2 we get the following bound.

Corollary 3. *Let G be an esp-digraph. Then, it holds that $\chi'_o(G) \leq 7$.*

We can also bound the oriented chromatic index of an esp-digraph G using the corresponding line digraph $LD(G)$ which is an msp-digraph. Then, by Observation 1(1.), Lemma 1 and Proposition 1 it holds that $\chi'_o(G) = \chi_o(LD(G)) \leq 7$.

The results of [7] even show that 7 is a tight upper bound for the oriented chromatic index of every orientation of series-parallel graphs. In order to show that this bound is also tight for the subclass of esp-digraphs, we use the esp-expression

$$\begin{aligned} X_3 = {} & (v_1, v_2) \times ((v_2, v_5) \cup (v_2, v_3) \times ((v_3, v_5) \cup (v_3, v_4) \times (v_4, v_5))) \\ & \times ((v_5, v_9) \cup ((v_5, v_7) \cup (v_5, v_6) \times (v_6, v_7)) \times ((v_7, v_9) \cup (v_7, v_8) \\ & \times (v_8, v_9))) \times ((v_9, v_{16}) \cup ((v_9, v_{13}) \cup ((v_9, v_{11}) \cup (v_9, v_{10}) \\ & \times (v_{10}, v_{11})) \times ((v_{11}, v_{13}) \cup (v_{11}, v_{12}) \times (v_{12}, v_{13}))) \times ((v_{13}, v_{16}) \\ & \cup ((v_{13}, v_{15}) \cup (v_{13}, v_{14}) \times (v_{14}, v_{15})) \times (v_{15}, v_{16}))) \times (v_{16}, v_{17}). \end{aligned}$$

Since $\chi'_o(\text{digraph}(X_3)) = \chi_o(LD(\text{digraph}(X_3))) = \chi_o(\text{digraph}(X_1)) = 7$, where X_1 is defined in Sect. 2, it holds that the bound of Corollary 3 is best possible.

By Proposition 2 and Observation 1(1.) we obtain the following result.

Theorem 1. *Let G be an esp-digraph. Then, the oriented chromatic index of G can be computed in linear time.*

Oriented Vertex Colorings. Since every esp-digraph is an orientation of a series-parallel graph by Proposition 3 we have the following bound.

Corollary 4. *Let G be an esp-digraph. Then, it holds that $\chi_o(G) \leq 7$.*

The proof of Proposition 3 given in [8] uses the color graph $H = (V_H, E_H)$ where $V_H = \{1, 2, 3, 4, 5, 6, 7\}$ and $E_H = \{(i, j) \mid j - i \equiv 1, 2, \text{ or } 4 \pmod 7\}$ which is built from the non-zero quadratic residues of 7. Next, we give an alternative proof of Corollary 4 using the recursive structure of esp-digraphs.

Theorem 2. *Let G be some esp-digraph. Then, it holds that $\chi_o(G) \leq 7$.*

Proof. Let $G = (V_G, E_G)$ be some esp-digraph. We use the color graph H built from the non-zero quadratic residues of 7 defined above to define an oriented 7-vertex-coloring $c : V_G \to \{1, \dots, 7\}$ for G.

First, we color the source of G by 1 and the sink of G by 2. Next, we recursively decompose G in order to color all vertices of G. In any step we keep the invariant that $(c(q), c(s)) \in E_H$, if q is the source of G and s is the sink of G.

- If G emerges from parallel composition $G_1 \cup G_2$, we proceed with coloring G_1 and G_2 on its own. Doing so, the color of the source and sink in G_1 and G_2 do not change.
- If G emerges from series composition $G_1 \times G_2$, let a be the color of the source and c be the color of the sink in G. For the sink of G_1 and the source of G_2 we choose color b, such that the arcs (a, b) and (b, c) are in color graph H.
- If G consists of a pair of vertices connected by a single arc, the coloring is given by our invariant. □

In order to verify the optimality of the shown bound, we consider the esp-expression

$$\begin{aligned} X_4 = &((v_1, v_4) \cup ((v_1, v_2) \times ((v_2, v_4) \cup ((v_2, v_3) \times (v_3, v_4))))) \\ &\times ((((v_4, v_6) \cup ((v_4, v_5) \times (v_5, v_6))) \times (v_6, v_7)) \cup (v_4, v_7)). \end{aligned}$$

Since $\chi_o(\text{digraph}(X_4)) = 7$ the bound of Theorem 2 is best possible.

In order to compute the oriented chromatic number of an esp-digraph $G = (V, E)$ defined by an esp-expression X, we recursively compute the set $F(X)$ of all triples (H, ℓ, r) such that H is a color graph for G, where ℓ and r are the colors of the source and sink, respectively, in G with respect to the coloring by H. By Theorem 2 and also by Corollary 4 we can conclude that $|F(X)| \leq 3^{7(7-1)/2} \cdot 7 \cdot 7 \in \mathcal{O}(1)$.

For two color graphs $H_1 = (V_1, E_1)$ and $H_2 = (V_2, E_2)$ we define $H_1 + H_2 = (V_1 \cup V_2, E_1 \cup E_2)$.

Lemma 2. *1. For every* $(u, v) \in E$ *it holds that* $F((u, v)) = \{((\{i, j\}, \{(i, j)\}), i, j) \mid 1 \leq i, j \leq 7,\ i \neq j\}$.

2. For every two esp-expressions X_1 *and* X_2 *we obtain* $F(X_1 \cup X_2)$ *from* $F(X_1)$ *and* $F(X_2)$ *as follows. For every* $(H_1, \ell_1, r_1) \in F(X_1)$ *and every* $(H_2, \ell_2, r_2) \in F(X_2)$ *such that graph* $H_1 + H_2$ *is oriented,* $\ell_1 = \ell_2$*, and* $r_1 = r_2$*, we put* $(H_1 + H_2, \ell_1, r_1)$ *into* $F(X_1 \cup X_2)$.

3. For every two esp-expressions X_1 *and* X_2 *we obtain* $F(X_1 \times X_2)$ *from* $F(X_1)$ *and* $F(X_2)$ *as follows. For every* $(H_1, \ell_1, r_1) \in F(X_1)$ *and every* $(H_2, \ell_2, r_2) \in F(X_2)$ *such that graph* $H_1 + H_2$ *is oriented, and* $r_1 = \ell_2$*, we put* $((V_1 \cup V_2, E_1 \cup E_2), \ell_1, r_2)$ *into* $F(X_1 \times X_2)$.

After performing the rules given in Lemma 2 on every sub-expression of X, we can solve our problem by $\chi_o(G) = \min\{|V| \mid ((V, E), \ell, r) \in F(X)\}$, which leads to the following result.

Theorem 3. *Let* G *be an esp-digraph. Then, the oriented chromatic number of* G *can be computed in linear time.*

4 Outlook

In our future work we want to analyze whether it is possible to compute oriented chromatic index and oriented chromatic number of orientations of series-parallel graphs efficiently in order to generalize Theorem 1 and Theorem 3.

References

1. Brandstädt, A., Le, V.B., Spinrad, J.: Graph Classes: A Survey. SIAM, Philadelphia (1999)
2. Courcelle, B.: The monadic second-order logic of graphs VI: on several representations of graphs by relational structures. Discrete Appl. Math. **54**, 117–149 (1994)
3. Gurski, F., Komander, D., Lindemann, M.: Oriented Coloring of msp-digraphs and oriented co-graphs (extended abstract). In: Wu, W., Zhang, Z. (eds.) COCOA 2020. LNCS, vol. 12577, pp. 743–758. Springer, Cham (2020). https://doi.org/10.1007/978-3-030-64843-5_50
4. Harary, F., Norman, R.: Some properties of line digraphs. Rend. Circ. Mat. Palermo **9**(2), 161–168 (1960). https://doi.org/10.1007/BF02854581
5. Klostermeyer, W., MacGillivray, G.: Homomorphisms and oriented colorings of equivalence classes of oriented graphs. Discrete Math. **274**, 161–172 (2004)
6. Ochem, P., Pinlou, A., Sopena, E.: On the oriented chromatic index of oriented graphs. J. Graph Theory **57**(4), 313–332 (2008)
7. Pinlou, A., Sopena, E.: Oriented vertex and arc coloring of outerplanar graphs. Inf. Process. Lett. **100**, 97–104 (2006)
8. Sopena, E.: The chromatic number of oriented graphs. J. Graph Theory **25**, 191–205 (1997)
9. Valdes, J., Tarjan, R., Lawler, E.: The recognition of series-parallel digraphs. SIAM J. Comput. **11**, 298–313 (1982)

The Student-Project Allocation Problem as Part of Timetabling in Project-Oriented Schools

Michael Hölscher(✉)

Chair of Business Management, esp. Industrial Management,
TU Dresden, 01069 Dresden, Germany
michael.hoelscher@tu-dresden.de

Abstract. This paper focusses on the student-project allocation problem (SPA) as part of the timetabling process for project-oriented schools. The goal is to find an optimal allocation of students to project groups in this specific learning environment. A multi-period integer linear programming (ILP) model based on students preferences is formulated. We compare the model with a decomposed formulation and examine the impact on computation time and solution quality. Two different objective functions and five different problem sizes are tested. We show that there is an incentive to use the decomposed version due to the significantly lower computing time especially for large-sized problem instances while maintaining acceptable solution quality.

Keywords: Integer linear programming · Student-project allocation

1 Introduction and Related Work

Timetabling in an educational context is a complex planning task and a well studied field of research. Regarding the university course timetabling a distinction is mainly made between curriculum-based course timetabling (CB-CTT) [4] and post-enrollment course timetabling (PE-CTT) [8]. The CB-CTT formulation takes into account a curriculum that reflects the courses to be fulfilled by the students and to which the constraints are oriented. In contrast, the PE-CTT formulation requires students to enroll in courses before the actual timetabling is done. In our case the decision which student participates in which project is not predetermined but the result of the solution process. That is because in this paper we focus especially on the isolated problem of allocating students to projects based on their preferences, the so-called student-project allocation problem (SPA). Very few contributions from the literature have considered student preferences in the context of timetabling at all (e.g. [6,10]). The SPA itself is usually relevant at universities, but because of the similarities we can transfer it to our case of project-oriented schools. In these schools there are no traditional classes like in regular schools but the students learn in small project groups that work together for several weeks. Two main variants of the SPA can be found

N. Trautmann and M. Gnägi (Eds.): OR 2021, LNOR, pp. 107–113, 2022.
https://doi.org/10.1007/978-3-031-08623-6_17

in the literature: In the first variant, only preferences of students over projects or courses are considered [2]. The other case involves also lecturer preferences over students [1], projects [9], or both [5]. The SPA for our case can be stated as a two-sided matching problem with only one-sided preferences because we only consider student preferences [7]. It is based on the real-life application case of the Universitätsschule Dresden, where such a concept of a project-oriented learning environment is being tested and evaluated.

2 Problem Formulation

In our case students need to be allocated to projects and work for several weeks in this constellation on their projects. This corresponds to one period in our upcoming planning approach. After that, a new allocation has to be found for the next period. The projects offered are divided into several categories in order to simplify the room planning that takes place in a succeeding planning step. At the beginning, each student expresses preferences for the projects offered. Since we define a maximization problem, higher preference values are associated with

Table 1. Notation

Sets			
C	Set of all categories of projects	J_i^{fix}	Set of projects already allocated to student i in previous iterations
I	Set of all students	J_s	Set of all projects of subject s
I_i^F	Set of all friends of student i	S	Set of all subjects
J	Set of all projects	T	Set of all periods
J_i^0	Set of projects for which the demand for the underlying subjects is met for student i		
Parameters			
λ_F	Friendship cost parameter	N^{proj}	Number of projects each student must attend in a period
λ_P	Preference cost parameter	$N^{proj,max}$	Maximum number of students allowed in a project
λ_S	Penalty cost parameter	$N^{proj,min}$	Minimum number of students required in a project
a_i^{fix}	Score of student i from previous iterations	N^{slot}	Number of remaining project allocation slots for each student
N_s^D	Demand for projects from subject s for each student	M_{ij}^{pref}	Matrix with preferences of student i over project j
N_{is}^{fix}	Number of allocated projects to student i of subject s in previous iterations	$M_{jc}^{proj,ctg}$	Matrix with allocation of project j to project category c
N_{is}^{need}	Number of projects of subject s that needs to be allocated to student i to fulfill demand		
Variables			
a_i	Score of student i calculated from preference and friendship value	y_{jt}	1, if project j is offered in period t, 0 otherwise
b_{is}	Penalty score of student i regarding subject s	z_{ifjt}	1, if student i is allocated with friend f to project j in period t, 0 otherwise
x_{ijt}	1, if student i is allocated to project j in period t, 0 otherwise		

higher satisfaction. In addition, a certain number of friends can be specified by the students, with whom they would like to be allocated to projects together. A minimum and maximum number of students per project is relevant for the allocation process. Each project has an underlying subject and over the entire planning period and a certain demand related to these subjects must be met for each student. This results in the special need for a multi-period approach which is new in the context of SPA. Table 1 presents the used notation, followed by the mathematical formulation of the optimization model.

$$\text{Obj 1:} \qquad \sum_{i \in I} a_i - \lambda_S \cdot \sum_{i \in I} \sum_{s \in S} b_{is} \rightarrow max \tag{1}$$

$$\text{Obj 2:} \qquad \min_{i \in I}(a_i) - \lambda_S \cdot \sum_{i \in I} \sum_{s \in S} b_{is} \rightarrow max \tag{2}$$

subject to:

$$a_i = a_i^{fix} + \lambda_P \cdot \sum_{j \in J} \sum_{t \in T} M_{ij}^{pref} \cdot x_{ijt} + \lambda_F \cdot \sum_{f \in I_i^F} \sum_{j \in J} \sum_{t \in T} z_{ifjt} \quad \forall i \in I \tag{3}$$

$$b_{is} \geq N_s^D - N_{is}^{fix} - \sum_{j \in J_s} \sum_{t \in T} x_{ijt} \qquad \forall i \in I, s \in S \tag{4}$$

$$\sum_{j \in J_i^0} x_{ijt} \leq N^{slot} - \sum_{s \in S} N_{is}^{need} \qquad \forall i \in I, t \in T \tag{5}$$

$$\sum_{j \in J_s} x_{ijt} \leq N^{slot} - \sum_{\substack{s^* \in S \\ s^* \neq s}} N_{is^*}^{need} \qquad \forall i \in I, s \in S, t \in T \tag{6}$$

$$\sum_{i \in I} x_{ijt} \leq y_{jt} \cdot N^{proj,max} \qquad \forall j \in J, t \in T \tag{7}$$

$$\sum_{i \in I} x_{ijt} \geq y_{jt} \cdot N^{proj,min} \qquad \forall j \in J, t \in T \tag{8}$$

$$\sum_{j \in J} x_{ijt} = N^{proj} \qquad \forall i \in I, t \in T \tag{9}$$

$$\sum_{j \in J} x_{ijt} \cdot M_{jc}^{proj,ctg} \geq 1 \qquad \forall c \in C, i \in I, t \in T \tag{10}$$

$$\sum_{t \in T} x_{ijt} \leq 1 \qquad \forall i \in I, j \in J \tag{11}$$

$$x_{ijt} = 0 \qquad \forall i \in I, j \in J_i^{fix}, t \in T \tag{12}$$

$$z_{ifjt} \leq x_{ijt} \qquad \forall i \in I, f \in I_i^F, j \in J, t \in T \tag{13}$$

$$z_{ifjt} \leq x_{fjt} \qquad \forall i \in I, f \in I_i^F, j \in J, t \in T \tag{14}$$

$$z_{ifjt} \geq x_{ijt} + x_{fjt} - 1 \qquad \forall i \in I, f \in I_i^F, j \in J, t \in T \tag{15}$$

$$a_i, b_{is} \in \mathbb{N}_0 \qquad \forall i \in I, s \in S \tag{16}$$

$$x_{ijt}, y_{jt}, z_{ifjt} \in \{0, 1\} \qquad \forall i \in I, f \in I_i^F, j \in J, t \in T \tag{17}$$

This ILP model can be used to solve the problem considering all periods of the planning horizon at once (multi-period) or it can be solved iteratively to represent the decomposed version of connected single-periods (with each $T = \{1\}$). Three different cases are tested, in reference to [3]: First, the overall satisfaction (*collective welfare*) is tried to be maximized with objective (1). In the second case we work with a maximin criterion to find an allocation where the worst score that a student receives is as high as possible (*individual welfare*) to increase fairness. Third, a lexicographic method is tested where the two criteria are combined in a multi-objective optimization. The model is solved with respect to the maximin criterion and afterwards we try to maximize the overall satisfaction without decreasing the objective value found in the first step. With constraints (3) the score for each student is calculated. It consists of a term for the preference values of the students over the projects and a term that reflects how often a student was allocated to projects together with his or her indicated friends. These two components are weighted by the integer factors λ_P (weight of student preferences) and λ_F (weight of friendship). For the decomposed version, the determined score a_i^{fix} is passed from one iteration to the next and thus accumulated. For the solution of the multi-period version this parameter is set to zero. Constraints (4) calculate how often the required demand for the subjects underlying the projects (N_s^D) is violated. Again for each student the number of already allocated projects belonging to subject s from previous iterations is stored in parameter N_{is}^{fix} while it is set to zero for the multi-period perspective. The issue of fulfilling the demand for underlying subjects is modeled as a soft constraint so the violation incurs a penalty in the objective functions by using the calculated b_{is} and an integer penalty cost factor λ_S. This factor only has a positive value in the multi-period version or in the last iteration of the decomposed version, otherwise it is zero. That is because the demand for the underlying subjects covers the complete planning horizon so for the decomposed version the calculation is only reasonable in the optimization of the last period. To avoid excessive violation of this soft constraint in the decomposed version the complementary constraints (5) and (6) are added to the model. Before each iteration the parameter N_{is}^{need} is calculated for each individual student, which results from the difference between the demand and already assigned projects of the corresponding subject. This parameter limits in combination with the remaining allocation slots the potential projects of the respective subjects allocated to each student. So for constraints (5) a subset J_i^0 is used that contains all the projects for which the demand for the respective underlying subject is already met for student i by assignments in previous iterations. The constraints ensure that any existing demands for all the other projects are taken into account and that not too many slots are filled with projects for which the need is already met anyway. Otherwise, the demands cannot be met over the entire planning period for certain subjects. Additionally we formulate constraints (6) from a subject perspective. When assigning projects within an iteration, sometimes projects from a particular subject must not be scheduled too often in that iteration because then again the fulfillment of demands from projects with other underlying subjects may be blocked. So the total demand for subjects to be completed and already made allocations are taken into account during every iteration. In addition, there are constraints (7) and (8) regarding the capacity of the projects. The projects are only offered for an attendance between $N^{proj,min}$ and $N^{proj,max}$ students. In each period, students should be allocated to a certain number of projects, which is determined through constraints (9). The projects are also divided into categories to simplify the subsequent room planning. Students should complete at least one project from each category per planning period as stated in constraints (10). Constraints (11) and (12) ensure that in both versions each project can only be allocated once to each student in the planning period. We need the constraints (13)–(15) to linearize the product of the two decision variables x_{ijt} and x_{fjt} to be able to reflect the friendship relationships. Therefore we use a standard linearization technique with which the product of two binary decision variables is replaced by an auxiliary variable and the usage of constraints (13)–(15). The domains of the decision variables are stated in constraints (16) and (17).

3 Computational Study

The presented solution approach is implemented in PYTHON and all computational tests are carried out on an Intel(R) Xeon(R) Gold 6136 with 3.0 GHz clock speed and 16 GB RAM. We use GUROBI 9.1.1 to solve the ILP-model. We created five instance classes, each with five randomly generated instances, to examine the effects of increasing problem size. The details about the different used parameters for the five configurations and the respective computation time limits are listed in Table 2. Apart from that, two parallel projects per student and period from two available project categories, a group size between four and six students, the possibility to specify one friend and three underlying subjects were planned for all instances. For the cost parameters in the objective functions we have chosen the weights $\lambda_P = 1$, $\lambda_F = 1$ and $\lambda_S = 50$.

Table 2. Design of problem instances

	$config1$	$config2$	$config3$	$config4$	$config5$
Number students $\|I\|$	15	30	45	60	75
Number projects $\|J\|$	12	18	24	30	36
Number periods $\|T\|$	2	3	4	5	6
Demand N_s^D	$\{1, 1, 1\}$	$\{2, 2, 2\}$	$\{2, 2, 2\}$	$\{3, 3, 3\}$	$\{4, 4, 4\}$
Time limit $[s]$	–	900	1800	2700	3600

Table 3 shows the results of the computational study. The figures given are in each case the average of the five instances per configuration. For a better comparison, the summed score over all periods and students as well as the lowest individual score are shown. "MP" stands for multi-period, "cSP" for connected single-periods (decomposed version).

Table 3. Results

		config1		*config2*		*config3*		*config4*		*config5*	
		MP	cSP	MP	cSP	MP	cSP	MP	cSP	MP	cSP
Obj. 1	$\sum_{i \in I} a_i$	313	307	867	828	1693	1635	2657	2583	–	3741
	$\min_{i \in I}(a_i)$	17.2	16	21	21	28.4	28.2	32.2	33.8	–	39
	Comp. time [s]	2.35	0.17	900	1.38	1800	6.11	2700	101.67	3600	332.40
	Opt.-Gap [%]	0	0	2.15	0	3.10	0	5.47	0	–	0
Obj. 2	$\sum_{i \in I} a_i$	301	276	814	729	1575	1431	2450	2191	3315	3173
	$\min_{i \in I}(a_i)$	19	16.2	26.2	22.8	34.4	31.2	40	35.4	41.6	41.6
	Comp. time [s]	11.52	0.45	900	3.01	1800	7.10	2700	15.32	3600	56.57
	Opt.-Gap [%]	0	0	6.12	0	8.17	0	12.53	0	28.01	0
Obj. 3.1/3.2	$\sum_{i \in I} a_i$	308	297	857	819	1660	1622	2603	2528	3577	3691
	$\min_{i \in I}(a_i)$	19	17.8	25.4	23.2	34	32	35	37.6	36	42.2
	Comp. time [s]	27.28	0.70	900	5.83	1800	16.93	2700	61.25	3600	442.17
	Opt.-Gap [%]	0	0	6.90	0	7.33	0	21.91	0	46.50	0

For each instance, the required demand for the underlying subjects could be met by means of the presented modeling, and therefore the term b_{is} is not taken into account in the results. We observe that the used computation time for the multi-period approach is on average more than 100 times higher than for the decomposed version. Nevertheless, it was not possible to solve the multi-period instances optimally for *config*2 to *config*5 within the respective time limit. The solution quality for the decomposed instances is on average 5% worse compared to the multi-period approach, but for the largest generated instances even better solutions are frequently found. The lexicographic optimization approach offers a good compromise between collective and individual welfare across both modeling variants and all problem classes. Within the periods of an instance, the score is more evenly distributed in the multi-periodic view, as expected. In the decomposed version, it decreases from period to period. That's because the periods are optimized sequentially one after the other and the solver tries to allocate projects with the highest preference values at the beginning of the planning horizon.

4 Conclusion

In this paper, a new model formulation for the SPA at project-oriented schools is presented. It enables us to solve multi-period variants of this problem and consider the fairness aspect and the present requirements over multiple allocation processes. For a flexible use and to obtain a solution within acceptable time, the decomposed version of the model is preferable. The next planning step would be to integrate the student-project allocations into the timetabling process and link them to the subsequent planning steps for space and personnel planning at the project-oriented school.

References

1. Abraham, D.J., Irving, R.W., Manlove, D.F.: Two algorithms for the student-project allocation problem. J. Discrete Algorithms **5**(1), 73–90 (2007)

2. Anwar, A.A., Bahaj, A.S.: Student project allocation using integer programming. IEEE Trans. Educ. **46**(3), 359–367 (2003)
3. Chiarandini, M., Fagerberg, R., Gualandi, S.: Handling preferences in student-project allocation. Ann. Oper. Res. **275**(1), 39–78 (2017). https://doi.org/10.1007/s10479-017-2710-1
4. Di Gaspero, L., McCollum, B., Schaerf, A.: The second international timetabling competition (ITC-2007): curriculum-based course timetabling (track 3). Technical report, Queen's University, Belfast, United Kingdom (2007)
5. El-Atta, A.H.A., Moussa, M.I.: Student project allocation with preference lists over (student, project) pairs. In: 2009 Second International Conference on Computer and Electrical Engineering, pp. 375–379. IEEE (2009)
6. Hoshino, R., Fabris, I.: Optimizing student course preferences in school timetabling. In: Hebrard, E., Musliu, N. (eds.) CPAIOR 2020. LNCS, vol. 12296, pp. 283–299. Springer, Cham (2020). https://doi.org/10.1007/978-3-030-58942-4_19
7. Kwanashie, A., Irving, R.W., Manlove, D.F., Sng, C.T.S.: Profile-based optimal matchings in the student/project allocation problem. In: Kratochvíl, J., Miller, M., Froncek, D. (eds.) IWOCA 2014. LNCS, vol. 8986, pp. 213–225. Springer, Cham (2015). https://doi.org/10.1007/978-3-319-19315-1_19
8. Lewis, R., Paechter, B., McCollum, B.: Post enrolment based course timetabling. Technical report, Cardiff University, Wales, United Kingdom (2007)
9. Manlove, D.F., O'Malley, G.: Student-project allocation with preferences over projects. J. Discrete Algorithms **6**(4), 553–560 (2008)
10. Van den Broek, J., Hurkens, C., Woeginger, G.: Timetabling problems at the TU Eindhoven. Eur. J. Oper. Res. **196**(3), 877–885 (2009)

Robust Optimization with Scenarios Using Belief Functions

Romain Guillaume[1], Adam Kasperski[2(✉)], and Paweł Zieliński[2]

[1] Université de Toulouse-IRIT, Toulouse, France
romain.guillaume@irit.fr
[2] Wrocław University of Science and Technology, Wrocław, Poland
{adam.kasperski,pawel.zielinski}@pwr.edu.pl

Abstract. In this paper a class of optimization problems with uncertain objective function coefficients is considered. The uncertainty is specified by providing a scenario set containing a finite number of parameter realizations, called scenarios. Additional knowledge about scenarios is modeled by specifying a mass function, which defines a belief function in scenario set. The generalized Hurwicz criterion is then used to compute a solution. Various computational properties of the resulting optimization problem are presented.

Keywords: Robust optimization · Belief function · Hurwicz criterion

1 Traditional Robust Problems with Scenarios

Consider the following deterministic optimization problem $\min\{\boldsymbol{c}^T\boldsymbol{x} : \boldsymbol{x} \in \mathbb{X}\}$, where $\boldsymbol{c} \in \mathbb{R}^n$ is a vector of objective function coefficients and $\mathbb{X}$ is a set of feasible solutions. For example, $\mathbb{X}$ can be a polyhedron in $\mathbb{R}^n$ which leads to the class of linear programming problems, or $\mathbb{X} \subseteq \{0,1\}^n$, which results in the class of combinatorial problems. In many applications the true realization of $\boldsymbol{c}$ is unknown before a solution must be determined. Assume that $\boldsymbol{c}$ is only known to belong to a given uncertainty (scenario) set $\mathcal{U} = \{\boldsymbol{c}_1, \dots, \boldsymbol{c}_K\} \subseteq \mathbb{R}^n$. Each vector $\boldsymbol{c}_i$, $i \in [K] = \{1, \dots, K\}$, is called a *scenario* and represents a possible realization of the objective function coefficients. This model of uncertainty is called a *discrete uncertainty representation* [9]. One common method of solving the problem with $\mathcal{U}$ is to apply the robust min-max criterion, which yields the following min-max problem:

$$\begin{array}{l} \min \max\limits_{k\in[K]} \boldsymbol{c}_k^T\boldsymbol{x} \\ \boldsymbol{x} \in \mathbb{X}. \end{array} \qquad (1)$$

A comprehensive description of computational properties of (1) for various classes of problems can be found in [1,8,9]. The solutions to (1) are known to be very conservative. Using the min-max criterion we assume an extreme risk aversion of decision maker. In order to soften (1) one can use the *Ordered Weighted Averaging Aggregation* (OWA) criterion, proposed in [14]. Let $\boldsymbol{w}$ be a

N. Trautmann and M. Gnägi (Eds.): OR 2021, LNOR, pp. 114–119, 2022.
https://doi.org/10.1007/978-3-031-08623-6_18

vector of nonnegative weights which sum up to 1. Given a solution $\boldsymbol{x} \in \mathbb{X}$, let σ be a permutation of $[K]$ such that $\boldsymbol{c}_{\sigma(1)}^T\boldsymbol{x} \geq \boldsymbol{c}_{\sigma(2)}^T\boldsymbol{x} \geq \cdots \geq \boldsymbol{c}_{\sigma(K)}^T\boldsymbol{x}$. Then

$$\mathrm{OWA}_{\boldsymbol{w}}(\boldsymbol{x}) = \sum_{k\in[K]} w_k \boldsymbol{c}_{\sigma(k)}^T\boldsymbol{x}.$$

It is easy to verify that choosing $\boldsymbol{w} = (1, 0, \ldots, 0)$ we get the maximum criterion. Hence, minimizing OWA generalizes the min-max problem (1). Choosing $\boldsymbol{w} = (\alpha, 0, \ldots, 0, 1-\alpha)$ for some $\alpha \in [0, 1]$, leads to the Hurwicz pessimism/optimism criterion. In general, the weight vector $\boldsymbol{w}$ can be used to model decision maker risk aversion. For a description of applications of OWA criterion to various optimization problems we refer the reader to [2,7,11,15].

2 Robust Problem with Belief Functions

In the traditional discrete uncertainty representation no additional knowledge in scenario set $\mathcal{U}$ is specified. This is equivalent to say that there is complete uncertainty in $\mathcal{U}$ and no evidence supports the claim that any subset of scenarios is more probable than other. In this section we propose a model of uncertainty which allows us to take such an additional knowledge in $\mathcal{U}$ into account. We will use some ideas from the Dempster and Shafer theory of evidence [4,12] and distributionally robust approach (see, e.g. [3]). In order to simplify the notations, we will identify each scenario $\boldsymbol{c}_k$ with its index $k \in [K]$

Let $m : 2^{[K]} \to [0, 1]$ be a mapping, called a *mass function*, such that $m(\emptyset) = 0$ and $\sum_{A\subseteq[K]} m(A) = 1$. The value of $m(A)$ supports the claim that the true scenario will belong to A. Set $F \subseteq [K]$ such that $m(F) > 0$ is called a *focal set.* We will use $\mathcal{F} = \{F_1, \ldots, F_\ell\}$ to denote the family of all focal sets in $[K]$. The mass function m induces a *belief function* in $2^{[K]}$ defined as follows:

$$\mathrm{B}(A) = \sum_{B\subseteq A} m(B),\ A \subseteq [K].$$

A probability distribution P in $[K]$ is said to be *compatible with* m (see, e.g., [5]) if $\mathrm{P}(A) \geq \mathrm{B}(A)$ for each event $A \subseteq [K]$. We will denote by $\mathcal{P}(m)$ the set of all probability distributions compatible with m. The set $\mathcal{P}(m)$ can be described by the following system of linear constraints:

$$\begin{array}{ll} \sum_{k\in A} p_k \geq \mathrm{B}(A) & A \subseteq [K] \\ p_1 + \cdots + p_K = 1 & \\ p_k \geq 0 & k \in [K], \end{array} \tag{2}$$

where p_k is a probability that scenario $k \in [K]$ will occur. It is well known that belief function B is supermodular (see, e.g., [10]). In this case, system (2) describes the core of a convex cooperative game. Because the cores of such games are nonempty [13], there is at least one probability distribution compatible with m, for any mass function m. Thus $\mathcal{P}(m) \neq \emptyset$.

Let $\tilde{\boldsymbol{c}}$ be a random vector with support $\mathcal{U}$ and a probability distribution $\mathrm{P} \in \mathcal{P}(m)$. In this paper, we study the following problem:

$$\begin{array}{l} \min \alpha \overline{E}_m[\tilde{\boldsymbol{c}}^T \boldsymbol{x}] + (1-\alpha)\underline{E}_m[\tilde{\boldsymbol{c}}^T \boldsymbol{x}], \\ \boldsymbol{x} \in \mathbb{X} \end{array} \tag{3}$$

where

$$\overline{E}_m[\tilde{\boldsymbol{c}}^T \boldsymbol{x}] = \max_{\mathrm{P} \in \mathcal{P}(m)} \mathbf{E}_{\mathrm{P}}[\tilde{\boldsymbol{c}}^T \boldsymbol{x}] = \sum_{F \in \mathcal{F}} m(F) \max_{k \in F} \boldsymbol{c}_k^T \boldsymbol{x}, \tag{4}$$

$$\underline{E}_m[\tilde{\boldsymbol{c}}^T \boldsymbol{x}] = \min_{\mathrm{P} \in \mathcal{P}(m)} \mathbf{E}_{\mathrm{P}}[\tilde{\boldsymbol{c}}^T \boldsymbol{x}] = \sum_{F \in \mathcal{F}} m(F) \min_{k \in F} \boldsymbol{c}_k^T \boldsymbol{x} \tag{5}$$

are *upper* and *lower* expectations with respect to the mass function m (see, e.g., [4,5]). The objective function of (3) is a *generalized Hurwicz criterion* and the parameter $\alpha \in [0,1]$ is the *optimism/pessimism* degree. In the following we will use $H_\alpha(\boldsymbol{x})$ to denote the objective function of (3). In general, defining a mass function requires providing up to 2^K numbers. We will assume that the size of (3) is polynomial in K. This assumption is satisfied, for example, when $|\mathcal{F}|$ is bounded by a polynomial in K.

Let us now briefly describe some special cases of (3) which are well known in literature. If $m([K]) = 1$, then (3) reduces to minimizing the traditional Hurwicz criterion. In this case the mass equal to 1 is assigned to the whole scenario set $\mathcal{U}$, which can be interpreted as a complete uncertainty in $\mathcal{U}$. If additionally $\alpha = 1$, then (3) becomes the robust minmax problem (1). If all the focal sets are singletons, i.e. $|F_i| = 1$ for each $F_i \in \mathcal{F}$, then (3) reduces to minimizing the expected solution cost. Indeed, in this case $\mathcal{P}(m)$ contains exactly one probability distribution such that $p_k = m(\{k\})$, $k \in [K]$. Let us provide yet another interesting interpretation of (3). Suppose $m(A) = 1/\binom{K}{l}$ for each $A \subseteq [K]$ such that $|A| = l \leq K$ and $m(A) = 0$ otherwise. Then (4) becomes the OWA criterion with nonincreasing weights $w_i = \binom{K-i}{l-1}/\binom{K}{l}$ for $i \leq K - l + 1$ and $w_i = 0$ otherwise. Similarly, (5) is then the OWA criterion with nondecreasing weights $w_i' = w_{K-i+1}$, $i \in [K]$. So, this particular structure of the mass function leads to OWA minimization.

Let us introduce binary variable $\delta_k(F)$ for each focal set $F \in \mathcal{F}$ and $k \in F$. Then (3) can be solved by using the following mixed integer program:

$$\begin{array}{ll} \min \sum_{F \in \mathcal{F}} m(F) y(F) & \\ y(F) \geq \alpha \boldsymbol{c}_k^T \boldsymbol{x} + (1-\alpha) u(F) & F \in \mathcal{F}, k \in F \\ u(F) \geq \boldsymbol{c}_k^T \boldsymbol{x} - M(1 - \delta_k(F)) & F \in \mathcal{F}, k \in F \\ \sum_{k \in F} \delta_k(F) = 1 & F \in \mathcal{F} \\ \boldsymbol{x} \in \mathbb{X} & \\ \delta_k(F) \in \{0,1\} & F \in \mathcal{F}, k \in F, \end{array} \tag{6}$$

where M is a large constant. The problem can be easier to solve if $\alpha = 1$. Using (4) we can then represent (3) as the following program:

$$
\begin{aligned}
\min \sum_{F\in\mathcal{F}} & m(F)y(F) \\
y(F) &\geq \boldsymbol{c}_k^T\boldsymbol{x} \qquad k\in F \\
\boldsymbol{x} &\in \mathbb{X}
\end{aligned} \tag{7}
$$

Notice that (7) is linear programming problem if $\mathbb{X}$ is a polyhedron in $\mathbb{R}^n$ described by a system of linear constraints. The next result demonstrates that the case $\alpha = 0$ can be much harder to solve.

Theorem 1 ([6]). *Problem (3) is NP-hard and not approximable if $\alpha = 0$ and $\mathbb{X}$ is a polyhedron in $[0,1]^n$.*

3 Application to Combinatorial Problems

In this section we will show some results for (3) if $\mathbb{X} \subseteq \{0,1\}^n$, i.e. for the class of combinatorial problems. We will assume that $\boldsymbol{c}_k \in \mathbb{R}^n_+$ for each $k \in [K]$, so the costs under all scenarios are nonnegative. Let us recall that (3) can be then NP-hard even if $m([K]) = 1$ and $\alpha = 1$, i.e. when (3) is the robust minmax problem (1) (see, e.g., [8,9]). This is contrary to the class of linear programming problems which can be solved by linear programming formulation (7). In the following we will propose a general approximation algorithm for (3).

Fix $r(\mathcal{F}) = \max_{F\in\mathcal{F}} |F|$ and $v_k = \sum_{F\in\mathcal{F}:k\in F} m(F)$, $k \in [K]$. If $\{F \in \mathcal{F} : k \in F\} = \emptyset$, then $v_k = 0$. Define also $\hat{\boldsymbol{c}} = \sum_{k\in[K]} v_k\boldsymbol{c}_k \in \mathbb{R}^n_+$. We can assume that $r(\mathcal{F}) \geq 2$. Indeed, the case $r(\mathcal{F}) = 1$ is equivalent to solving the deterministic problem with the cost vector $\hat{\boldsymbol{c}}$. We can also assume that $|F| \geq 2$ for each focal set $F \in \mathcal{F}$. Indeed, if a positive mass is assigned to a single scenario $\boldsymbol{c}_k$, then we add a copy $\boldsymbol{c}'_k$ of this scenario to $\mathcal{U}$ and fix $m(\{k,k'\}) := m(\{k\})$ and $m(\{k\}) := 0$. This transformation does not change the values of $\overline{E}_m[\tilde{\boldsymbol{c}}^T\boldsymbol{x}]$ ad $\underline{E}_m[\tilde{\boldsymbol{c}}^T\boldsymbol{x}]$ (see Eqs. (4) and (5)) and does not increase $r(\mathcal{F})$.

Theorem 2. *Assume that the deterministic problem $\min\{\hat{\boldsymbol{c}}^T\boldsymbol{x} : \boldsymbol{x} \in \mathbb{X}\}$ is polynomially solvable. Then problem (3) is approximable within $r(\mathcal{F})$ for $\alpha \in [0.5, 1]$ and within $\frac{1-\alpha}{\alpha}r(\mathcal{F})$ for $\alpha \in (0, 0.5]$.*

Proof. Let $\boldsymbol{x}^*$ be an optimal solution to (3) and $\hat{\boldsymbol{x}}$ be an optimal solution to the deterministic problem for the cost vector $\hat{\boldsymbol{c}}$. We will show first that for each $\alpha \in [0,1]$ the following inequality holds:

$$
H_\alpha(\boldsymbol{x}^*) \geq \frac{\alpha}{r(\mathcal{F})} \sum_{F\in\mathcal{F}} m(F) \sum_{k\in F} \boldsymbol{c}_k^T\hat{\boldsymbol{x}}. \tag{8}
$$

Using Eqs. (4) and (5) we get

$$
\begin{aligned}
H_\alpha(\boldsymbol{x}^*) &\geq \alpha \sum_{F\in\mathcal{F}} m(F) \max_{k\in F} \boldsymbol{c}_k^T\boldsymbol{x}^* \geq \alpha \sum_{F\in\mathcal{F}} m(F) \frac{1}{|F|} \sum_{k\in F} \boldsymbol{c}_k^T\boldsymbol{x}^* \\
&\overset{(a)}{\geq} \alpha \sum_{F\in\mathcal{F}} m(F) \frac{1}{r(\mathcal{F})} \sum_{k\in F} \boldsymbol{c}_k^T\boldsymbol{x}^* = \frac{\alpha}{r(\mathcal{F})} \sum_{k\in[K]} \sum_{\{F\in\mathcal{F}:k\in F\}} m(F)\boldsymbol{c}_k^T\boldsymbol{x}^* \\
&= \frac{\alpha}{r(\mathcal{F})} \sum_{k\in[K]} v_k\boldsymbol{c}_k^T\boldsymbol{x}^* \overset{(b)}{\geq} \frac{\alpha}{r(\mathcal{F})} \sum_{k\in[K]} v_k\boldsymbol{c}_k^T\hat{\boldsymbol{x}} = \frac{\alpha}{r(\mathcal{F})} \sum_{F\in\mathcal{F}} m(F) \sum_{k\in F} \boldsymbol{c}_k^T\hat{\boldsymbol{x}}.
\end{aligned}
$$

Inequality (*a*) follows from the fact that $r(\mathcal{F}) \geq |F|$ for each $F \in \mathcal{F}$. In the inequality (*b*) we use the optimality of $\hat{\boldsymbol{x}}$ under $\hat{\boldsymbol{c}} = \sum_{k\in[K]} v_k \boldsymbol{c}_k$.

If $\alpha \in [0.5, 1]$ and $|F| \geq 2$, then $\alpha \sum_{k\in F} \boldsymbol{c}_k^T \hat{\boldsymbol{x}} \geq \alpha \max_{k\in F} \boldsymbol{c}_k^T \hat{\boldsymbol{x}} + (1-\alpha) \min_{k\in F} \boldsymbol{c}_k^T \hat{\boldsymbol{x}}$ and (8) implies $H_\alpha(\boldsymbol{x}^*) \geq \frac{1}{r(\mathcal{F})} H_\alpha(\hat{\boldsymbol{x}})$. On the other hand, if $\alpha \in (0, 0.5]$ and $|F| \geq 2$, then $(1-\alpha) \sum_{k\in F} \boldsymbol{c}_k^T \hat{\boldsymbol{x}} \geq \alpha \max_{k\in F} \boldsymbol{c}_k^T \hat{\boldsymbol{x}} + (1-\alpha) \min_{k\in F} \boldsymbol{c}_k^T \hat{\boldsymbol{x}}$ and (8) implies $H_\alpha(\boldsymbol{x}^*) \geq \frac{\alpha}{(1-\alpha) r(\mathcal{F})} H_\alpha(\hat{\boldsymbol{x}})$. □

Observe that $r(\mathcal{F}) \leq K$ for $\alpha = 1$. Hence the approximation algorithm generalizes the known approximation result for the robust minmax problem (see, e.g., [1]). The result from Theorem 2 can also be applied when the deterministic problem is NP-hard, but admits a ρ-approximation algorithm for some $\rho \geq 1$. An easy modification of the proof yields then $\rho r(\mathcal{F})$ and $\rho \frac{1-\alpha}{\alpha} r(\mathcal{F})$-approximation algorithms for the cases $\alpha \in [0.5, 1]$ and $(0, 0.5]$, respectively. Observe that the approximation ratio tends to infinity as $\alpha \to 0$. The next result shows that the boundary case $\alpha = 0$ is not at all approximable.

Theorem 3. *If $\alpha = 0$, then the problem is not at all approximable when $\mathbb{X}' = \{(\boldsymbol{x}, \overline{\boldsymbol{x}}) \in \{0,1\}^{2n} : x_i + \overline{x}_i = 1, i \in [n]\}$.*

Proof. The proof is the same as the proof of [6, Theorem 1]). It is enough to replace the polyhedron $\mathbb{X}$ in the proof of [6, Theorem 1]) with the set $\mathbb{X}'$.

The deterministic $\min\{\boldsymbol{c}^T \boldsymbol{x} : \boldsymbol{x} \in \mathbb{X}'\}$ is a very simple selection problem, which has a trivial $O(n)$-time algorithm. It is not difficult to show that it is a special case of basic network problems such as the shortest path and the minimum spanning tree. Let $G = (V, A)$ be a chain composed of $|V| = n$ nodes and $2n$ arcs of the form $a_i = (i, i+1)$ and $\overline{a}_i = (i, i+1)$ for $i = 1, \ldots, n-1$. Then $\mathbb{X}'$ is the set of characteristic vectors of the simple paths from 1 to n and the spanning trees in G. So, the negative result from Theorem 3 remains true for many basic combinatorial problems.

4 Conclusions

The introduction of a mass function in scenario set $\mathcal{U}$ allows us to refine the model of uncertainty. In many practical situations, we are able to identify subsets of scenarios which are more probable to occur than others, even if the true probability distribution in $\mathcal{U}$ is unknown. However, this additional information restricts the set of possible probability distributions. We can then compute a solution minimizing the total expected cost under a combination of the worst and the best probability distribution which can occur. The general problem can be hard to solve and also hard to approximate. In this paper we have described some special cases which can be solved efficiently or for which efficient approximation algorithms exists. Our results are general, so better algorithms can exist for particular problems.

Acknowledgements. Romain Guillaume was partially supported by the project caasc ANR-18-CE10-0012 of the French National Agency for Research. Adam Kasperski and Paweł Zieliński were supported by the National Science Centre, Poland, grant 2017/25/B/ST6/00486.

References

1. Aissi, H., Bazgan, C., Vanderpooten, D.: Min-max and min-max regret versions of combinatorial optimization problems: a survey. Eur. J. Oper. Res. **197**(2), 427–438 (2009)
2. Chassein, A.B., Goerigk, M.: Alternative formulations for the ordered weighted averaging objective. Inf. Process. Lett. **115**(6–8), 604–608 (2015)
3. Delage, E., Ye, Y.: Distributionally robust optimization under moment uncertainty with application to data-driven problems. Oper. Res. **58**, 595–612 (2010)
4. Dempster, A.: Upper and lower probabilities induced by a multivalued mapping. Ann. Math. Stat. **38**, 325–339 (1967)
5. Denoeux, T.: Decision-making with belief functions: a review. Int. J. Approximate Reasoning **109**, 87–110 (2019)
6. Guillaume, R., Kasperski, A., Zieliński, P.: Robust optimization with scenarios using random fuzzy sets. In: 2021 IEEE International Conference on Fuzzy Systems (FUZZ-IEEE), pp. 1–6 (2021)
7. Kasperski, A., Zieliński, P.: Combinatorial optimization problems with uncertain costs and the OWA criterion. Theor. Comput. Sci. **565**, 102–112 (2015)
8. Kasperski, A., Zieliński, P.: Robust discrete optimization under discrete and interval uncertainty: a survey. In: Doumpos, M., Zopounidis, C., Grigoroudis, E. (eds.) Robustness Analysis in Decision Aiding, Optimization, and Analytics. ISORMS, vol. 241, pp. 113–143. Springer, Cham (2016). https://doi.org/10.1007/978-3-319-33121-8_6
9. Kouvelis, P., Yu, G.: Robust Discrete Optimization and Its Applications. Kluwer Academic Publishers (1997)
10. Kwuida, L., Schmidt, S.E.: Valuations and closure operators on finite lattices. Discrete Appl. Math. **159**, 990–1001 (2011)
11. Ogryczak, W., Śliwiński, T.: On solving linear programs with the ordered weighted averaging objective. Eur. J. Oper. Res. **148**(1), 80–91 (2003)
12. Shafer, G.: A Mathematical Theory of Evidence. Princeton University Press, Princeton (1976)
13. Shapley, L.S.: Cores of convex games. Internet J. Game Theory **1**, 11–26 (1971). https://doi.org/10.1007/BF01753431
14. Yager, R.R.: On ordered weighted averaging aggregation operators in multi-criteria decision making. IEEE Trans. Syst. Man Cybern. **18**(1), 183–190 (1988)
15. Yager, R.R., Kacprzyk, J., Beliakov, G. (eds.): Recent Developments in the Ordered Weighted Averaging Operators: Theory and Practice. Springer, Heidelberg (2011). https://doi.org/10.1007/978-3-642-17910-5

Optimal Numbers, Locations, and Configurations of Tower Cranes on Construction Sites

Thomas Ihor Maindl[1](✉) and Jannik Vogel[1,2]

[1] SDB Science-Driven Business Ltd., Larnaca, Cyprus
thomas.maindl@sdb.ltd
[2] Heidelberg Analytics, Heidelberg, Germany
jannik.vogel@hd-analytics.com

Abstract. Choosing the right number, configurations, and locations of tower cranes can significantly impact building schedules and reduce construction cost. These decisions also depend on the choice of material supply points where the material is stored before being lifted to the given demand points. To ensure that the capacity of the cranes is not exceeded, load charts that specify the maximum weight that can be lifted at a certain boom radius need to be considered. We present a MILP model implemented in GAMS that solves this decision problem minimizing the crane operating, installation and rental costs. In addition to the optimization model, a heuristic in closed form is presented. For a realistic case of a construction site of mid-rise buildings, we compare solution quality, performance, and scalability of both approaches.

Keywords: Tower crane · Construction sites · Software · MILP

1 Introduction and Positioning in Related Literature

Tower cranes are widely used on construction sites to lift material from supply to demand points. In this context, decisions on crane selection, crane location, and lift planning have to be made [1].

Typically, a variety of cranes in different configurations is available to the decision maker [10]. They differ for example in their mast/hook heights, jib lengths, and hook speeds. The capacity of a tower crane needs to be considered during its operation. In practice, it is given by load charts that specify the maximum weight that can be lifted at a certain boom radius [11]. Different crane configurations lead to significant differences in rental and operating costs [11].

For placing a tower crane, several factors need to be considered including the geometric layout of the construction site, ground conditions, and existing site obstacles [5,7]. Selecting the right supply areas to store material before lifting it to the demand points is closely integrated with choosing crane locations [5,14]. The integrated planning of the crane configuration and locations as well as the

N. Trautmann and M. Gnägi (Eds.): OR 2021, LNOR, pp. 120–126, 2022.
https://doi.org/10.1007/978-3-031-08623-6_19

locations of supply points may lead to significant savings in crane rental and operating costs [3].

Publications relevant to this manuscript are presented in the following. Several existing studies determine the location of a single tower crane and supply locations [5,13,14]. [18] consider multiple tower cranes by determining disjunct groups of tasks and applying a single tower crane location model to each task group in isolation. They, however, only decide on the tower crane locations and not on the supply point locations. [16] determine crane and supply point locations in a multi crane model with a given number of two cranes.

Another stream of literature integrates a decision on the crane configuration. [6] present a heuristic to determine the number and configurations of cranes as well as the crane locations in continuous space. [11] consider a single crane and choose a crane configuration that does not exceed capacity for each possible crane location before the optimization. Their capacity constraints are based on load moments, i.e., the load weight multiplied with the maximum boom radius. However, in practice load charts do not lead to constant load moments for different operation radii. [8] present a MILP model similar to ours, however, they consider a fixed number of tower cranes and the objective function considers operating costs and a penalty for increased height and reach of the cranes. [2] and [3] decide on the number of cranes, their configurations, and the crane and supply locations. Their objective is to minimize crane installation and rental costs. [17] integrate decisions on the installation of tower cranes over the course of the construction project and use simulated annealing to solve the problem.

The contribution of this manuscript is the development of a MILP model for determining the optimal number of cranes, their configurations and locations, as well as supply locations in an integrated way. The model considers capacity of tower cranes via load charts and an objective function that includes the operating, installation and rental costs.

2 Integrated Determination of the Number of Cranes, Crane Locations, and Their Configurations

An a priori unknown number of tower cranes N^{c} is to be placed at candidate locations $k \in \mathcal{K}$. The set $\mathcal{K}$ is determined by considering the factors described in the last section, such as the geometric layout of the construction site or ground conditions, see [7]. Tower crane configurations $c \in \mathcal{C}$ differ with respect to their operating cost C^{o}_c, their dimensions (height and boom reach), trolley velocities in radial, angular, and vertical direction, and configuration-specific load charts. Configuration-dependent installation and rental costs add up to the fixed costs C^{f}_c. Eligible crane configurations for each crane location k are indicated by Ξ_{kc}.

As it is typically assumed in the literature, several supply points $i \in \mathcal{I}$ with corresponding coordinates have been identified on the construction site. Demand points $j \in \mathcal{J}$ are known and characterized by their coordinates, the quantity Q_j of demand units, and the weight of one demand unit (e.g., a concrete bucket) W_j. Δ_{jl} indicates whether demand j is for material $l \in \mathcal{L}$. All locations have associated cartesian xyz coordinates with z being the vertical component.

We implement load charts via the indicator Γ_{ijkc} for feasible lift operations: $\Gamma_{ijkc} = 1$ if and only if the maximum lifting weight at a boom radius of the maximum distance between crane location k and supply point i and crane location k and the demand point j is at least the weight of demand j.

A set of coordination factors determines simultaneous radial-angular and vertical-horizontal hook travel, in addition to the hook control difficulty, see also [8]. These factors are influenced by many aspects including crane operator skills and experience, terrain, visibility, obstacles, and weather conditions. We follow the formulation in [8] assuming all factors depend on the crane location.

3 Mixed-Integer Linear Programming Model

Each demand j needs to be fulfilled exactly once, tracked by binary variable δ_{ijkcl} which indicates whether material l is moved from supply point i to demand j by a crane at location k in configuration c,

$$\sum_{ikcl} \delta_{ijkcl} = 1 \quad \forall j \, . \tag{1}$$

There will be no lift if demand j is not for material l (i.e., $\Delta_{jl} = 0$) or if it is infeasible (load chart violated, $\Gamma_{ijkc} = 0$),

$$\delta_{ijkcl} \leq \Delta_{jl} \quad \forall ijkcl \, , \quad \delta_{ijkcl} \leq \Gamma_{ijkc} \quad \forall ijkcl \, . \tag{2}$$

The binary variable ξ_{kc} indicates whether a crane at location k is in configuration c. There cannot be lift operations for non-utilized $k - c$ combinations,

$$\sum_{c} \xi_{kc} \leq 1 \quad \forall k \, , \quad \delta_{ijkcl} \leq \xi_{kc} \quad \forall ijkcl \, . \tag{3}$$

Also, if there is a crane at location k, its configuration c will be unique and chosen from the eligible configurations given by Ξ_{kc},

$$\xi_{kc} \leq \Xi_{kc} \quad \forall kc \, . \tag{4}$$

Another binary variable restricts material movement: $\chi_{il} = 1$ if and only if location i is a supply point for material l, subject to "at most one product per supply point" and "on average, at most one material supply point per crane",

$$\sum_{c} \delta_{ijkcl} \leq \chi_{il} \quad \forall ijk, \quad \sum_{l} \chi_{il} \leq 1 \quad \forall i, \quad \sum_{i} \chi_{il} \leq \sum_{kc} \xi_{kc} \quad \forall l \, . \tag{5}$$

Materials without demand will not have a supply point,

$$\chi_{il} \leq \sum_{j} \Delta_{jl} \quad \forall il \, . \tag{6}$$

The objective function z sums the operating, installation and rental costs,

$$z = \sum_{ijkcl} \delta_{ijkcl} T_{ijkc} Q_j C^{\mathrm{o}}_c + \sum_{kc} C^{\mathrm{f}}_c \xi_{kc} \, . \tag{7}$$

To determine the total hook travel times T_{ijkc}, we use the angular, radial, and vertical trolley travel time components between supply and demand points. It is based on [8] and extended by configuration-dependent trolley velocities. Note that as pointed out in [15], due to a sign error in applying the law of cosines, some other literature sources state incorrect expressions for the angular trolley travel times.

4 Heuristic Approach

The problem described above consists of several decisions that depend on each other. In the following, a heuristic is presented that can be applied in closed form. Therefore, the individual decisions are made successively.

The number of cranes is investigated in the range $N^c = 1, 2, ..., \min(|\mathcal{K}|, |\mathcal{J}|)$ successively. For a fixed N^c, the heuristic goes through the following steps:

1. *Demand groups:* We separate the demand locations into N^c groups depending on its coordinates, e.g., by using a *KMeans* clustering algorithm [9]. Hence, the *KMeans* model is trained on the coordinates of the demand locations and, subsequently, the demand location clusters are predicted which represent the demand groups.
2. *Crane locations:* For each demand group, choose the crane location with the lowest maximum distance to the demand points in the demand group. Only consider crane locations that are not already used by another demand group as there can only be one crane located at a specific crane location $k \in \mathcal{K}$.
3. *Supply points:* For each demand group and material required in this demand group, determine a supply point. Therefore, choose the supply point with the lowest maximum distance to the crane location and the demand points that require this material in the demand group. To fulfill the model constraints, only consider supply points without any material assigned to it.
4. *Crane configurations:* For each demand group and its corresponding crane and supply point location, choose the feasible crane configuration with the lowest fixed crane costs C^f_c that can fulfill all the demands of the demand group using the supply locations of the last step. Checking the feasibility of a crane configuration consists of verifying that (i) the weight of the material to be lifted at the distance between the crane location and the supply points as well as the demand points is not exceeded, and (ii) the maximum height and reach of the crane configuration is attained. It may occur that this step does not find a feasible crane configuration to fulfill the demand.

Finally, from all feasible solutions generated for different values of N^c, the solution with the lowest objective value (total cost) is chosen.

5 Realistic Test Case and Conclusion

We demonstrate the solution of the MILP model and the heuristic for a real case of the construction of a mid-rise building complex. The demand, possible supply locations, and the crane data and operating cost are adopted from [8]. Figure 1a depicts the layout of the construction site. There are 18 demand points for three materials (concrete, plywood, and rebar), each with a quantity Q_j between 10 and 20, corresponding to one level. The rental costs and installation costs are adopted from [4] and [12], respectively. Note that the individual cost structure will depend on the location, jurisdiction, and the specific project's details. The computations are performed on a Lenovo X1 Carbon with an Intel Core i7-10510U and 16 GB RAM. The solver CPLEX 12.10 is used in GAMS, the heuristic is implemented in Python 3.8.[1]

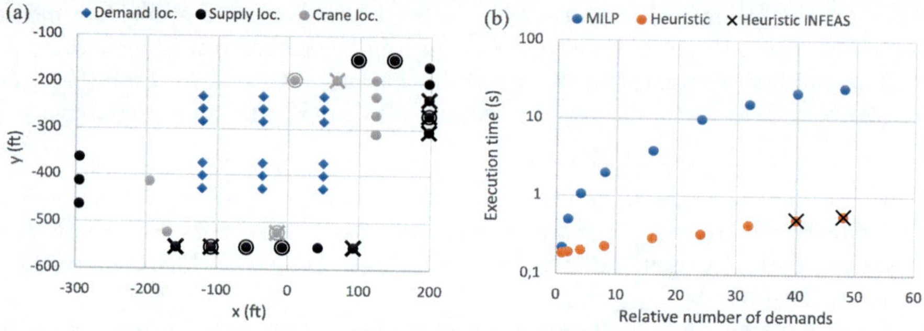

Fig. 1. **a)** Location map: crosses denote locations selected by the MILP model, circles those selected by the heuristic. **b)** MILP model and heuristic performance scaling; 'relative number of demands' refers to multiples of the 18 'one level' demands in (a).

The MILP model solves to optimality in 0.2 s and yields 106,151 \$ total costs, of which 351 \$ are crane operating cost ('opt' solution). It selects two cranes.[2] The heuristic yields total costs of 106,212 \$ (+0.06% compared to 'opt') including 412 \$ (+17.38%) operating cost. See Fig. 1a for the selected locations. Compared to the 'worst case' (i.e., maximum cost) solution with two cranes, the 'opt' solution improves total costs by 6.67% and operating cost by 21.27%.

In the following, we investigate an increasing number of demand locations as they occur on bigger construction sites for multi-level buildings. Therefore, we assume that the same demand of the base case repeats on higher levels (larger z coordinate values). We find that the optimization model always solves the model to a proven optimal solution. The objective values of the heuristic are very close to the objective values of the MILP model:

[1] In the GAMS model, we implemented the bounds (2), (4), and (6) by directly restricting the domain of the affected variables to those combinations of $ijkcl$ and its subsets that are allowed by the values of Γ_{ijkc}, Δ_{jl}, and Ξ_{kc}, respectively. While the resulting model is equivalent to the one defined in Sects. 2 and 3, it is smaller and in many cases easier to solve.

[2] Due to costs related to the number of cranes exceeding operating costs by >2 orders of magnitude, this decision may be singled out in practical applications e.g., by imposing a constraint on $\sum_{kc} \xi_{kc}$ (cf. Sect. 3).

#demand sets:	1	2	4	8	16	24	32	40	48
Gap opt-heur	0.06%	0.11%	0.19%	0.32%	0.47%	0.55%	0.6%	∞	∞
Gap opt-worst	6.67%	6.78%	6.97%	7.23%	7.52%	7.67%	6.04%	6.16%	6.21%

For the investigated case, the 'opt-heur' gap increases with increasing demand. Finally, the heuristic may lead to infeasible solutions for large demands (high number of levels). This originates from the heuristic making the individual decisions sequentially as opposed to the integrated approach in the MILP model. Figure 1b shows the computation times for the MILP and the heuristic. While the heuristic executes orders of magnitude faster, execution times of the optimization model are still in an acceptable range, even for larger numbers of lifts.

To conclude, the MILP model was found to be accurate, robust in the number of demand points, and delivered fast solutions. A heuristic that is based on closed form solutions and, thus, can be applied without solving an optimization model, delivered even faster solutions but may be infeasible, especially for a high number of demand points.

Acknowledgment. We thank our reviewer for useful & constructive comments improving this work.

References

1. Briskorn, D., Dienstknecht, M.: Comput. Oper. Res. **92**, 194–207 (2018). https://doi.org/10.1016/j.cor.2017.11.012
2. Briskorn, D., Dienstknecht, M.: Eur. J. Oper. Res. **273**(1), 160–174 (2019). https://doi.org/10.1016/j.ejor.2018.07.033
3. Briskorn, D., Dienstknecht, M.: Omega **97**, 1–18 (2020). https://doi.org/10.1016/j.omega.2019.102114
4. Editors: 2020 rental rate survey. Cranes & access, Vertikal Press Ltd. **22**(9), 23–35 (2021). https://vertikal.net/en/cranes-and-access/issue/383
5. Huang, C., Wong, C.K., Tam, C.M.: Autom. Constr. **20**(5), 571–580 (2011). https://doi.org/10.1016/j.autcon.2010.11.023
6. Irizarry, J., Karan, E.P.: Electron. J. Inf. Technol. Constr. **17**, 351–366 (2012)
7. Ji, Y., Leite, F.: Autom. Constr. **93**, 78–90 (2018). https://doi.org/10.1016/j.autcon.2018.05.003
8. Ji, Y., Leite, F.: J. Constr. Eng. Manag. **146**(3) (2020). https://doi.org/10.1061/(asce)co.1943-7862.0001781
9. Lloyd, S.P.: IEEE Trans. Inf. Theory **I**(2), 129–137 (1982)
10. Marzouk, M., Abubakr, A.: Autom. Constr. **61**, 1–15 (2016). https://doi.org/10.1016/j.autcon.2015.09.008
11. Nadoushani, Z.S.M., Hammad, A.W.A., Akbarnezhad, A.: J. Constr. Eng. Manag. **143**(1), 04016,089 (2017). https://doi.org/10.1061/(asce)co.1943-7862.0001215
12. Rasmussen, G.: How are tower cranes built? (2021). https://www.craneblogger.com/crane-resource-library/how-are-tower-cranes-built/

13. Tam, C.M., Tong, T.K.: Constr. Manag. Econ. **21**(3), 257–266 (2003). https://doi.org/10.1080/0144619032000049665
14. Tam, C.M., Tong, T.K.L., Chan, W.K.W.: J. Constr. Eng. Manag. **127**(4), 315–321 (2001). https://doi.org/10.1061/(asce)0733-9364(2001)127:4(315)
15. Trevino, C., Abdel-Raheem, M.: J. Constr. Eng. Manag. **144**, 07018,001 (2018). https://doi.org/10.1061/(ASCE)CO.1943-7862.0001557
16. Wang, J., et al.: Autom. Constr. **59**, 168–178 (2015). https://doi.org/10.1016/j.autcon.2015.05.006
17. Wu, K., García de Soto, B., Zhang, F.: Autom. Constr. **111**, 1–17 (2020). https://doi.org/10.1016/j.autcon.2019.103060
18. Zhang, P., Harris, F.C., Olomolaiye, P.O., Holt, G.D.: J. Constr. Eng. Manag. **125**(2), 115–122 (1999). https://doi.org/10.1061/(asce)0733-9364(1999)125:2(115)

A Heuristic-Based Reduction for the Temporal Bin Packing Problem with Fire-Ups

John Martinovic(✉) and Nico Strasdat

Institut für Numerische Mathematik, Technische Universität Dresden, Dresden, Germany
{john.martinovic,nico.strasdat}@tu-dresden.de

Abstract. Given the ever-increasing role of data centers in global energy consumption, the problem of assigning jobs (items) to servers (bins) in an energy-efficient manner has become more and more important in recent years. In that respect, the temporal bin packing problem with fire-ups (TBPP-FU) takes into account not only the number of servers used but also the switch-on processes required to operate them. Due to the challenging size of the resulting ILP models, various tailored reduction strategies can be applied to obtain more tractable formulations. In this article, we show how the information from a heuristic solution can be used to further improve these exact approaches, extending a theoretical result that was previously proven only for some very restrictive cases in the literature. Finally, the benefits of this new reduction procedure will be demonstrated based on computational tests.

Keywords: Cutting and packing · Temporal bin packing · Fire-ups · Heuristics · Reduction

1 Introduction and Preliminaries

The *temporal bin packing problem* (TBPP) extends the well-known *bin packing problem* [7,11] with regard to an additional time dimension. Given $n \in \mathbb{N}$ *items* (*jobs*), it requires to find the minimum number of bins (*servers*) of capacity $C \in \mathbb{N}$ needed to accommodate them. Here, each job is described by a *size* (*resource demand*) $c_i \in \mathbb{N}$ and a *lifespan* (*activity interval*) $[s_i, e_i)$, $i \in I := \{1, \dots, n\}$, and an allocation is feasible if and only if the capacity is respected at any instant of time, see Fig. 1.

N. Trautmann and M. Gnägi (Eds.): OR 2021, LNOR, pp. 127–133, 2022.
https://doi.org/10.1007/978-3-031-08623-6_20

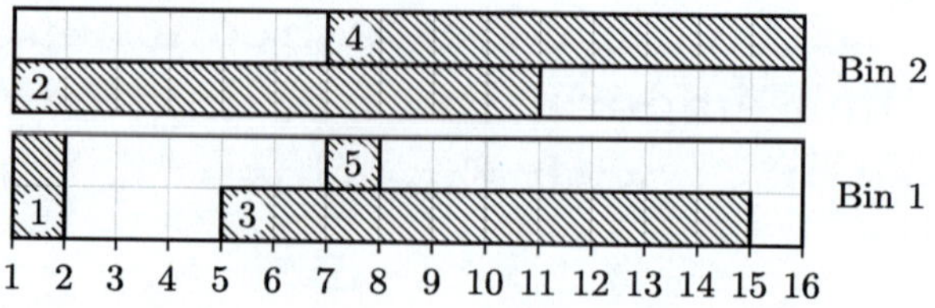

Fig. 1. A feasible assignment of five jobs to two servers.

The TBPP owes its practical importance to the ever-growing energy demand of the IT industry [1,8] and, accordingly, has been introduced in an application-oriented article dealing with efficient workload management in data centers [5]. Besides, it naturally generalizes the *temporal knapsack problem* [3,4] and can be solved by branch-and-price algorithms exploiting numerous heuristics [6]. Recent publications suggest to also include the operating mode of the servers into the objective function leading to the *temporal bin packing problem with fire-ups* (TBPP-FU) [2]. In that scenario, an inactive server can be temporarily deactivated (to save energy), but must later be "awakened" again (causing a *fire-up*) if required. Addressing the two objectives by a weighted-sum method (with $\gamma > 0$ scaling the number of fire-ups), the resulting ILP formulations (called M1 and M2 in [2]) turn out to be challenging in size. To this end, in the same article, a *constructive look-ahead heuristic* (CLH) was proposed which is able to strongly support the ILP models with heuristic information whenever $\gamma \leq 1/n$ holds. Very recently, in [9] a variety of variable and constraint reduction methods to improve these ILP models (for arbitrary values of γ) has been derived, and they were shown to lead to substantial performance gains of the exact approaches. As a result, the "optimized" version of M1 has been identified as the (on average) most competitive formulation.

In this article, we combine the two central ideas of the previous publications. More precisely, we show how heuristic information for arbitrary (!) values of γ can be used to further improve the already refined approaches from [9]. By that, we close an open question in the field of cutting and packing and succeed in solving several benchmark instances from the literature for the first time. Due to limited space, we only consider model M1 here; further information and more detailed explanations (also for M2) can be found in the full-length article [10].

2 The Basic Version of Model M1

In addition to the terminology mentioned in Sect. 1, let $K := \{1, \ldots, n\}$ denote the set of servers and let $T := \bigcup_{i \in I}\{s_i, e_i\}$ and $T_S := \bigcup_{i \in I}\{s_i\}$ collect the relevant instants of (starting) times. Moreover, we assume the items to be ordered with respect to non-decreasing starting times (where ties are broken arbitrarily), and we use $I_t := \{i \in I \mid t \in [s_i, e_i)\}$ to indicate active jobs at time $t \in T$. To state M1, let us consider the following four types of binary variables: (i) $z_k = 1$

iff. server $k \in K$ is used, (ii) $x_{ik} = 1$ iff. job $i \in I$ runs on server $k \in K$, (iii) $y_{tk} = 1$ iff. server $k \in K$ is active at time $t \in T$, (iv) $w_{tk} = 1$ iff. server $k \in K$ is activated at time $t \in T_S$. Following [2], we then obtain

Model 1 (M1)

$$\sum_{k \in K} \left(z_k + \gamma \sum_{t \in T_S} w_{tk} \right) \rightarrow \min$$

$$\text{s.t.} \quad y_{tk} \leq \sum_{i \in I_t} c_i x_{ik} \leq y_{tk} C, \qquad k \in K, t \in T, \qquad (1)$$

$$\sum_{k \in K} x_{ik} = 1, \qquad i \in I, \qquad (2)$$

$$x_{ik} \leq y_{s_i,k}, \qquad i \in I, k \in K, \qquad (3)$$

$$y_{tk} \leq z_k, \qquad k \in K, t \in T, \qquad (4)$$

$$y_{tk} - y_{t-1,k} \leq w_{tk}, \qquad k \in K, t \in T_S, \qquad (5)$$

$$x_{ik} \in \{0,1\}, \qquad i \in I, k \in K, \qquad (6)$$

$$y_{tk} \in \{0,1\}, \qquad k \in K, t \in T, \qquad (7)$$

$$w_{tk} \in \{0,1\}, \qquad k \in K, t \in T_S, \qquad (8)$$

$$z_k \in \{0,1\}, \qquad k \in K. \qquad (9)$$

The objective function minimizes a weighted sum of the number of servers used and the fire-ups required to operate them. Constraints (1) ensure respecting the server capacity (right hand side), and switch off empty servers (left hand side). Conditions (2) demand that any job is assigned exactly once, while coupling the different types of variables is done by Restrictions (3)–(5). In particular, (5) is responsible for detecting a fire-up exactly in those cases where the server under consideration is currently active but was inactive at the preceding time instant (referred to as $t-1$ in a symbolic way).

In [9], techniques for avoiding symmetry, introducing valid inequalities, and lifting item sizes have already partly eliminated the assignment-based disadvantageous properties of M1. However, the methods proposed so far have not yet contributed to the reduction of the quantity $|K|$, which has a crucial impact on the model size, since the index k occurs in each of the four variable types of M1. Thus, tailoring the set K to the actual or approximate number of servers needed in an optimal solution is an important preprocessing challenge.

3 A New Theoretical Result

To move to a more appropriate set K, heuristic information can be used. However, this has so far only been achieved for very small values of γ in the literature:

Theorem 1 ([2]). *Let $\gamma \leq 1/n$. Then, the number of servers required for the TBPP-FU is equal to the number of servers in an optimal solution to the TBPP.*

The preceding result allows either to compute the optimal size of K beforehand by solving a somewhat simpler auxiliary problem (i.e., the TBPP), or at least to limit the number of servers to be considered to a reasonably small value obtained by a heuristic solution. However, as reported in [2, Example 2.2], such an approach cannot be extended in a straightforward way to arbitrary scaling parameters γ. To address this deficiency, the following result provides a very simple way of using heuristic information in the general case, too.

Theorem 2. *Let z_{heu} be the objective value obtained by any heuristic for the TBPP-FU. Then, the number of servers required in an optimal solution is at most $k^\star := \lfloor z_{heu}/(1+\gamma) \rfloor$.*

Proof. If the assertion were false, we would need at least $k^\star + 1$ servers in an optimal solution. Since each server is switched on at least once, this would lead to an objective value of at least $(k^\star+1)+\gamma \cdot (k^\star+1) = (1+\gamma)\cdot(k^\star+1) > z_{heu}$, which yields the contradiction since the heuristic would have to be better than the optimal solution. □

This result permits to replace the set $K = \{1, \dots, n\}$ at all positions in M1 by a (strongly) reduced set $K^\star := \{1, \dots, k^\star\}$ leading to an improved model hereinafter referred to as *M1heu*. For that purpose, we make use of the constructive look-ahead heuristic (CLH) from [2], see also Algorithm 1. While in [2] the look-ahead parameter was simply fixed to $q = 3$, we will study the influence of this value in a bit more detail in the next section, to then pass the best possible compromise between effort and benefit to M1.

Algorithm 1. CLH with look-ahead parameter q

Input: Item list ordered by non-decreasing starting times s_i, parameter $q \in \mathbb{N}$.
1: Initialize the "empty" assignment $\mathcal{A}(0) := \emptyset$.
2: **for** $i \in I$ **do**
3: Assign item i to any open server of $\mathcal{A}(i-1)$ that can accommodate it and (as another alternative) also to a new empty server. By that we obtain the assignments $\mathcal{A}_1(i), \dots, \mathcal{A}_p(i)$.
4: Add the next q items to each of these allocations in a best-fit fashion and obtain the (updated) assignments $\widetilde{\mathcal{A}}_1(i), \dots, \widetilde{\mathcal{A}}_p(i)$.
5: Choose one assignment (from $\widetilde{\mathcal{A}}_1(i), \dots, \widetilde{\mathcal{A}}_p(i)$) with the lowest objective value and define it as $\mathcal{A}(i)$, that is, the starting point of the next iteration.
6: **end for**
Output: heuristic solution with objective value z_{heu}.

4 Computational Results

For our numerical calculations, we coded the M1 and M1heu in Python (version 3.9.2) and used its Gurobi (version 9.1.1) interface to solve the resulting ILP formulations with default settings and a time limit of 30 minutes per instance.

All the experiments were run on an AMD A10-5800K processor with 16 GB RAM. We consider a benchmark set from [6] originally designed for the ordinary TBPP (without fire-ups). However, possessing all the required input data, these instances can easily be interpreted in the context of the TBPP-FU, too. Given the somewhat more difficult scenario associated with adding fire-ups, we limit our investigations to $\gamma = 1$ and a reasonable subset of the 1700 instances from [6], that is, the moderate instance sizes with $|T| \in \{10, 20, \ldots, 50\}$ time steps, as similarly suggested in [9]. Note that, for a fixed choice of $|T|$, the data set consists of 100 instances that are divided into ten classes (numbered by I-X) of different difficulty level. Due to limited space, we will not go into further detail about these classes in our results and discussions, so that we just focus on complete packages of 100 instances each.

Let us first consider the results of CLH for different choices of the look-ahead parameter q in Table 1. Although there is no strict monotonicity, it can be seen that looking further into the future tends to lead to better heuristic results. For completeness, we also record the average of the respective best heuristic values (column 'best') and that of the LP bounds (column 'LP') for each instance block. The latter shows that the deviation of 'best' from the actual optimum is $\leq$15% on average, but increases with the instance size. On average, we observe that $q = 20$ leads to the best heuristic results, but larger look-ahead parameters can be slightly better for specific values of $|T|$. In addition, the times required average only 0.5 s for $q = 20$, while they are already about four to five times larger for $q \in \{n/2, n\}$, but without providing any noticeable gains. Hence, we will use the heuristic value obtained from $q = 20$ to define $k^\star$ according to Theorem 2, and pass the corresponding feasible solution to the solver to give it a warm start.

Table 1. Heuristic value for different look-ahead parameters q. (The column n_{avg} states the average number of items in the respective subclass.)

$\|T\|$	n_{avg}	1	2	3	5	10	20	$n/2$	n	Best	LP
10	54.90	40.9	40.0	39.8	39.1	38.1	37.5	37.3	**37.0**	36.6	34.8
20	88.39	46.9	45.8	45.4	44.2	42.4	41.4	41.2	**41.1**	39.9	36.7
30	121.43	52.3	50.9	50.5	48.5	46.1	44.9	**44.7**	45.0	43.2	37.6
40	154.10	57.5	55.9	55.1	52.7	49.6	**47.8**	48.0	48.8	46.2	38.2
50	186.70	62.9	60.9	59.9	57.0	53.4	**51.2**	52.8	53.4	50.0	39.0
Avg value	102.10	52.1	50.7	50.1	48.3	45.9	**44.5**	44.8	45.0	43.2	37.3

First of all, observe that using the heuristic information leads to ILP formulations of much smaller size. Based on Table 2, the average savings are about 72% for both, the numbers of variables and constraints. Remarkably, the percentage of reduction even increases for larger values of $|T|$. These observations also translate to the number of instances solved exactly and the computation times required. While for small values of $|T|$ both models still show very similar performance results, M1heu proves to be clearly superior for larger instances. Overall, the improved model was able to solve 17 additional benchmark instances and required about a quarter less computation time on average.

Table 2. Comparison between M1 and M1heu in terms of model size (left table), as well as instances solved to optimality and times required (right table).

$\|T\|$	#variables M1	#variables M1heu	#constraints M1	#constraints M1heu
10	2.7	**1.4**	3.7	**1.8**
20	7.2	**2.7**	9.8	**3.6**
30	13.8	**4.2**	18.7	**5.6**
40	22.3	**5.8**	30.1	**7.9**
50	33.0	**7.6**	44.3	**10.4**
Avg	15.8	**4.3**	21.3	**5.9**

$\|T\|$	M1 t	M1 opt	M1heu t	M1heu opt
10	28.1	(99)	**22.7**	**(99)**
20	244.8	(91)	**239.5**	**(91)**
30	483.7	(81)	**399.4**	**(82)**
40	711.2	(70)	**499.8**	**(78)**
50	858.5	(65)	**596.7**	**(73)**
Avg/Sum	465.3	(406)	**351.6**	**(423)**

5 Conclusions

In this article, we examined the TBPP-FU and, as a main contribution, showed how the information of a heuristic solution can be used to reduce the model size. On the one hand, this closed an open theoretical question, but at the same time led to much more convincing numerical results. Among other things, the number of variables and constraints could be reduced by about 72% leading to 25% lower computation times on average. These beneficial properties also enabled the optimal solution of 17 previously unsolved benchmark instances. Future research will focus primarily on finding improved heuristics or alternative modeling approaches.

References

1. Andrae, A.S.G., Edler, T.: On global electricity usage of communication technology: trends to 2030. Challenges **6**(1), 117–157 (2015)
2. Aydin, N., Muter, I., Ilker Birbil, S.: Multi-objective temporal bin packing problem: an application in cloud computing. Comput. Oper. Res. **121**, 104959 (2020)
3. Caprara, A., Furini, F., Malaguti, E.: Uncommon Dantzig-Wolfe reformulation for the temporal Knapsack problem. INFORMS J. Comput. **25**(3), 560–571 (2013)
4. Clautiaux, F., Detienne, B., Guillot, G.: An iterative dynamic programming approach for the temporal Knapsack problem. Eur. J. Oper. Res. **293**(2), 442–456 (2021)
5. de Cauwer, M., Mehta, D., O'Sullivan, B.: The temporal bin packing problem: an application to workload management in data centres. In: Proceedings of the 28th IEEE International Conference on Tools with Artificial Intelligence, pp. 157–164 (2016)
6. Dell'Amico, M., Furini, F., Iori, M.: A branch-and-price algorithm for the temporal bin packing problem. Comput. Oper. Res. **114**, 104825 (2020)

7. Delorme, M., Iori, M., Martello, S.: Bin packing and cutting stock problems: mathematical models and exact algorithms. Eur. J. Oper. Res. **255**, 1–20 (2016)
8. Jones, N.: How to stop data centres from gobbling up the world's electricity. Nature **561**, 163–166 (2018)
9. Martinovic, J., Strasdat, N., Selch, M.: Compact integer linear programming formulations for the temporal bin packing problem with fire-ups. Comput. Oper. Res. **132**, 105288 (2021)
10. Martinovic, J., Strasdat, N., Valério de Carvalho, J., Furini, F.: Variable and constraint reduction techniques for the temporal bin packing problem with fire-ups. Optim. Lett. (2021). https://doi.org/10.1007/s11590-021-01825-x
11. Scheithauer, G.: Introduction to Cutting and Packing Optimization - Problems, Modeling Approaches, Solution Methods. International Series in Operations Research & Management Science, vol. 263. Springer, Cham (2018). https://doi.org/10.1007/978-3-319-64403-5

Check for updates

Segmentation and Repetitive Scheduling of Linear Continuous Construction Projects

Michael Moos(✉), Sebastian Velten, and Christian Weiß

Fraunhofer Institute for Industrial Mathematics ITWM, Kaiserslautern, Germany
{michael.moos,sebastian.velten,christian.weiss}@itwm.fraunhofer.de

Abstract. We examine a structuring problem that arises in the first phase of linear repetitive projects which can be found in the area of, e.g., highway construction. The tasks are performed consecutively along an elongated construction site which can be split into segments. This leads to a repetitive project scheduling problem where the same tasks are performed in a repetitive way in every segment. In this paper, we assume the segmentation can be different for each task and that the split of a task into task segments incurs costs for hiring more resources/subcontractors as well as for handling the resulting communication overhead. The goal is to optimize the tradeoff between the project's makespan and total cost. We develop a constraint programming model and use it to compute pareto optimal solutions.

Keywords: Project scheduling · Repetetive scheduling · Constraint programming

1 Introduction

In this paper we investigate a planning problem posed in the early phase of linear construction projects. During this phase, decisions are made regarding the structure of the construction activities and division of work among project participants. The task of the planner(s) is to define segments and to structure necessary construction steps into work packages executed on these segments. Finally, these work packages are assigned to participating subcontractors.

Formally, assume there are n tasks $T_1, \dots, T_n$ with duration $d_1, \dots, d_n \in \mathbb{N}$ that need to be executed in sequence along an elongated construction site (e.g. highway construction). The total length of the construction site is $L \in \mathbb{N}$. Let $0 = A_0 < A_1 < \dots < A_m < A_{m+1} = L$ be equidistant points along the construction site with $A_1, \dots, A_m$ being the locations where segmentation of tasks is possible. We set $l = A_{j+1} - A_j \in \mathbb{N}$, $j = 0, \dots, m$ as the distance of two points on the site which implies $L/l = m + 1$.

There is given a minimum delay $a_i \in \mathbb{N}$ for consecutive tasks T_i and T_{i+1}, $i = 1, \dots, n-1$. For every point on the construction site, task T_{i+1} has to be a_i

N. Trautmann and M. Gnägi (Eds.): OR 2021, LNOR, pp. 134–140, 2022.
https://doi.org/10.1007/978-3-031-08623-6_21

time units behind task T_i. Without segmentation, this is modelled by start-after-start and end-after-end relations with delay of a_i between consecutive tasks.

For execution of the tasks, subcontractors $C_1, \ldots, C_d$ can be hired. Each subcontractor C_u is specialized on a single task T_i and has several resource crews $R_{u1}, \ldots, R_{ur_u}$, which can handle one segment of T_i in parallel. There are three types of costs associated with hiring subcontractor C_u: the base cost $b_u \in \mathbb{N}$ paid once for hiring the subcontractor, the resource cost $w_u \in \mathbb{N}$ paid for every resource crew hired, and cost $h_u \in \mathbb{N}$ for every l units of length the subcontractor works on. Since splitting a task into smaller segments at a point A_j leads to communication overhead between the participating project partners, an additional cost $z_i \in \mathbb{N}$ for every split of task T_i is introduced.

The goal is to simultaneously minimize the makespan M, i.e. the latest finish time of any segment of task T_n, and the total cost Γ, i.e. the sum of all costs incurred by hiring of resources and splitting of tasks. Since these are competing objectives, our goal is to compute the tradeoff between the two objectives by enumerating all pareto-optimal solutions. To achieve this we construct a constraint program which is repeatedly applied to compute the tradeoff in full.

To the best of our knowledge, simultaneous scheduling and task segmentation has not been considered before in the literature. Mostly, segmentation of tasks is either predefined or proposed in such a way that choosing the smallest possible segmentation is always optimal. For a review see [6]. In [1] the authors consider minimizing the interruption time of resource crews, i.e. the sum of idle times scheduled between consecutive work parts assigned to the same crew. Minimizing this interruption time leads to a reduction of segmentation, which makes it in some way similar to our model. In [8], the authors study the tradeoff among three dimensions, namely time, cost and quality in continuous repetitive projects but they do not consider segmentation as a degree of freedom. Authors [10] analyzed the time-cost tradeoff of continuous projects by developing a constraint programming model, again assuming segmentation is fixed in advance.

2 Constraint Programming Model

In order to solve the optimization problem, we developed a constraint programming model and implemented it using the SAT-Solver of Google OR Tools [4].

An optional interval variable $X = (S, E, P)$ is a tuple of three integer variables $S, E \in \mathbb{N}$, $P \in \{0, 1\}$, where S represents the start time of the interval, E the end time, and P its presence, i.e. whether the interval is used in the solution. An optional interval variable is given a duration $d \in \mathbb{N}$ and ensures that $E = S + d$ if $P = 1$, i.e. if the interval is present. For a collection of optional interval variables $X_1, \ldots, X_\omega$ the constraint noOverlap$(X_1, \ldots, X_\omega)$ ensures no two present intervals of the collection may intersect. Similar variables and constraints can be found in other solvers, such as ILOG [7]. The noOverlap constraint is similar to the one dimensional version of the diffn constraint and disjunctive constraint [2].

For an arbitrary given constraint Θ and boolean variables $v_1, \ldots, v_\omega$, using onlyEnforceIf$(\Theta \mid v_1, \ldots, v_\omega)$ adds the constraint Θ with the additional condition that it is not enforced unless all variables $v_1, \ldots, v_\omega$ are true. For a more in depth explanation on the propagation of such *reified* constraints, see [9]. The term onlyEnforceIf$(\Theta \mid v_1, \ldots, v_w)$ is equivalent to the conditional expression $\neg(\wedge_{i=1}^{\omega} v_i) \vee (\Theta \text{ is satisfied})$.

In what follows, we denote by $[n]$ the set $\{1, 2, \ldots, n\}$ for some $n \in \mathbb{N}$ and further use the notation $[n]_0 = [n] \cup \{0\}$. Moreover, we denote by $\mathcal{C}(T_i)$ the set of subcontractors C_u specialized in task T_i. Finally, we denote by $T_{\max} \in \mathbb{N}$ the largest allowed finish time of any task. Note that an upper bound is given by $T_{\max} = \sum_{i=1}^{n}(d_i + a_i)$.

Variables

$X_{uv}^{ij} = (S_{uv}^{ij}, E_{uv}^{ij}, P_{uv}^{ij})$	optional interval variable with duration d_i for $i \in [n]$, $j \in [m]_0$, $C_u \in \mathcal{C}(T_i)$, $v \in [r_u]$; present, if and only if resource crew R_{uv} is used to perform task T_i between points A_j and A_{j+1}
$F_{uv} \in \{0,1\}$	availability of the vth resource crew of subcontractor C_u
$Z^{ij} \in \{0,1\}$	segmentation of task T_i at A_j, $j \in [m]$
$B_u \in \{0,1\}$	hiring of subcontractor C_u
$W_u \in [r_u]_0$	# resource crews available from subcontractor C_u
$H_u \in [m+1]_0$	# parts of length l on which subcontractor C_u works on

Note that we introduce optional interval variables for every possible smallest segment of every task. However, in the final solution, two consecutive intervals belonging to the same task may still be seen as belonging to a common segment, see the "task split constraints" (6) and (7) below. So, in the final solution, not every present interval variable X_{uv}^{ij} necessarily defines its own segment.

We denote the total cost variable, consisting of the resource cost and the splitting cost, with

$$\Gamma = \sum_{u=1}^{d} (B_u \cdot b_u + W_u \cdot w_u + H_u \cdot h_u) + \sum_{i=1}^{n} z_i \cdot \left(\sum_{j=1}^{m} Z^{ij} \right),$$

and the makespan, equal to the maximum completion time of the segments of the last task T_n, with

$$M = \max_{\substack{j \in [m]_0, \\ C_u \in \mathcal{C}(T_n), \\ v \in [r_u]}} E_{uv}^{nj}.$$

Objective. We consider both the cost C and the makespan M, as defined above, to be minimized simultaneously,

$$\min(\Gamma, M),$$

and determine all pareto optimal solutions using the epsilon-constraint method described in [3, p. 285].

Constraints

Scheduling Constraints

We consider start-after-start and end-after-end precedences for consecutive tasks with a time lag of a_i. These need to be ensured for every start point and the combination of resource crews which execute the associated tasks:

$$\begin{aligned}&\text{onlyEnforceIf}\left(S^{ij}_{u_1v_1}+a_i\leq S^{(i+1)j}_{u_2v_2}\mid P^{ij}_{u_1v_1},P^{(i+1)j}_{u_2v_2}\right),\\&i\in[n-1],\ j\in[m]_0,\ S_{u_1}\in\mathcal{C}(T_i),\ S_{u_2}\in\mathcal{C}(T_{i+1}),\ v_1\in[r_{u_1}],\ v_2\in[r_{u_2}],\end{aligned}\tag{1}$$

$$\begin{aligned}&\text{onlyEnforceIf}\left(E^{ij}_{u_1v_1}+a_i\leq E^{(i+1)j}_{u_2v_2}\mid P^{ij}_{u_1v_1},P^{(i+1)j}_{u_2v_2}\right),\\&i\in[n-1],\ j\in[m]_0,\ S_{u_1}\in\mathcal{C}(T_i),\ S_{u_2}\in\mathcal{C}(T_{i+1}),\ v_1\in[r_{u_1}],\ v_2\in[r_{u_2}].\end{aligned}\tag{2}$$

We introduce no overlap constraints for the interval variables associated with one resource crew:

$$\text{noOverlap}\left(X^{i0}_{uv},X^{i1}_{uv},\ldots,X^{im}_{uv}\right),\quad i\in[n],\ C_u\in\mathcal{C}(T_i),\ v\in[r_u].\tag{3}$$

The following constraints set the end of every interval variable to zero, when it is not present:

$$\text{onlyEnforceIf}\left(E^{ij}_{uv}=0\mid\neg P^{ij}_{uv}\right),\quad i\in[n],\ j\in[m]_0,\ C_u\in\mathcal{C}(T_i),\ v\in[r_u].\tag{4}$$

We add exclusive constraints:

$$\sum\nolimits_{C_u\in\mathcal{C}(T_i)}\sum\nolimits_{v=1}^{r_u}P^{ij}_{uv}=1,\quad i\in[n],\ j\in[m]_0.\tag{5}$$

Task Split Constraints

In order to determine if a task T_i is split at a point A_j these constraints check if the same resource crew is hired for the left and right task part of T_i at A_j and if the completion and start time coincide:

$$\begin{aligned}&\text{onlyEnforceIf}\left(P^{i(j-1)}_{uv}=P^{ij}_{uv}\mid\neg Z^{ij}\right),\\&i\in[n],\ j\in[m]_0,\ C_u\in\mathcal{C}(T_i),\ v\in[r_u],\end{aligned}\tag{6}$$

$$\begin{aligned}&\text{onlyEnforceIf}\left(E^{i(j-1)}_{uv}=S^{ij}_{uv}\mid\neg Z^{ij}\right),\\&i\in[n],\ j\in[m]_0,\ C_u\in\mathcal{C}(T_i)\,v\in[r_u].\end{aligned}\tag{7}$$

Resource and Subcontractor Constraints

Since all resource crews of one subcontractor are interchangeable we impose an order on the resource crew variables to break symmetries. Further, assuming a resource crew is hired for a task segment, availability has to be ensured:

$$F_{uv}\geq F_{u(v+1)},\quad u\in[d],\ v\in[r_u-1],\tag{8}$$

$$P^{ij}_{uv}\leq F_{uv},\quad i\in[n],\ j\in[m]_0,\ C_u\in\mathcal{C}(T_i),\ v\in[r_u].\tag{9}$$

We set the auxiliary variables associated with a subcontractor C_u. Since we fixed the order for crew resources, checking the hiring status of a subcontractor simplifies. The other two sets of constraints sum up the relevant boolean variables:

$$B_u = F_{u0}, \quad u \in [d], \tag{10}$$

$$W_u = \sum_{v=1}^{r_u} F_{uv}, \quad u \in [d], \tag{11}$$

$$H_u = \sum_{v=1}^{r_u} \sum_{j=0}^{m} P_{uv}^{ij}, \quad i \in [n], C_u \in \mathcal{C}(T_i). \tag{12}$$

3 Results

We consider a highway construction project of the form introduced in [5]. Further, we set the length of the construction site to be equal to 4 units of length. We assume segmentation of tasks is possible at $(A_1, A_2, A_3) = (1, 2, 3)$. For every task of the project, two subcontractors are available to complete it. Subcontractors differ only in the base cost, where the base cost is zero for the first subcontractor and five for the second subcontractor. In order to introduce different scenarios, we keep the number of resource crews per subcontractor variable. The parameters are given by:

Task no.	Task description	d_i	a_i	z_i	w_u	h_u
1	Clearing and grubbing	20	1	1	0	2
2	Earth moving	24	3	2	5	5
3	Subbase	12	2	2	2	4
4	Base	8	1	2	2	4
5	Paving	24	3	5	7	8
6	Finish shoulders	16	–	1	0	3

We apply the introduced constraint programming model to the four scenarios: 1. One subcontractor with one resource crew per task, 2. One subcontractor with two resource crews per task, 3. Two subcontractors with one and two resource crews per task and, 4. Two subcontractors with two resource crews per task.

We calculate the overlapping pareto frontiers for all scenarios which are shown in Fig. 1. Given these solutions, it is interesting to note that increasing the number of resource crews does not decrease the makespan for previous cost bounds. It is, however, usually possible to enlarge the front by finding solutions for smaller makespan and higher cost bounds when more resource crews are available. The only time this does not hold, is when the crew number is increased from two to three, which can be observed in Fig. 1. This is due to the fact that having only three resource crews available implies that there are always two task segments left, which cannot be performed in parallel. Note, however, that it may be beneficial to split tasks into more segments than resource crews available. See, e.g., Fig. 2, which presents the lowest makespan solution of Scenario 2 where two resources are considered.

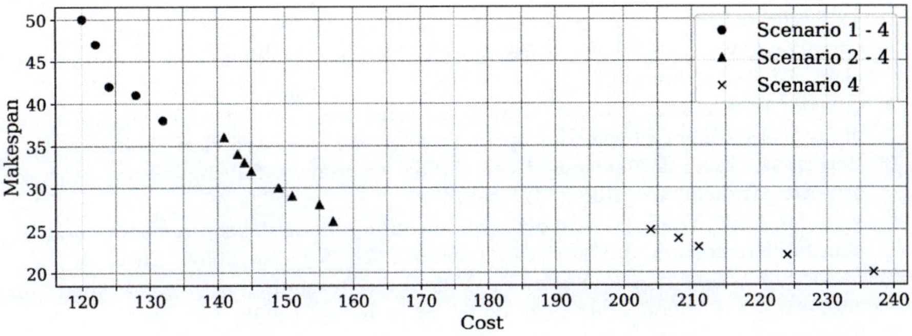

Fig. 1. Pareto frontiers for all scenarios.

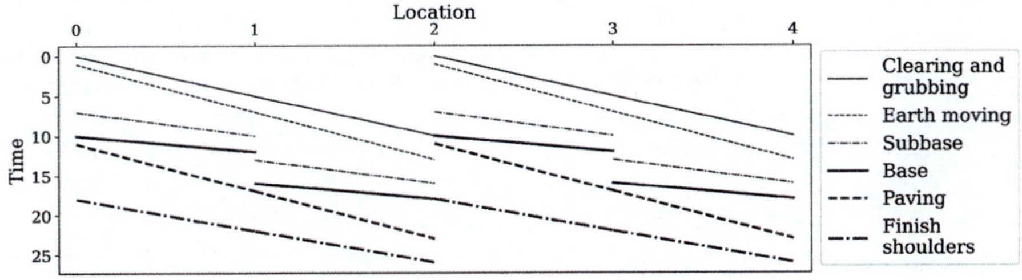

Fig. 2. Time minimized solution for Scenario 2.

4 Future Work

For future studies, a natural question to ask is how the proposed constraint programming model can be applied to more complex projects. This includes taking into consideration non-linear task graphs and including subcontractors or resources which can handle more than one task. Furthermore, it is possible to include additional optimization criteria such as the minimization of the idle time of resource crews as in [1]. Finally, another interesting direction would be the introduction of more (or even variable) splitting points, where splitting points no longer need to be equidistant or where only a lower bound on the segment length is given but splits can otherwise happen arbitrarily.

Acknowledgements. We thank Kristina Heim at TU Kaiserslautern, FGBBW, for initial inspiration and useful discussions. The research presented in this paper is partially supported by the Federal Ministry of Transport and Digital Infrastructure (BMVI) project "Infra-Bau 4.0" (grant number DG 25 - 836.7/12).

References

1. Altuwaim, A., El-Rayes, K.: Optimizing the scheduling of repetitive construction to minimize interruption cost. J. Constr. Eng. Manag. **144**(7), 04018051 (2018)
2. Beldiceanu, N., Carlsson, M., Rampon, J.X.: Global constraint catalog, (revision a). Technical report T2012:03, Swedish Institute of Computer Science (2012)
3. Chankong, V., Haimes, Y.: Multiobjective Decision Making: Theory and Methodology. North Holland (1983)
4. Google Developers: Google OR-Tools Reference. Google Developers (2021). https://developers.google.com/optimization/reference. Accessed 22 July 2021
5. Herbsman, Z.J.: Evaluation of scheduling techniques for highway construction projects. Transp. Res. Rec. (1126) (1987)
6. Ioannou, P.G., Yang, I.T.: Repetitive scheduling method: requirements, modeling, and implementation. J. Constr. Eng. Manag. **142**(5), 04016002 (2016)
7. Laborie, P., Rogerie, J., Shaw, P., Vilím, P.: Reasoning with conditional time-intervals. Part II: an algebraical model for resources (2009)
8. Nguyen, D.T., Doan, D.T.V., Tran, N.N.C., Tran, D.H.: A novel multiple objective whale optimization for time-cost-quality tradeoff in non-unit repetitive projects. Int. J. Constr. Manag. (2021). https://doi.org/10.1080/15623599.2021.1938939
9. Rossi, F., Van Beek, P., Walsh, T. (eds.): Handbook of Constraint Programming. Elsevier, Amsterdam (2006)
10. Zou, X., Zhang, L.: A constraint programming approach for scheduling repetitive projects with atypical activities considering soft logic. Autom. Constr. **109**, 102990 (2020)

Distributed Solving of Mixed-Integer Programs with COIN-OR CBC and Thrift

Jakob Görner, Ramon Janssen, and Jochen Rethmann(✉)

Faculty of Electrical Engineering and Computer Science,
Niederrhein University of Applied Sciences, 47805 Krefeld, Germany
jochen.rethmann@hs-niederrhein.de

Abstract. We present a distributed solver for mixed-integer problems. Our aim is to offer an alternative to commercial solvers so that even small companies come to the benefit of an optimization suite. The solver utilizes the COIN-OR Branch-and-Cut solver (CBC) for solving subproblems. Interprocess communication is achieved by using the remote procedure call library Thrift. Using ordinary office hardware, we evaluate our solution on instances from the MIPLIB and multidimensional knapsack instances both in deterministic and non-deterministic mode.

1 Introduction

Parallel MIP solvers for distributed memory computing environments often use a sequential MIP solver as a black box. Algorithmic improvements in the sequential solver can thereby be exploited directly. Examples of this approach are ParaSCIP by Shinano et al. [8] and GAMS in combination with Condor and CPLEX by Bussieck et al. [2]. Such frameworks require much knowledge from the user and adjustment of many parameters is needed in order to solve specific classes of problems efficiently. In contrast to that we would like to build a powerful, yet easy-to-use solver by using open-source software running on ordinary office hardware.

In [4] a distributed MIP solver is presented in which a sequential solver based on GLPK is used as a black box. The sequential solver carries out its own branch and bound procedure on a subproblem for a limited amount of time. If it times out, new subproblems are created.

We carry on this approach using CBC instead of GLPK. CBC is among the fastest open source MIP solvers, see H.D. Mittelmann's Web-Page. Using CBC in the distributed setup leads to performance gains and we compare the runtimes to those of Gurobi and multi-threaded CBC. Furthermore, we introduce a mechanism for enforcing determinism and assess its impact on solver runtimes.

2 Branch and Bound

A MIP in standard form defines a problem in which a linear objective function $\boldsymbol{c}^T\boldsymbol{x}$ is to be minimized subject to linear constraints $\boldsymbol{A}\boldsymbol{x} \leq \boldsymbol{b}$, where some of the

N. Trautmann and M. Gnägi (Eds.): OR 2021, LNOR, pp. 141–147, 2022.
https://doi.org/10.1007/978-3-031-08623-6_22

variables of $\boldsymbol{x}$ are required to take on integer values. The LP-relaxation of the problem is given by dropping the integrality constraints. The solution to the LP relaxation is a *lower bound* on the MIP solution, whereas a feasible solution is a valid *upper bound.*

Most MIP solvers use branch and bound to recursively subdivide the MIP problem into subproblems until either a solution is found or infeasibility is established. The LP relaxation is usually employed to create subproblems. A common approach is to select an integer variable x_i with a fractional value $\tilde{x}_i$ in the LP relaxation to define two new bounds: $x_i \leq \lfloor \tilde{x}_i \rfloor$ and $x_i \geq \lceil \tilde{x}_i \rceil$. By adding these bounds to the subproblem, two new subproblems are created.

In line with [4], we consider variables for branching in decreasing order of the absolute value of their coefficients in the objective function vector. Moreover, we branch on k variables at a time. All combinations of the new upper and lower bounds of the variables are generated, leading to 2^k new subproblems. The parameter k scales with the number of processors in our distributed architecture to prevent scalability issues.

By repeatedly branching on variables, a *branching tree* is created. The purpose of the *bounding* operation in branch and bound is to constrain the size of this branching tree. A subproblem may be *fathomed*, i.e. discarded, if its lower bound is worse than any solution to the MIP.

An important ingredient of a branch and bound procedure is the search strategy. It determines the order in which subproblems are processed. We use a *best-first search* which chooses the candidate subproblem with the smallest lower bound. While this strategy tends to minimize the number of subproblems processed, incremental improvements to the upper bound are often few and far between. This may reduce the effectiveness of the bounding operation [7]. We rely on heuristics in the sequential solver to quickly find improvements to the upper bound in the early phases of the algorithm.

3 Architecture

We use the farmer-worker architecture since load balancing and termination detection are easy, and the architecture must not be high scalable, since in an office of mid-sized companies there will be no more than a few tens of computers.

Farmer. The farmer loads the problems from file and first applies *presolve reductions* using CBC. The presolved problem is then passed to the workers. The farmer then creates initial subproblems and maintains active subproblems in a priority queue.

Worker. The worker receives the new variable bounds of a subproblem to solve and, if available, the best optimum found so far. The bounds are applied to the problem to reconstruct the subproblem. The best optimum is added to the subproblem model as a cutoff. This often enables the CBC solver to establish infeasibility quickly. CBC ships with a highly fine-tuned version of its solver. Due to technical difficulties, we use a much simpler solver in our benchmarks.

We let the CBC solver carry out its own branch and bound procedure for 25 seconds since this works well in our setup. If infeasibility is established by the CBC solver, the subproblem is fathomed. In the case of a timeout, the worker creates and returns new subproblems as well as the best solution found in the timed out subproblem.

Determinism. Non-deterministic behaviour is the possibility that an algorithm will follow different execution paths at different times on the same input due to external effects, see [6]. As a result, solver times may vary and different solutions may be reported. This type of behaviour is undesirable in practice.

A deterministic parallel branch and bound procedure comes at the cost of reduced efficiency and scalability due to the synchronization required [6]. We use a *barrier* synchronization scheme to realize a deterministic execution mode in our distributed setup. In between two barrier synchronization points, the workers finish a subset S of the active subproblems. The results of these problems, i.e. new subproblems and upper bounds, are processed after all the problems of subset S have been finished. The order in which the problems of a subset are processed may still vary, but this does not affect the overarching search order.

The size of subset S is important in this synchronization scheme: While increasing the size reduces the amount of idle time, it may reduce the number of subproblems being fathomed, since the best upper bound is only updated after all problems of a subset have been processed. The optimal subset size depends both on the problem instance as well as the number of workers. We get satisfactory results when the subset size is about twice the number of workers.

4 Computational Results and Evaluation

Benchmark Data. We evaluate our distributed CBC-solver on some multidimensional knapsack instances from Chu and Beasley [3] and on some instances from the MIPLIB 2003 [1] and 2010 [5].

Setup. Our solver is running on a small cluster of 12 machines, each equipped with an Intel i7-8700 CPU, 16 GB RAM, running 4 worker processes. We use CBC version 2.10.5, GLPK version 4.65, Thrift version 0.13.0 and Gurobi version 9.1.0. The benchmarks for Gurobi shown in Table 1 and 2 were carried out using its default settings on a single machine equipped with an Intel i7-8700K CPU and 16 GB RAM.

GLPK has presolving enabled and uses gomory, mixed integer rounding, mixed cover and clique cuts. The feasibility pump is used as heuristic and hybrid pseudo-cost branching is used as variable selection rule. In line with [4], GLPK uses a timeout of 10 seconds in the benchmarks that follow. CBC is running with gomory, probing, knapsack cover, clique, mixed integer rounding and flow cover cuts. The feasibility pump, rounding and greedy heuristics are used.

Scalability. To assess the scalability of a parallel implementation, its *speed-up* is often regarded. The speed-up $S_N := T_1/T_N$ is the ratio between the runtime T_1 when using 1 worker and the runtime T_N when using N workers. To empirically assess the speed-up, instances need to be solvable in a reasonable amount of time with the sequential solver. Unfortunately, such instances may not be difficult enough when using a large number of workers. We therefore use the speed-up from 16 to 32 and 48 workers as a proxy for the actual speed-up in our setup. To combat different sources of variability, we run each instance multiple times and compute the *average* value of these runtimes.

Comparison to Distributed GLPK and Gurobi. Table 1 gives runtimes in seconds on some multidimensional knapsack instances. Runtimes greater than 3600 s are indicated by a bar. Our distributed CBC-solver (labeled d-CBC) consistently outperforms the distributed GLPK-solver on these instances, see columns 2 and 3. Furthermore, using 32 workers, the distributed CBC-solver is able to outperform Gurobi on the majority of instances, see columns 4 and 6.

Table 1. Runtimes on some multidimensional knapsack instances.

Instance	Non-deterministic				Gurobi	Deterministic	
	GLPK	d-CBC				d-CBC	
	4 × 4	4 × 4	8 × 4	12 × 4	1 × 12	4 × 4	8 × 4
10 × 250–0.75_1	2633	273	136	95	209	346 (+27%)	253 (+86%)
10 × 250–0.75_2	—	626	349	251	641	750 (+20%)	517 (+48%)
10 × 250–0.75_3	2288	238	127	95	155	313 (+32%)	234 (+84%)
10 × 250–0.75_4	1407	146	101	95	103	262 (+80%)	192 (+90%)
10 × 250–0.75_5	3259	432	246	186	230	398 (–8%)	302 (+23%)
10 × 250–0.75_9	1321	112	77	63	77	219 (+96%)	182 (+136%)
10 × 250–0.25_6	—	326	178	127	278	465 (+43%)	315 (+77%)
30 × 100–0.25_9	1954	376	169	149	237	387 (+3%)	303 (+79%)
30 × 100–0.50_2	1024	191	108	77	118	356 (+86%)	250 (131%)
30 × 100–0.50_3	1470	228	141	113	181	470 (+106%)	321 (128%)

Table 2 summarizes the runtimes on some instances from the MIPLIB. While the distributed CBC-solver outperforms the distributed GLPK-solver on most instances, see columns 2 and 3, Gurobi's performance appears to be out of reach on most instances, see columns 5 and 6.

Gurobi and CBC are not able to simplify the instances of OR-LIB in Table 1 using presolve reductions. Furthermore, the available cutting plane techniques appear to be less effective on these instances. Such instances, which inevitably require the exploration of a large number of subproblems, naturally favour our distributed solution in which more computational power can be utilized.

Impact of Determinism. Tables 1 and 2 also summarize the results on the impact of determinism. The numbers in parentheses indicate the slow-down due to deterministic calculation.

Let us now evaluate the results by some comparisons to Gurobi and multi-threaded CBC (labeled mt-CBC). We ran the above instances on these two solvers with 4, 8, 12, and 16 threads, measured the runtimes and summarized

Table 2. Runtimes in seconds on some MIPLIB instances.

Instance	Non-deterministic				Gurobi	Deterministic	
	GLPK	d-CBC				d-CBC	
	4 × 4	4 × 4	8 × 4	12 × 4	1 × 12	4 × 4	8 × 4
mas74	1915	106	61	54	73	235 (+122%)	204 (+234%)
danoint	530	747	403	275	309	1080 (+45%)	657 (+63%)
qiu	241	93	45	41	0	375 (+303%)	162 (+260%)
noswot	1023	756	346	251	13	1768 (+134%)	1331 (+285%)
neos-1616732	—	1757	920	625	347	2159 (+23%)	1263 (+37%)
ns1830653	—	677	326	250	30	1432 (+112%)	1155 (+254%)
pigeon-10	—	—	2169	1970	0	—	—
ran14 × 18	1453	574	269	251	92	1084 (+83%)	776 (+189%)
ran14 × 18-disj-8	—	958	494	381	398	1867 (+95%)	1325 (+168%)
reblock67	2687	740	277	323	34	1766 (+138%)	1142 (+312%)
rmine6	1413	501	243	234	61	845 (+69%)	753 (+201%)
zib54-UUE	—	1306	656	567	109	—	1983 (+203%)

the geometric mean values for the different benchmark sets. The benchmarks ran on a VM on a single machine.

Efficiency of Parallelization. Table 3 gives the results of a comparison between workers in d-CBC and threads in mt-CBC, both in non-deterministic mode using fixed parameterization. The table entries are calculated according to *runtime of d-CBC divided by runtime of mt-CBC.* Our form of parallelism seems to be effective for OR-LIB, on instances of MIPLIB it seems to be improvable.

Table 3. mt-CBC using threads versus d-CBC using workers.

Benchmark set	4W:4T	8W:8T	12W:12T	16W:16T
OR-LIB	1.35	1.14	1.28	1.25
MIPLIB	2.35	2.90	3.46	3.48

Scalability of CBC and Gurobi. Table 4 summarizes the results for the speed-ups, where the values are calculated according to *runtime using 4 threads divided by runtime using n threads* for values $n = 8, 12, 16$. Our type of parallelization seems to scale well. Gurobi seems to scale slightly worse than multi-threaded CBC, probably because the instances are not hard enough.

Table 4. Comparison of speed-ups on different solvers.

Instances	Gurobi			mt-CBC			d-CBC	
of set	4:8T	4:12T	4:16T	4: 8T	4:12T	4:16T	16:32W	16:48W
OR-LIB	1.82	2.23	2.10	1.94	2.49	2.84	1.76	2.27
MIPLIB	1.60	2.47	2.21	1.96	2.18	2.55	2.01	2.38

Influence of Determinism. Table 5(a) summarizes the results for the slow-down due to deterministic calculation according to *runtime deterministic divided by runtime non-deterministic*. Since Gurobi is inherently deterministic, no comparison to Gurobi is possible. Our form of deterministic calculation by barrier synchronization seems to be effective.

Table 5. Slow-down due to deterministic calculation and fixed parameterization.

(a)

instances of set	mt-CBC			d-CBC	
	4T	8T	12T	16W	32W
OR-LIB	2.71	2.38	2.10	1.43	1.85
MIPLIB	1.84	1.83	2.26	1.96	2.74

(b)

instances of set	mt-CBC			
	4T	8T	12T	16T
OR-LIB	3.33	3.73	3.31	3.50
MIPLIB	2.44	2.73	2.74	3.10

Influence of Fixed Parameterization. Table 5(b) summarizes the results for the slow-down due to fixed parameterization instead of a parameterization specific to the problem instance using mt-CBC. The table entries are calculated according to *runtime fixed divided by runtime specific*. The performance degration is remarkable.

5 Conclusion

Using CBC instead of GLPK yield significant performance gains in the distributed architecture. A deterministic mode for the distributed solver based on a simple barrier synchronization scheme was introduced and shown to be effective. We could show that our form of parallelism is effective, too. In a multi-threaded environment, a lot of time can be lost due to synchronization and caching effects. In such an environment, our distributed system may have advantages because the worker processes can run truly in parallel. Although we had to choose a fixed parameterization to make CBC run reliable, good results were achieved.

References

1. Achterberg, T., Koch, T., Martin, A.: MIPLIB 2003. Oper. Res. Lett. **34**(4), 361–372 (2006)
2. Bussieck, M.R., Ferris, M.C., Meeraus, A.: Grid-enabled optimization with GAMS. INFORMS J. Comput. **21**(3), 349–362 (2009)
3. Chu, P.C., Beasley, J.E.: A genetic algorithm for the multidimensional knapsack problem. J. Heurist. **4**(1), 63–86 (1998)
4. Gurski, F., Rethmann, J.: Distributed solving of mixed-integer programs with GLPK and thrift. In: Fink, A., Fügenschuh, A., Geiger, M.J. (eds.) Operations Research Proceedings 2016. ORP, pp. 599–605. Springer, Cham (2018). https://doi.org/10.1007/978-3-319-55702-1_79
5. Koch, T., et al.: MIPLIB 2010. Math. Program. Comput. **3**(2), 103–163 (2011)

6. Maher, S.J., Ralphs, T., Shinano, Y.: Assessing the Effectiveness of (Parallel) Branch-and-bound Algorithms. Tech. Rep. 19-03, ZIB (2019)
7. Ralphs, T.K.: Parallel Branch and cut. Parallel Comb. Optim. **58**, 53–101 (2006)
8. Shinano, Y., Achterberg, T., Berthold, T., Heinz, S., Koch, T.: ParaSCIP: a parallel extension of SCIP. In: Competence in High Performance Computing 2010, pp. 135–148. Springer, Heidelberg (2011). https://doi.org/10.1007/978-3-642-24025-6

The Balanced Maximally Diverse Grouping Problem with Attribute Values and Varying Group Sizes

Arne Schulz(✉)

Institute of Operations Management, Universität Hamburg, Moorweidenstraße 18, 20148 Hamburg, Germany
arne.schulz@uni-hamburg.de

Abstract. The balanced maximally diverse grouping problem with attribute values and varying group sizes (BMDGPAVVG) searches for the best balanced solution amongst all optimal solutions of the corresponding instance of the maximally diverse grouping problem with attribute values and varying group sizes (MDGPAVVG). We describe the set of optimal solutions of MDGPAVVG which allows us to use a short integer program presented in the literature to solve the BMDGPAVVG.

Keywords: Combinatorial optimization · Integer programming · Assignment · Grouping

1 Introduction

The maximally diverse grouping problem (MDGP) is a well-investigated combinatorial optimization problem, which is applied for example in the assignment of students to learning groups. The MDGP is the problem of assigning items, $i, j \in I$, to groups, $g \in G$, such that the sum of pairwise distances $d_{ij}^a \geq 0$ of item pairs which are assigned to the same group is maximized over all attributes $a \in A$. The problem is either formulated with equal-sized groups or the number of assigned items is required to be within a certain range for each group.

In the general variant, d_{ij}^a is arbitrary. However, in common applications, e.g. the assignment of students to groups, distances are absolute differences of attribute values av_i^a, i.e. $d_{ij}^a = |av_i^a - av_j^a|$. Assignments of students to groups have been extensively investigated in the literature (e.g. assignment of students to project groups [1], study groups [4], or work groups [2]). In the assignment of students to learning groups, teachers usually aim at equally strong groups with respect to the students' academic achievements. Thus, av_i^a can be the average grade of student i. It can also be the grade in a certain course or another performance measure. Moreover, the teacher might also aim at well-balanced groups regarding the gender or wishes that international students are equally distributed over the groups. Then, av_i^a is a zero-one attribute which indicates whether the student is female/male and international, respectively.

N. Trautmann and M. Gnägi (Eds.): OR 2021, LNOR, pp. 148–153, 2022.
https://doi.org/10.1007/978-3-031-08623-6_23

Recently, Schulz [5] investigated the MDGP with attribute values and found that the set of optimal solutions can be described by a set of equalities, called block constraints, if this set has a feasible solution. He proposed that it is desirable to search for the best balanced solution amongst all optimal solutions of the MDGP and introduced the balanced maximally diverse grouping problem with attribute values. However, the paper considers only the case with equal-sized groups.

In this paper, we generalize the approach by Schulz [5] to a variant where the number of items is required to be within a certain range for each group. We adapt the construction of the block constraints proposed by Schulz [5] to describe the set of optimal solutions of the MDGP with attribute values and varying group sizes if the block constraints have a feasible solution.

The paper is constructed as follows: First, we give a formal introduction of the problem setting and the prior work by Schulz [5] (Sect. 2). Then, we present our adaptation for the variant with varying group sizes (Sect. 3). The paper closes with a conclusion in Sect. 4.

2 Problem Description

We consider a set of items $i \in I$, a set of groups $g \in G$, and a set of attributes $a \in A$ as given. Each item has according to each attribute an attribute value $av_i^a \in \mathbb{Q}_{\geq 0}$. With them we define the distance among each item pair as $d_{ij}^a = |av_i^a - av_j^a|$. Each group must get at least l_g and at most u_g items assigned.

The MDGP is generally formulated as follows (compare e.g. [3,6]):

$$\max \sum_{a \in A} \sum_{g \in G} \sum_{i \in I} \sum_{j \in I: j > i} d_{ij}^a x_{ig} x_{jg} \tag{1}$$

with the constraints

$$\sum\nolimits_{g \in G} x_{ig} = 1 \qquad \forall i \in I \tag{2}$$

$$l_g \leq \sum\nolimits_{i \in I} x_{ig} \leq u_g \qquad \forall g \in G, \tag{3}$$

$$x_{ig} \in \{0, 1\} \qquad \forall i \in I, g \in G \tag{4}$$

Thereby, x_{ig} is a binary variable which equals to one if item i is assigned to group g and is zero otherwise. Objective function (1) maximizes the pairwise differences between each pair of items assigned to the same group according to all attributes. Constraints (2) ensure that each item is assigned to exactly one group while Constraints (3) take care that all groups get at least l_g and at most u_g items assigned. Constraints (4) are the binary constraints for the x variables.

Schulz [5] found another representation of the objective function (1). Therefore, he introduced blocks $k \in K$ with $|K| = |I|/|G|$ (equal-sized groups) and assigned the items for each attribute according to the following assumption to the blocks:

Assumption 1. *Let $|G|$ be the number of groups. Then, the $|G|$ items with the largest attribute values according to the considered attribute are assigned to the first block, the $|G|$ items with the next largest attribute values according to the considered attribute are assigned to the second block, and so on.*

The item to block assignment is depicted by a binary parameter b^a_{ki} which is one if item i is according to attribute a in block k and zero else. Schulz [5] introduced the block constraints

$$\sum\nolimits_{i\in I:b^a_{ki}=1} x_{ig} = 1 \qquad \forall a \in A, g \in G, k \in K \qquad (5)$$

and proved that the set of optimal solutions for (1)–(2), (4) and

$$\sum_{i\in I} x_{ig} = |I|/|G| \qquad \forall g \in G,$$

equals the set of feasible solutions for (2), (4), and (5) if a feasible solution exists, blocks are determined according to Assumption 1, and the assignment to blocks is unique in the sense that no two items with identical attribute value are assigned to different blocks regarding the corresponding attribute. If $|A| = 1$, it is sufficient that the blocks are determined according to Assumption 1. By this, we get

$$\sum_{a\in A}\sum_{g\in G}\sum_{i\in I}\sum_{j\in I:i<j} d^a_{ij} x_{ig} x_{jg} = \sum_{a\in A}\sum_{k\in K}\sum_{i\in I} b^a_{ki} c_k a v^a_i \qquad (6)$$

with

$$c_k = \begin{cases} \bar{c}_k & \text{if } k \leq \left\lceil \frac{|K|}{2} \right\rceil \\ -\bar{c}_k & \text{else,} \end{cases} \qquad (7)$$

where

$$\bar{c}_k = 2 \cdot \max\left(\frac{|K|}{2} - k, k - \frac{|K|}{2} - 1\right) + 1 \qquad (8)$$

if $|K|$ is even and

$$\bar{c}_k = 2 \cdot \left| k - \left\lceil \frac{|K|}{2} \right\rceil \right| \qquad (9)$$

if $|K|$ is odd. Due to the page restriction, we refer the reader to [5] for a proof. Since the set of all optimal solutions for the MDGP can be described in the case with attribute values, the procedure by Schulz [5] searches for the best balanced solution amongst them, i.e. minimizes

$$\sum_{a\in A}\sum_{g,g'\in G:g'>g} \left| \sum_{i\in I}\sum_{j\in I:i<j} d^a_{ij} x_{ig} x_{jg} - \sum_{i\in I}\sum_{j\in I:i<j} d^a_{ij} x_{ig'} x_{jg'} \right|$$

with the constraints (2)–(5) and $u_g = l_g = |I|/|G|$ for all $g \in G$. In the next section, we use the structure of the right side of (6) to adapt this approach to varying group sizes.

3 Solution Approach for Varying Group Sizes

In this section, we present our solution approach. First, we consider fixed but not necessarily equal group sizes, i.e. $l_g = u_g$ for all $g \in G$ (Subsect. 3.1). Afterwards, we generalize the approach to the case $l_g \leq u_g$ for all $g \in G$ but $l_g \geq l_{g'}$ and $u_g \geq u_{g'}$ for all $g < g'$ (Subsect. 3.2). As the number of items per group varies, we replace c_k by c_{gk}, whereat $c_{gk} = c_k$ with $|K|$ equal to the number of items in group g in (7)–(9).

3.1 Fixed Group Sizes

If we know the group sizes, c_{gk} values are fixed. Besides, av_i^a is fixed. Hence, we have to set b_{ki}^a values to one (each item is assigned to exactly one block per attribute) such that the right side of (6) is maximized. The following theorem sets b_{ki}^a values for each attribute such that the right side of (6) is maximized.

Theorem 1. *Let $a \in A$ be fixed. Let two lists with the c_{gk} values as well as the attribute values av_i^a, both in decreasing order, be given. Let p_i be the position of av_i^a in the list of attribute values. Then, the right side of (6) is maximized by multiplying av_i^a with the c_{gk} value at position p_i of the c_{gk}-list and summing up the products over all $i \in I$.*

Proof. Let $a, b, c, d \in \mathbb{Q}$ with $a \geq b$ and $c \leq d$. Then,

$$\begin{aligned} a \cdot c + b \cdot d &\leq a \cdot c + b \cdot d + (a - b) \cdot (d - c) \\ &= a \cdot c + a \cdot (d - c) + b \cdot d - b \cdot (d - c) \\ &= a \cdot d + b \cdot c. \end{aligned}$$

Thus, the sum over the products $c_{gk} \cdot av_i^a$ is maximized if the largest c_{gk} value is multiplied with the largest av_i^a, the second largest c_{gk} value with the second largest av_i^a, and so on, which proves the theorem.

Let $\bar{K} = \{c_{gk} : g \in G, k \in K\}$. We repeat the procedure in Theorem 1 for each attribute. Set $b_{\bar{k}i}^a = 1$, $\bar{k} \in \bar{K}$, if av_i^a is matched with a $c_{gk} = \bar{k}$ for the corresponding $a \in A$ in Theorem 1 and introduce the new block constraints

$$\sum_{i \in I : b_{\bar{k}i}^a = 1} x_{ig} = 1 \qquad \forall a \in A, \bar{k} \in \bar{K}, g \in G : \exists k \in K : c_{gk} = \bar{k} \tag{10}$$

Then, a feasible solution for (2), (4), and (10) maximizes the right side of (6) and must therefore be optimal for (1)–(4) and we can search for the best balanced optimal solution of the MDGP with attribute values with the procedure by Schulz [5].

3.2 Bounded Group Sizes

If $l_g < u_g$, c_{gk} cannot be predetermined but are endogenous. Thus, we have to choose the number of items per group n_g with $l_g \leq n_g \leq u_g$ for all $g \in G$ such that c_{gk} values are maximized. Since c_{gk} values count the number of connections between blocks k and $|K| + 1 - k$ visiting at most one further block in between (compare [5]), all c_{gk} values of the group increase by one if a further item is assigned to it. Thus, it is, if $l_g \geq l_{g'}$ and $u_g \geq u_{g'}$ for all $g < g'$, optimal to assign a further item always to the group with the largest number of items assigned but $n_g < u_g$, i.e. the group with the smallest index g for which $n_g < u_g$ holds.

The following procedure finds optimal c_{gk} values: First, we set $n_g = l_g$ for all $g \in G$. If this is not possible, i.e. $\sum_{g \in G} l_g > |I|$, (1)–(4) has no feasible solution. Second, we increase n_1 until either $n_1 = u_1$ or $\sum_{g \in G} n_g = |I|$. If the first criterion is reached first, we repeat the same with the next group (2, 3, 4, ..., $|G|$) until the second criterion is reached. This procedure maximizes the counted c_{gk} values over all groups (remember that we assume $l_g \geq l_{g'}$ and $u_g \geq u_{g'}$ for all $g < g'$) if $\sum_{g \in G} l_g \leq |I| \leq \sum_{g \in G} u_g$. Otherwise there is no feasible solution. Algorithm 1 illustrates this procedure.

Algorithm 1

1: *Set* $n_g = l_g$ *for all* $g \in G$ *and* $\bar{I} = |I| - \sum_g l_g$
2: *Set* $counter = 0$
3: ***while*** $\bar{I} > 0$ *and* $counter < |G|$ ***do***
4: *Set* $counter = counter + 1$
5: *Set* $h = \min(u_{counter} - n_{counter}, \bar{I})$
6: *Set* $n_{counter} = n_{counter} + h$
7: *Set* $\bar{I} = \bar{I} - h$
8: ***end while***

Apply Theorem 1 for all $a \in A$ with the c_{gk} values determined by the optimal n_g (Algorithm 1) and set $\bar{K}$ and $b^a_{\bar{k}i} = 1$ as described in Subsect. 3.1. Then, a feasible solution for (2), (4), and (10) must be optimal for (1)–(4) and we can search for the best balanced optimal solution of the MDGP with attribute values with the procedure by Schulz [5].

Note that it is for general d^a_{ij} values not always optimal to assign a further item to the group with the larger number of assigned items. Let g_1 and g_2 be two groups whereat g_1 has a higher number of items assigned than g_2. Let further item i be unassigned. With general d^a_{ij} values it is possible that $\sum_{j \in g_1} d^a_{ij} < \sum_{j \in g_2} d^a_{ij}$ although the first sum contains more addends.

The following example shows that $l_g \geq l_{g'}$ and $u_g \geq u_{g'}$ for all $g < g'$ is a necessary restriction (beside symmetry). Consider four groups with $l_1 = 1$, $l_2 = l_3 = l_4 = 2$, $u_1 = 4$, and $u_2 = u_3 = u_4 = 3$, and ten items. As seven items have to be assigned to ensure that each group has at least l_g items assigned, three items are variable. If we assign them always to the group g with $g = \text{argmax}_{g' \in G}\{n_g : n_g < u_g\}$, we assign the remaining items to groups 2, 3, and 4. Then, at most three items are assigned to the same group (i.e. largest three c_{gk}

values are 2). If we assign all three variable items to group 1, it contains four items (i.e. largest c_{gk} value is 3, the second and third largest is 1). Let further $|A| = 1$, $av_1^1 = 1$ and $av_i^1 = 0$ for all $i = 2, ..., 10$. Then, the only non-zero addend on the right side of (6) is 1 (av_1^1) multiplied with the largest c_{gk} value. Thus, the second assignment is better. However, if instead $av_1^1 = av_2^1 = av_3^1 = 1$ and $av_i^1 = 0$ for all $i = 4, ..., 10$, then the right side of (6) has three non-zero addends each multiplying 1 (av values) with one of the three largest c_{gk} values. So, the first assignment is better.

4 Conclusion

The paper generalizes the findings by Schulz [5] for the balanced MDGP with attribute values to groups with varying group sizes. The MDGP with attribute values has many practical applications as shown in the introduction. Since it has potentially a large number of optimal solutions, it is beneficial to search for the best balanced one amongst them. This paper presents a way how to do this.

The block constraints (10) are always fulfilled if $|A| = 1$, as all block constraints are independent. Let $n_1 \geq n_g + 2$ for all $1 < g \in G$. Then, the two items with the smallest attribute values regarding the first attribute are assigned to the first group. If they are in the same block regarding the second attribute, there is no feasible solution for (2), (4), and (10). So, there is not necessarily a feasible solution for $|A| \geq 2$. Nonetheless, our findings lead to an upper bound for the MDGP with attribute values, as Theorem 1 can still be applied to every single attribute to compute the right side of (6), which is a relaxation of (10).

References

1. Beheshtian-Ardekani, M., Mahmood, M.A.: Education development and validation of a tool for assigning students to groups for class projects. Decis. Sci. **17**(1), 92–113 (1986)
2. Caserta, M., Voß, S.: Workgroups diversity maximization: a metaheuristic approach. In: Blesa, M.J., Blum, C., Festa, P., Roli, A., Sampels, M. (eds.) HM 2013. LNCS, vol. 7919, pp. 118–129. Springer, Heidelberg (2013). https://doi.org/10.1007/978-3-642-38516-2_10
3. Gallego, M., Laguna, M., Martí, R., Duarte, A.: Tabu search with strategic oscillation for the maximally diverse grouping problem. J. Oper. Res. Soc **64**(5), 724–734 (2013)
4. Krass, D., Ovchinnikov, A.: Constrained group balancing: why does it work. Eur. J. Oper. Res. **206**(1), 144–154 (2010)
5. Schulz, A.: The balanced maximally diverse grouping problem with block constraints. Eur. J. Oper. Res. **294**(1), 42–53 (2021)
6. Singh, K., Sundar, S.: A new hybrid genetic algorithm for the maximally diverse grouping problem. Int. J. Mach. Learn. Cybernet. **10**(10), 2921–2940 (2019). https://doi.org/10.1007/s13042-018-00914-1

Finding Minimum Balanced Separators - An Exact Approach

William Surau(✉), Ralf Borndörfer, and Stephan Schwartz

Zuse Institute Berlin, Takustr. 7, 14195 Berlin, Germany
{surau,borndoerfer,schwartz}@zib.de

Abstract. Balanced separators are node sets that split the graph into size bounded components. They find applications in different theoretical and practical problems. In this paper we discuss how to find a minimum set of balanced separators in node weighted graphs. Our contribution is a new and exact algorithm that solves Minimum Balanced Separators by a sequence of Hitting Set problems. The only other exact method appears to be a mixed-integer program (MIP) for the edge weighted case. We adapt this model to node weighted graphs and compare it to our approach on a set of instances, resembling transit networks. It shows that our algorithm is far superior on almost all test instances.

Keywords: Node separators · Integer programming

1 Balanced Separators

The definition of balanced separators is not consistent in the literature. Here we use the following:

Definition 1. *Let $G = (V, E)$ be a graph with node weights $w \in \mathbb{R}^V_{\geq 0}$, $X \subseteq V$, and $\alpha \in [0, 1]$. We call X an* **α-balanced separator** *if any connected component C in $G - X$ satisfies $w(C) \leq \alpha \cdot w(G)$.*

Other definitions only consider unit node weights or require that X separates G into two disjoint vertex sets $A, B \subseteq V$ such that no edge between a vertex in A and B exists, $V = A \cup B \cup X$, and $|A|, |B| \leq \alpha \cdot |V|$. In the latter definition we call X an α-balanced biseparator. Note that for $\alpha \geq \frac{2}{3}$ both definitions intersect. We consider the problem of finding balanced separators of minimum cardinality.

Definition 2. *Let $G = (V, E)$ be graph with node weights $w \in \mathbb{R}^V_{\geq 0}$ and $\alpha \in [0, 1]$. The* **Minimum Balanced Separators** *problem is to find a set $X \subseteq V$ such that X is an α-balanced separator and $|X|$ is minimum.*

Observe that Minimum Balanced Separators is $\mathcal{NP}$-hard. By setting $w \equiv 1$ and $\alpha = \frac{1}{|V|}$ we basically search for a minimum vertex cover, which is one of the classical $\mathcal{NP}$-hard problems.

Solving Minimum Balanced Separators is usually done by approximation algorithms [3] or even pseudo approximation algorithms with relaxed upper

N. Trautmann and M. Gnägi (Eds.): OR 2021, LNOR, pp. 154–159, 2022.
https://doi.org/10.1007/978-3-031-08623-6_24

bounds [9]. In [5] the authors found kernels for unweighted graphs. The problem is also studied for different graph classes like planar graphs [10] or graphs with maximum degree 3 [4]. In [13] they propose an exact approach to find minimum balanced biseparators.

Balanced separators have a variety of applications. They are used in approximation algorithms to find tree decompositions [7]. In many graph algorithms based on the divide and conquer paradigm [11] finding balanced separators of small size is important. Separators are also used to indicate bottlenecks in communication systems [1].

Our work is motivated by covering or partitioning a graph with connected subgraphs subject to lower and upper bounds. These problems occur in many districting [12], waste collection [6] or toll control enforcement [2] problems and can be solved by compact MIPs as presented in [8,14]. Enforcing connectivity in MIPs is crucial and can be done by declaring vertices as roots from which connected subgraphs are formed. A naive approach is to allow every vertex to operate as a root. Since every considered root increases the number of variables and constraints, we aim to minimize the set of roots. Here we can exploit the lower weight bound L of each connected subgraph. If we set $\alpha = \frac{L-\varepsilon}{w(G)}$ (for ε sufficiently small) then each connected subgraph S with $w(S) \geq L$ shares at least one vertex with any α-balanced separator. Therefore, finding a minimum set of roots reduces to MINIMUM BALANCED SEPARATORS.

2 Compact Flow Formulation

Elijazyfer [8] proposes an exact method for finding α-balanced separators in edge weighted graphs. For the sake of completeness we present the adapted version to the node weighted case. The MIP is based on a flow formulation, which requires directed graphs. To turn G into a directed graph we define A as the bidirected arc set of the edges in E. The idea is to search for an arc partition such that each component induces a connected subgraph and its cumulative weight does not exceed $\alpha \cdot w(G)$. The set of vertices occurring in more than one component constitutes an α-balanced separator. Therefore, the objective is to minimize the cardinality of this vertex set. We introduce variables $x \in \{0,1\}^V$ for the indication of the α-balanced separator, variables $y \in \{0,1\}^{V \times A}$ for the arc partition, variables $q \in \mathbb{N}_0^{V \times A}$ for flows ensuring connectivity of the components, and variables $s \in \{0,1\}^{V \times V}$ and $c \in \{0,1\}^{V \times V}$ to determine the weight of a component.

Constraints (1b) partition the arcs and (1c) force both directions of an arc to be in the same component. In (1d) it is ensured that both endpoints of an arc are in the same component as the arc. Lines (1e) and (1f) specify a flow to ensure connectivity, for more details see [2]. If a vertex is in more than one subgraph it has to be a separator, which is forced by (1g). The upper bound of each component is set in (1h). The variable c_v^r indicates if vertex v in the component rooted at r is excluded in the cumulative weight. This is only possible if $s_v^r = 1$ (1i) and v is a separator (1j).

$$\min_{x,s,c,y,q} \quad \sum_{v \in V} x_v \tag{1a}$$

$$\text{s.t.} \quad \sum_{r \in V} y_a^r = 1 \qquad \forall\, a \in A, \tag{1b}$$

$$y_{(u,v)}^r = y_{(v,u)}^r \qquad \forall\, r \in V\ \forall\, (u,v) \in A, \tag{1c}$$

$$2 \cdot y_{(u,v)}^r \le s_u^r + s_v^r \qquad \forall\, r \in V\ \forall\, (u,v) \in A, \tag{1d}$$

$$q_a^r \le y_a^r \cdot (|V| - 1) \qquad \forall\, r \in V\ \forall\, a \in A, \tag{1e}$$

$$\sum_{v \in \delta^+(v)} q_a^r - \sum_{v \in \delta^-(v)} q_a^r \ge s_v^r \qquad \forall\, r \in V\ \forall\, v \in V \setminus \{r\}, \tag{1f}$$

$$1 + x_v \cdot (|V| - 1) \ge \sum_{r \in V} s_v^r \qquad \forall\, v \in V, \tag{1g}$$

$$\sum_{v \in V} (s_v^r - c_v^r) \cdot w_v \le \alpha \cdot w(G) \qquad \forall\, r \in V. \tag{1h}$$

$$c_v^r \le s_v^r \qquad \forall\, r \in V\ \forall\, v \in V, \tag{1i}$$

$$c_v^r \le x_v \qquad \forall\, r \in V\ \forall\, v \in V. \tag{1j}$$

We can see that the model contains a large number of variables and constraints and uses many big M constraints. In particular, for the connectivity it considers a flow emerging from every vertex, while our main motivation is to avoid this necessity.

3 Exact Algorithm

We propose a different exact approach that circumvents these issues. While our formulation still has a large number of constraints we present a separation routine that dynamically generates necessary constraints. To this end, we define $\mathcal{S}$ as the set of all connected subgraphs of G with a cumulative weight greater than $\alpha \cdot w(G)$. The master problem has the following form:

$$\min_{x} \quad \sum_{v \in V} x_v \tag{2a}$$

$$\text{s.t.} \quad \sum_{v \in V(S)} x_v \ge 1 \qquad \forall\, S \in \mathcal{S}, \tag{2b}$$

$$x_v \in \{0,1\} \qquad \forall\, v \in V. \tag{2c}$$

The variable x_v indicates if v is part of the balanced separator. Our objective is to minimize the cardinality of the balanced separator (2a). Constraints (2b) ensure that each subgraph in $\mathcal{S}$ contains at least one separating vertex.

Model (2) solves Minimum Balanced Separators. Let $X \subseteq V$ be a solution of (2), then every connected component C in $G - X$ fulfills $w(C) \le \alpha \cdot w(G)$, otherwise X would not be feasible since $C \in \mathcal{S}$. Let X^* be a solution of Minimum Balanced Separators. Any subgraph $S \in \mathcal{S}$ has to have a common vertex with X^*, otherwise a connected component $C \supseteq S$ in $G - X^*$ exists with $w(C) > \alpha \cdot w(G)$. Therefore, model (2) is correct. Observe that this model is the well known Hitting Set, which is defined as:

Definition 3. *Let $\mathcal{U}$ be a set and $\mathcal{S} \subseteq \mathcal{P}(\mathcal{U})$. The* **Hitting Set** *problem is to find a set $X \subseteq \mathcal{U}$ such that for any $S \in \mathcal{S}$ holds $X \cap S \neq \emptyset$ and $|X|$ is minimum.*

The algorithm starts by solving Hitting Set$(V, \emptyset)$ and adds iteratively violated constraints, corresponding to the connected components exceeding the upper bound.

Algorithm 1. Minimum Balanced Separators

Input: $G = (V, E), w \in \mathbb{R}^V_{\geq 0}, \alpha \in [0, 1]$
Output: $X \subseteq V$ s.t. X is an α-balanced separator and $|X|$ is minimum.

$S, X \leftarrow \emptyset$
while in $G - X$ exists a connected component C with $w(C) > \alpha \cdot w(G)$ **do**
  for all connected component C in $G - X$ **do**
    if $w(C) > \alpha \cdot w(G)$ **then**
      $S \leftarrow S \cup \{V(C)\}$
  $X \leftarrow$ Hitting Set(V, S)
return X

Proposition 1. *Algorithm 1 is correct.*

Proof. Let X^* be a minimum α-balanced separator and X be a solution from Algorithm 1. Since $G - X$ contains no connected component C with $w(C) > \alpha \cdot w(G)$, X is an α-balanced separator. If $|X^*| < |X|$, then X is not an optimal solution of Hitting Set, because at any iteration X^* is a feasible solution.

In each iteration one or more subgraphs from $\mathcal{S}$ are added to S, otherwise we stop. Since at least one vertex of every subgraph in S is in X no subgraph gets added twice. Therefore, and since $|S|$ is finite, we conclude that Algorithm 1 terminates. □

We have improved Algorithm 1 by only adding minimal connected subgraphs that are violating the upper bound. Let $\mathcal{S}' \subseteq \mathcal{S}$ be the set of all connected subgraphs, such that no proper connected subgraph is in $\mathcal{S}$ as well. Let $X \subseteq V$ be the current solution and C be a connected component in $G - X$ with $w(C) > \alpha \cdot w(G)$. We search for an arbitrary covering of C with subgraphs of $\mathcal{S}'$. This can be done heuristically by a simple breadth first search (BFS) on C. Starting at any vertex we construct a subgraph following the BFS. As soon as the current subgraph violates the upper bound we start to construct a new subgraph from an uncovered vertex. We achieve another improvement by bounding the objective value of Hitting Set in each iteration. Let us say we are at the i-th iteration of the while loop. Let $X' \subseteq V$ be the solution of Hitting Set in the previous iteration. We know that the current solution $X \subseteq V$ fulfills $|X'| \leq |X|$. Therefore, we can add the constraint $\sum_{v \in V} x_v \geq |X'|$.

4 Computational Study

We ran the experiments on machines equipped with Intel Xeon E3-1234 CPUs with 3.7 GHz and 32 GB RAM. The code is written in Python 3.6 and to solve MIPs Gurobi 9.1 is used. The time limit is set to 2 h.

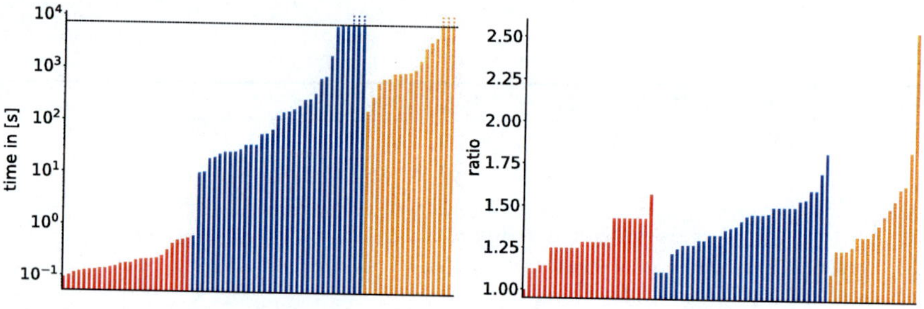

(a) running time of RG

(b) ratio between best objective from MIP and RG

Fig. 1. Computational results of RG and MIP on `tree_lg` (red), `voronoi_medium_lg` (blue), and `voronoi_large_lg` (orange). The order of the instances is sorted independently.

We consider test instances that resemble transit networks of different complexity. All test instances are described and can be found in a GitHub repository[1], and our instance-wise computational results in an online supplement[2]. We only use the node weighted instances labeled with "`_lg`" (for line graph). Inspired by a real world problem from [2], we set parameter α such that $\alpha \cdot w(G) = 100 - \varepsilon$, where $\varepsilon = 10^{-4}$. This leads to values of α ranging from 0.12 to 0.21. The instances are split into three groups, `tree_lg`, `voronoi_medium_lg`, and `voronoi_large_lg`. By MIP we denote model (1) and RG refers to Algorithm 1.

In Fig. 1a we see the running time of RG. The time limit is marked by a horizontal dashed line. If an instance exceeds the time limit, the corresponding bar ends with a dotted line. Our algorithm finds an optimal solution within the time limit for all but 6 instances. Note that the scale is logarithmic and the running times lie in a broad range, even within the same group. The MIP,

1 https://github.com/stephanschwartz/vertex_covering_with_capacitated_trees.
2 https://github.com/williamsurau/finding_minimum_balanced_separators-an_exact_approach.

on the other hand, exceeds the time limit on all 75 instances. Our algorithm is not only much faster but also provides better solutions. In Fig. 1b we see that the incumbent separator set in MIP after 2 h is significantly larger than the optimal solution from RG. Again, the instances are ordered by ratio and grouped. While MIP only finds an optimal solution for one instance it could not find a corresponding dual bound better than 0. In fact, the dual bound for MIP is still 0 on all instances after 7200 s. Moreover, we ran MIP on several selected instances with a time limit of 24 h, and it was not able to close the gap for any of those. Once again, this illustrates the superiority of the presented approach.

References

1. Bhatt, S.N., Leighton, F.T.: A framework for solving VLSI graph layout problems. J. Comput. Syst. Sci. **28**(2), 300–343 (1984)
2. Borndörfer, R., Schwartz, S., Surau, W.: Vertex Covering with Capacitated Trees. Eng. Tech. rep. 21–14. Takustr. 7, 14195 Berlin: ZIB (2021)
3. Brandt, S., Wattenhofer, R.: Approximating small balanced vertex separators in almost linear time. In: WADS 2017. LNCS, vol. 10389, pp. 229–240. Springer, Cham (2017). https://doi.org/10.1007/978-3-319-62127-2_20
4. Nguyen Bui, T., Jones, C.: Finding good approximate vertex and edge partitions is NP-hard. Inf. Process. Lett. **42**(3) 153–159 (1992)
5. Casel, K., Friedrich, T., Issac, D., Niklanovits, A., Zeif, Z.: Balanced crown decomposition for connectivity constraints. arXiv:2011.0452, arXiv.preprint (2020)
6. Clautiaux, F., Guillot, J., Pesneau, P.: Exact approaches for solving a covering problem with capacitated subtrees. Comput. Oper. Res. **105**, 85–101 (2019)
7. Cygan, M., et al.: Parameterized Algorithms, vol. 5. 4. Springer, Cham (2015). https://doi.org/10.1007/978-3-319-21275-3
8. Ziena Elijazyfer. "Längenbeschränkte Teilgraphenbildung zur Maut-Kontrollstreckenoptimierung". Masterthesis. Freie Universität Berlin, 2018
9. Feige, U., Taghi Hajiaghayi, M., Lee, J.R.: Improved approximation algorithms for minimum weight vertex separators. SIAM J. Comput. **38**(2), 629–657 (2008)
10. Holzer, M., Prasinos, G., Schulz, F., Wagner, D., Zaroliagis, C.: Engineering planar separator algorithms. In: Brodal, G.S., Leonardi, S. (eds.) ESA 2005. LNCS, vol. 3669, pp. 628–639. Springer, Heidelberg (2005). https://doi.org/10.1007/11561071_56
11. Rosenberg, A.L., Heath, L.S.: Graph Separators, with Applications. Springer, NewYork (2001). https://doi.org/10.1007/b115747
12. Salazar-Aguilar, M., Róos-Mercado, R.Z., Ríos, M.C.: New models for commercial territory design. Netw. Spatial Econ. **11**(3), 487–507 (2011)
13. de Souza, C.C., Cavalcante, V.F.: Exact algorithms for the vertex separator problem in graphs. Networks **57**(3) 212–230 (2011)
14. Surau, W.: Das homogene längenbeschränkte zusammenhängende Teilgraphenüberdeckungsproblem. Freie Universität Berlin, Bachelorthesis (2020)

Global Pricing and Feasible Directions in Linear Programming

Biressaw Chali Wolde(✉) and Torbjörn Larsson

Department of Mathematics, Linköping University, 58183 Linköping, Sweden
{biressaw.wolde,torbjorn.larsson}@liu.se

Abstract. We present a linear programming approach based on a global pricing function and feasible directions. It is embedded in the framework of the simplex method through the use of external columns, which are combinations of original columns. The global pricing function is composed by the pricing function of the simplex method, which captures the objective's behaviour over the cone of feasible directions, and an exterior penalty function that captures information about the topology of the entire feasible set. Given a non-degenerate basic feasible solution, a global pricing problem yields a non-edge improving feasible direction, which is translated into an external column that enters the basis.

Preliminary computational results indicate that the global pricing principle may have a significant advantage over the ordinary pricing of the simplex method. Further, our new approach allows for several computational strategies, which need to be investigated in future research in order to explore its full potential.

Keywords: Linear program · Pricing · Feasible direction · External pivoting

1 Derivation

Let the matrix $A \in \mathbb{R}^{m\times n}$, with $n > m$, have full rank, let the vectors $b \in \mathbb{R}^m$ and $c \in \mathbb{R}^n$, and consider the Linear Program (LP)

$$\begin{aligned} z^\star = \min\ & z = c^{\mathrm{T}}x \\ \text{s.t. } & Ax = b \\ & x \geq 0. \end{aligned}$$

Assume that LP is feasible and that a basic feasible solution is at hand. It corresponds to a column partitioning $A = (B, N)$, with $B \in \mathbb{R}^{m\times m}$ being the non-singular basis matrix and $N \in \mathbb{R}^{m\times(n-m)}$ being the matrix of non-basic columns. We further partition $c = (c_B^{\mathrm{T}}, c_N^{\mathrm{T}})^{\mathrm{T}}$ and $x = (x_B^{\mathrm{T}}, x_N^{\mathrm{T}})^{\mathrm{T}}$. Then the basic solution is $x_B = B^{-1}b \geq 0$ and $x_N = 0$, the complementary dual solution is $u^{\mathrm{T}} = c_B^{\mathrm{T}}B^{-1}$, and the vector of reduced costs is $\bar{c}_N^{\mathrm{T}} = c_N^{\mathrm{T}} - u^{\mathrm{T}}N$. Letting I_m denote the identity matrix of size m, problem LP is equivalent to

N. Trautmann and M. Gnägi (Eds.): OR 2021, LNOR, pp. 160–166, 2022.
https://doi.org/10.1007/978-3-031-08623-6_25

$$\begin{aligned} z^\star = \min\ & z(x_N) = c_B^{\mathrm{T}} B^{-1} b + \bar{c}_N^{\mathrm{T}} x_N \\ \text{s.t. } & I_m x_B + B^{-1} N x_N = B^{-1} b \\ & x_B,\ x_N \geq 0, \end{aligned}$$

or

$$\begin{aligned} z^\star = \min\ & z(x_N) = c_B^{\mathrm{T}} B^{-1} b + \bar{c}_N^{\mathrm{T}} x_N \\ \text{s.t. } & x_B(x_N),\ x_N \geq 0, \end{aligned} \tag{1}$$

where $x_B(x_N) = B^{-1}b - B^{-1}Nx_N$.

By further using the indicator function $\sigma : \mathbb{R}^m \mapsto \mathbb{R}_+ \cup \{+\infty\}$, with $\sigma(\xi) = 0$ if $\xi \in \mathbb{R}_+^m$ and $\sigma(\xi) = +\infty$ otherwise, problem LP can be equivalently stated as

$$z^\star = \min_{x_N \geq 0} z(x_N) = c_B^{\mathrm{T}} B^{-1} b + \bar{c}_N^{\mathrm{T}} x_N + \sigma(x_B(x_N)).$$

This problem is clearly computationally intractable. The indicator function is therefore approximated with an exterior penalty function $p : \mathbb{R}^m \mapsto \mathbb{R}_+$ which is everywhere continuously differentiable and convex and takes values $p(\xi) = 0$ if $\xi \in \mathbb{R}_+^m$ and $p(\xi) > 0$ otherwise. An example of such a function is the well-known quadratic penalty function $p(\xi) = \frac{\rho}{2} \sum_{i=1}^m (\min\{0, \xi_i\})^2$, where $\rho > 0$ is a penalty parameter.

Definition 1. *Let u be the complementary dual solution to a basic feasible solution to* LP *and let p be a penalty function, as stated above. Then $P : \mathbb{R}^{n-m} \mapsto \mathbb{R}$ with $P(x_N) = \bar{c}_N^{\mathrm{T}} x_N + p(x_B(x_N))$ is a* global pricing function *for* LP. □

The global pricing function includes the ordinary linear pricing function of the simplex method, which captures the behaviour of the objective function over the cone of feasible directions from the given extreme point of the feasible set, and a nonlinear penalty function which captures approximate global information about the topology of the feasible polyhedron of the problem. Based on the global pricing function, we next define a pricing mechanism that takes global information about the feasible set into account.

Definition 2. *Let P be a global pricing function for* LP. *Then the* global pricing problem *for* LP *is given by $P^* = \min_{x_N \geq 0} P(x_N) = \bar{c}_N^{\mathrm{T}} x_N + p(x_B(x_N))$.* □

Note that $P^* \leq P(0) = 0$. Further, since $p(\xi) \geq 0$ for all $\xi \in \mathbb{R}^m$ and $p(\xi) = 0$ for all $\xi \in \mathbb{R}_+^m$, it follows that $P^* = -\infty$ if and only if $z^* = -\infty$.

Our approach relies on four results. Their proofs are straightforward and therefore omitted. We first establish that the global pricing problem is a relaxation of problem LP and that a non-degenerate basic feasible solution is optimal in LP exactly when it also solves the global pricing problem.

Theorem 1. $P^* + c_B^{\mathrm{T}} B^{-1} b \leq z^*$. □

Theorem 2. *A non-degenerate basic feasible solution to problem* LP *is optimal if and only if* $0 \in \arg\min_{x_N \geq 0} P(x_N)$. □

Remark 1. In the non-degenerate case, optimality holds exactly when $\bar{c}_N \geq 0$ and every optimal solution to the global pricing problem is non-zero exactly when $\bar{c}_N \ngeq 0$. In the degenerate case, optimality may hold even though $\bar{c}_N \ngeq 0$ and the global pricing problem has a non-zero optimal solution. □

The global pricing problem can be used to calculate a near-optimal solution to problem LP, by using a sufficiently coercive penalty function, but we do not consider this to be a viable computational strategy. It will instead be used for computing a non-edge feasible direction of descent for LP. Such a direction can be used within the framework of the simplex method by translating it into an *external column* [2]; this is an auxiliary column that is a non-negative linear combination of the original columns of LP. Our approach is related to the works in [4] and [2], which both translate feasible non-edge descent directions into auxiliary variables. The feasible directions constructed in these works are however not based on a direction-finding problem, but on *ad hoc* rules.

We next establish that, under non-degeneracy, any $x_N^* \geq 0$ with $P(x_N^*) < 0$ gives a feasible direction of descent for problem (1), and thus also for LP. We here use the notation $x(\theta) = (x_B(\theta)^{\mathrm{T}}, x_N(\theta)^{\mathrm{T}})^{\mathrm{T}}$ with $x_B(\theta) = B^{-1}b - \theta B^{-1}Nx_N^*$ and $x_N(\theta) = \theta x_N^*$ for some given $x_N^* \geq 0$.

Theorem 3. *Let the basic feasible solution $x_B = B^{-1}b$ and $x_N = 0$ be non-degenerate and non-optimal. Let $x_N^* \geq 0$ be such that $P(x_N^*) < 0$ and $\theta > 0$. Then $\bar{c}_N^{\mathrm{T}} x_N^* < 0$, the solution $x(\theta)$ has objective value $c_B^{\mathrm{T}} B^{-1}b + \theta \bar{c}_N^{\mathrm{T}} x_N^* < c_B^{\mathrm{T}} B^{-1}b$, and it is feasible in* LP *if θ is sufficiently small.* □

Remark 2. Under degeneracy, a descent direction given by an $x_N^* \geq 0$ with $P(x_N^*) < 0$ may be infeasible. Hence, in this case the situation is the same as in the simplex method, where the direction corresponding to an entering variable may be infeasible. □

Theorem 4. *Let the basic feasible solution $x_B = B^{-1}b$ and $x_N = 0$ be non-degenerate and non-optimal, and let $x_N^* \geq 0$ be such that $P(x_N^*) < 0$. Assume that $\left(B^{-1}Nx_N^*\right)_i > 0$ holds for some $i \in \{1, \ldots, m\}$ and define*

$$\theta^* = \min_{i \in \{1,\ldots,m\}} \left\{ \frac{(B^{-1}b)_i}{(B^{-1}Nx_N^*)_i} : \left(B^{-1}Nx_N^*\right)_i > 0 \right\}.$$

Then $\theta^* > 0$, $x(\theta)$ is feasible in LP *if and only if $\theta \in [0, \theta^*]$, θ^* minimizes $z(\theta x_N^*)$ over $\theta \in [0, \theta^*]$, and $x(\theta^*)$ is on the boundary of the feasible set of* LP. □

If $P(x_N^*) < 0$ and $\left(B^{-1}Nx_N^*\right)_i \leq 0$ holds for all $i \in \{1, \ldots, m\}$, then LP clearly has an unbounded optimal value. If the given basic feasible solution is non-degenerate and non-optimal, and x_N^* is optimal in the global pricing problem, then x_N^* is infeasible in LP and $\theta^* < 1$; this is because the penalty function p is exterior and continuously differentiable. If the penalty function is sufficiently coercive and the global pricing problem is solved to near-optimality, then θ^* will be close to one and the solution $x(\theta^*)$ will be near-optimal.

The solution $x(\theta^*)$ is on the boundary of the feasible set but typically not an extreme point of this set, and therefore not associated with a basic feasible solution for LP. To be able to remain within the framework of the simplex method, the feasible descent direction is therefore expressed as a non-negative linear combination of the original columns in LP, called an external column [2].

For an $x_N^* \geq 0$ with $P(x_N^*) < 0$, the external column is given by $c_{n+1} = c_N^{\mathrm{T}} x_N^*$ and $a_{n+1} = N x_N^*$, and LP is replaced by the augmented, but equivalent, problem

$$\begin{aligned} z^* = \min\ & z = c_B^{\mathrm{T}} B^{-1} b + \bar{c}_N^{\mathrm{T}} x_N + \bar{c}_{n+1} x_{n+1} \\ \text{s.t. } & I_m x_B + B^{-1} N x_N + B^{-1} a_{n+1} x_{n+1} = B^{-1} b \\ & x_B,\ x_N,\ x_{n+1} \geq 0, \end{aligned}$$

where $\bar{c}_{n+1} = c_{n+1} - u^{\mathrm{T}} a_{n+1} = c_N^{\mathrm{T}} x_N^* - u^{\mathrm{T}} N x_N^* = \bar{c}_N^{\mathrm{T}} x_N^* \leq P(x_N^*) < 0$ and x_{n+1} is the external variable. Letting the external column enter the basis clearly corresponds to following the feasible direction in problem LP. Further, the external variable will then take the value θ^*. If the external column is basic in an optimal solution to the augmented problem, then an optimal solution to the original problem can easily be recovered [2].

2 Numerical Illustrations

To make a first evaluation of the potential advantage of using global pricing and external columns, we made a straightforward MATLAB implementation of the revised simplex method. Since pivots on external columns lead to solutions that are not extreme points in the feasible set, such pivots are combined with ordinary simplex pivots. We choose to simply generate a single external column at the initial basic feasible solution, and thereafter use the standard simplex method with the Dantzig entering variable criterion. The test problem instances used are randomly generated as described in [2]; they are of maximization type, have positive objective coefficients, and are inequality-constrained with positive right-hand-sides. The simplex method is started with a slack basis.

The global pricing problem can produce external columns that are of high quality compared to the non-basic columns, and thereby reduce the number of simplex iterations, but it is computationally demanding. For larger problem instances it should therefore be of interest to consider restricted versions of the global pricing problem. Letting $\mathcal{N}$ be the index set of the non-basic columns, we consider a $J \subseteq \mathcal{N}$ and define $x_J = (x_j)_{j \in J}$, $\bar{c}_J = (\bar{c}_j)_{j \in J}$, and $N_J = (a_j)_{j \in J}$. The restricted global pricing problem is $\min_{x_J \geq 0} P(x_J) = \bar{c}_J^{\mathrm{T}} x_J + p(x_B(x_J))$, where $x_B(x_J) = B^{-1} b - B^{-1} N_J x_J$. (The restricted problem is clearly not certain to provide a lower bound to the optimal value of problem LP; cf. Theorem 1). The computational cost of finding the external column by solving the restricted global pricing problem and the reduction in simplex computations that it may lead to will clearly depend of the size of J and how it is generated.

The set J is selected in two ways, called *Dantzig selection* and *steepest-edge selection*. Letting $k = |J|$, they select the k best non-basic columns according to the Dantzig and the steepest-edge (e.g., [3]) entering variable criteria, respectively. The latter selection is more computationally expensive but should provide the restricted global pricing problem with a better collection of non-basic columns. The pricing problem is formulated as a quadratic program (by using some auxiliary variables) and solved by the built-in MATLAB solver **quadprog**.

We compare the standard simplex method and our global pricing strategy for various values of k and the two selection criteria. Three instances, with 1000 constraints and 2000, 3000, and 4000 variables, respectively, are used. Their optimal basic solutions include 335, 408, and 428 original variables, respectively. We compare the number of iterations and the runtimes needed to reach optimality for $\rho \in \{10^{-8}, 10^{-6}, 10^{-4}\}$. (The values of ρ of interest are small because of the uneven scalings of the objective and constraints in these test instances). Figure 1 shows convergence histories for the instance of size 1000×2000 for the choice $\rho = 10^{-6}$, when using the two selection criteria and different values of k, as percentages of the number of original columns. The results for the three instances, for the same values of k as used above, are shown in Table 1. We have also tried to use $k > 0.50 \times n$, but the results then obtained are very close to those for $k = 0.50 \times n$.

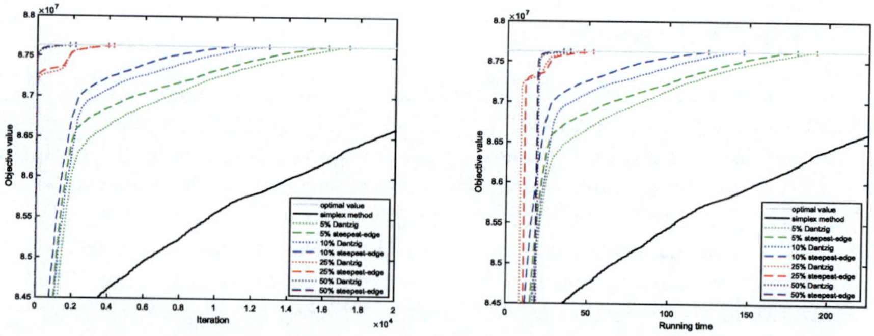

Fig. 1. Objective values versus simplex iterations and runtimes

We first note that the value of the penalty parameter is not very crucial for the overall performance of the global pricing strategy; even relatively small values yield external columns of sufficient quality to considerably reduce iterations and runtimes. We further note that the steepest-edge criterion mostly gives better performance than the Dantzig criterion, which is as expected, and that the most important factor for the overall performance is the value of k. External columns with highest quality are typically obtained when using a relatively large penalty parameter value, the steepest-edge selection criterion, and a large enough restricted global pricing problem. Further, for external columns of high quality the value of the maximal step θ^* is very close to one, and such columns remain in the basis until the late simplex iterations.

Table 1. Simplex iterations and runtimes. Here, (m, n) is the test problem size and e_{iter} is the number of iterations with the external column in the basis

Size		Std simplex		Parameters		Dantzig selection				Steepest-edge selection			
m	n	iterations	time	k/n (%)	ρ	iterations	time	e_{iter}	θ^*	iterations	time	e_{iter}	θ^*
1000	2000	35,181	410.2	5	10^{-8}	15,811	173.5	1,869	0.742308	15,383	170.0	2,363	0.730709
				5	10^{-6}	17,386	192.3	2,291	0.989642	16,225	180.0	2,473	0.991153
				5	10^{-4}	19,380	217.6	2,528	0.999895	17,169	193.1	2,354	0.999914
				10	10^{-8}	14,961	167.2	2,225	0.750753	8,804	101.4	2,123	0.781021
				10	10^{-6}	12,907	146.4	2,483	0.991447	10,977	124.2	2,340	0.993612
				10	10^{-4}	11,832	137.8	2,285	0.999913	10,064	117.6	2,245	0.999935
				25	10^{-8}	5,628	66.7	1,869	0.829007	5,847	66.8	1,942	0.824768
				25	10^{-6}	3,978	47.0	1,970	0.995734	4,281	52.7	1,921	0.995644
				25	10^{-4}	4,306	52.1	2,079	0.999957	3,861	47.5	2,181	0.999956
				50	10^{-8}	5,536	77.2	1,895	0.829990	5,536	76.6	1,895	0.829990
				50	10^{-6}	2,125	38.5	1,622	0.996242	1,802	34.2	1,759	0.996211
				50	10^{-4}	2,313	40.0	1,840	0.999962	1,789	35.4	1,762	0.999962
1000	3000	47,647	762.1	5	10^{-8}	20,154	300.9	3,175	0.585040	15,364	220.9	3,087	0.836747
				5	10^{-6}	20,471	304.5	3,117	0.997036	18,273	262.6	3,043	0.996217
				5	10^{-4}	18,385	271.3	2,707	0.999970	20,712	309.0	3,436	0.999961
				10	10^{-8}	13,318	194.2	2,951	0.874610	9,228	131.1	3,022	0.882759
				10	10^{-6}	12,944	188.2	3,367	0.997960	10,961	157.2	3,069	0.997723
				10	10^{-4}	13,286	195.3	3,028	0.999980	9,995	143.0	3,067	0.999977
				25	10^{-8}	4,514	65.8	2,418	0.914427	4,773	68.9	2,528	0.915966
				25	10^{-6}	4,222	63.2	2,567	0.998881	3,282	49.6	2,464	0.998833
				25	10^{-4}	4,141	62.7	2,824	0.999989	3,908	58.2	2,801	0.999988
				50	10^{-8}	4,199	68.8	2,430	0.918132	4,199	68.5	2,430	0.918132
				50	10^{-6}	2,349	45.2	2,334	0.998820	2,349	46.1	2,334	0.998820
				50	10^{-4}	2,337	46.9	2,336	0.999988	2,337	48.1	2,336	0.999988
1000	4000	57,572	1,126.7	5	10^{-8}	23,919	435.8	3,928	0.912434	20,648	374.8	3,817	0.908613
				5	10^{-6}	25,337	464.1	3,930	0.998762	25,115	462.6	3,962	0.998729
				5	10^{-4}	26,017	477.1	3,997	0.999988	22,753	415.9	3,693	0.999987
				10	10^{-8}	11,928	211.4	3,431	0.929075	9,998	174.9	3,355	0.933425
				10	10^{-6}	12,960	211.6	3,848	0.998898	11,309	198.7	3,641	0.999058
				10	10^{-4}	15,180	267.7	4,167	0.999989	12,808	226.1	3,884	0.999991
				25	10^{-8}	5,379	97.4	2,921	0.947856	5,640	104.6	2,996	0.947902
				25	10^{-6}	4,243	80.4	3,222	0.999363	3,927	76.2	2,946	0.999365
				25	10^{-4}	4,074	77.8	3,033	0.999994	3,817	75.1	3,004	0.999993
				50	10^{-8}	5,834	112.9	3,175	0.948566	5,834	113.5	3,175	0.948566
				50	10^{-6}	2,810	60.4	2,793	0.999309	2,810	65.4	2,793	0.999309
				50	10^{-4}	2,880	64.4	2,879	0.999993	2,880	65.1	2,879	0.999993

3 Conclusions

Our results indicate that the use of the global pricing principle and external columns have the potential to considerably improve the performance of the simplex method with respect to both iterations and computing times; these findings demand for continued research. This includes questions regarding how often external columns should be generated, the choice of size of the restricted global pricing problem, tailored algorithms for this problem (e.g. based on the projected Newton method [1]), and the accuracy to which it should be solved. Further, the global pricing problem must be modified to properly handle degeneracy.

Finally, the strategy of using the global pricing function and external columns should be generalized to a column generation setting.

References

1. Bertsekas, D.P.: Projected Newton methods for optimization problems with simple constraints. SIAM J. Control Optim. **20**, 221–246 (1982)
2. Eiselt, H.A., Sandblom, C.-L.: Experiments with external pivoting. Comput. Oper. Res. **17**, 325–332 (1990)
3. Murty, K.G.: Linear Programming. John Wiley and Sons, New York (1983)
4. Murty, K.G., Fathi, Y.: A feasible direction method for linear programming. Oper. Res. Lett. **3**, 121–127 (1984)

Tight SDP Relaxations for Cardinality-Constrained Problems

Angelika Wiegele and Shudian Zhao(✉)

Institut für Mathematik, Alpen-Adria-Universität Klagenfurt,
Universitätsstraße 65-67, 9020 Klagenfurt, Austria
{angelika.wiegele,shudian.zhao}@aau.at

Abstract. We model the cardinality-constrained portfolio problem using semidefinite matrices and investigate a relaxation using semidefinite programming. Experimental results show that this relaxation generates tight lower bounds and even achieves optimality on many instances from the literature. This underlines the modeling power of semidefinite programming for mixed-integer quadratic problems.

Keywords: Semidefinite programming · Cardinality-constrained problem · Mixed-integer nonlinear programming

1 Introduction

The cardinality-constrained optimization problem is widely applied in areas of finance such as the portfolio optimization problem where the number of stocks to be traded is bounded. It is even NP-complete to test the feasibility of an optimization problem with cardinality constraints [1].

There are various previous studies about cardinality-constrained problems. Frangioni and Gentile [4] and Zheng et al. [11] work with the approaches that use diagonalizations and perspective cuts to improve the continuous relaxation bound in order to deal with the cardinality-constrained problems with lower bounds for nonzero stocks. Burdakov et al. [3] have proposed nonconvex relaxations that still have the same solutions (in the sense of global minima), and then tools for minimization problems in continuous variables are applied to solve the relaxations. This problem also lies in the family of quadratic programming problems with complementary constraints, where Braun and Mitchell [2] have introduced heuristics via semidefinite programming to generate upper bounds for this problem.

Semidefinite programming belongs to the field of convex optimization. A semidefinite programming (SDP) problem can be solved in polynomial time. SDP has shown advantages in approximating integer or mixed-integer nonlinear

This project has received funding from the European Union's Horizon 2020 research and innovation programme under the Marie Skłodowska-Curie grant agreement MINOA No 764759.

N. Trautmann and M. Gnägi (Eds.): OR 2021, LNOR, pp. 167–172, 2022.
https://doi.org/10.1007/978-3-031-08623-6_26

programming problems. In particular, quadratic problems can be quite naturally modeled by semidefinite programming. For quadratically constrained quadratic programming (QCQP), the optimal solution of the SDP relaxation is tight when the QCQP problem is convex [10]. Interior point methods (IPMs) are commonly used to solve SDP problems and implemented in SDP solvers such as Mosek [9]. IPMs can solve SDP problems to high precision as long as the number of constraints and the size of the matrices is reasonable.

Notation. The notation $[n]$ stands for the set of integers $\{1, \ldots, n\}$. We denote by $\mathcal{S}^n$ the set of symmetric $n \times n$ matrices. The operation $\operatorname{diag}(M)$ maps the diagonal entries of matrix M to a vector. We denote by $\langle \cdot, \cdot \rangle$ the trace inner product. That is, for any $M, N \in \mathbb{R}^{m \times n}$, we define $\langle M, N \rangle := \operatorname{trace}(N^\top M)$. We write $X \succeq Y$ if matrix $X - Y$ is positive semidefinite.

2 The Cardinality-Constrained Portfolio Optimization Problem

Given n stocks and the covariance matrix Q, a profit vector μ and the minimum expected return ρ. The objective is to find a portfolio that minimizes the risk while a minimum expected return is achieved. A mathematical programming formulation is as follows.

$$\begin{aligned} \min\ & x^\top Q x \\ \text{s.t. } & \mu^\top x \geq \rho, \\ & e^\top x = 1, \\ & 0 \leq x_i \leq u_i, \ \forall i \in [n], \end{aligned} \tag{1}$$

where x indicates the weights for each stock of a portfolio which is nonnegative and bounded by u, $x^\top Q x$ is the standard deviation of the portfolio, which is used to measure the risk of that portfolio, and $\mu^\top x$ is the expected return.

One commonly used constraint for a portfolio problem is the cardinality constraint, in other words, the maximum number of stocks to be chosen.

$$\begin{aligned} \min\ & x^\top Q x \\ \text{s.t. } & \mu^\top x \geq \rho, \\ & e^\top x = 1, \\ & 0 \leq x_i \leq u_i, \ \forall i \in [n], \\ & \mathbf{Card}(x) \leq \aleph, \end{aligned} \tag{2}$$

where $\mathbf{Card}(x)$ indicates the number of nonzero components of x.

Problem (2) can be formed as a mixed-integer nonlinear programming problem as follows [3].

$$\begin{aligned}\min_{x,y}\ & x^\top Q x \\ \text{s.t.}\ & \mu^\top x \geq \rho, \\ & e^\top x \leq 1, \\ & e^\top y \geq n - \aleph, \\ & x_i y_i = 0,\ \forall i \in [n], \\ & y_i \in \{0,1\},\ \forall i \in [n], \\ & 0 \leq x_i \leq u_i,\ \forall i \in [n],\end{aligned} \tag{3}$$

where the slack variables $y \in \{0,1\}^n$ are integer and the complementary constraints $x_i y_i = 0$ relates x and y. Hence, x_i and y_i cannot be both positive, and the constraint $e^\top y \geq n - \aleph$ requires that y has at least $n - \aleph$ nonzero elements, in return, then x has at most $\aleph$ nonzero elements.

This problem is NP-hard [7]. When $\aleph$ is small, an optimal solution can be found easily by global search, while it gets more complicated to solve when $\aleph$ increases.

3 A Semidefinite Programming Relaxation

Before introducing a relaxation using semidefinite programming, note that in problem (3) we have $x^\top Q x = \langle x, Qx\rangle = \langle Q, xx^\top\rangle$. Moreover, $y_i \in \{0,1\}$ implies $y_i^2 = y_i$ and hence the diagonal of matrix $yy^\top$ must be equal to y. The condition $x_i y_i = 0$ translates to the diagonal of matrix $xy^\top$ being 0.

We now introduce matrices X and Y, substitute the term $xx^\top$ by the symmetric matrix X, i.e., $X = xx^\top$, and relax this condition to $X \succeq xx^\top$. Similarly, we relax $Y = yy^\top$ to $Y \succeq yy^\top$.

Then, with the Schur complement we have

$$X \succeq xx^\top,\ Y \succeq yy^\top \iff \exists Z \in \mathcal{S}^n \text{ s.t. } \begin{pmatrix} 1 & x^\top & y^\top \\ x & X & Z \\ y & Z^\top & Y \end{pmatrix} \succeq 0.$$

Hence, the final SDP relaxation is given as

$$\begin{aligned}\min_{\bar{X}}\ & \langle Q, X\rangle \\ \text{s.t.}\ & \mu^\top x \geq \rho, \\ & e^\top x \leq 1, \\ & e^\top y \geq n - \aleph, \\ & \operatorname{diag}(Z) = \mathbf{0}, \\ & \operatorname{diag}(Y) = y, \\ & 0 \leq x_i \leq u_i,\ \forall i \in [n], \\ & \bar{X} \succeq 0,\end{aligned} \tag{4}$$

where $\bar{X} = \begin{pmatrix} 1 & x^\top & y^\top \\ x & X & Z^\top \\ y & Z & Y \end{pmatrix}$. The positive semidefinite matrix in (4) is of dimension $2n + 1$.

Madani et al. [8] proved that any polynomial optimization problem can be reformulated as a quadratically constrained quadratic problem (with certain sparsity properties) with a corresponding SDP relaxation having on optimal solution with rank at most two. Therefore, we can expect that the SDP relaxation (4) will yield strong bounds. In the next section we will evaluate the quality of the bounds of the SDP relaxation on problem (3).

4 Numerical Results

We performed numerical experiments in order to evaluate the quality of the lower bounds that we obtain from the SDP relaxation (4). We used python and ran the tests on a ThinkPad-X1-Carbon-6th with 8 Intel(R) Core(TM) i7-8550U CPU @ 1.80 GHz.

The model data Q, μ, ρ, and upper bounds u are from the instances in the paper of Frangioni and Gentile [4] and can be found at the website [5].

We use Gurobi 9.1.2 [6] to solve the MIQP problem (3). We report the gap between the upper and lower bounds found by Gurobi after a time limit of 90 s. The lower bounds from the SDP relaxation (4) are obtained by Mosek [9].

The code can be downloaded from https://github.com/shudianzhao/cardinality-SDP.

Table 1. Average results for Gurobi solving (3) within a time limit of 90 s and Mosek for solving (4) on data with $n \in \{200, 300, 400\}$; 30 instances are tested for each pair of n and $\aleph$

$\aleph$	n	Gap (MIQP (3))			Gap (SDP (4))			
		Min %	Avg %	Max %	Min %	Avg %	Max %	Time (s)
5	200	63.92	84.05	91.05	0.00	0.02	0.23	4.05
	300	0.00	86004	93.55	0.00	0.03	0.48	13.27
	400	58.81	91.62	95.09	0.00	0.00	0.03	30.11
10	200	63.40	75.78	83.98	0.00	0.02	0.15	4.53
	300	13.71	80.30	88.25	0.00	0.02	0.47	12.30
	400	56.62	85.83	90.57	0.00	0.01	0.05	27.92
20	200	4163	59.38	70.53	0.00	0.01	0.06	4.25
	300	0.00	67.37	78.60	0.00	0.01	0.13	12.05
	400	33.21	74.85	82.74	0.00	0.01	0.03	27.74

Table 1 shows the overall performance on instances orl$n$-005- and pard$n$-005- with $n \in \{200, 300, 400\}$ and $\aleph \in \{5, 10, 20\}$. We test 30 instances for each pair of n and $\aleph$.

The optimality gap is measured as difference between the lower bound (lb) and the best found upper bounds (ub). We compute the relative gap as $(ub - lb)/ub$, where the upper bounds ub are found by Gurobi and the lower bounds lb are obtained from relaxations solved in Gurobi and the SDP relaxation (4) solved by Mosek.

When the time limit is reached, the gap for Gurobi solving (3) is typically between 60% and 80%. There is only one instance, namely orl300-05-i, that can be solved by Gurobi within 90 s for $\aleph \in \{5, 20\}$.

In contrast to that, the lower bounds from the SDP relaxation (4) are exceptional. The gaps between the SDP lower bounds and the best upper bounds found by Gurobi are closed for most of the instances.

Actually, 96% of the SDP solutions have rank one, and hence the optimal solutions for problem (3) can be built from it. The computation times for solving the SDP problems are less than 40 s, even for large instances where $n = 400$ and hence the size of the SDP matrix is 801.

These results are another evidence of the strong modeling power of semidefinite programming, in particular when quadratic functions and binary variables are present.

5 Conclusion

Numerical results illustrate the strong modeling power of semidefinite programming for solving the cardinality-constrained portfolio optimization problem. This shows a promising direction for solving combinatorial optimization problems arising from the area of finance.

References

1. Bienstock, D.: Computational study of a family of mixed-integer quadratic programming problems. Math. Program. **74**(2), 121–140 (1996)
2. Braun, S., Mitchell, J.E.: A semidefinite programming heuristic for quadratic programming problems with complementarity constraints. Comput. Optim. Appl. **31**(1), 5–29 (2005)
3. Burdakov, O.P., Kanzow, C., Schwartz, A.: Mathematical programs with cardinality constraints: reformulation by complementarity-type conditions and a regularization method. SIAM J. Optim. **26**(1), 397–425 (2016)
4. Frangioni, A., Gentile, C.: SDP diagonalizations and perspective cuts for a class of nonseparable MIQP. Oper. Res. Lett. **35**(2), 181–185 (2007)
5. Frangioni, A., Gentile, G.: Mean-variance problem with minimum buy-in constraints (2017). data and documentation. http://www.di.unipi.it/optimize/Data/MV.html
6. Gurobi Optimization, LLC: Gurobi optimization reference manual. Version 9.1 (2021). https://www.gurobi.com
7. J Júdice, J., Faustino, A.: The linear-quadratic bilevel programming problem. INFOR: Inf. Syst. Oper. Res. **32**(2), 87–98 (1994)

8. Madani, R., Fazelnia, G., Lavaei, J.: Rank-2 matrix solution for semidefinite relaxations of arbitrary polynomial optimization problems. Technical report, Columbia University, New York (2014)
9. MOSEK ApS. MOSEK Optimizer API for Python. Version 9.1.13. (2020). http://docs.mosek.com/9.1.13/toolbox/index.html
10. Park, J., Boyd, S.: General heuristics for nonconvex quadratically constrained quadratic programming. arXiv preprint, https://doi.org/10.48550/arXiv.1703.07870 (2017)
11. Zheng, X., Sun, X., Li, D.: Improving the performance of MIQP solvers for quadratic programs with cardinality and minimum threshold constraints: a semidefinite program approach. INFORMS J. Comput. **26**(4), 690–703 (2014)

Energy and Environment

Optimal Trading of Flexible Power Consumption on the Day-Ahead Market

Neele Leithäuser(✉), Till Heller, Elisabeth Finhold, and Florian Schirra

Fraunhofer ITWM, 67663 Kaiserslautern, Germany
{neele.leithaeuser,till.heller,elisabeth.finhold,
florian.schirra}@itwm.fraunhofer.de

Abstract. As power generators shift from inert power plants to more volatile renewable sources of electricity, variable loads allow energy intensive businesses to monetize their flexibility in the electricity market. In the aluminum industry, engineering innovations have enabled such flexibility. A virtual battery can balance changing power inflows over a period of a few days. We have modeled the optimization problem of when and how to use this flexibility on the day-ahead energy market based on hourly price forecasts as a linear program. Besides the expected revenue, we also consider technical costs which occur in an unsteady energy schedule. In order to account for realistic trading settings, we embed the optimization problem in a daily rolling horizon planning framework. In this work, we generalize the specific planning problem to a broader range of applications and analyze the main influences on a profitable trading strategy. For a typical parametrization, we present a constructive heuristic that does not require an LP-Solver.

1 Introduction

In the past, the majority of power plants was inert and the variation of loads was expensive. As a result, grid operators were keen for energy-intensive industry to keep their load constant. However, due to the steady expansion of renewable energies as part of the energy transition, more volatile and uncontrollable generators such as wind or solar power plants will dominate the power supply. Industries that are able to flexibilize their loads according to price signals (which is known as *demand side management* (DSM[1])) assist the grid stability, which is rewarded by yielding profits on the various energy markets.

While energy is traded on different future markets, in this work, we solely focus on the German DA market at the European Power Exchange (EPEX SPOT SE), which is a single auction with hourly products and a market clearing prices that all successful bidders will pay or receive, respectively. While more complex bids are possible (the detailed rule set can be found in [2]), for the rest of this

[1] For a general introduction to DSM, we refer to a survey by Zhang and Grossmann [1].

N. Trautmann and M. Gnägi (Eds.): OR 2021, LNOR, pp. 175–181, 2022.
https://doi.org/10.1007/978-3-031-08623-6_27

work, we only consider the common strategy of mere singular stepwise single-contract orders[2] bids in our model.

However, the optimal trading problem on the DA market is a complex problem, which is well-studied in its general form (e.g. see [3]).

The TRIMET Aluminium SE has successfully flexibilized their formerly steady process by enabling the variation of loads: the virtual energy storage allows for balancing loads to a certain extent [4]. However, calling this flexibility comes at the price of higher maintenance costs and reduced process efficiency. The trade optimization of this setting has been studied in an offline approach in [5].

Unlike related research for this setting [5], we will explore the profit potential of this setting in a rolling horizon approach, and, more importantly, we will generalize the specific problem and perform a parameter study for alternative industrial processes with similar conditions. For the sake of brevity, we neglect the explicit consideration of price risks here, however, their existence plays a major role in the long term profit, which is analyzed in the rolling horizon framework.

2 Model

We consider a general process with a time-invariant baseline load of b, from which the load may deviate by l_t, u_t within an interval $[b - l_t, b + u_t]$ at hour t. Whenever the actual load is above (below) the baseline value, the virtual storage (also referred to as battery) is charged (discharged) proportionally. The battery has a given capacity range of $[\kappa_{\min}, \kappa_{\max}]$ which must not be exceeded. We assume switching costs (emissions, maintenance fees, process complexity) c_{sw} that are proportional to the absolute change in load. Further, the decrease in efficiency is modeled via costs c_{ef} proportional to the absolute charging level deviation of the battery w.r.t. to its balanced base level 0.

Given a price prognosis $[p_t]_{t=1}^{T}$ for a certain lookahead time of T hours, transaction costs proportional to the traded volume θ, and an initial battery load of κ_0, we model the deterministic problem by the linear program **LP 1** (Fig. 1).

Here, v_t is the volume of power that is bought (if positive) or sold (if negative) at hour t. Since we assume a liquid market, prices for buying and selling are symmetric. Equations (1c) model the loss or profit of the trades reduced by the transaction costs. Equations (1d) ensure that the battery level remains within the feasible capacity limits. Equations (1e) is the optional constraint to force the battery level at the end of the planning time back to the base level ($L = 0$) or another predefined level L. Equations (1f) model the efficiency cost induced by hours where the battery is not on the base level. Equations (1g) model the switching costs caused by changing load levels. In order to resolve the absolute values, due to the structure of the equations, one can reformulate the equations as linear inequalities with positive and negative parts.

[2] In this way, bidding prices will be low/high enough in order to accept all reasonable clearing prices with one price-invariant amount.

$$
\begin{aligned}
\min \quad & C^{\mathrm{trd}} + C^{\mathrm{eff}} + C^{\mathrm{swt}} && \text{(1a)} \\
\text{s.t.} \quad & l_t \le v_t \le u_t \qquad \forall t = 1, \ldots, T && \text{(1b)} \\
& \sum_{t=1}^{T} \left(p_t v_t + \theta \left| v_t \right| \right) \le C^{\mathrm{trd}} && \text{(1c)} \\
& \kappa_{\min} \le \kappa_0 + \sum_{t=1}^{k} (v_t - b) \le \kappa_{\max} \qquad \forall k = 1, \ldots, T && \text{(1d)} \\
& \left[\sum_{t=1}^{T} (v_t - b) = L \right] && \text{(1e)} \\
& \sum_{t=1}^{T} \left(c_{\mathrm{ef}} \left| \sum_{k=1}^{t} (v_k - b) \right| \right) \le C^{\mathrm{eff}} && \text{(1f)} \\
& c_{\mathrm{sw}} \sum_{t=1}^{T-1} \left| v_t - v_{t+1} \right| \le C^{\mathrm{swt}} && \text{(1g)} \\
& v_t \in \mathbb{R} && \text{(1h)}
\end{aligned}
$$

Fig. 1. Linear program **LP 1** for the deterministic optimal trading problem for a fixed lookahead.

3 Methods

In the basic approach, to which we restrict ourselves here, we solve problem **LP 1** with an LP-solver[3] on a rolling basis: every day, we calculate an optimal schedule for a fixed planning period with a daily updated price forecast. However, only the bids of the first day are actually set on the day-ahead market. The remaining parts of the solution are discarded as they only serve as a preview of the relative development of the price level to better evaluate the prices of the next day. Note that the initial battery level gets updated every day with the first day's final level. Also note that in general there is no proof of optimality or competitiveness for the resulting schedule after a number of iterations with respect to the *ex post* offline schedule for the total timespan, even if the price forecasts never change.

3.1 Quantile Heuristic

In our setting, the trading profit dominates the technical costs. Note that otherwise, remaining at the baseline and not trading at all would be a dominant strategy. Analyzing the trading schedules in detail reveals a hence unsurprising pattern: although the trading amount is modeled as a continuous variable, almost all equations (1b) become tight in an optimal solution.

We further observe that the optimal solution contains frequent intervals over which a simple threshold price divides the hours into *selling* or *buying* hours.

[3] We use Gurobi 9.0.

Neglecting the technical costs, we exploit this behavior in the so-called *quantile heuristic*. For this, let κ_0 be the current battery level and $r_{bat} = \frac{\kappa_0 - \kappa_{min}}{\kappa_{max} - \kappa_{min}}$ the ratio of the battery fill level. As a threshold, we now use the r_{bat}-quantile of the hourly price forecasts over the next seven days. In an hour with price forecast above the threshold, we buy the maximum possible volume (under the given technical restrictions). If the forecast is below the threshold, we sell the maximum feasible volume. The idea behind this approach is that if we already are at a high battery level, we want to use the remaining capacity to buy in the hours that yield highest expected profits. For example, if the battery is already 80% charged, we require our price forecast to be among the lowest 20% to buy.

3.2 Price Forecasts

In the present study, we use our own machine learning based forecast model for the generation of electricity price prognosis. We train a neural network with five years of electricity price data as well as forecasts of wind and solar energy infeed. Analogous to meteorological forecasts, for each forecast, we evaluate the model on a daily updated ensemble forecast of the renewable energy shares and compute their mean value. For more details on our model see [6].

4 Results

In the following, we will give a brief idea on how the choice of parameters and strategies influences the longterm profit on the German DA market. We first specify a reference scenario as in Table 1. Note that the values are inspired by the TRIMET use case, but do not reflect their real cost structure.

Table 1. Baseline scenario parameter settings.

Parameter	Variable	Value	Unit
Flexibility	$[l_t, u_t]$	[68,112]	MW
Initial battery level	κ_0	0	MWh
Battery capacity	$[\kappa_{min}, \kappa_{max}]$	[−1056,1056]	MWh
Baseline load	b	90	MW
Prognosis prices	$[[p_t^i]_{t=1}^{14}]_{i=1}^n$	forecast	EUR
Efficiency costs	c_{sw}	0.1	EUR
Switching costs	c_{ef}	1	EUR
Transaction costs	θ	0.05	EUR
Number of iterations	n	182	-

We use the time period of 2020-01-01 until 2020-06-30 for the following results. The evolution of profit for this scenario is plotted in Fig. 2 in comparison to the profits that are achieved by the *quantile heuristic* from Sect. 3.1.

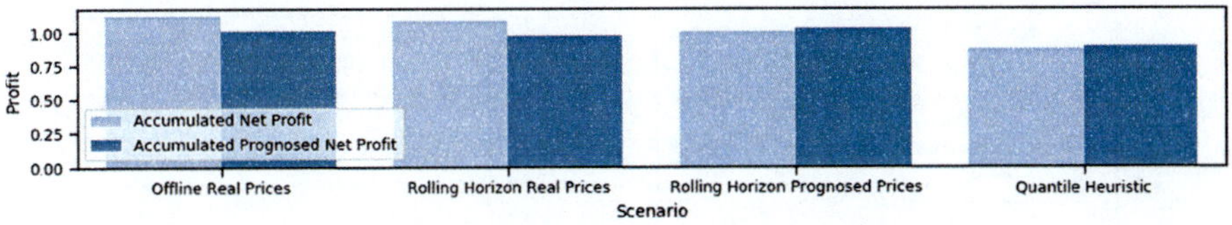

Fig. 2. Comparison of accumulated profit over six months for different algorithmic approaches. The two scenarios with real prices only serve to estimate the theoretical optimum, since these are not yet known at the time of the decision.

We evaluate the *accumulated net profit* as the sum of hourly trade profit reduced by technical costs on the realized clearing prices. Analogously, the *accumulated predicted net profit* is calculated on the first days of the respective price forecasts. To simplify comparability, we state all profits relative to the realized profit of the baseline scenario. Starting from this scenario, we vary one of the following

- the lookahead from 1 to 14 days in steps of 1 day.
- the flexibility from $[89, 91]$ MW to $[65, 115]$ MW in 1 MW steps.
- the battery size from $[0, 0]$ MWh to $[-1450, 1450]$ MWh in 50 MWh steps.
- the efficiency costs from 0 EUR to 1 EUR in steps of 0.02 EUR.
- the switching costs from 0 EUR to 10 EUR in steps of 0.2 EUR.

We also distinguish three battery strategies, namely *AllDev*, *NoDev*, and *FixDev*, which refer to omitting the battery restrictions (1e), setting $L = 0$, or $L = \kappa_0$, respectively. Note that the value of κ_0 changes in every iteration. Hence, for the strategy *FixDev*, one always tries to return to the start level of the battery, while in the *NoDev* scenario, one always aims for returning to the baseload level. The results can be seen in Fig. 3. Note that the profit hardly differs for the individual battery strategies for a lookahead of at least seven days. Since we keep the value of 14 days as a fixed value in all other parameter studies, we restrict ourselves only to the *AllDev* scenario there for the sake of clarity. The study shows that the accumulated profit based on the solution evaluation on the realized prices (darker blue) is not too far away from the ones evaluated on the concatenated prognosis prices of the respective next day prognosis (lighter blue), but the latter clearly tends to be too optimistic.

5 Discussion and Outlook

As displayed in Fig. 3, it is crucial to have a lookahead period that covers at least a full week in order to account for weekly patterns. However, with very large lookaheads, the profit will not increase anymore due to the inaccuracy of the forecasts. Similarly, the battery capacity is only a bottleneck if it is very small with respect to the flexibility. A further increase of battery capacity does not help, since the process off the baseline is rather expensive due to the efficiency costs. In the analyzed spectrum, the flexibility range has a linear effect on the

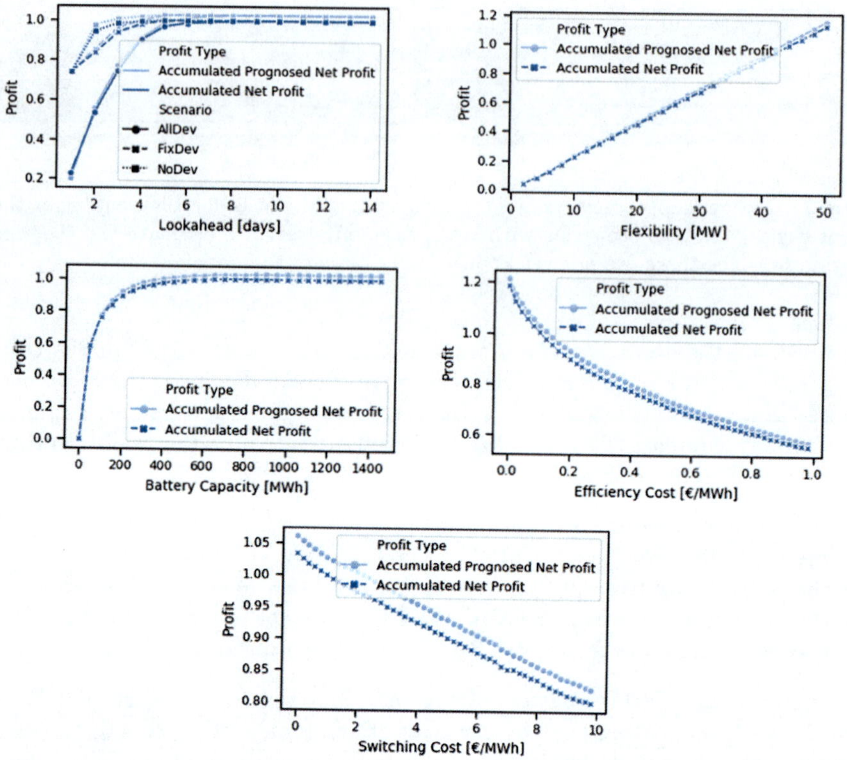

Fig. 3. Analysis of the change in profitability for parameter variations (cet. par.), w.r.t. the baseline scenario setting over six months.

profit. The technical cost parameters have different effects on the net profit. Although the switch costs are much higher, the effect on the net profit is much less. Due to the lack of space, we do not discuss the explicit consideration of price risk due to incorrect forecasts nor the combined optimization on different markets (cross-market optimization) nor the multi-objective optimization of the individual cost factors here. However, we do currently study all of these aspects in order to increase the practical value of the model.

References

1. Zhang, Q., Grossmann, I.E.: Enterprise-wide optimization for industrial demand side management: fundamentals, advances, and perspectives. Chem. Eng. Res. Des. **116**, 114–131 (2016)
2. EPEX Spot Operational Rules. Accessed 14 July 2021
3. Faria, E., Fleten, S.-E.: Day-ahead market bidding for a Nordic hydropower producer: taking the Elbas market into account. Comput. Manag. Sci. **8**(1–2), 75–101 (2011)

4. Depree, N., Düssel, R., Patel, P., Reek, T.: The virtual battery — operating an aluminium smelter with flexible energy input. In: Williams, E. (ed.) Light Metals 2016, pp. 571–576. Springer, Cham (2016). https://doi.org/10.1007/978-3-319-48251-4_96
5. Schäfer, P., Westerholt, H.G., Schweidtmann, A.M., Ilieva, S., Mitsos, A.: Model-based bidding strategies on the primary balancing market for energy-intense processes. Comput. Chem. Eng. **120**, 4–14 (2019)
6. Wagner, A., Ramentol, E., Schirra, F., Michael, H.: Short- and long-term forecasting of electricity prices using embedding of calendar information in neural networks. J. Commod. Markets 100246 (2022). ISSN 2405-8513. https://doi.org/10.1016/j.jcomm.2022.100246, https://www.sciencedirect.com/science/article/pii/S2405851322000046

Blending Hydrogen into Natural Gas: An Assessment of the Capacity of the German Gas Grid

Jaap Pedersen[1(✉)], Kai Hoppmann-Baum[1,2], Janina Zittel[1], and Thorsten Koch[1,2]

[1] Zuse Institute Berlin, Berlin, Germany
pedersen@zib.de
[2] Technical University Berlin, Berlin, Germany

Abstract. In the transition towards a pure hydrogen infrastructure, repurposing the existing natural gas infrastructure is considered. In this study, the maximal technically feasible injection of hydrogen into the existing German natural gas transmission network is analysed with respect to regulatory limits regarding the gas quality. We propose a transient tracking model based on the general pooling problem including linepack. The analysis is conducted using real-world hourly gas flow data on a network of about 10,000 km length.

Keywords: Hydrogen blending · Linear programming · Energy policy and planning

1 Introduction

With its National Hydrogen Strategy (NHS) [1], the German government defines a first general framework for a future hydrogen economy. In the transition phase towards a pure hydrogen infrastructure, there is a growing interest in blending hydrogen into the existing natural gas grid, ensuring a guaranteed outlet, and incentivizing hydrogen production [2]. To determine the maximal feasible amount of hydrogen that can be injected into the existing natural gas grid, we propose a transient hydrogen propagation model based on the pooling problem [3], that takes different regulatory and technical limits into account. The assessment is performed a posteriori on measured hourly gas flow data of one of Europe's largest transmission network operators.

2 Hydrogen Propagation Model

Let $G = (V, A)$ be a directed graph representing the gas network, where V denotes the set of nodes and A the set of arcs. The set of nodes V consists of entry nodes V^+, exit nodes V^-, and inner nodes V^0, i.e. $V = V^+ \dot{\cup} V^- \dot{\cup} V^0$. The set of arcs A consists of the set of pipelines $A^{\text{pi}} \subseteq A$ and the set of non-pipe elements

N. Trautmann and M. Gnägi (Eds.): OR 2021, LNOR, pp. 182–187, 2022.
https://doi.org/10.1007/978-3-031-08623-6_28

$A^{\mathrm{np}} := A \backslash A^{\mathrm{pi}}$. Moreover, we further distinguish the set of compressing arcs, a subset of the non-pipe elements, i.e. $A^{\mathrm{cs}} \subseteq A^{\mathrm{np}}$. For a discrete-time horizon $T := \{t_1, t_2, \dots, t_k\}$, where $\tau_t = t_t - t_{t-1}$ denotes the elapsed time between two subsequent time steps t and $t-1$, measured gas flow data is provided for each time step $t \in T$. In particular, each entry $i \in V^+$ has a given inflow $s_{i,t} \in \mathbb{R}_{\geq 0}$, and each exit $j \in V^-$ has a given outflow $d_{j,t} \in \mathbb{R}_{\geq 0}$ for each time step $t \in T$. For each pipeline $a = (l, r) \in A^{\mathrm{pi}}$, the flow into a at l and out of a at r is given and denoted by $f^l_{a,t}$, $f^r_{a,t} \in \mathbb{R}$ at time $t \in T$, respectively. The amount of gas stored in each pipeline $a \in A^{\mathrm{pi}}$, also known as *linepack*, is denoted by $F_{a,t}$ for each time step $t \in T$. For each non-pipe element $a \in A^{\mathrm{np}}$, a single flow value $f_{a,t}$ is given at time $t \in T$. Let A^{ind}_i be the set of arcs $a \in A$ incident to node i. W.l.o.g, we assume that each entry $i \in V^+$ has exactly one adjacent arc with flow greater or equal than zero. Additionally, a subset of entry nodes $V^+_{H_2} \subseteq V^+$ is selected where hydrogen injection is possible. The variable $\tilde{w}_{i,t} \in [0, q^{UB}_{i,t}]$ denotes the hydrogen mass fraction for each node $i \in V$ at time $t \in T$, where $q^{UB}_{i,t} \in [0, 1]$ represents an upper bound on it. Additionally, let $w_{a,t}, w^l_{a,t}, w^r_{a,t} \in [0, 1]$ denote the hydrogen mass fraction in the pipe, at l, and at r for each pipeline $a = (l, r) \in A^{\mathrm{pi}}$ at time $t \in T$, respectively. Throughout this paper, the term *fraction* refers to the mass fraction. However, it is easily translated to the *volume fraction* (vol.-%), which is usually used to describe regulatory and technical limits, and vice versa.

2.1 Mixing in Nodes

The mixing process is described as follows: The hydrogen fraction at a node is the amount of hydrogen flowing into the node divided by the total amount of gas entering it. The amount of hydrogen flowing into a node i is given by the product of the flow $f^{in}_{i,a,t}$ and its hydrogen fraction $w^{in}_{i,a,t}$ into node i over arc a at time step t. In particular, we formulate the mixing process as follows

$$\tilde{w}_{i,t} \sum_{a \in A^{ind}_i} f^{in}_{i,a,t} = \sum_{a \in A^{ind}_i} w^{in}_{i,a,t} f^{in}_{i,a,t} \qquad \forall i \in V \backslash V^+, \forall t \in T \tag{1}$$

$$f^{in}_{i,a,t} = \begin{cases} f_{a,t}, & a = (j,i) \in A^{\mathrm{np}} \wedge f_{a,t} > 0 \\ -f_{a,t}, & a = (i,j) \in A^{\mathrm{np}} \wedge f_{a,t} < 0 \\ f^i_{a,t}, & a = (j,i) \in A^{\mathrm{pi}} \wedge f^i_{a,t} > 0 \\ -f^i_{a,t}, & a = (i,j) \in A^{\mathrm{pi}} \wedge f^i_{a,t} < 0 \\ 0, & \text{otherwise} \end{cases} \qquad \forall a \in A^{ind}_i \tag{2}$$

$$w^{in}_{i,a,t} = \begin{cases} \tilde{w}_{j,t}, & a = (j,i) \in A^{\mathrm{np}} \wedge f_{a,t} > 0 \\ \tilde{w}_{j,t}, & a = (i,j) \in A^{\mathrm{np}} \wedge f_{a,t} < 0 \\ w^i_{a,t}, & a = (j,i) \in A^{\mathrm{pi}} \wedge f^i_{a,t} > 0 \\ w^i_{a,t}, & a = (i,j) \in A^{\mathrm{pi}} \wedge f^i_{a,t} < 0 \\ 0, & \text{otherwise.} \end{cases} \qquad \forall a \in A^{ind}_i \tag{3}$$

2.2 Mixing and Linepack in Pipelines

To keep track of the amount of hydrogen in the network over time, we need to consider the amount of gas stored in each pipeline $a = (l, r) \in A^{\text{pi}}$, i.e., the linepack. We assume instant mixing, i.e., the hydrogen fraction is always equal over the whole length of the pipeline and adapts immediately. In particular, the hydrogen fraction in the pipeline is described by

$$w_{a,t}F_{a,t} = w_{a,t-1}F_{a,t-1} + w^l_{a,t}f^l_{a,t}\tau_t - w^r_{a,t}f^r_{a,t}\tau_t \quad \forall a = (l, r) \in A^{\text{pi}}, \forall t \in T \quad (4)$$

$$w^r_{a,t} = \begin{cases} w_{a,t}, & f^r_{a,t} \geq 0 \\ \tilde{w}_{r,t}, & f^r_{a,t} < 0 \end{cases} \quad \forall a = (l, r) \in A^{\text{pi}}, \forall t \in T \quad (5)$$

$$w^l_{a,t} = \begin{cases} \tilde{w}_{l,t}, & f^l_{a,t} \geq 0 \\ w_{a,t}, & f^l_{a,t} < 0 \end{cases} \quad \forall a = (l, r) \in A^{\text{pi}}, \forall t \in T, \quad (6)$$

where (4) describes the change of linepack over time as well as the mixing process in the pipeline and (5) and (6) consider the flow direction w.r.t the hydrogen fraction at l and r of pipe $a = (l, r) \in A^{\text{pi}}$.

2.3 Objective and Complete Model

As our goal is to assess the hydrogen capacity of the gas network, the objective is to maximize the technically feasible hydrogen injection, i.e. hydrogen fraction $\tilde{w}$ multiplied by the gas inflow s, for the given set of entry nodes $V^+_{H_2} \subseteq V^+$ over the considered time horizon T. However, to obtain a smooth operation, we introduce additional weights $\mu_{i,t}$ and penalize changes in the hydrogen fraction at $i \in V^+_{H_2}$ between $t-1$ and t. Hence, our hydrogen propagation model is defined as the following linear program

$$\textbf{HPM:} \quad \max_{\tilde{w}, w} \sum_{t \in T} \sum_{i \in V^+_{H_2}} \tilde{w}_{i,t}s_{i,t} - \mu_{i,t}|\Delta\tilde{w}_{i,t}| \quad (7)$$

$$\text{s.t. } (1) - (6)$$

$$0 \leq \tilde{w}_{i,t} \leq 1 \quad \forall i \in V^+_{H_2}, \forall t \in T \quad (8)$$

$$0 \leq \tilde{w}_{i,t} \leq q^{UB}_{i,t} \quad \forall i \in V \backslash V^+, \forall t \in T \quad (9)$$

$$\tilde{w}_{i,t} = 0 \quad \forall i \in V^+ \backslash V^+_{H_2}, \forall t \in T, \quad (10)$$

where $|\Delta\tilde{w}_{i,t}| = |\tilde{w}_{i,t} - \tilde{w}_{i,t-1}|$ can be linearized using common techniques. The mixing process is represented by (1)–(6). At hydrogen entries 100% hydrogen is allowed, see (8). For all non-entry nodes an hydrogen upper bound is defined in (9), and for all non-hydrogen entries the hydrogen fraction is set to zero in (10).

3 Sequential Hydrogen Propagation Model

Even though the hydrogen propagation model in Subsect. 2.3 is an LP, difficulties occur when the model is solved for large networks and long time periods.

Thus, the time period and the input data are split into smaller overlapping time horizons, which are solved iteratively. For each iteration, we define its starting state based on the previous solution. However, hydrogen bounds may be violated when multiple successive iterations are considered as future flows and bounds are not available, e.g., constraints (9). Thus, to ensure feasibility across multiple iterations, slack variables for the hydrogen limits on critical elements, i.e., exits and compressor stations, are introduced. Let $\sigma^d_{i,t} \in \mathbb{R}_{\geq 0}$ be the slack variable on the hydrogen upper bound of exit $i \in V^-$ at time t. As the hydrogen bound of a compressor station $a = (i,j) \in A^{\text{cs}}$ is imposed on its outgoing node i, let $\sigma^a_{i,t} \in \mathbb{R}_{\geq 0}$ be the slack variable on the hydrogen upper bound of this node i at time t. To penalize the usage of slack variables, let $\kappa_{i,t}$ and $\gamma_{i,t}$ be penalty weights for using slack $\sigma^d_{i,t}$ on exit $i \in V^-$ and slack $\sigma^a_{i,t}$ on the outgoing node i of compressor $a = (i,j) \in A^{\text{cs}}$ at time t, respectively. Thus, **HPM** is extended to

$$\textbf{sHPM:} \max_{\substack{\tilde{w},w,\\ \sigma^d,\sigma^a}} \sum_{t \in T} \left(\sum_{i \in E} \tilde{w}_{i,t} s_{i,t} - \mu_{i,t} |\Delta \tilde{w}_{i,t}| - \sum_{i \in V^-} \kappa_{i,t} \sigma^d_{i,t} - \sum_{a=(i,j) \in A^{cs}} \gamma_{i,t} \sigma^a_{i,t} \right)$$

$$\begin{aligned} \text{s.t. } & (1) - (6), (8), (10) \\ & 0 \leq \tilde{w}_{i,t} - \sigma^d_{i,t} \leq q^{UB}_{i,t} && \forall i \in V^-, \forall t \in T \\ & 0 \leq \tilde{w}_{i,t} - \sigma^a_{i,t} \leq q^{UB}_{i,t} && \forall a = (i,j) \in A^{\text{cs}}, \forall t \in T \end{aligned}$$

Finally, as it is more critical to stay technically feasible at compressor stations than at exit nodes[1], a *sequential hydrogen propagation algorithm* is formulated, taking this hierarchy into account. If no feasible solution is found for **HPM**, **sHPM** is solved in the next stage where only exit slacks are allowed to be nonzero. If still no feasible solution is found, **sHPM** is solved again with both exit and compressor slacks in the final stage.

4 Case Study

Our analysis of the hydrogen capacity is conducted on a major part of the German gas grid using hourly measured gas flow data from the time period April to December 2020. There are frequent changes in the network's topology in this period, i.e., over 50 network configurations lasting from one hour up to 30 days. Hence, we apply the sequential hydrogen propagation model from Sect. 3 here. A three-day rolling horizon with a two-day overlap is chosen. The network consists of 8600 nodes, with 56 entry and over 1000 exit nodes, and 10000 arcs. For hydrogen injection, 20 entries in Northwestern Germany are chosen in consultation with experts at the TSO, taking the general location of green hydrogen projects from the NEP Gas 2020–2030 [5] into account.

Besides technical restrictions regarding compressor stations, an important measure is the gas interchangeability represented by the Wobbe-Index *WI* at

[1] Compressor parts have to be replaced with hydrogen volume fraction above 10% [4].

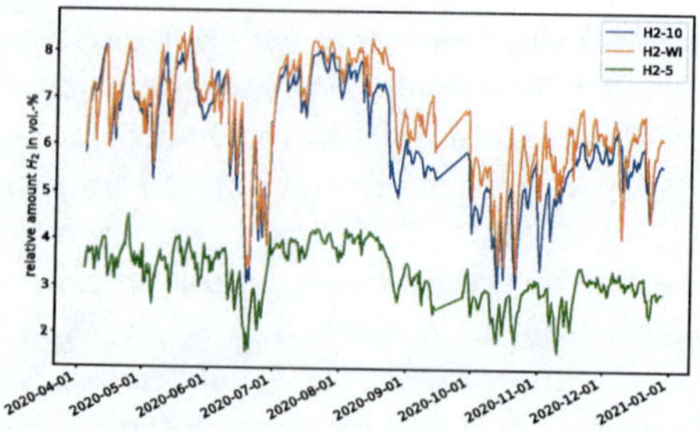

Fig. 1. Relative amount of hydrogen stored in the network for `H2-10` (blue), `H2-WI` (orange), `H2-5` (green) for the time period April to December 2020

exit nodes. Based on regulatory limits for *WI* and its behaviour for hydrogen/gas mixtures defined in [6,7], the permitted hydrogen volume fraction at exit nodes is determined a priori and ranges between 7–18 vol.-%. In this study, an hydrogen limit of 10 vol.-% is imposed on compressor stations, cf. [4]. For all inner nodes not adjacent to a compressor station we set the hydrogen limit to 1.0. In the following, we consider three scenarios, which differ in the hydrogen limit on exit nodes. `H2-10`: At each exit $i \in V^-$ and time t, $q_{i,t}^{UB}$ is determined by the limits on *WI* and must not be greater than 10 vol.-%. `H2-WI`: At each exit $i \in V^-$ and time t, $q_{i,t}^{UB}$ is only based on *WI*. `H2-5`: Three groups of exits are introduced, *country borders* with $q_{i,t}^{UB} = 10$ vol.-%, *industry-like exits* with $q_{i,t}^{UB} = 5$ vol.-%, and all other exits with limits based on the *WI*.

5 Results and Conclusion

The amount of hydrogen injected into the network is 11.2, 12.2, and 5.7 TWh for `H2-10`, `H2-WI`, and `H2-5`, respectively. As comparison, the amount of hydrogen given by a constant injection of 10 vol.-% at each entry $i \in V_{H_2}^+$ in each time step $t \in T$ without respecting any bounds results in 12 TWh and the planned capacity for green hydrogen in Germany is 14 TWh by 2030 according to [1].

The relative amount of hydrogen stored in the network at time t is shown in Fig. 1. `H2-WI` shows the highest amount of hydrogen in the network peaked in May with over 8 vol.-%, followed by a slightly smaller amount in `H2-10`. By imposing the much smaller bounds on *industry-like* exits in `H2-5`, the overall level of injection reduces by roughly 50% compared to the other two scenarios. In September, the overall gas level increases with gas injection from Eastern parts of the network. Therefore, the hydrogen level drops in all scenarios.

The distribution of the hydrogen fraction, shown in the left of Fig. 2, shows that in `H2-10` the H_2 entries can inject 10 vol.-% hydrogen most of the time. This is expected as this is the upper bound given for both exits and compressor stations. By removing the 10 vol.-% limit in `H2-WI`, the distribution is shifted to the right between 10 and 20 vol.-%. In `H2-5`, the distribution is centered at 5 vol.-% hydrogen representing the tightest bound given by *industry-like exits*. The

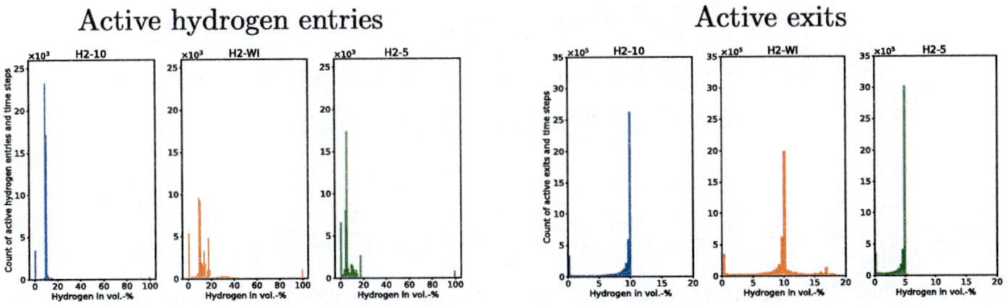

Fig. 2. Histogram of hydrogen fraction at active hydrogen entries (left) and active exits (right) over the whole time period for all scenarios.

peaks at 0% hydrogen represent the time steps in which slack was needed forcing hydrogen injection to zero. The bounds on the exits in these three scenarios are recognized in the histogram of active exits, shown in the right of Fig. 2.

With the proposed hydrogen propagation model, the hydrogen capacity of an existing gas network can be assessed. The analysis is performed on historical gas flow data using several hydrogen injection locations and imposing different bounds on exits and compressor stations.[2] In the future, we plan to integrate hydrogen tracking into the operation.

Acknowledgement. The work for this article has been conducted in the Research Campus MODAL funded by the German Federal Ministry of Education and Research (BMBF) (fund numbers 05M14ZAM, 05M20ZBM).

References

1. Federal Ministry for Economic Affairs and Energy: The National Hydrogen Strategy (2020)
2. Quarton, C.J., Samsatli, S.: Should we inject hydrogen into gas grids? Practicalities and whole-system value chain optimisation. Appl. Energy **275**, 115–172 (2020). https://doi.org/10.1016/j.apenergy.2020.115172
3. Haverly, C.A.: Studies of the behavior of recursion for the pooling problem, ACM SIGMAP Bull. **25** (1978). https://doi.org/10.1145/1111237.1111238
4. Siemens Energy, Gascade Gastransport GmbH, Nowega GmbH. Wasserstoffinfrastruktur - tragende Säule der Energiewende (2020). https://www.get-h2.de/wp-content/uploads/200915-whitepaper-h2-infrastruktur-DE.pdf
5. FNB-Gas. Netzentwicklungsplan Gas 2020-2030 (2021). https://www.fnb-gas.de/netzentwicklungsplan/netzentwicklungsplaene/netzentwicklungsplan-2020/
6. DVGW-Arbeitsblatt G 260: Gasbeschaffenheit. Bonn. März 2013
7. DVGW-Arbeitsblatt G 262: Nutzung von Gasen aus regenerativen Quellen in der öffentlichen Gasversorgung. September 2011

[2] More detailed preprint: http://nbn-resolving.de/urn:nbn:de:0297-zib-82838.

Statistical Analysis and Modeling for Detecting Regime Changes in Gas Nomination Time Series

Milena Petkovic(✉), Nazgul Zakiyeva, and Janina Zittel

Applied Algoritmic Intelligence Methods Department, Zuse Institute Berlin, Takustraße 7, 14195 Berlin, Germany
{petkovic,zakiyeva,zittel}@zib.de

Abstract. As a result of the legislation for gas markets introduced by the European Union in 2005, separate independent companies have to conduct the transport and trading of natural gas. The current gas market of Germany, which has a market value of more than 54 billion USD, consists of Transmission System Operators (TSO), network users, and traders. Traders can nominate a certain amount of gas anytime and anywhere in the network. Such unrestricted access for the traders, on the other hand, increase the uncertainty in the gas supply management. Some customers' behaviors may cause abrupt structural changes in gas flow time series. In particular, it is a challenging task for the TSO operators to predict gas nominations 6 to 10 h-ahead. In our study, we aim to investigate the regime changes in time series of nominations to predict the 6 to 10 h-ahead of gas nominations.

Keywords: Regime switching · Nonlinear time series · Gas nomination forecast

1 Introduction

Natural gas plays as key energy source in Europe. Due to the legislation for gas markets introduced by the European Union in 2005, gas traders can independently nominate a certain amount of gas in any entry or exit point in the network. The amount of gas, either input or off-take, can be recorded or changed by the trader anytime up to 2 h before the realization. Such unrestricted access for the traders increases the uncertainty in the gas network management [2]. In particular, abrupt pattern changes of some customers' behaviors make it challenging for the gas network operators to predict 6 to 10 h-ahead gas nominations.

From the point of view of our industry partner, one of Germany's biggest TSOs, Open Grid Europe [3], the objective is to accurately predict the final gas nominations at 6am to 10am starting at midnight. It is the most challenging period of prediction horizon in gas nomination forecast, as the final gas nomination for 6am can abruptly change around 2am or 3am. Autoregression based time series models are popular in energy prediction due to its interpretability

N. Trautmann and M. Gnägi (Eds.): OR 2021, LNOR, pp. 188–193, 2022.
https://doi.org/10.1007/978-3-031-08623-6_29

and accurate forecasting compared to black-box machine learning methods [1,7]. In our study we apply Markov Switching Autoregression model with Exogenous (MS-ARX) to study the regime changes in customer's behaviour pattern and predict 6 to 10 h ahead final gas nominations. MS-ARX detects 3 regimes and shows superior multi-step ahead prediction performance compared to the benchmark and linear models. 93% of days in the out-of-sample period, MS-ARX provides effective predictions for 6am to 10am predictions.

2 Data

We study the regime changes in gas nomination time series in the German gas pipeline network. We consider the dataset of one customer nominating gas amount in seven different entry points, which creates the most complex dynamics for multi-step ahead prediction among others. The dataset covers 678 days (22 months and 6 days) from March Y1 to January Y2 with 16,272 observations, where Y1 and Y3 denotes the years of the dataset.

Figure 1a displays the aggregated final gas nomination time series of one customer at 7 entry points. The dynamic pattern of the time series shows sudden changes within the first 6 to 10 h from midnight. The short period of sudden increases and drops in time series may challenge the modeling and multi-step-ahead predictions of final gas nomination. It motivates us to consider regime changes in the time series modeling, as the customer's behaviour in nominating gas may be dependent on the unobservable discrete state variables.

Gas nomination history is an additional time series dataset which contains nomination history for every final hour in a day. Figure 1b illustrates the gas nomination history up to 10 h of offset for daily final gas nominations for 6am from March Y1 and January Y3. The graph shows the amount of the gas nominated for 6am 1 to 10 h in advance every day. It also has abrupt time series changes around three hours before 6am. Figure 2 shows the cross-dependence between the final nominations at 24 h and their history up to 10 h of offset. Although the cross-dependence between hourly final nominations and their history show high correlation, 6 am final nomination and its history start decreasing after 4 h of offset, which is the nomination set at 2am. And similar pattern is observed for 7am to 12pm.

3 Method

Let $y_t, t = 1, \dots, T$ denote a time series of T univariate observations. Observable stochastic process y_t is

$$y_t = \mu_{S_t} + \phi_{S_t} x + \epsilon_t, \qquad \epsilon_t \sim N(0, \sigma^2_{S_t}), \tag{1}$$

where x is a vector observed explanatory variables and μ_{S_t} is an intercept term, ϕ_{S_t} is a parameter vector and $\sigma^2_{S_t}$ is the variance of error terms, depending on hidden discrete stochastic process S_t, [4,5]. The probability of moving from

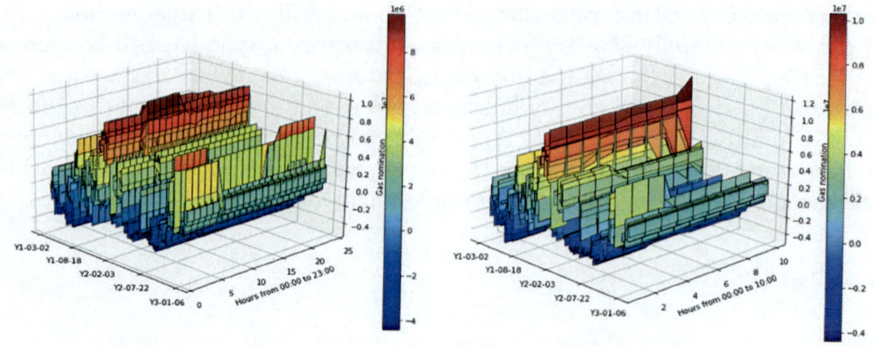

(a) Final gas nomination time series. (b) Gas nomination history for 6am.

Fig. 1. Final gas nomination (W) time series and gas nomination history between March 2 Y1 and January 6 Y3, where Y1 and Y3 denotes the years of the dataset.

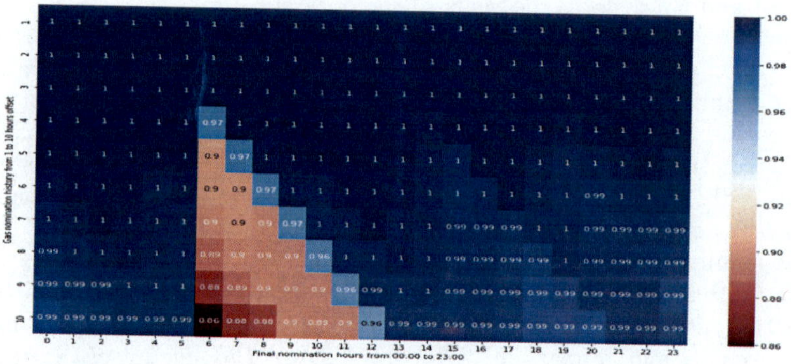

Fig. 2. Correlation between hourly final gas nomination and the gas nomination history offset hours.

regime i to regime j: $P(S_t = j|S_{t-1} = i) = p_{ij},\ i,j = 0,1,\ldots,K-1$ called transition probabilities, with $p_{i,0} + p_{i,1} + \cdots + p_{i,K-1} = 1,\ i = 0,1,\ldots,K-1$. In this process the current regime depends only on the regime before and it is called a Markov process.

With given past observations $F_{t-1} = \{y_{t-1}, y_{t-2}, \ldots\}$, we estimate the transition probabilities p_{ij} by maximizing the likelihood function with respect to the unknown parameters $\theta = (p_{i,j}, \phi_{S_t}, \mu_{S_t}, \sigma^2_{S_t}),\ i,j = 0,\ldots,K-1$ and the transition probabilities $p_{i,j}$. Taking the log of the likelihood function $l(\theta) = f(\theta|F_{t-1})$ and using the chain rule for the conditional probability, we have the log-likelihood function as follows,$l(\theta) = \sum_{t=1}^{T} l_t(\theta) = \sum_{t=1}^{T} \left(ln \sum_{S_t=0}^{m_s} \sum_{S_{t-1}=0}^{m_s} f(y_t, S_t, S_{t-1}| F_{t-1})\right) = \sum_{t=1}^{T} \left(ln \sum_{S_t=0}^{m_s} \sum_{S_{t-1}=0}^{m_s} f(y_t|S_t, S_{t-1}, F_{t-1}) P(S_t, S_{t-1}|F_{t-1})\right)$. The conditional probability density function for the observations y_t given the state variables S_t, S_{t-1} and the previous observations

$$f(y_t|S_t, S_{t-1}, F_{t-1}) = \frac{1}{\sqrt{2\pi\sigma_{S_t}^2}} \exp\left\{ -\frac{[y_t - \mu_{S_t} - \phi_{S_t} x]^2}{2\sigma_{S_t}^2} \right\},$$

as $\epsilon_t = y_t - \mu_{S_t} - \phi_{S_t} x \sim N(0, \sigma_{S_t}^2)$. Using the chain rule for conditional probabilities and the Markov property $P(S_t|S_{t-1}, F_{t-1}) = P(S_t|S_{t-1})$, we rewrite the conditional joint probability $P(S_t, S_{t-1}|F_{t-1})$ as

$$P(S_t, S_{t-1}|F_{t-1}) = P(S_t|S_{t-1}, F_{t-1})P(S_{t-1}|F_{t-1}) = P(S_t|S_{t-1})P(S_{t-1}|F_{t-1}).$$

The time dependent state probabilities $P(S_{t-1}|F_{t-1})$ are updated by Kim algorithm [6]. Finally, with the estimated parameters, the h step-ahead prediction is calculated as the weighted average of the predictions in each regime, $\hat{y}_{t+h} = \sum_{j=0}^{K-1} (\mu_{S_{t+h}} + \phi_{S_{t+h}} x) P(S_{t+h} = j|F_t)$. $P(S_{t+h} = j|F_t)$ is a smoothed probability that observation at time $t + h$ is in state j given the history $F_t = \{y_t, \dots, y_1\}$.

4 Forecast Setup and Evaluation

We predict final gas nomination time series described in Sect. 2. We make 6 to 10 h-ahead out-of-sample forecast in real time starting from 00:00 on January 1 Y2 to January 6 Y3, a total of 373 days. In the training-validation sample we choose the optimal hyperparameters for the training the model with sliding window method. Then we estimate the model weights at each point in the out-of-sample period.

For an h-step ahead forecast, where h = 6, 7, 8, 9, 10, we apply the MS-ARX model with 3 regimes and 3 variables including the y_{t_0}, y_{t_0-24+h} and x_{t_0}, denoting the last observed gas nomination, 24 h before gas nomination and history of gas nomination at time t_0, respectively, where t_0 is the last observed point. Updating the model using the whole observed train sample, we move forward predicting the rest of the test sample one period at a time.

As alternative models, we apply three benchmark methods namely, Base I (y_{t_0}), Base II (y_{t_0-24+h}), and Reference (x_{t_0}). Furthermore, we consider Autoregression model with the exogenous (ARX) using the three variables. We train the ARX model using the last 7 days of rolling window, as it gave the optimal out-of-sample prediction results in the training-validation period.

We evaluate the relative forecast accuracy according to average root mean squared error (RMSE), mean absolute error (MAE), and mean absolute percentage error (MAPE) for the h-step-ahead predictions. The smaller the average error, the better accuracy is obtained by the forecast model. We obtain hourly h-step-ahead predicted gas flow series $\hat{Y}_{t+h}$. The h-step-ahead prediction performance is evaluated over the forecasting period. The RMSE, MAE, and MAPE are obtained as RMSE $= [\sum_{t\in T_2} (\hat{y}_{t+h} - y_{t+h})^2/|T_2|]^{1/2}$, MAE $= \sum_{t\in T_2} |\hat{y}_{t+h} - y_{t+h}|/|T_2|$, and MAPE $= \sum_{t\in T_2} \left|(\hat{y}_{t+h} - y_{t+h})/y_{t+h}\right|/|T_2|$, for $h = 6, \cdots, 10$, where $|T_2|$ is the length of forecasting period T_2.

Furthermore, we apply traffic light evaluation criteria. As discussed in Sect. 2, the extreme sudden changes in the final gas nomination time series occur between 2am to 3am. It complicates predicting the 6am to 10am final gas nominations using the observations only until 12am. The proportion of green days are evaluated with MAE $< 10^6$W and MAPE < 0.1, yellow days with MAE $< 2 * 10^6$W or MAPE < 0.2, and red days with MAE $> 10^6$W and MAPE > 0.1 in predicting 6am to 10am final gas nominations trained with the time series until 12am.

5 Forecast Results

We demonstrate the multi-step ahead forecasting results of final gas nomination of a difficult customer in German gas transmission network. Table 1 shows the average forecast errors, denoted as aRMSE, aMAE and aMAPE, by the MS-ARX and the alternative models for the 6 to 10 h-ahead gas nomination prediction. Table 1 compares the out-of-sample predictions done at 12am for 6am to 10am using the traffic light criteria. Overall, the results demonstrate the advantages of the regime changes assumption in improving the multi-step ahead final gas nomination prediction. The MS-ARX performs as the most accurate forecast model with the smallest average forecast errors for multi-step ahead predictions. It illustrates that the assumption of the regime changes in the final gas nomination improves the forecast errors of the linear ARX model from aRMSE 5.18, aMAE 2.35 and aMAPE 1.65 to aRMSE 3.85, aMAE 1.48 and aMAPE 0.67 for 6 to 10 h-ahead gas nomination forecast. Compared to the Reference (x_{t_0}), which is the best performing model within the benchmark models, the MS-ARX improves its prediction errors from aRMSE 5.88, aMAE 2.03 for 6 to 10 h-ahead predictions. Table 1 shows the MS-ARX model outperforms all the alternative models with the maximum percentage of green days (93.01%) and minimum percentage of the red days (2.42%). It is worth mentioning that the MS-ARX shows not only the accurate prediction results on overall h-step-ahead forecasts, but its significant advantage is shown in predicting the final nomination in the most difficult hours (6am to 10am). In summary, it is useful to consider the regime changes assumption for accurate and stable multi-step ahead final gas nominations forecasts.

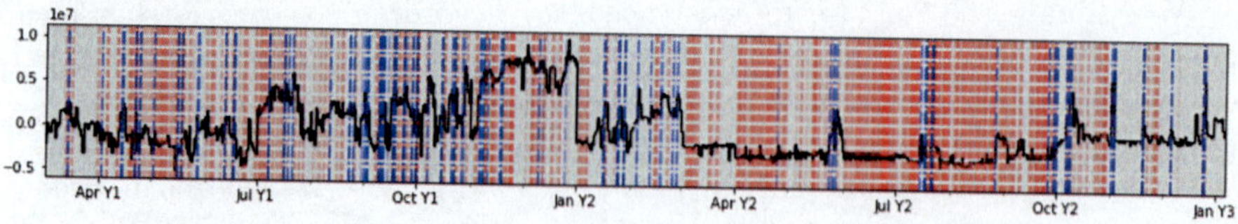

Fig. 3. Aggregated hourly final gas nomination in 3 estimated regimes, regimes 1, 2 and 3 are shown in grey, blue, red, respectively.

Table 1. Average forecast errors for h-step ahead forecast, $h = 6, 7, 8, 9, 10$, and traffic light criteria for 6am to 10am nomination forecast in particular. Best forecast result is marked in bold-face.

	MS-ARX	ARX	Base I (y_{t_0})	Base II (y_{t_0-24+h})	Reference (x_{t_0})
aRMSE (10^6W)	**3.85**	5.18	7.66	11.86	5.88
aMAE (10^6W)	**1.48**	2.35	2.82	6.13	2.03
aMAPE	0.67	1.65	1.43	4.66	**0.55**
Green	**93.01%**	87.37%	82.26%	81.99%	72.31%
Yellow	**4.57%**	7.80%	10.48%	10.75%	20.70%
Red	**2.42%**	4.84%	7.26%	7.26%	6.99%

6 Conclusion

We apply the MS-ARX model to predict 6 to 10 h-ahead final gas nominations in the German high-pressure gas pipeline network. It demonstrates the 3 regime periods in the hourly gas nomination time series and shows a superior out-of-sample forecast performance over linear and benchmark prediction models.

Acknowledgement. The work for this article has been conducted within the Research Campus Modal funded by the German Federal Ministry of Education and Research (fund numbers 05M14ZAM, 05M20ZBM).

References

1. Brabec, M., et al.: A nonlinear mixed effects model for the prediction of natural gas consumption by individual customers. Int. J. Forecast. **24**(4), 659–678 (2008)
2. Fügenschuh, A.: Mathematical optimization for challenging network planning problems in unbundled liberalized gas markets. Energy Syst. **5**(3), 449–473 (2013). https://doi.org/10.1007/s12667-013-0099-8
3. Open Grid Europe GmbH. https://www.oge.net/
4. Hamilton, J.D.: A new approach to the economic analysis of nonstationary time series and the business cycle. Econometrica 47, 357–384 (1989)
5. Hamilton, J.D.: Regime switching models. In: Durlauf, S.N., Blume, L.E. (eds.) Macroeconometrics and Time Series Analysis. TNPEC, pp. 202–209. Palgrave Macmillan UK, London (2010). https://doi.org/10.1057/9780230280830_23
6. Kim, C.-J.: Dynamic linear models with Markov-switching. J. Econ. **60**(1–2), 1–22 (1994)
7. Kurek, T., Wojdan, K., Swirski, K.: Long-term prediction of underground gas storage user gas flow nominations. J. Power Technol. **99**(4) 272–280 (2020)

A Conceptual Framework for Determining the Economically Optimal Level of Microgrid Resilience

Sammit Vartak[1] and Reinhard Madlener[2,3](✉)

[1] RWTH Aachen University, Aachen, Germany
Sammit.Vartak@rwth-aachen.de
[2] Institute for Future Energy Consumer Needs and Behavior (FCN), School of Business and Economics/E.ON Energy Research Center, RWTH Aachen University, Aachen, Germany
RMadlener@eonerc.rwth-aachen.de
[3] Department of Industrial Economics and Technology Management, Norwegian University of Science and Technology (NTNU), Trondheim, Norway
http://fcn.eonerc.rwth-aachen.de

Abstract. Microgrids are considered to be an effective measure for resilience due to their ability to operate in island mode during a power outage and to ensure continuity of supply. Current standard frameworks for microgrid resilience evaluation overlook the economic parameters, which are important for decision-making about investments in resilience enhancement. To bridge this gap, a new methodology is proposed in order to evaluate the economically optimal level of microgrid resilience. An availability-based resilience evaluation framework is used for quantification of the resilience using both a deterministic and a stochastic approach. The quantified resilience values are monetized using the economic indicator Value of Lost Load. The economically optimal resilience level is evaluated using the Net Present Value capital budgeting method.

Keywords: Resilience · Microgrid · Economic evaluation

1 Introduction

Storms, hurricanes, floods, wildfires, and other natural disasters may result in major disruptions to the power supply. The increased frequency and severity of natural disasters in recent years have necessitated the enhancement of the resilience of power grids. The losses due to electrical outages have a large negative economic impact. Hence it is important to mitigate these impacts in the future to a reasonable extent. In this context, the resilience of the power grid comes into the picture. The resilience of power grids is the ability of a grid to adapt and recover from extreme events.

Microgrids are considered to be one of the most effective measures to increase resilience during natural disasters because of their ability to isolate themselves

N. Trautmann and M. Gnägi (Eds.): OR 2021, LNOR, pp. 194–199, 2022.
https://doi.org/10.1007/978-3-031-08623-6_30

from the main grid during the occurrence of an outage and to maintain continuity of supply to the local load [1]. Microgrids are a part of the power grid, having their own local distributed generation resources and being able to operate autonomously to supply local loads and thus reducing the impact of any disaster. While planning a microgrid for resilience purposes, it is important to quantify the level of resilience of the microgrid in order to evaluate its preparedness for extreme events. Since the enhancement of microgrid resilience involves significant capital investment, it is important to determine the economic value of the resilience in order to decide about the optimal level of resilience that a microgrid should possess. Currently, there is no standard to measure the economic value of microgrid resilience, which is one of the challenges when designing a microgrid for resilience enhancement purposes [2].

This study aims to propose a new methodology in order to evaluate the economically optimal level of resilience for a microgrid. First, the concept of availability-based resilience evaluation is discussed in Sect. 2. The quantification of microgrid resilience using a deterministic versus a Markov-chain-based stochastic approach is presented in Sect. 3. In Sect. 4, it is discussed how the measured resilience value can be translated into an economic metric using the Value of Lost Load (VoLL) [5]. A methodology is then proposed to evaluate the optimal resilience enhancement option which justifies the cost of resilience enhancement. In Sect. 5, the applicability of the proposed methodology is discussed, conclusions drawn, and some ideas for future research outlined.

The proposed methodology evaluates the resilience of a microgrid only during its islanded operation and not in grid-connected mode. But the suggested framework can easily be extended to include the grid-connected operation of the microgrid as well. Also, it is assumed that during a disaster, the microgrid is not physically damaged and that the microgrid resources are able to serve the load. Partial supply of load is not considered yet in the proposed approach.

2 Resilience Evaluation Framework

This work uses the so-called availability-based resilience measurement framework for measuring microgrid resilience, as proposed in [3]. US Presidential Policy Directive (2013) on Critical Infrastructure Security and Resilience defines Resilience as 'the ability to prepare for and adapt to changing conditions and withstand and recover rapidly from disruptions'. According to the proposed framework, the resilience of a microgrid can be measured as a quantity corresponding to the availability. Hence, the resilience level R of the microgrid can be calculated as

$$R = \frac{T_U}{T_D + T_U}, \tag{1}$$

where T_U is the uptime of the microgrid during the outage and T_D is the downtime. The sum of T_D and T_U denotes the total resilience evaluation time T. This equation gives a suitable measure of resilience within the context of the definition provided above. The downtime T_D is related to the recovery speed and is influenced by both the available infrastructure and to human-related activities such as repairs and maintenance policies. The uptime T_U is directly dependent on the withstanding capability of the microgrid to the given event. Its value is mostly related to infrastructure characteristics and grid design.

Although preparation/planning capacity and adaptation capability are not directly reflected in the resilience measurement, they influence the resilience metric indirectly. For example, they will be reflected in the resilience values of two different scenarios. Scenarios may differ in the available generation capacity or storage capacity of the microgrid and will give two different values for resilience. This difference will indicate a change in resilience when adopting different technological or infrastructure improvements in the microgrid. A comparison between the two resilience values will represent the planning capacity and adaptability of scenarios. This quantifiable difference can be used in conjunction with a cost analysis and a probabilistic evaluation to decide the priority, based on the resilience level and the costs of implementing different scenarios.

3 Resilience Quantification

The uptime for the microgrid is considered to be the time when the microgrid can satisfy the load demand by using its own resources. The downtime is the rest of the time when the microgrid is not able to serve the loads. Let $G[t]$ be the fitness function indicating the ability of the microgrid to serve the loads. At every time instance t, $G[t]$ measures the total difference between the capacity of the microgrid resources available (energy supply + energy storage) and the load demand, i.e.

$$G[t] = X[t] + B[t] - L[t], \tag{2}$$

where $X[t]$ is the energy supplied by the source in time interval t, $L[t]$ is the energy load during time interval t, and $B[t]$ is the energy in the battery storage at time interval t. Function $G[t]$ indicates the ability of the energy source and the battery to serve the load at time interval t. Function $G[t] > 0$ indicates that there is surplus energy in the grid, whereas $G[t] < 0$ indicates that there is an energy deficiency. If $f_X[t]$ is the probability distribution of $X[t]$, $f_L[t]$ is the probability distribution of $L[t]$, and $f_B[t]$ is the probability distribution of $B[t]$ at time t, then the probability distribution of $G[t]$, i.e. $f_G[t]$, can be calculated using Eq. (2). The resilience of the microgrid $R[t]$ at time t is given as

$$R[t] = Pr(G[t] \geq 0) = \sum_{g \geq 0} f_{G[t]}, \tag{3}$$

i.e. the resilience $R[t]$ is equal to the probability that the microgrid resources can satisfy the load demand at time instance t (probability of $G[t] \geq 0$). To find the probability distribution of $G[t]$, i.e. $f_G[t]$, the probability distributions of $f_X[t]$, $f_L[t]$ and $f_B[t]$ are required. The probability distribution of battery storage $f_B[t]$ depends upon the probability distributions of the energy source and the load, i.e. $f_X[t]$ and $f_L[t]$.

Deterministic Computational Approach: The deterministic approach evaluates resilience by calculating the availability of the microgrid. This method can be used to analyze the best- and the worst-case scenario. Figure 1 shows the algorithm developed for an example microgrid having a PV system and Li-ion battery storage as its only resources. This algorithm takes into account bidirectional energy flows and maintains a minimum state of charge in the battery during normal operation. The battery is allowed to discharge fully in the case of an outage. Since the deterministic approach provides a range of resilience values possessed by the microgrid (based on the best- and the worst-case scenario), it has limitations and does not provide any generalized value that can be used directly for resilience planning and investment decisions.

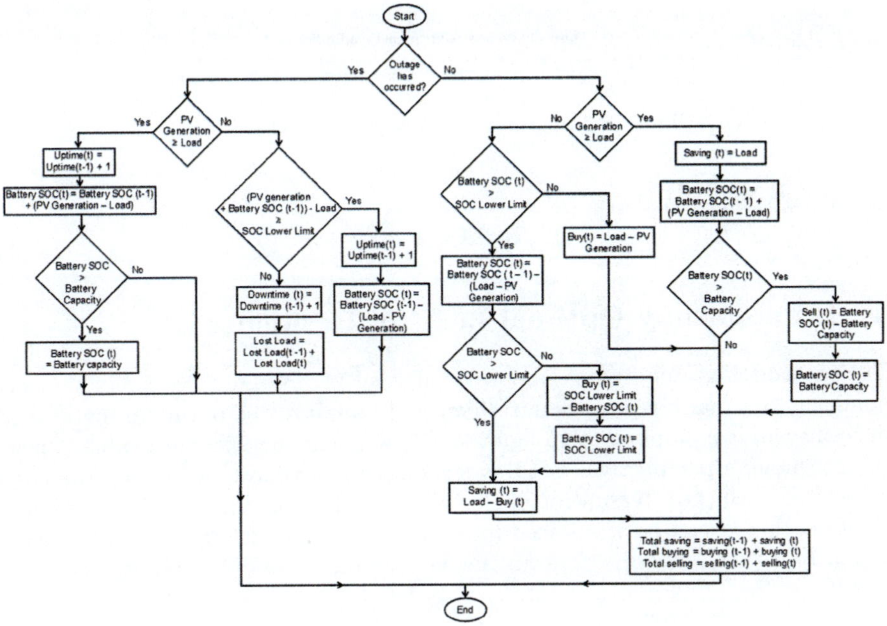

Fig. 1. Flowchart of the computational method adopted

Markov-Chain-Based Approach: The Markov-chain-based approach provides generalized value for resilience of the microgrid by calculating the probability of a microgrid being available to supply the load demand considering a large amount of possible values of load and generation. The stochastic data is generated using the Monte Carlo method to acquire more accurate availability results. The Markov chain model is designed according to the model proposed in [4]. The Markov chain model for microgrid resilience quantification is summarized in a flowchart in Fig. 2. Once π_E is known, the availability A and resilience

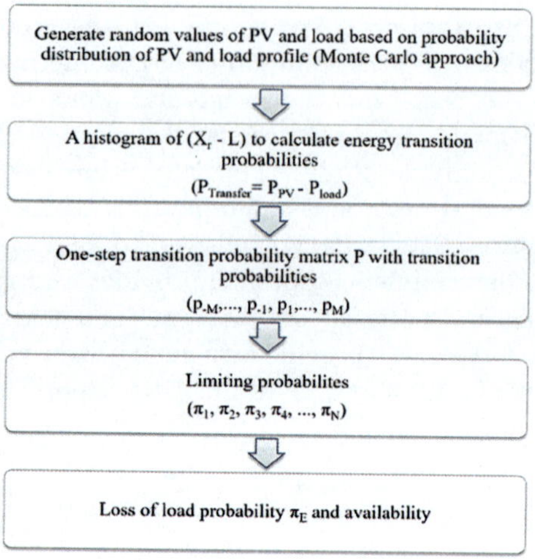

Fig. 2. Summary of Markov-chain-based method

R can be calculated as

$$R = A = 1 - \pi_E \tag{4}$$

4 Economically Optimal Level of Resilience

The methodology for finding the optimal level of resilience considers the economically optimal investment in resilience. To analyze the optimum investment in resilience, it is important to figure out how much the resilience value is from the consumer's perspective. VoLL is a monetary indicator expressing the costs associated with the interruption of electrical supply in €$/kWh$ [5]. The resilience values obtained can be translated into economic values using the VoLL specification in Eq. (5), where h denotes the average duration of the outage.

$$VoLL_{Total} = h \times (1 - R) \times (VoLL_{PerUnit} \times Avg.Load) \tag{5}$$

VoLL, as an economic metric to monetize resilience, is complementary to the availability-based resilience measurement framework. The greater the availability of the system, the less the economic impact due to the lost load is. The most resilient case is the one resulting in the least $VoLL_{Total}$.

The investment needed to obtain a certain level of resilience and the economic benefits due to resilience are compared with each other in order to find the economically optimal level of resilience by using the Net Present Value approach. The benefits G obtained from resilience are calculated as follows.

$$G_{Resilience} = VoLL_{Total(BaseCase)} - VoLL_{Total(ResilientCase)} \tag{6}$$

The methodology is further detailed in [9] for the case of a stylized microgrid (incl. results).

5 Conclusion and Outlook

In this paper, we have presented a new methodology for evaluating the economically optimal level of resilience in a microgrid using an availability-based resilience framework. Deterministic and stochastic approaches for resilience measurement are explained. The main aim of this work is to introduce an economic evaluation framework which monetizes the resilience value in order to facilitate the decision-makding process involved in the identification of optimum technologies for enhanced resilience.

The methodology proposed is also useful to evaluate different economic and policy instruments, such as feed-in tariffs and subsidies, and to observe the change in economic performance due to these interventions. In future research, the methodology can be extended to take into account the stochastic nature of the duration of an outage and change in load demand. This methodology can also be used in conjunction with more sophisticated optimization techniques with the objective of minimizing the total Value of Lost Load.

References

1. Kezunovic, M., Overbye, T.J.: Off the beaten path: resiliency and associated risk. IEEE Power Energy Mag. **16**(2), 26–35 (2018). https://doi.org/10.1109/MPE.2017.2780961
2. Asmus, P.: Unpacking the value of resiliency that Microgrids offer (2019). https://www.navigantresearch.com/news-and-views/unpacking-the-value-of-resiliency-that-microgrids-offer. Accessed 31 July 2021
3. Krishnamurthy, V., Kwasinski, A.: Effects of power electronics, energy storage, power distribution architecture, and lifeline dependencies on Microgrid resiliency during extreme events. IEEE J. Emerg. Sel. Topics Power Electron. **4**(4), 1310–1323 (2016). https://doi.org/10.1109/JESTPE.2016.2598648
4. Song, J., Krishnamurthy, V., Kwasinski, A., Sharma, R.: Development of a Markov-Chain-based energy storage model for power supply availability assessment of photovoltaic generation plants. IEEE Trans. Sustain. Energy **4**(2), 491–500 (2013). https://doi.org/10.1109/TSTE.2012.2207135
5. Schröder, T., Kuckshinrichs, W.: Value of lost load: an efficient economic indicator for power supply security? A Liter. Rev. Front. Energy Res. **3**, 55 (2015). https://doi.org/10.3389/fenrg.2015.00055
6. Rosen, C., Madlener, R.: Regulatory options for local reserve energy markets: implications for prosumers, utilities, and other stakeholders. Energy J. **37**(SI2), 39–50 (2016). https://doi.org/10.5547/01956574.37.SI2.cros
7. Mercurio, A.: Micrograds and energy security: the business case. International Association for Energy Economics (IAEE). Energy Forum, Fourth Quarter, 43–45 (2013). http://www.iaee.org/documents/2013EnergyForum4qtr.pdf
8. Vonsien, S., Madlener, R.: Cost-effectiveness of Li-Ion battery storage with a special focus on photovoltaic systems in private households. J. Energy Storage **29**(June), 101407 (2020). https://doi.org/10.1016/j.est.2020.101407
9. Vartak, S., Madlener, R.: On the optimal level of Microgrid resilience from an economic perspective. FCN Working Paper No. 5/2020, Institute for Future Energy Consumer Needs and Behavior, RWTH Aachen University, April 2020

Modeling and Forecasting Gas Network Flows with Multivariate Time Series and Mathematical Programming Approach

Nazgul Zakiyeva(✉) and Milena Petkovic

Zuse Institute Berlin, Berlin, Germany
{zakiyeva,petkovic}@zib.de

Abstract. With annual consumption of approx. 95 billion cubic meters and similar amounts of gas just transshipped through Germany to other EU states, Germany's gas transport system plays a vital role in European energy supply. The complex, more than 40,000 km long high-pressure transmission network is controlled by several transmission system operators (TSOs) whose main task is to provide security of supply in a cost-efficient way. Given the slow speed of gas flows through the gas transmission network pipelines, it has been an essential task for the gas network operators to enhance the forecast tools to build an accurate and effective gas flow prediction model for the whole network. By incorporating the recent progress in mathematical programming and time series modeling, we aim to model natural gas network and predict gas in- and out-flows at multiple supply and demand nodes for different forecasting horizons. Our model is able to describe the dynamics in the network by detecting the key nodes, which may help to build an optimal management strategy for transmission system operators.

Keywords: Forecasting · Mathematical programming · Natural gas

1 Introduction

Germany is the largest market for natural gas in the European Union. Natural gas is the second most used energy source in Germany, with 25% share in primary energy consumption [1] and plays an important role in the German "Energiewende", especially in the transition from fossil fuels to renewables in the power sector. Recently, the natural gas market is becoming more and more competitive and is moving towards more short-term planning, e.g., day-ahead contracts, which makes the dispatch of natural gas even more challenging [4]. Despite these challenges, TSOs have to fulfill all transport demands in a cost-efficient way. Since the average gas pipe velocity is only 25 km/h [3], a high-accuracy and high-frequency forecasting of supplies and demands in natural gas network is essential for efficient and safe operation of the complex natural gas transmission networks and distribution systems.

N. Trautmann and M. Gnägi (Eds.): OR 2021, LNOR, pp. 200–205, 2022.
https://doi.org/10.1007/978-3-031-08623-6_31

This work is part of a joint project with one of Germany's biggest transmission system operators, Open Grid Europe [2]. In order to provide a comprehensive understanding of the network dynamic, together with OGE, we develop a Network Autoregressive Linear model with Balance constraint (NAR-LB) model that detects influential nodes in the network (the nodes that demonstrate a strong effect on the future flows of other nodes and drive the movement of the network over time) and provides high-precision, multi-step ahead hourly forecasts for more than 200 nodes in the gas network. These nodes have very different statistical characteristics, as they represent a variety of different source and consumption points, ranging from connections to other gas networks or countries over industrial consumers to storage facilities. By addressing the key challenges with high dimensionality and balance constraint simultaneously as in Chen et al. [3], we aim to enhance the computational efficiency of the high dimensional constrained network problem by integrating recent advances in optimization and statistical modeling techniques. As the size of the network grows, the number of parameters is controlled by regularizing the only significant dependencies, which highlights the influential nodes in the network. Compared to the existing prediction models in machine learning and statistics, the NAR-LB model provides more accurate prediction results for the whole network and interpret the complex dynamics in the network to help reduce financial and technical risks.

2 Methodology

Let N denote the number of nodes in a large-scale complex gas transmission network. We denote with $q_{t,i}$ the gas flow time series at node i, where $t = 1, ..., T+1$ and $i = 1, ..., N$. Following the model proposed in Chen et al. [3] we define Network Autoregressive Linear model with Balance constraint, to capture the network effect of different nodes and determine influential nodes for each node in the network, where the total gas in-flows and out-flows need to be balanced, as follows:

$$\begin{aligned} q_{t,i} &= \sum_{j=1}^{N} q_{t-1,j} \cdot w_{j,i}, \qquad i = 1, ..., N, t = 2, ..., T+1 \\ \text{s.t.} \sum_{i=1}^{N} \sum_{j=1}^{N} q_{t-1,j} &\cdot w_{j,i} = 0, \qquad t = 2, ..., T+1 \end{aligned} \tag{1}$$

The parameter $w_{i,i}$ model the autoregressive dependence for each node while parameter $w_{j,i}$ when $i \neq j$ model the influence of j-th node to the i-th node i.e. influence of the past value of the j-th node on the current value of the i-th node. Let us define $T \times N$ matrix $Q = (q_1, .., q_N)$ containing T previous flow values of N network nodes $q_i = (q_{1,i}, ..., q_{T,i})^\intercal$, matrix $X = (x_1, ..., x_N)$ containing previous values of Q and matrix W representing $N \times N$ matrix of unknown parameters $w_{i,j}$ that mutually model the influence of the network nodes.

While autoregressive dependence is modeled by diagonal elements of W, the non-diagonal elements define the weighted adjacency matrix where row vector $W_j = (w_{j,1}, ..., w_{j,N})$ represents the influence of the j-th node on the future values of the other nodes in the network. We assume that the weighted adjacency matrix is sparse but we have no prior knowledge of the sparse structure in terms of number of significant elements and their location. To detect the influential nodes in the dynamic network, we adopt the Lasso type regularization in estimation. Since in any point in time only a small number of nodes have a significant effect to the network dynamic and each of influential nodes only have an influence on the future flow of small number of nodes, we impose 2 layers of sparsity: groupwise and inner group sparsity on the weighted adjacency matrix. We observe matrix W in N groups (W_j) and estimate the parameters by applying Sparse-Group Lasso penalty [3].

$$
\begin{gathered}
\min_{W} |E| + \sum_j \lambda_g \left\|W_j\right\|_2 + \lambda \sum_{i \neq j} |w_{i,j}| \\
\text{s.t. } E = Q - XW \\
XWI = 0 \\
lb \leq w_{i,j} \leq ub, \qquad i = 1, ..., N, j = 1, ..., N
\end{gathered} \tag{2}
$$

where I is an $N \times N$ identity matrix, $0 \leq \lambda \leq 1$ is a Lasso tuning parameter for the individual weight $w_{i,j}$ when $i \neq j$ and $\lambda_g = \sqrt{N}\lambda$ is a tuning parameter for group W_j. The tuning parameters are chosen to optimize the forecasting performance via the rolling window technique [3]. In addition we set the lower and upper bounds for the weights to $lb = -2$ and $ub = 2$, respectively.

We use the estimated weighted adjacency matrix to choose the most influential lagged flows for each node in the network as features for multi-step ahead forecast. For each node $i \in 1, .., N$ we select F features with the highest absolute value of $w_{j,i}, j = 1, \ldots, N$. Further, we approximate future gas flow with weighted sum of features:

$$
\hat{q}_{t,i} = \sum_{j=1}^{F} q_{t-1,j} \cdot f_{j,i}, \qquad i = 1, ..N \tag{3}
$$

where $\hat{q}_{t,i}$ is the approximated gas flow for node i at the time t, $q_{t-1,j}, j = 1, \ldots, F$ are previous flows of F nodes with the highest influence to node i and $f_{j,i}$ are corresponding weights. If we define the error of approximation as:

$$
e_{t,i} = q_{t,i} - \sum_{j=1}^{F} q_{t-1,j} \cdot f_{j,i}, \qquad i = 1, ..., N \tag{4}
$$

then the optimal weights are calculated by minimizing the sum of absolute errors:

$$
\begin{gathered}
\min_{f} \sum_{i=1}^{N} |e_{t,i}| \\
\text{s.t. } q_{t,i} - \sum_{j=1}^{F} q_{t-1,j} \cdot f_{j,i} = e_{t,i}, \qquad i = 1, ..., N, t = 2, ..., T+1 \\
\sum_{i=1}^{N} \sum_{j=1}^{F} q_{t-1,j} \cdot f_{j,i} = 0, \qquad i = 1, ..., N, t = 2, ..., T+1 \\
lb \leq f_{i,j} \leq ub, i = 1, ..., N, j = 1, ..., F
\end{gathered} \tag{5}
$$

To optimize the forecasting performance we choose the tuning parameters using the rolling window technique with window size of 120 h over the training set T_{train}. We minimize average mean square forecast error (MSE), $MSE = \sum_{t \in T_{train}} \sum_{h=0}^{H-1} (q_{t+h} - \hat{q}_{t+h})^2 / (H * |T_{train}|)$, for each node by performing the grid search for $\lambda \in \{0, ..., 1\}$, where $q_{t,h}$ and $\hat{q}_{t,h}$ are the real and forecasted values of the natural gas flows at hour t and H is a forecasting horizon.

3 Experimental Setup and Results

In this paper we investigate the dynamic patterns of natural gas flows in the high-pressure gas pipeline network of OGE [2]. The dataset consists of demand and supply flows with an hourly time resolution for a period of 21 months (625 days). The network contains 1029 nodes in total. We consider 210 nodes with mean daily flow of more than 25 MW and less than 95% of zero flows. In the observed data set we have 14 supply nodes (labeled S1–S14), 9 storages (can change a behavior and have both positive and negative flows over time, labeled ST1–ST9) and 187 demand nodes (labeled D1–D187). Furthermore, we add an artificial node to the network to represent the contribution of the nodes that are less important in terms of volume and active time.

All gas flow data are normalized with mean zero and unit variance. Figure 1 illustrates the temporal dependence among the observed 210 nodes. As it can be seen in the diagonal, there is a strong positive autocorrelation of each node with its own past values. Off the diagonal, the cross-correlations represent the dynamic flow of gas from one node to another. While all of the supply nodes and most of the demand nodes contain their own predictive information, the dynamic dependence in the network is sparse and is driven by small number of nodes. Among the supply nodes, only the 4 seem to be active, demonstrating a strong positive cross-correlation within the supply group and strong negative cross-correlation to some demand nodes and very limited dependence with the storage nodes. This is also the case for the demand nodes. The temporal dependence suggests that only a small number of nodes may have a significant effect on the future gas flows in the network.

We use training-validation sample of 260 days (T_{train}) to choose optimal hyperparameters for the sparsity estimation (see Sect. 2). Further, we use the chosen parameters to estimate the large-scale weighted adjacency matrix W at each point in the testing period T_{test} consisting of 365 days (T_{test}). With a rolling window size of 120 h, we move forward one period at a time to update adjacency matrix and calculate the forecast for three different forecasting horizons (1 h, 6 h and 12 h ahead), until we reach the end of the sample. We compare the performance of NAR-LB model with several well-known benchmarks: Baseline forecast (repeating last known value for the same hour in the day), ARIMA and LSTM. We determine the best ARIMA models for a univariate time series of 210 nodes according to a Akaike information criterion (AIC) using 28 days of rolling window. The LSTM is implemented using one LSTM hidden layer with an optimized number of hidden neurons for each node (32–128) and learning rate for each individual node (0.0001–0.001), a dropout of 0.1 followed by a fully connected output layer with the number of neurons equal to the forecast horizon and trained for 100 epochs. The parameters are optimized based on the performance on the training set (T_{train}).

The performance of NAR-LB model is measured and quantified by calculating the forecast accuracy for individual nodes, as well as the mean for the entire network. We use root mean squared error (RMSE), as well as normalized mean absolute percentage error (nMAPE) defined as:

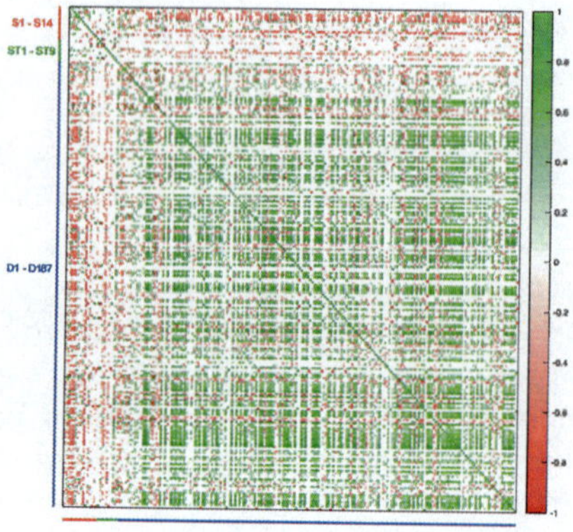

Fig. 1. Sample cross-correlation heatmap for 210 nodes in gas network.

$RMSE = [\sum_{t \in T_{test}} \sum_{h=0}^{H-1} (q_{t+h} - \hat{q}_{t+h})^2 / (H * |T_{test}|)]^{1/2}$ and $nMAPE = (\sum_{t \in T_{test}} \sum_{h=0}^{H-1} |(q_{t+h} - \hat{q}_{t+h}) / \max(q)|) / (H * |T_{test}|)$, where $q_{t,h}$ and $\hat{q}_{t,h}$ are the real and forecasted values of the natural gas flows at hour t and H is a forecasting horizon.

Table 1. Performance comparison

	RMSE				nMAPE			
H	NAR-LB	BAS	ARIMA	LSTM	NAR-LB	BAS	ARIMA	LSTM
1	0.249	0.534	0.408	0.387	0.095	0.261	0.150	0.177
6	0.443	0.534	0.503	0.462	0.185	0.261	0.281	0.205
12	0.513	0.534	0.559	0.559	0.220	0.261	0.299	0.299

In Table 1 we report an average $RMSE$ and $nMAPE$ for three considered forecasting horizons and comparing to three alternative benchmark models. The results show that NAR-LB consistently outperforms all benchmark models. The difference is most significant for the shorter horizons where mean $nMAPE$ is improved for 37% comparing to second best alternative model, while for the longer horizon (12 h) NAR-LB perform similar to LSTM model with the improvement of 7.6%. It is clear that the proposed model benefits from modeling temporal dependencies in the network. Figure 2 illustrates the estimated weighted adjacency matrix $\hat{W}$ in different periods of the year. It can be seen that during the winter time there is much more dynamic in the network while during the summer time, the number of influential nodes is significantly smaller.

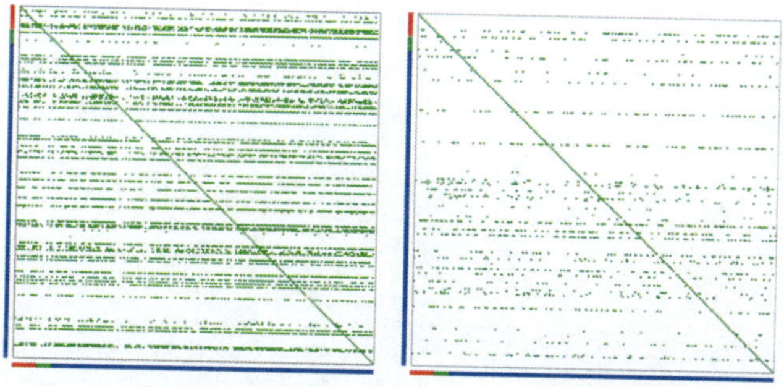

(a) Winter (b) Summer

Fig. 2. Illustration of the estimated weights matrix $\hat{W}$ in out-of-sample forecast

4 Conclusion

In this paper we propose the Network Autoregression Linear model with Balance constraint for identifying the influential nodes in the large-scale complex gas transmission network and multi-step ahead forecasting of gas flows in the network. The obtained results show that NAR-LB model consistently outperforms the alternative models by at least 7.6% giving accurate multi-step forecast results for more than 200 nodes in gas network simultaneously providing the information about temporal dynamic of the network flow.

Acknowledgement. The work for this article has been conducted within the Research Campus Modal funded by the German Federal Ministry of Education and Research (fund numbers 05M14ZAM, 05M20ZBM).

References

1. Arbeitsgemeinschaft Energiebilanzen: Evaluation tables on the energy balance for the Federal Republic of Germany 1990 to 2019 (2019)
2. Open Grid Europe GmbH. https://oge.net/. Accessed 15 June 2021
3. Chen, Y., Koch, T., Zakiyeva, N., Zhu, B.: Modeling and forecasting the dynamics of the natural gas transmission network in Germany with the demand and supply balance constraint. Appl. Energy **278** (2020). ISSN: 0306-2619
4. Chen, Y., Chua, W.S., Koch, T.: Forecasting day-ahead high-resolution natural-gas demand and supply in Germany. Appl. Energy **228** (2018). ISSN: 0306-2619
5. Petkovic, M., et al.: A hybrid approach for high precision prediction of gas flows, ZIB Report 19-26 (2019)

Health Care Management

Classifying Ready-for-Transfer Patients in the Intensive Care Unit Based on Clinical Data

Franz Ehm[1(✉)], Volkmar Franz[2], Maic Regner[2], Udo Buscher[1], Hanns-Christoph Held[3], and Peter Spieth[2]

[1] Department of Industrial Management, Technische Universiät Dresden, Dresden, Germany
franz.ehm@tu-dresden.de
[2] Department of Anesthesiology and Critical Care Medicine, Universitätsklinikum Carl Gustav Carus Dresden, Dresden, Germany
[3] Surgical Intensive Care Unit, Universitätsklinikum Carl Gustav Carus Dresden, Dresden, Germany

Abstract. In the intensive care unit (ICU), a common task for clinicians is to choose patients who are ready-for-transfer to a lower ward in order to make limited capacity available for new arrivals. To support this process, we build three predictive models based on historical data from more than 25,000 ICU cases to evaluate patients according to their actual medical state. The decision is modeled as a classification problem to predict the chance of adverse patient outcome defined by ICU-readmission within 72 h or readmission with subsequent exitus. In addition to a screening method based on critical criteria, we propose logistic regression models relying on critical parameter counts and metrical features from measurements, scores, and patient characteristics, respectively. Performance testing using ICU data demonstrates the ability of our approach to assist the process of patient selection for transfer.

Keywords: Intensive care · Patient transfer · Machine learning

1 Introduction

While the intensive care unit (ICU) is crucial to treat the most severely sick patients its operation is expensive and resources are limited. In order to enable new admissions to the ICU, clinicians have to regularly choose patients who are ready for transfer to a lower ward. The decision is subject to a variety of medical and operational factors compounded by uncertainty, stress or fatigue. Data-based tools offer support to complement clinical expertise and existing guidelines on discharge policies. In this paper, we propose three classification models to predict a patients risk of subsequent readmission or mortality following transfer based on historical ICU data. After discussing previous research on this topic in Sect. 2, we describe the design of the proposed classifiers and analyze their performance using data from a case study in Sect. 3. Conclusions are drawn in Sect. 4.

N. Trautmann and M. Gnägi (Eds.): OR 2021, LNOR, pp. 209–215, 2022.
https://doi.org/10.1007/978-3-031-08623-6_32

Table 1. Confusion matrix and accuracy metrics ($RFT = 0$, $NRFT = 1$)

	Predicted 1	Predicted 0			n=25,179		n=24,051
Actual 1	*True positive* (TP)	*False negative* (FN)	$Sensitivity = \frac{TP}{TP+FN}$	w/ exitus	2,523	no exitus	1,395
Actual 0	*False positive* (FP)	*True negative* (TN)	$Specificity = \frac{TN}{TN+FP}$		22,656		22,656
	$Accuracy = \frac{TP+TN}{TP+FP+TN+FN}$						

2 Patient Classification for ICU Admission and Discharge

The relevance of sound admission and discharge decisions for ICU planning is pointed out by Bai et al. [1]. The authors review operations research and management science in the context of ICU management. Classification models assist in finding optimal admission or discharge policies. Kim et al. estimate patient outcome based on physiological and operational variables [2]. Different models for prediction of a patients discharge status from clinical information at the time of admission were studied by Roumani et al. [3].

At the other end, several researchers proposed models to identify ready-for-transfer patients by predicting patient outcome following discharge. Since therapeutic benefit is generally hard to estimate (see [4]) readiness is indirectly assessed by evaluating adverse patient outcome in the retrospective of an ICU stay. McWilliams et al. defined patients as non-ready-for-discharge if they had to be readmitted to ICU or died in hospital following transfer. Vital parameters and patient data were used to design random forest tree and logistic regression classifiers and compare against clinical guidelines [5]. In similar approaches, multivariate logistic regression [6], transfer learning models [7], and genetic feature weighting [8] were proposed. Rosa et al. compared a variety of ICU related illness scores to predict readmission within 48 h or death following discharge [9].

In this study, we deploy (N)RFT to denote patients as (non-)ready-for-transfer. The label of interest, NRFT, is characterized by 'readmission within 72 h' and/or 'readmission with subsequent death' at ICU. Another issue is whether or not patients who die during their current stay should be included in the classification [6]. We consider both options by differentiating scenarios with and without exitus. Table 1 displays relevant definitions and metrics to assess predictive quality in binary classification. Accuracy metrics reflect the share of predicted labels which are derived from the estimated class probabilities using a cutoff value. As shown in Table 1, we face unbalanced data meaning positives are relatively rare among the observed classes. Although false predictions within this smaller part of data do not greatly compromise overall accuracy, the cost of misjudging rare critical cases is substantial. This is a typical problem in medical classification and can be addressed by shifting the cutoff point via a misclassification cost ratio to emphasize true positives (TP) at the expense of false positive predictions (FP) [3].

3 Design and Comparison of Classifiers

3.1 ICU Data

Clinical data was extracted from *Intensive Care Manager (ICM)*, a patient data management software supplied by *Dräger*. For the considered anesthetic ICU 25,552 cases were reported between 2003-05-15 and 2020-09-01. Each case denotes an episode of stay at ICU, i.e. one patient might relate to multiple cases in the event of readmission. Data was structured in separate sets that account for transfer history and patient characteristics (*Stays*), vital and laboratory parameters (*Measurements*), categories in the "Simplified Acute Physiology Score" (*SAPS*) and other medication or therapy related information. In the course of preprocessing, data was stripped from cases with missing, unrealistic or non-conform entries. In some cases, data was replaced by imputing values from other sources of information. Further elimination of cases was required to synchronize differences in the availability and time-wise resolution of data. As shown in Fig. 1, the final data sets for single- and multi-variate classification comprise 25,179 and 10,743 cases, respectively.[1] Technically, classifiers rely on training data to estimate model parameters and predict labels of unseen (test) data. Most studies presented in Sect. 2 deployed k-fold cross validation to minimize statistic noise by creating k split samples. We adopt this practice by using five folds to cross validate our classifiers.

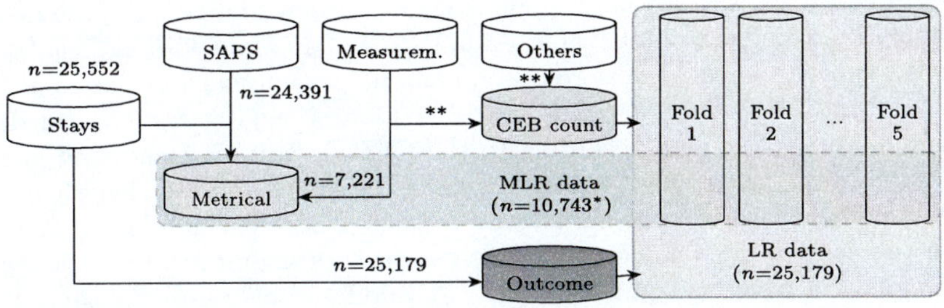

Fig. 1. Processing singlevariate (LR) and multivariate (MLR) feature data for 5-fold cross validation with exitus cases included as NRFT and CEB count computed according to the approach described in Sect. 3.2 (*fewer cases with incomplete data after feature selection, **missing values were considered as non-critical)

3.2 Clinical Expertise Based Classifier

Expert knowledge from physicians at Uniklinikum Dresden (UKD) is used to establish a first classifier. This so called clinical expertise based (CEB) classifier combines practical expertise and common clinical rules to form criteria for ICU discharge. Aside from knock-out criteria (e.g. ventilation), CEB defines a set

[1] This refers to the scenario with exitus cases included as NRFT.

Table 2. CEB parameters by critical upper (UB), lower (LB) or two-sided bounds (2B)

I_{UB} :	Temperature, Lactat, C-reactive protein, Procalcitonin, Urea, Creatinin, Bilirubin, FiO_2, ALAT, ASAT, O_2 (intake), red blood cells (intake)
I_{LB} :	Hemoglobin, arterial blood pressure, pH, Quick, GCS (score), Albumin, PaO_2, Extubation time
I_{2B} :	Heart rate, Kalium, Natrium, Glucose, Leucocytes, RASS (score)

of relevant parameters I which are considered critical if their values v exceed certain upper (UB), lower (LB) or two-sided bounds (2B) as shown in Table 2. We use the term *CEB-t* to denote a classifier which labels a case as NRFT or $y = 1$ if the critical parameter count x_{CEB} reaches a certain threshold t, i.e.

$$y = 1, \quad if \quad t \leq x_{CEB} := \sum_{i \in I_{UB} \cup I_{2B}} ||v_i > UB|| + \sum_{i \in I_{LB} \cup I_{2B}} ||v_i < LB|| \tag{1}$$

where v_i is an aggregate measurement for parameter i 24 h prior to discharge.

3.3 Logistic Regression

Single-Variate Model: Logistic regression (LR) uses training data to fit an s-shaped probability function to the binary classification data. In singlevariate LR the estimate y_{prob} is based on a single feature x. Binary labels y_{pred} are derived by applying a defined cutoff value $\in [0, 1]$ to y_{prob}. Time-wise aggregation of data is required to synchronize resolution among parameters and was conducted by computing daily *min*, *max* or *avg*. Using these aggregates, we can determine daily critical counts according to the CEB definitions mentioned above.[2] To enable joint processing of all ICU cases regardless of their length of stay (LOS), daily critical counts have to be further condensed into single features. From the various possible combinations two feature definitions were chosen: 1) *total_critical_0* which is equivalent to x_{CEB} and 2) *total_critical_weighted* being the sum of daily critical counts smoothed by a weigthing factor $1/(1+offset\_day)$. This was done to reflect time progress without a change in criticality thereby rewarding prolonged stays to "stabilize" non-critical patients with regard to clinical application of the tool. The two resulting models *LR-0* and *LR-w* are trained using the 'liblinear' solver with *l1* regularization from the *scikit-learn* library in Python.

Multi-variate Model: Condensing variables into a single feature means valuable information of individual clinical parameters might be lost. The proposed multivariate model (MLR) features an ensemble of 52 metrical variables mainly derived from time-wise aggregates of 21 CEB parameters. For each parameter,

[2] Time windows for aggregation were defined by offset-days, i.e. multiples of 24-h prior to the date of discharge.

we considered the average over LOS (*avg_los*). We further included the reported maximum (*max_0*) and/or minimum value (*min_0*) from the last 24 h prior to discharge in accordance to the CEB bounds. The resulting 47 features were complemented by SAPS values (*saps_avg_los*, *saps_0*) and patient characteristics: *gender*, *age*, and *body mass index*. Feature selection was performed by backward elimination of variables which 1) showed a low significance, i.e. $p \geq 0.05$ or 2) a high variance inflation factor, i.e. $vif > 5$ in any stage of repeated model fitting. This way the number of variables in the model was narrowed down to 18 and 11 in the scenarios with and without exitus, respectively.

For the training of the MLR model we relied on the binomial 'General Linear Models' solver from the *StatsModels* library in Python. Inconveniently, the use of metrical variables means a significant amount of cases that has to be excluded from analysis due to incomplete data. The MLR model could thus only be trained and validated on respective subsets of the cross fold splits used for CEB and LR as shown in Fig. 1. Still, feature selection allowed some of the previously dropped cases to be reconsidered for final model training and testing.

Cutoff Variation: Predictive outcome of the (M)LR models is highly sensitive to the choice of cutoff. Specificity increases with higher cutoff while sensitivity decreases. In this study, we aim for the "sweet spot" between both measures by varying the threshold probability until specificity reaches the level of sensitivity.

Table 3. Comparing performance metrics of classifiers: CEB with $t = 1, 2$, LR with *total_critical_0*, *total_critical_weighted*, and MLR using mean values (SD) from 5-fold cross validation for NRFT with exitus

	CEB-1	CEB-2	LR-0	LR-w	MLR
AUROC	0.649 (n.a.)	0.600 (n.a.)	0.687 (0.018)	0.734 (0.020)	0.771 (0.018)
Accuracy	0.803 (n.a.)	0.891 (n.a.)	0.505 (0.024)	0.651 (0.039)	0.659 (0.047)
Sensitivity	0.457 (n.a.)	0.235 (n.a.)	0.752 (0.026)	0.694 (0.046)	0.702 (0.057)
Specificity	0.841 (n.a.)	0.964 (n.a.)	0.478 (0.027)	0.646 (0.046)	0.654 (0.057)
Cutoff	1.000 (n.a.)	1.000 (n.a.)	0.090 (0.000)	0.070 (0.000)	0.068 (0.004)

3.4 Classifier Performance

From the results in Table 3 we note CEB classifiers to show the highest accuracy albeit at the cost of sensitivity. CEB with threshold $t = 1$ performs better than $t = 2$. In contrast, (M)LR models yield reasonably high sensitivity which is desired from a clinical perspective. This can be attributed to the sensitive choice of relatively small cutoff values. It also confirms that accuracy is misleading as a performance measure when facing imbalanced data. As illustrated by the ROC curves in Fig. 2, the best predictive quality is achieved by MLR. At the same time, LR-w outperforms any other single-variate classifier justifying the weighting method in the critical count definition. We also note that the inclusion of exitus data makes for an easier distinction of labels. This result was also noted in by Badawi/Breslow [6] and seems logical given the multimorbid condition of near exitus patients.

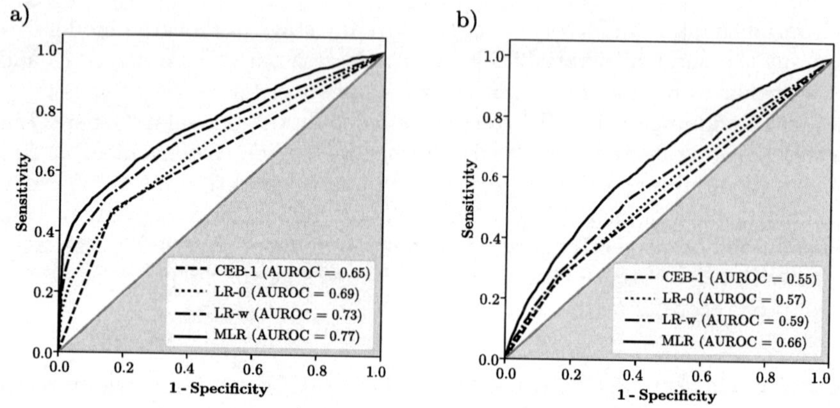

Fig. 2. Receiver operating characteristic (ROC) curves for different classifiers with a) exitus cases considered as NRFT, b) exitus cases excluded from analysis

4 Conclusions

In this study, three approaches were presented to classify ready-for-transfer patients at the ICU. Results for NRFT w/o exitus suggest a bias in the data considering that historical transfer decisions were already taken by experts. We can thus use the results to reflect on the effectiveness of previous clinical decisions which seem to be in accordance with CEB definitions. Nevertheless, we note the promising potential especially of the (M)LR models to assist in the identification of RFT patients to minimize risk of adverse outcome. While MLR yields the best overall performance LR-w represents a valid and robust alternative for daily clinical use since it requires less complete data entries for classification.

Acknowledgement. This study was funded by Else-Kröner-Fresenius Zentrum (EKFZ) for Digital Health. The authors are grateful for the valuable comments by the reviewer.

References

1. Bai, J., et al.: Operations research in ICU management: a literature review. Health care Manage. Sci. **21**(1), 1–24 (2018)
2. Kim, S.H., et al.: ICU admission control: an empirical study of capacity allocation and its implication for patient outcomes. Manage. Sci. **61**(1), 19–38 (2015)

3. Roumani, Y.F., et al.: Classifying highly imbalanced ICU data. Health Care Manage. Sci. **16**(2), 119–128 (2013)
4. Skowronski, G.A.: Bed rationing and allocation in the intensive care unit. Current Opin. Crit. Care **7**(6), 480–484 (2001)
5. McWilliams, C.J., et al.: Towards a decision support tool for intensive care discharge: machine learning algorithm development using electronic healthcare data from MIMIC-III and Bristol. BMJ Open. **9**, e025925 (2019)
6. Badawi, O., Breslow, M.J.: Readmissions and death after ICU discharge: development and validation of two predictive models. PLoS One **7**, e48758 (2012)
7. Desautels, T., et al.: Prediction of early unplanned ICU readmission in a UK tertiary care hospital: a crosssectional machine learning approach. BMJ Open **7**, e017199 (2017)
8. Kramer, A.: A novel method using vital signs information for assistance in making a discharge decision from the ICU. Med. Res. Arch. **5**(12) (2017)
9. Rosa, R.G., et al.: Comparison of unplanned intensive care unit readmission scores: a prospective study. PLoS ONE **10**(11), e0143127 (2015)

Leveling ICU Bed Occupancy Using a Quota System: An Applied Heuristic for Operating Theater Managers

Steffen Heider[1,2](✉)

[1] Chair of Health Care Operations/Health Information Management, Faculty of Business and Economics, University of Augsburg, Universitätsstraße 16, 86159 Augsburg, Germany

[2] Unit of Digitalization and Business Analytics, Universitätsklinikum Augsburg, Stenglinstraße 2, 86156 Augsburg, Germany

steffen.heider@uni-a.de

Abstract. The operating theater, as well as the intensive care unit, are both one of the most expensive departments within a hospital but also one of the largest revenue drivers. Extensive planning on multiple strategic levels is necessary, to guarantee patient safety, workload leveling as well as profitability of a hospital. Patients are scheduled for each department individually but also jointly across departments when resources are shared. In research, many papers focus on optimizing the utilization of each department individually but also on the patient flow from one department to the other. However, few papers focus on the development of scheduling heuristics that can be used by operating theater managers without knowledge of mathematical optimization. We present an operating theater quota system for elective intensive care patients that minimizes the expected maximum bed demand in the intensive care unit. We develop a heuristic that can be easily understood and applied by operating theater managers. We use multiple instances to show that the heuristic can achieve near-optimal results by comparing the heuristic with an optimal approach.

Keywords: Health care · Scheduling

1 Introduction

Scheduling surgeries in the operating theater (OT) is a challenging task for physicians and OT managers. When scheduling, not only the capacity within the OT needs to be considered, also the capacity of downstream units, e.g., the intensive care unit (ICU) or regular ward stations. Especially the ICU is often seen as a bottleneck where poor planning can lead to canceled surgeries or early discharges, which show a higher readmission rate compared to intended discharges, all harming patient recovery [1].

In literature, many papers focus on surgery scheduling considering one or multiple up- or downstream units to adequately model the patient flow through the

N. Trautmann and M. Gnägi (Eds.): OR 2021, LNOR, pp. 216–222, 2022.
https://doi.org/10.1007/978-3-031-08623-6_33

hospital. For a general overview of literature with a focus on multiple departments including the OT and the ICU see [5], for a recent overview focused on OT scheduling in general see [6]. [4] use a simulation model and show that the introduction of ICU quotas for elective patients drastically reduces the number of last-minute cancellations of those patients. [3] use a combined optimization and simulation approach to show that a central scheduling approach of ICU patients across departments can reduce ICU bed utilization variability by up to 17.5%.

Even though these works cover interesting conclusions for hospitals and OT managers, practical applicability is limited, especially without mathematical optimization knowledge. We present an easy-to-implement scheduling heuristic for ICU quotas in the OT - a system that reduces the number of canceled elective ICU surgeries and the bed utilization variability in the ICU (i.e., [2–4]) - aimed at OT managers and practitioners in hospitals. The quota system determines ICU slots in the OT to limit and control the daily number of ICU patients for each specialty to implicitly level the average bed utilization. The quota system is in active use at University Hospital of Augsburg, one of the largest hospitals in Germany [2]. We use multiple instances of small, medium-sized, and large OTs with varying patient demand and length of stay (LOS) distributions to show that our heuristic produces near-optimal results compared to a mixed-integer program (MIP).

2 Data Preparation and Mathematical Model

The heuristic requires little data which can be determined using historical data. For each day $t \in T$ in the planning horizon, i.e. 7 or 14 days, we need to know the maximum number of possible ICU patients or slots $C_{s,t}$ for each specialty $s \in S$. In practice, there are usually not more elective ICU patients than rooms per specialty and day, which means $C_{s,t}$ is equal to the number of rooms for each specialty s on day t. The second part of data consists of the elective ICU patient demand P_s for each specialty s in the planning horizon rounded to the next integer. The last part of data is the convolved LOS distribution $L_{s,t}$ for each specialty s on day t in the planning horizon. This distribution depicts the probability that an ICU patient still needs a bed t days after surgery. Table 1 shows an exemplary calculation of the convolved complementary cumulative distribution function (convolved CCDF) for one specialty.

Table 1. Preprocessing of convolved cumulative distribution function

t	0	1	2	3	4	5	6	7	8	9	Sum
Historical observation [n]	35	229	49	28	26	19	6	11	8	4	415
PMF [%]	8.4	55.2	11.8	6.7	6.3	4.6	1.4	2.7	1.9	1.0	100
CCDF [%]	0.0	91.6	36.4	24.6	17.8	11.6	7.0	5.5	2.9	1.0	
Convolved CCDF [%]	5.5	94.5	37.3	24.6	17.8	11.6	7.0	-	-	-	

In the first step, historically observed LOS values rounded to the next integer need to be counted. In this example 229 patients stayed one day in the ICU, 415 patients were observed in total. The probability mass function (PMF) determines the relative probability that a patient stayed exactly t days, in this case 229 out of 415 patients or 55.2% for $t = 1$. The CCDF is the probability that a patient is still in the ICU after t days, which can be calculated by the sum of the PMF of day t to the last day. The CCDF of day 0 is an exception and is equal to 0, since patients with a rounded LOS of 0 are neglected. Lastly, the convolved CCDF is used to account for the cyclical approach of this model. This is necessary to model LOS distribution in cases where the LOS exceeds the planning horizon. The convolved LOS has to have the same length as the planning horizon - in this case 7 days - and can be calculated by summing up all CCDF values that exceed the planning horizon to the corresponding days. For example, the convolved CCDF value for 1 day after surgery is the sum of the CCDF of day 1 and day 8 or 91.6% and 2.9%, respectively. Hereby we model the fact that if the quotas are the same every week, there is a 2.9% probability that a patient from last week is still in the ICU, resulting in 94.5% total. If the CCDF would exceed two weeks in total, the convolved CCDF for day 1 would include the CCDF value of day 15 if the planning horizon is still 7 days.

The goal of the mathematical model is the minimization of the maximum bed utilization b^{max} in the planning horizon by determining the number of ICU slots $x_{s,t}$ for each specialty s on each day t and therefore to minimize the bed utilization variability. The mathematical model is shown in the following.

$$\text{Min } b^{max} \tag{1}$$

$$\text{subject to: } x_{s,t} \leq C_{s,t} \qquad \forall\, s \in S,\ t \in T \tag{2}$$

$$\sum_{t \in T} x_{s,t} = P_s \qquad \forall\, s \in S \tag{3}$$

$$\sum_{s \in S} \sum_{k \in T} L_{s,k} \cdot x_{s,(t-k) mod |T|} \leq b^{max} \qquad \forall\, t \in T \tag{4}$$

$$x_{s,t} \in \mathbb{N}_0 \qquad \forall\, s \in S,\ t \in T \tag{5}$$

$$b^{max} \geq 0 \tag{6}$$

The objective function in (1) minimizes the maximum bed utilization determined in (4). Constraints (2) limit the maximum number of ICU slots for each specialty on each day. With Constraints (3) the number of slots that need to be distributed for each specialty in the planning horizon is set. Constraints (4) determine the resulting bed utilization resulting from the ICU slots in the OT and therefore the maximum bed utilization in the planning horizon (see [3] for more details). Constraints (5) and (6) determine the domain of the decision variables.

3 ICU Quota System Heuristic

The goal of the heuristic is to have a close to optimal solution of the presented mathematical model that is easy to understand for OT managers and requires few steps. The developed heuristic can be executed with pen and paper but implementation in a spreadsheet, e.g., Microsoft Excel is recommended. The number of calculations depends on the planning horizon $|T|$, the number of ICU slots I, and the number of days the OT is open O. The number of required steps by hand is equal to $|T| \cdot I \cdot O$, which can be reduced to $I \cdot O$ if done with formulas in a spreadsheet. The heuristic works similar to a bin packing heuristic and we use four different sorting algorithms to determine which specialty should be scheduled first: mean LOS - which is equal to the sum of all CCDF values - of each specialty in ascending and descending order, as well as the mean LOS multiplied by the number of ICU slots of each specialty also in ascending and descending order. A pseudo code of the heuristic is shown in the following.

Procedure Heuristic

```
sort specialties defined by rule
for s = 1 → |S| do
    for i = 1 → |P_s| do
        for t = 1 → O do
            if one additional slot is possible then
                add LOS distribution to ICU utilization
                save b^max of t
                revert LOS distribution from ICU utilization
        schedule ICU slot on day t where b^max is minimum
```

The central idea of the heuristic is to start with no slots and schedule one additional ICU slot on each day in the planning horizon where possible and save the resulting maximum utilization in the ICU for each day. The additional slot should be added where the new maximum bed utilization in the planning horizon b^{max} is minimum. To calculate the resulting bed utilization that results from adding a new slot, the convolved CCDF has to be added to the current bed utilization starting from day 0 of the convolved CCDF on the day of the added slot. This means that if a new slot is added on a Wednesday in a 7-day planning horizon, the convolved CCDF from 0 to 4 has to be added to Wednesday to Sunday and the values of 5 and 6 to Monday and Tuesday, respectively.

4 Computational Study

To test if the developed heuristic is able to deliver good results in various environments, we create multiple instances for this computational study. First, we generate LOS distributions based on a lognormal (σ, μ) distribution with σ ranging from 1 to 1.5 and μ ranging from 0 to 1, both incremented by 0.1, resulting in 66 distributions in total. The distributions are truncated after 70 days and are convolved as previously explained. We calculate the mean LOS of every distribution and build 3 clusters - short, medium, and long - containing 22 distributions each, grouped by the mean LOS. As a second step, we create different OT setups. There are 3 criteria determining the OT instances: the number of specialties, the number of operating rooms, and the allocation of specialties to operating rooms. We use either 3, 6, or 9 specialties. The number of rooms is either equal to the number of specialties or twice that number. If the allocation of specialties to rooms is equally distributed, each specialty has the same number of rooms available in the planning horizon which is set to 7 days. Since we only consider elective patients, there are no rooms available on weekend days. If the allocation of specialties to rooms is not equally distributed, the available total blocks, which is the number of rooms multiplied by the number of days the OT is opened for elective patients, are assigned to the specialties in ascending order. Each specialty gets assigned a random number between 0 and half of the remaining blocks in the planning horizon. The last specialty in the assignment process gets all remaining blocks. The block to day assignment follows in a second step. Here, starting from the first specialty, every block of each specialty is assigned in steps of one to the consecutive day. If specialty 1 would have 4 blocks available, these blocks would be assigned from Monday to Thursday. Specialty 2 would then start with the first assigned block on Friday and all remaining blocks would be assigned starting from Monday again. By assigning the blocks in this way, the number of blocks per specialty each day is equally distributed and results in a balanced workload for surgeons in the OT. Lastly, the total number of ICU slots in the planning horizon is twice the number of specialties. The number of ICU slots per specialty is set in relation to the number of blocks per specialty. In total 12 different OT instances are generated. An excerpt of the generated OT instances can be seen in Table 2.

Table 2. Excerpt of generated OT instances

Instance	Specialties	Rooms	Equal OT	ICU slots	OT blocks			ICU slots		
					S1	S2	S3	S1	S2	S3
1	3	3	No	6	4	2	9	2	1	3
2	3	3	Yes	6	5	5	5	2	2	2
3	3	6	No	12	6	2	22	2	1	9
4	3	6	Yes	12	10	10	10	4	4	4
...	...	...	...	...	...	...	...	...	...	...

In the final step, the generated LOS distributions and the OT instances are combined. We use the 12 OT instances to generate a total of 48 instances divided into 4 groups. 3 groups combine the 12 OT instances with randomly assigned LOS distributions from only the short, medium, and long group, respectively. The last group evenly distributes randomly picked LOS distributions from the short, medium, and long group in relation to the number of specialties. Within one instance, no LOS distribution is assigned twice. For the computational study, the mathematical model is implemented in IBM ILOG CPLEX 12.9 and all 48 instances are solved to optimality. The heuristic is implemented in Microsoft Excel using Visual Basic for Applications (VBA). We compare the relative gap to optimality of all instances using the four different sorting algorithms to test if one algorithm outperforms others in some environments. The results are shown in Table 3. In general, all four sorting algorithms perform very well with the heuristic. The optimality gap ranges from 0.1% to 8.7%, the average optimality gap over all instances from 1.3% to 1.6%. All heuristics perform slightly better with longer LOS distributions compared to short LOS distributions. There is no clear sign if equal or unequal distribution of blocks to specialties performs better or not. In total, the ascending sorting algorithm where the LOS is multiplied with the number of ICU slots performs best on average and has the smallest range from worst to best optimality gap. Therefore, we recommend using this sorting algorithm with the developed heuristic.

Table 3. Optimality gap of the heuristic for grouped instances - mean [min – max]

Instances	LOS·n des %	LOS·n asc %	LOS des %	LOS asc %
All	1.5 [0.3–5.4]	1.3 [0.1–3.6]	1.3 [0.1–5.5]	1.6 [0.2–8.7]
Unequal OT	1.9 [0.3–5.4]	1.3 [0.1–3.6]	1.5 [0.1–5.5]	1.9 [0.2–8.7]
Equal OT	1.1 [0.4–2.5]	1.3 [0.7–2.4]	1.1 [0.4–2.5]	1.3 [0.7–2.4]
Short	2.1 [0.3–5.4]	2.0 [0.5–3.6]	1.7 [0.1–3.7]	2.3 [0.9–5.1]
Unequal OT	2.6 [0.3–5.4]	2.2 [0.5–3.6]	1.9 [0.1–3.7]	2.7 [0.9–5.1]
Equal OT	1.6 [0.9–2.5]	1.8 [1.1–2.4]	1.6 [0.9–2.5]	1.8 [1.1–2.4]
Medium	1.1 [0.3–2.8]	0.9 [0.1–1.5]	1.4 [0.1–5.5]	1.8 [0.3–8.7]
Unequal OT	1.1 [0.3–2.8]	0.8 [0.1–1.2]	1.7 [0.1–5.5]	2.4 [0.3–8.7]
Equal OT	1.1 [0.4–1.8]	1.1 [0.7–1.5]	1.1 [0.4–1.8]	1.1 [0.7–1.5]
Long	0.9 [0.3–2.3]	0.9 [0.1–1.5]	0.8 [0.3–1.5]	0.9 [0.2–1.6]
Unequal OT	1.0 [0.3–2.3]	0.6 [0.1–1.3]	0.8 [0.3–1.2]	0.7 [0.2–1.6]
Equal OT	0.9 [0.6–1.5]	1.1 [0.9–1.5]	0.9 [0.6–1.5]	1.1 [0.9–1.5]
Mixed	1.9 [0.6–4.7]	1.4 [0.9–2.2]	1.2 [0.4–2.5]	1.5 [0.4–4.0]
Unequal OT	2.8 [1.5–4.7]	1.4 [1.0–2.2]	1.5 [0.4–2.5]	1.7 [0.4–4.0]
Equal OT	1.0 [0.6–1.3]	1.3 [0.9–1.7]	1.0 [0.6–1.3]	1.3 [0.9–1.7]

References

1. Baker, D.R., Pronovost, P.J., Morlock, L.L., Geocadin, R.G., Holzmueller, C.G.: Patient flow variability and unplanned readmissions to an intensive care unit. Crit. Care Med. **37**(11), 2882–2887 (2009)
2. Heider, S., Schoenfelder, J., Koperna, T., Brunner, J.O.: Balancing control and autonomy in master surgery scheduling: benefits of ICU quotas for recovery units. Health Care Manag. Sci. **25**, 311–332 (2022)
3. Heider, S., Schoenfelder, J., McRae, S., Koperna, T., Brunner, J.O.: Tactical scheduling of surgeries to level bed utilization in the intensive care unit. IISE Trans. Healthcare Syst. Eng. **10**(4), 229–242 (2020)

4. Kim, S.C., Horowitz, I.: Scheduling hospital services: the efficacy of elective-surgery quotas. Omega **30**(5), 335–346 (2002)
5. Vanberkel, P.T., Boucherie, R.J., Hans, E.W., Hurink, J.L., Litvak, N.: A survey of health care models that encompass multiple departments. Int. J. Health Manage. Inf. **1**(1), 37–69 (2010)
6. Wang, L., et al.: Operating room planning and scheduling for outpatients and inpatients: a review and future research. In: Working Papers of Department of Decision Sciences and Information Management, Leuven (2021)

The Effects of a Boarding Area on Patient Waiting Time in the Emergency Department Using a Discrete-Event Simulation

Jakob Heins[1,2](✉)

[1] Chair of Health Care Operations/Health Information Management, Faculty of Business and Economics, University of Augsburg, Universitätsstraße 16, 86159 Augsburg, Germany
[2] University Hospital Augsburg, Stenglinstraße 2, 86156 Augsburg, Germany
jakob.heins@wiwi.uni-augsburg.de

Abstract. Growing patient volume combined with limited capacities in emergency departments represents a challenge for hospital management to maintain a high quality of work without an increase in waiting times. Often patients cannot be directly admitted to the hospital after treatment in an emergency department due to limited bed capacities at the intensive care units or regular ward stations. Various approaches to solving this problem were proposed in recent years including ideas for active bed management within the emergency department or different allocation rules for specific boarding areas. This study develops a discrete-event simulation approach to simulate the patient flow of an emergency department based on real data. Furthermore, the set up of different patient boarding area systems are compared to other mechanisms such as adding more treatment beds to evaluate the effects on patients length of stay and waiting times in a scenario analysis.

Keywords: Discrete-event simulation · Healthcare · Emergency department · Boarding area

1 Introduction

In the last decades, growing pressures on hospital emergency departments (EDs) were creating major problems for hospital management. Overcrowded waiting areas in an ED not only lead to a reduced quality of care but also patient dissatisfaction [2]. The increasing demand from patients who need emergency care, the temporary use of bed capacity for intensive care patients, or the lack of response from regular wards for a fast and efficient admission after a patient has received initial treatment intensify the underlying problem [3].

The last few years have seen a variety of comprehensive literature reviews focusing on simulation modeling of EDs [5,8]. A particular field of simulation

N. Trautmann and M. Gnägi (Eds.): OR 2021, LNOR, pp. 223–228, 2022.
https://doi.org/10.1007/978-3-031-08623-6_34

studies in the ED concerns discrete-event simulation (DES). Authors [1] present a DES analysis study to investigate the effect of inpatient boarding on the ED showing significant improvements by using inpatient boarding areas. A boarding area is an area to hold patients due to the lack of free capacity of downstream units accepting new patients. Authors [6] investigate how patient waiting time can be reduced in the ED and conclude that only increasing the number of treatment beds can (though it does not necessarily) reduce that time.

Even though these papers offer promising results, there are often gaps in the data presented. As a consequence, more general assumptions are applied which effect the overall conclusions significantly. We present a DES to support in decision-making for a medium sized German hospital, located in South-Bavaria. Hereby, we use a data set based on real data from that hospital and simulate one week of patient flow throughout its ED. We also use different personnel resource capacities in combination with a certain shift plan. Furthermore, we determine the value of separate boarding areas for inpatients and outpatients. In particular, we give advise on the number of additional resources that should be implemented to provide a good patient care. We also determine the validation of the simulation model based on statistical hypotheses testing. As a result, shifting resources and establishing a boarding area for inpatients can be seen to have positive effects on the average length of stay (LOS) and waiting time for each patient category, whereas merely adding more treatment beds to the ED does not show significant improvements.

2 Emergency Department Simulation Model

In the simulation study presented here, a DES is used to replicate the ED of a medium sized German hospital, located in South-Bavaria, with 400 beds. After arrival, each patient is triaged into a specific severity group with the help of the Manchester Triage System (MTS). The MTS is a system that supports emergency nurses to categorize patients into different groups based on their underlying urgency levels [7]. Depending on the patient's urgency level, the maximum allowed waiting time differs among each group. Whereas the level *immediate* (red) means a patient definitively needs isolation and direct treatment, the second level *very urgent* (orange) allows a maximum waiting time of 10 min. Depending on the hospital, the maximum time limits for the other three categories *urgent* (yellow), *standard* (green), and *non-urgent* (blue) may differ. In the data set provided, the limits are set to 30, 90, and 120 min, respectively. Furthermore, patients are treated by a specialized physician. We consider the three different medical specialties to be general surgery, neurology, and internal medicine. The distribution for these values are drawn out according to the provided data set. Besides the previously mentioned physicians, nurses do also take care of the patients in the simulation model. These two resource types work according to a predefined shift plan to simulate reality. A simplified process flow can be seen in Fig. 1.

After triage, patients with the most urgent MTS group might use a shock room, whereas the other patients move into the waiting room until further treatment. The waiting room queue follows a strict priority-based order and depends

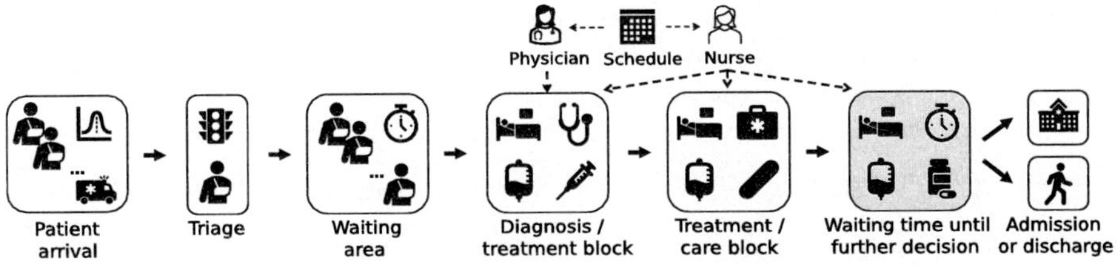

Fig. 1. Simplified process flow of the underlying problem

on the patient's urgency level. After the waiting room follows the first diagnosis/treatment block, in which a physician, nurse, and a treatment room are assigned. Afterwards, the patient may need further nurse care which is implemented by a second treatment block. After completion of the second care block, the patient is either discharged from or admitted to the hospital. As highlighted in Fig. 1, there is a waiting time after finishing the second treatment block until the discharge or admission is executed. During that waiting time of a patient, resources such as treatment bed or nurses are occupied. Afterwards, the treatment bed is freed up once more to be used for another patient.

The available data set provides patient data for one complete year with around 24,000 patient arrivals. During the day, the arrivals follow a specific arrival schedule with individual arrival times for each hour of a day with a peak of up to 5.2 patients between 10:00 am–11:00 am and the lowest arrival rate of 0.59 patients between 04:00 am–05:00 am. Furthermore, the distributions of the different MTS categories, medical specialties, and discharge or admission probabilities can be obtained. Within the simulation, the treatment times are stochastically derived from the data set. However, the exclusive physician treatment time is not given. We have therefore approximated the treatment time using a research study provided by [4] in combination with a triangular distribution using the given mean value as the mode and ±15 min as the interval borders, which enables to model a stochastic distribution for the physician treatment times. Due to variations in the arrival rate during the day, the shift schedules of the physicians and nurses are adapted accordingly based on real data. The number of available nurses, physicians, and treatment rooms for any particular patient volume is also provided by the data set.

3 Verification and Validation of the Simulation Model

The simulation model was implemented and verified using AnyLogic 8.7.5. One of the main measurable output parameters to operationally validate the simulation model is the average LOS for each patient urgency level. Figure 2 depicts this validation procedure graphically. The pairwise comparison of the three largest MTS groups for the real and simulated data shows similar interval borders. The data set contains some extreme outliers. For the sake of clarity, the outliers are not shown in the graphic. Furthermore, the two MTS groups *immediate* and *not-urgent* were not taken into account in the validation process since they only amount to 3% of the total patient arrivals in the data set and the stochastic inaccuracy arising from this might distort the results.

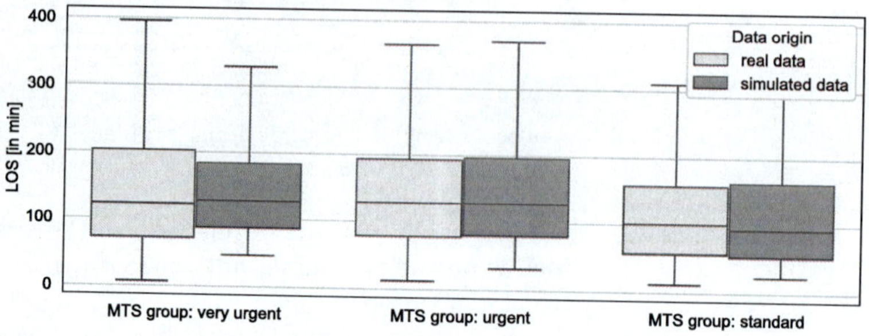

Fig. 2. Data distribution of the average LOS for three different MTS groups

In addition to the graphical validation, hypothesis testing can be used as another technique to compare means, variances, or distributions [9]. The Kol-mogorov-Smirnov test (or KS test) and the independent sample t-test are applied.

The sample data presents a timespan of one week. Applying the arrival distributions of the data set it follows that 46, 273, and 129 samples are randomly chosen representing the three largest MTS categories *very urgent*, *urgent*, and *standard*. The real data contains more than 24,000 patient arrivals for one year, the simulation generates 1,000 one-week (approx. 462 arrivals) replications. Assuming a significance level (α) of 0.05, the p-values of the KS test for the three MTS categories *very urgent*, *urgent*, and *standard* are 0.300, 0.579, 0.328, respectively. The p-values for the independent sample t-test are 0.366, 0.598, and 0.200, respectively. None of the H_0 hypotheses of the KS test as well as the t-test can be rejected. This underlines the successful operational validation [9,10].

4 Scenario Analysis

In this section, seven different scenarios are developed to discuss their effects on the average LOS as well as the waiting time of the patients. These two measurements are connected to each other. Hereby, Scenario 1 refers to the original simulated data. The first adjustments are performed in Scenarios 2 and 3, in which one and two more treatment rooms, each offering one treatment bed, are added to the DES. As showing in Fig. 3, the average LOS as well as the percentile of patients treated within its MTS category threshold could be slightly improved. Nevertheless, out of all scenarios, adding more treatment beds has the lowest rate of improvement.

Based on the data set, the outpatients might also posses a certain boarding time after completing treatment which is currently dismounted in the treatment

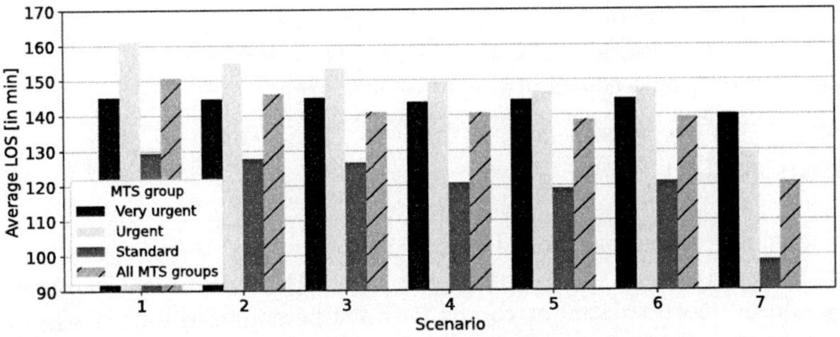

(a) *Average LOS*

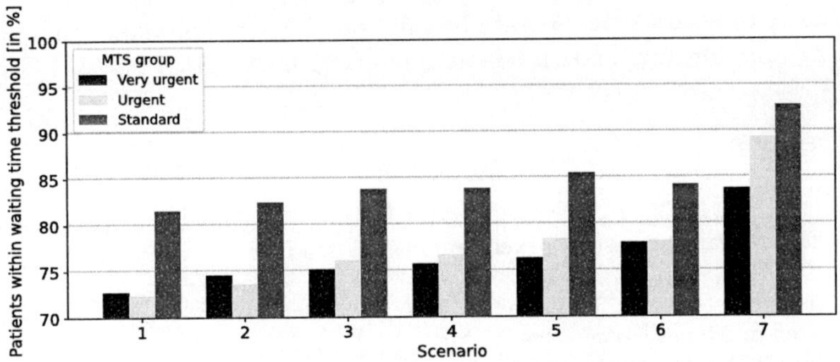

(b) *Percentile waiting time threshold*

Fig. 3. Results of the seven scenarios for the three MTS groups

room (Scenario 1). In the Scenarios 4 and 5, the newly generated room capacity is now no longer used as single treatment beds anymore, but rather as a boarding area for outpatients. Hereby, one newly built room can provide space for up to three boarding beds. Since these outpatients do not require further care, no additional nurse is required to be used for patient monitoring. Comparing Scenarios 4 and 5 with Scenarios 2 and 3, shows a slight better result for the waiting time and LOS using an outpatient boarding area.

An adaption of Scenario 5 is shown in Scenario 6 in which one of the two boarding rooms is used as an boarding area for inpatients and the other one as a boarding area for outpatients. Different from the outpatient boarding area, the inpatient boarding does certainly require patient monitoring. Hereby, one nurse from the early and late shift will be assigned from the current shift schedule to work full time at this inpatient boarding area. Due to the low staffing at night, a boarding area cannot be maintained during the night shift. Even though one nurse is missing two out of three shifts a day, the results in Fig. 3 assume similar treatment time as in Scenarios 3, 4, or 5.

In the final Scenario 7, the same concept as already introduced in Scenario 6 is employed. In addition to that, the monitoring of the inpatient boarding area does not require additional resources but are rather provided by the hospital management. A reduction of the average LOS throughout all MTS groups is noticeable and can be especially seen for less urgent MTS groups.

5 Conclusion

If management representatives plan to reconfigure EDs, they might also consider adding more treatment beds (rooms) to the ED environment. Unfortunately, merely adding more treatment rooms without personnel resource adaption, e.g. one treatment bed per room, does not significantly alter the average LOS experienced by patients in the different categories of urgency (see Fig. 3). A better approach is to establish a separate boarding area for outpatients and a separate boarding area for inpatients full-time monitored by an extra nurse.

References

1. Bair, A.E., Song, W.T., Chen, Y.C., Morris, B.A.: The impact of inpatient boarding on ED efficiency: a discrete-event simulation study. J. Med. Syst. **34**(5), 919–929 (2010)
2. Bernstein, S.L., et al: The effect of emergency department crowding on clinically oriented outcomes. Acad. Emerg. Med. **16**(1), 1–10 (2009)
3. Derlet, R.W.: Overcrowding in emergency departments: increased demand and decreased capacity. Ann. Emerg. Med. **39**(4), 430–432 (2002)
4. Gries, A., Michel, A., Bernhard, M., Martin, J.: Personnel planning in the emergency department. Der Anaesthesist **60**(1), 71–78 (2011)
5. Gul, M., Guneri, A.F.: A comprehensive review of emergency department simulation applications for normal and disaster conditions. Comput. Indus. Eng. **83**, 327–344 (2015)
6. Khare, R.K., Powell, E.S., Reinhardt, G., Lucenti, M.: Adding more beds to the emergency department or reducing admitted patient boarding times: which has a more significant influence on emergency department congestion? Ann. Emerg. Med. **53**(5), 575–585 (2009)
7. Mackway-Jones, K., Marsden, J., Windle, J.: Emergency Triage: Manchester Triage Group. John Wiley & Sons, Oxford (2014)
8. Salmon, A., Rachuba, S., Briscoe, S., Pitt, M.: A structured literature review of simulation modelling applied to emergency departments: current patterns and emerging trends. Oper. Res. Health Care **19**, 1–13 (2018)
9. Sargent, R.G.: Verification and validation of simulation models. In: Proceedings of the 2010 Winter Simulation Conference, pp. 166–183 (2010)
10. Stewart, W.J.: Probability, Markov Chains, Queues, and Simulation. Princeton University Press, Princeton (2009)

Artificial Intelligence-Based Decision Support in Laboratory Diagnostics

Alexander Scherrer(✉), Michael Helmling, Christian Singer, Sinan Riedel, and Karl-Heinz Küfer

Fraunhofer Institute for Industrial Mathematics (ITWM), Fraunhofer-Platz 1, 67663 Kaiserslautern, Germany
alexander.scherrer@itwm.fraunhofer.de

Abstract. This research work introduces a solution approach for detecting infectious diseases in modern laboratory diagnostics. It combines an artificial intelligence (AI)-based data analysis by means of random forest methods with decision support based on intuitive information display and suitable planning functionality. The approach thereby bridges between AI-based automation and human decision making. It is realized as a prototypical diagnostic web service and demonstrated for the example of Covid-19 and Influenza A/B detection.

Keywords: Laboratory diagnostics · Artificial intelligence · Decision support

1 Introduction

Many diseases like cancer or viral infections cause early characteristic value changes in blood counts. Laboratory medicine thus allows for an efficient screening and preliminary diagnosis [5], but also features high-throughput and widely manual processing of blood counts. This research work introduces an approach for efficient and reliable large-scale lab diagnostics featuring data analysis with artificial intelligence (AI) and mathematical decision support. Section 2 explains the underlying mathematics, Sect. 3 presents the corresponding web service and obtained numerical results and Sect. 4 provides a conclusion for this work.

2 Material and Methods

2.1 Concept

Figure 1 depicts the concept of AI-based decision support in laboratory diagnostics. The *existing lab diagnostics environment* supports the manual workflow of laboratory diagnostics. This workflow starts with the access to a blood count administrated in a lab information system (LIS), continues with its manual diagnostic assessment and ends with a documentation of the diagnosis in the LIS.

N. Trautmann and M. Gnägi (Eds.): OR 2021, LNOR, pp. 229–235, 2022.
https://doi.org/10.1007/978-3-031-08623-6_35

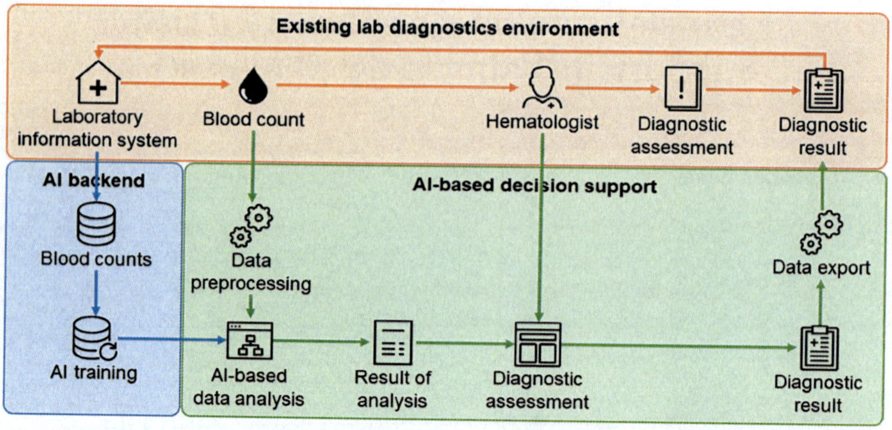

Fig. 1. Workflows of laboratory diagnostics: classical workflow based on the existing lab diagnostics environment and advanced workflow with AI-based decision support based on an AI back end (Icons: banzaitokyo.com, icons8.com, www.iconsmind.com)

The advanced workflow uses a diagnostic web service featuring *AI-based decision support*. Blood counts obtained from the LIS undergo a data preprocessing to enable successful data analysis. The preprocessed data enters an AI-based data analysis, which yields preliminary diagnoses for the blood counts. The results of analysis are provided to the hematologist in a decision support module for diagnostic assessment. The diagnostic results are exported for further processing in the LIS. The AI-based decision support builds upon an *AI back end* for training the random forest (RF) method on reference data from the LIS.

2.2 AI-Based Decision Support and AI Back End

Data Preprocessing and Export: These steps connect the advanced AI-based workflow to the existing environment. The data preprocessing comprises the import of a blood count from the LIS, its reduction to the crucial parameter values $c(p)$ as determined in the AI training and a check for value consistency [1]. The data export provides the diagnostic results to the LIS for further processing.

AI-Based Data Analysis - Decision Trees: The AI-based data analysis uses an RF binary classification method for assigning a positive or negative preliminary diagnosis to a blood count [2]. The binary decision tree T can be considered as a connected directed graph, which has a unique root vertex without incoming edges and for each other vertex v exactly one incoming and either no (for leafs) or two (for nodes) outgoing edges $(v, v'), (v, v'')$ leading into other vertices. At each node, a decision is done which child vertex v', v'' the blood count $\mathbf{c}$ is assigned to. A decision considers some parameter p and checks whether $c(p)$ exceeds some split value $s(p)$ or not. Each leaf represents a preliminary diagnosis for the blood count. The decision at v based on p for assigning $\mathbf{c}$ to v' or v'' is evaluated with the *Gini impurity*

$$\mathrm{GI}_p(v) = 1 - \mathrm{Prob}^2(\mathbf{c} \in v') - \mathrm{Prob}^2(\mathbf{c} \in v'') \tag{1}$$

where $\mathrm{Prob}(\mathbf{c} \in v')$ denotes the probability for assigning $\mathbf{c}$ to v'. High GI indicates high selectivity of the decision for separating positive from negative blood counts. The *Gini gain* of a split at v based on p and leading to v', v''

$$\mathrm{GG}_p(v, v', v'') = \mathrm{GI}_p(v) - \mathrm{GI}_p(v') \cdot |\{\mathbf{c} \in v'\}|^{-1} - \mathrm{GI}_p(v'') \cdot |\{\mathbf{c} \in v''\}|^{-1} \tag{2}$$

quantifies the achieved improvement in separating positive from negative blood counts on the way towards preliminary diagnoses [6]. High GG indicates major benefit from a node split in the sense of a precise preliminary diagnosis.

AI-Based Data Analysis - Quality Measures: The decision tree used for AI-based data analysis is created in the AI back end as described in the corresponding section. The obtained decision tree imitates the approach of a hematologist as decision maker of considering parameter values one after the other in a suitable order before formulating the medical diagnosis. The *tree variable importance*

$$\mathrm{Imp}(p, T) = \Big(\sum_v \mathrm{GG}_p^2(v, v', v'')\Big)^{\frac{1}{2}} \tag{3}$$

measures the impact of p in a decision tree T for classifying blood counts. The quality of preliminary diagnostic decisions is measured with

$$TP(T), \quad TN(T), \quad F_1(T) \tag{4}$$

The true-positive rate TP and true-negative rate TN are the percentages of correctly diagnosed positive and negative blood counts respectively. The F_1 measure is the harmonic mean of the true-positive rate and the percentage of positive blood counts among the positively diagnosed blood counts. The defect of possibly few available data can be dealt with *k-fold cross-validation* [7]. This approach distributes the data into k equally sized parts, trains the AI method on $k-1$ of them, tests its on the renaming part and averages the test results over all parts.

AI-Based Data Analysis - Random Forest: In AI-based decision making, single decision trees may turn out highly sensitive to the training data and parameter selections. RF methods therefore create a forest $\mathcal{T}$ of multiple decision trees and aggregate the single decisions to a majority decision of higher quality [2]. Let $\mathrm{Dec}(\mathbf{c}, \mathcal{T})$ denote the majority decision of $\mathcal{T}$ for the blood count $\mathbf{c}$. The *validity* of a joint preliminary diagnostic decision is computed as

$$\mathrm{Val}(\mathrm{Dec}(\mathbf{c}, \mathcal{T})) = |\mathcal{T}|^{-1} \cdot |\{T \in \mathcal{T} : \mathrm{Dec}(\mathbf{c}, T) = \mathrm{Dec}(\mathbf{c}, \mathcal{T})\}| \tag{5}$$

The mean values of the measures (4) over $\mathcal{T}$ yield the average measures

$$TP(\mathcal{T}), \quad TN(\mathcal{T}), \quad F_1(\mathcal{T}) \tag{6}$$

The transparency of a single tree is lost with the use of multiple trees and averaging of results. But the impact of a parameter p on the classification of blood counts with the RF $\mathcal{T}$ can still be quantified with the *RF variable importance*

$$\mathrm{Imp}(p,\mathcal{T}) = \Big(|\mathcal{T}|^{-1} \cdot \sum_{T\in\mathcal{T}} \mathrm{Imp}^2(p,T)\Big)^{\frac{1}{2}} \tag{7}$$

AI Back End: Creation of a single tree takes place on a training collection of blood counts with known diagnoses. Training essentially comprises the iterative addition of vertices to the tree with parameters p and split values $s(p)$, which provide maximal GG (2). This addition ends with in case of a small gain obtained with a further split of a vertex or few blood counts contained in a vertex. The multiple trees used by RF methods also make use of the training collection and crucial parameters p. In order to obtain maximally uncorrelated trees, they are created by randomization of the underlying training collection with *bootstrapping* and selection of split parameters from random parameter subsets [2].

Diagnostic Assessment: This step presents the results of analysis to the hematologist for formulating the final diagnostic decisions. The results of analysis consist of the preliminary diagnostic results and the values for (5), (6) and (7). The decision maker can exploit this context information for the diagnostic decision with filtering and sorting features.

3 Results and Discussion

3.1 Blood Counts

Functionality and performance of the diagnostic web service were examined for two collections of blood data with tests for Covid-19 or Influenza A/B infections. For each collection, all cases with existing test results and sufficient homogeneity in terms of documented parameters were selected. The publicly available collection [4] gave 105 cases with Covid-19 test results, which are described in terms of 15 crucial parameters and have a positive test rate of 12.4%. The second collection was provided by a commercial lab diagnostic service provider and thus remains undisclosed. It gave 93 blood counts with Covid-19 test results, which are described in terms of 8 crucial parameters and have a positive rate of 82.8%. This collection also gave 407 cases with Influenza A/B test results, which are described in terms of 8 crucial parameters and have a positive rate of 93.1%.

3.2 AI Training and Numerical Results

An RF method with 2000 trees was trained and tested on these blood counts with 20 runs of a 5-fold cross-validation [8]. In case of the first collection with Covid-19 test results, the obtained quality measures are

$$TP(\mathcal{T}) = 0.524, \quad TN(\mathcal{T}) = 0.934, \quad F_1(\mathcal{T}) = 0.521 \tag{8}$$

The comparably small number of available blood counts and the clear bias towards negative test results yield a rather low TP and F_1 measure and a very good TN. Computation of the RF variable importance (7) gave values partly

Table 1. AI training: crucial blood parameters with highest variable importance for preliminary diagnosis of Covid-19 infections.

p	$\mathrm{Imp}(p, \mathcal{T})$	p	$\mathrm{Imp}(p, \mathcal{T})$
Leukocytes	0.25929131	Monocytes	0.10817097
Platelets	0.18836478	Lymphocytes	0.07070973
Eosinophils	0.12277543	Mean platelet volume	0.06415193

listed in Table 1. Indeed, the decision trees used the top ranked parameters with high GG (2) prominently. Hence, (7) is a good means for adding transparency to the RF method. The dependency on the available data is also reflected by the results obtained for the second data collection with its very high positive test rates. The Covid-19 cases from that collection gave a much higher $TP(\mathcal{T}) = 0.975$ and $F_1(\mathcal{T}) = 0.910$ and a comparable $TN(\mathcal{T}) = 0.837$. The Influenza A/B cases from that collection gave equally high $TP(\mathcal{T}) = 0.944$ and $F_1(\mathcal{T}) = 0.961$, but low $TN(\mathcal{T}) = 0.235$ because of the very few available negative test results.

3.3 Software Realization

The solution concept and underlying methods for AI-based data analysis and decision support were implemented as a diagnostic web service, which can be used in combination with an existing lab diagnostics environment like in Fig. 1. The graphical user interface of this web service is shown in Fig. 2. The menu

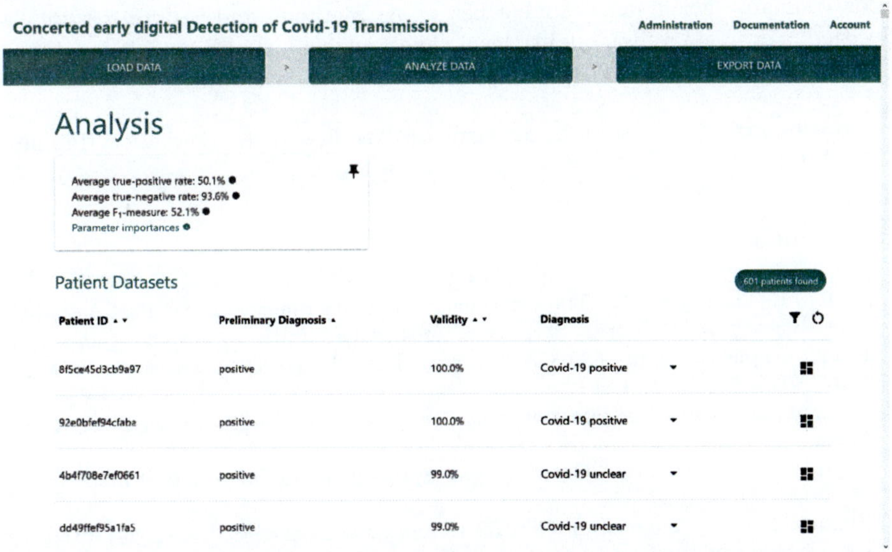

Fig. 2. GUI of the diagnostic web service: menu bar with workflow steps and display of analytic results with planning features

bar in the upper part guides the hematologist through the main work steps of data preprocessing, data analysis and data export. The lower part shows the corresponding detailed information view, in this case during assessment of the preliminary diagnoses.

AI-Based Decision Making: In this view, the hematologist first obtains basic information about the reliability of the data analysis method in terms of the quality measures (8). A click on that text line reveals the importance (7) of the considered blood values partly listed in Table 1. The table of results contains a column with the quality indices (5) for each preliminary diagnosis. These information displays allow for a direct use of AI-based quality measures in human decision making. The blood values as most detailed information can be accessed with the functional icons to the right. The hematologist then takes all this information into account for the final diagnostic decisions for the considered blood counts and documents the results in the column to the right. The columns enable a reordering in the sense of lexicographic optimization and are equipped with filtering and sorting features [3]. In combination, these features allow for an efficient use of AI results in human decision making. For example, they can be used for quick access to all not yet diagnosed blood counts with strong positive preliminary diagnostic decisions as shown in Fig. 2.

4 Conclusion

The solution approach presented herein combines AI methods for data analysis and decision support concepts for software-assisted lab diagnostics. It thereby allows for an efficient and reliable use of AI methods and results for human decision making on a high quality level. The prototypical web service provides a compatible amendment to existing lab diagnostics environments.

Acknowledgments. This work was funded by the Fraunhofer Innovation Hub program Anti-Corona.

References

1. Acuna, E., Rodriguez, C.: The treatment of missing values and its effect on classifiers accuracy. In: McMorris, F.R., et al. (eds.) Classification, Clustering, and Data Mining Applications, pp. 639–647. Springer, Heidelberg (2004). https://doi.org/10.1007/978-3-642-17103-1_60
2. Breiman, L., et al.: Classification and Regression Trees. CRC Press, Boca Raton (1984)
3. Ehrgott, M.: Multicriteria Optimization. Springer, Heidelberg (2005). https://doi.org/10.1007/3-540-27659-9
4. Einstein Data4u: Diagnosis of COVID-19 and its clinical spectrum (2020). https://www.kaggle.com/einsteindata4u/covid19
5. Grewatta, P., et al.: Algorithm-based large-scale screening for blood cancer. PLoS ONE (2019)

6. Hasti, T., et al.: The Elements of Statistical Learning. Springer, New York (2001). https://doi.org/10.1007/978-0-387-21606-5
7. Kohavi, R.: A study of cross-validation and bootstrap for accuracy estimation and model selection. In: Proceedings of the 14th IJCAI, vol. 14, no. 2, pp. 1137–1145 (1995)
8. Singer, C.: Digitally assisted data analysis and decision making in laboratory diagnostics. Bachelor's thesis, TU Kaiserslautern (2021)

Heuristics, Metaheuristics and Matheuristics

The Picking and Packing Problem in Buy-Online-Pick-up-in-Store Retailing

Nicola Ognibene Pietri[1], Xiaochen Chou[1], Dominic Loske[2], Matthias Klumpp[2], Jafar Jamal[1], and Roberto Montemanni[1(✉)]

[1] University of Modena and Reggio Emilia, Reggio Emilia, Italy
roberto.montemanni@unimore.it

[2] Georg-August-University of Göttingen, Göttingen, Germany

Abstract. With the rapid increase of digitization and desire for contactless shopping during the COVID-19 pandemic, online grocery sales keep growing fast. Correspondingly, optimized policies for order picking are nowadays central in omnichannel supply chains, not only within dedicated warehouses but also in grocery stores while processing online orders. In this work, we apply the Buy-Online-Pick-up-in-Store concept and optimize the in-store picking and packing procedure.

The approach we propose, which is based on two mathematical programming models, guides pickers on how to organize articles into bags while collecting items. In this way bags are filled up evenly and they are ready to be handled to the customers at the end of each picking task, with no further rearrangement needed.

Keywords: In-store order picking and packing · Omnichannel grocery retailing · Mathematical programming

1 Introduction

In the past decade we have witnessed a rapid increase of digitization, that has also been transforming our behaviour as consumers. Online grocery purchases have gradually entered our lives as a convenient option for shopping. This process has been further accelerated by the restrictions and desire for contactless shopping during the COVID-19 pandemic. In such a scenario, traditional brick-and-mortar (B&M) retailers have been greatly incentivized to develop omnichannel solutions [1]. Among the three classical ways for the design of an omnichannel model [2], we are interested in the mode of grocery retailers using their existing B&M structures to fulfill both online and offline demands. The corresponding Buy-Online-Pick-up-in-Store concept (BOPS) is the focus of the remainder of this paper.

The BOPS model has been extensively studied in terms of market strategies [3,4], but approaches for operations management of in-store order picking are scarce. In the BOPS model, a picker travels through the store to pick all the items on a given shopping list and checks out at the cashier like a regular customer. From the information made available by retailers, these picking operations are

N. Trautmann and M. Gnägi (Eds.): OR 2021, LNOR, pp. 239–244, 2022.
https://doi.org/10.1007/978-3-031-08623-6_36

commonly not optimized. Employees who are not (yet) familiar with the store may need to go back and forth for a picking task. Even skilled pickers may return to the previously visited shelves, as a consequence of lack of correlation between the articles' order given on a shopping list and the layout of the supermarket.

Since human workforce is considered a major cost-driver in omnichannel retailing, optimizing the picking and packing process is crucial in cost reduction. An effective way to improve the efficiency of the picking would be reordering the articles in the shopping list by a simple shortest route analysis [5]. The problem is treated as a Travelling Salesman Problem (TSP): the picker follows the shortest route to visit each zone of the store only once, and picks up all the articles on the list at each zone. This is a very effective way to reduce the operating time. However, the B&M stores layout are commonly designed to maximize revenue [2,6] and customer satisfaction [7], and the main product attributes being considered are visibility, variety, availability and position to maximize impulsive purchases. Product attributes such as size and fragility are less considered, thus picking the articles following the shortest path may lead to potential product damaging and a substantial rearrangement overhead at the cashier. Since manpower is consumed in the packing process as well, determining a good packaging during the picking process would be a great benefit, especially in presence of portable devices that can be used by the employees to produce the bill already while picking. In a previous work [8], a scoring model that gives each product a priority value defined by its characteristics has been proposed. By adding precedence constraints on top of the shortest route analysis, the model is solved as a Sequential Ordering Problem (SOP) [9]. In this paper, the solutions are devised by the heuristic algorithms described in [10]. The picking route is then optimized to reduce the time of moving back and forth in the store while not damaging the articles due to the order in which they are placed. Such an approach provides a good trade-off between efficiency and customer satisfaction. Nevertheless, this model has a drawback considering actual situations. When more than one bag is required for a given shopping list, there is a high probability that we will get one full bag of heavy stuffs (e.g. beverages) and another bag full of light products (e.g. potato chips), since packing in multiple bags is not considered by the optimizer.

To close the gap, in this work we propose a two-step process that estimates the number of bags required for a given shopping list and prepares for the picker an ordered list with each article associated with a certain bag. The articles are conveniently arranged in bags concurrently. In such a way, multiple packages are ready for the customers at the end of each picking task, that further improves the overall efficiency of the BOPS model.

2 Problem Statement

The purpose of this paper is to expand and complete the research conducted in a previous work [8], in which the in-store picking problem was optimized as a SOP to find a trade-off between shorter route and safe pickings that avoid damages to the articles.

In this work, we further optimize the picking and packing process when multiple shopping bags are considered. When the picker uses an ordered list with the previous solving method, it is impossible to put the articles in a reasonable and even manner simultaneously in each bag, as it is difficult for a picker to directly estimate the exact number of shopping bags needed by taking a look at a list of products. To solve this issue, we further optimize the shopping list with a mathematical model that estimates the minimum number of bags required for a given shopping list, and assigns each article in a shopping list to a certain bag in order to have also a balanced packing.

3 Methodology

First of all, we consider the basic problem (BASE) to determine the minimum number of bags required b_{min} for a given shopping list.

BASE. We define $P = \{1, 2..., m\}$ as the set of products and $B = \{1, 2..., n\}$ as the set of available homogeneous bags. v_i is the volume of product $i \in P$, w_i is the weight of product $i \in P$. The maximum volume of a bag is V_{max} and the maximum weight of a bag is W_{max}. In the model we define two binary variables: x_{ij} and y_j. If product i is in bag j, x_{ij} takes the value 1, otherwise $x_{ij} = 0$. Similarly, $y_j = 1$ if bag j is used, 0 otherwise. The basic problem can be written as the following binary linear programming problem:

$$b_{min} = \min \quad \sum_{j \in B} y_j \tag{1}$$

$$s.t \quad \sum_{i \in P} w_i x_{ij} \leq W_{max} y_j \qquad j \in B \tag{2}$$

$$\sum_{i \in P} v_i x_{ij} \leq V_{max} y_j \qquad j \in B \tag{3}$$

$$\sum_{j \in B} x_{ij} = 1 \qquad i \in P \tag{4}$$

$$x_{ij} \in \{0, 1\} \qquad i \in P, j \in B \tag{5}$$

$$y_j \in \{0, 1\} \qquad j \in B \tag{6}$$

The objective function (1) is to minimize the total number of bags. The constraints (2) and (3) restricts the articles in each bag from exceeding the maximum weight and volume. Constraints (4) enforce each article to be in one and only one bag. The optimal solution to the base problem is the minimum number of bags required b_{min}. If $b_{min} = 1$, the solution is equivalent to the SOP solution discussed in [8] and no further optimization is required. When $b_{min} > 1$, the next step is to look for the optimal solution that assign each product into a shopping bag, based on some heuristically approximated characteristics of the good, with the aim of having a balanced packing. To achieve this, we solve a Min-Max problem that ensures that the total volume and weight of the contents in each shopping bag are almost identical.

Bag 1				Bag 2				Bag 3			
Item	Quantity	Volume	Weight	Item	Quantity	Volume	Weight	Item	Quantity	Volume	Weight
KEIMOEL	1	0,75	0,77	ORIGINAL PALMOLIVE	1	0,75	0,77	FRZ.WUERZMITTEL PET FL	1	1,00	1,02
RAPSOEL	4	0,50	0,51	RAPSOEL	1	0,50	0,51	REINES RAPSOEL	1	0,75	0,77
PIZZATOMATEN	2	0,38	0,40	SONNENBLUMENOEL	3	0,50	0,51	LINSENEINTOPF	2	0,60	0,80
TOMATEN STUECKIG	1	0,38	0,40	BURGER SAUCE JALAPENOS	1	0,30	0,31	BBQ SAUCE ORIGINAL	1	0,30	0,31
GOLDMAIS	1	0,43	0,33	BAKED BEANS I.TOMATENS	1	0,40	0,42	BURGER SAUCE AIOLI	1	0,30	0,31
MAIS 425ML	1	0,43	0,33	BEANZ O. ZUCKERZUSATZ	3	0,40	0,42	KICHERERBSEN	1	0,21	0,23
BOLOGNESE SAUCE	1	0,38	0,40	GULASCHSUPPE	1	0,40	0,42	FEINE GUERKCHEN	1	0,58	0,51
FEINE GUERKCHEN	1	0,58	0,51	KOKOSMILCH	1	0,40	0,42	MAISKOELBCHEN	1	0,37	0,40
LINGUINE AL BRONZO	1	0,50	0,50	MANGO-SCHEIB. LEI GEZ	1	0,43	0,45	PESTO VERDE	1	0,15	0,19
LINGUINE HARTWEIZEN	2	0,50	0,50	TOMATENCREMESUPPE	2	0,40	0,42	KAPERN SURFINES	3	0,04	0,06
VOLLKORN-SPAGHETTI	2	0,50	0,50	SAUERKIRSCHEN	1	0,72	0,75	ARTISCH.VIERT. GEGRIL.	4	0,25	0,28
TOMATENMARK BIO	1	0,20	0,20	ARTISCH.VIERT. GEGRIL.	1	0,25	0,28	COCKTAIL-CORNICHONS	1	0,21	0,24
HUEHNERSUPPE M.NUDELN	1	0,75	0,06	SOSSE ZUM BRATEN	1	0,75	0,25	DIJON-SENF ORIGINAL	1	0,20	0,23
INST. NUDELGER. HUHN	1	0,20	0,06	ROESTZWIEBELN	1	1,00	0,15	COLLEZIONE LASAGNE	2	1,40	0,50
								SENF MITTELSCHARF	1	0,20	0,25
	20	9,35	8,40		19	9,40	8,35		22	9,39	8,36

Fig. 1. Example of a shopping list optimized by the two-step model

MIN-MAX. With respect to the basic problem, we define $B_m = \{1, ..., b_{min}\}$ as the set of bags, as calculated with problem BASE, z as the minmax variable, d_w^+, d_w^-, d_v^+ and d_v^- as the distance variables that are dummy variables required to minimize the difference between the volume and weight of a pair of bags. The problem can be written as follows:

$$\min \quad z \tag{7}$$

$$s.t \quad (2) - (6)$$

$$\sum_{i \in P} w_i x_{ij} - \sum_{i \in P} v_i x_{ik} - d_w^+ + d_w^- = 0 \qquad j, k \in B_m, j < k \tag{8}$$

$$\sum_{i \in P} v_i x_{ij} - \sum_{i \in P} v_i x_{ik} - d_v^+ + d_v^- = 0 \qquad j, k \in B_m, j < k \tag{9}$$

$$0 \leq d_w^+, d_w^-, d_v^+, d_v^- \leq z \tag{10}$$

With constraint (8) and (9), each shopping bag contains products that sum up to a similar volume and weight. The distance variables d_w^+,d_w^-,d_v^+ and d_v^- are non negative, and the minmax constraints for the distance variables are shown in (10). An example of an optimized shopping list by the two-step model is shown in Fig. 1. Volumes are expressed in litres (l) and weights in kilograms (kg).

4 Experimental Simulation

In this section we present the experimental results based on layout and orders made available by a German retailer. Simulation experiments are conducted on 10 real historical BOPS orders that contains 48 items on average with a standard deviation of 5. All models were solved with the Excel Solver performed on a Windows 10 virtual machine with dual-core Intel core i5 processor 2.3 GHz and 4GB of RAM.

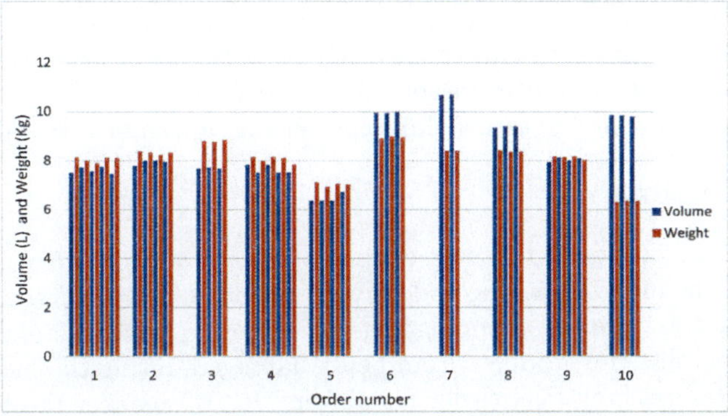

Fig. 2. Simulation results for 10 order picking and packing tasks. For each bag used, the volume (blue) and weight (red) are reported

By adopting the two-step model proposed in this work, the space of each packaging bag should be used to the greatest extent and the articles are reasonably stacked and will not be damaged. Therefore the articles can be directly scanned during the picking process as they are ready-to-go at the end of a picking trip. Consider that the reality picking is fast but rearranging and repacking articles at the cashier is time consuming. We conducted a simulation for a comprehensive understanding on the total manpower required by the picking, scanning, and packing processes: the travel time for moving between the shelves are derived from [8]. During the picking task, the time to pick up an item is estimated as 7 s if the optimization methods we propose, that adapt to the scan-as-you-pick model is applied, and 5 s otherwise. At the cashier, the time to scan and place the products in the final bag is estimated as 8 s per product when the scan-as-you-pick protocol is not applied, of which 5 for the picker and 3 for the cashier [11] (the picker loads the article and packs, the cashier scans).

The result shows a remarkable difference: an average time of 177 s is saved when the optimization methods that adapt to the scan-as-you-pick model is applied, that roughly equals to 23% of the total time required. Such an improvement can be achieved only when the picker is completely free of distractions. Deviations could exist in actual operation due to the number of articles, the pickers' experience, and even the articles' attributes, therefore the estimation is only marginal. Nevertheless, the optimization method proposed in this work reduces the complexity of the task, thereby reduces the pressure on employees.

The solutions of the model are reported in Fig. 2. For each of the 10 orders, we show the estimated number of bags and the total volume and weight of the articles in each shopping bag. The maximum volume and weight of a shopping bag are set to 12 l and 9 kg respectively.

In Fig. 2 we can observe that the articles are divided evenly in each shopping bag (each pair of blue/red columns in the chart represents a bag). The maximum difference in terms of volume and weight between all the shopping bags for a certain shopping list is 0.363 l and 0.315 kg. We can conclude that the model estimates the number of shopping bags required correctly and allocates the articles in each bag with respect to the attributes reasonably. In general, an significant advantage of the optimization model in terms of manpower saving can be observed.

5 Conclusion

In this work we propose a two-step optimization approach to increase the overall efficiency of the in-store picking and packing problem in BOPS retailing. Articles are assigned into balanced shopping bags according to its characteristics. Such an approach prevents the products from being damaged during the picking process, and makes the packing task easier for pickers. It also reduces the time required to process an order when packaging time is also considered.

References

1. Wang, Y., Xu, R., Schwartz, M., Chen, X.: COVID-19 and retail grocery management: insights from a broad-based consumer survey. IEEE Eng. Manag. Rev. **48**(3), 202–211 (2020)
2. Wollenburg, J., Hübner, A., Kuhn, H., Trautrims, A.: From bricks-and-mortar to bricks-and-clicks. Int. J. Phys. Distrib. Logist. Manag. **48**(4), 415–438 (2018)
3. Li, Z., Yang, W., Jin, H.S., Wang, D.: Omnichannel retailing operations with coupon promotions. J. Retail. Consum. Serv. **58**, 102324 (2021)
4. Kong, R., Luo, L., Chen, L., Keblis, M.F.: The effects of BOPS implementation under different pricing strategies in omnichannel retailing. Transp. Res. Part E: Logist. Transp. Rev. **141**, 102014 (2020)
5. Chou, X., Loske, D., Klumpp, M., Gambardella, L.M., Montemanni, R.: In-store picking strategies for online orders in grocery retail logistics. In: Cerulli, R., Dell'Amico, M., Guerriero, F., Pacciarelli, D., Sforza, A. (eds.) Optimization and Decision Science. ASS, vol. 7, pp. 181–189. Springer, Cham (2021). https://doi.org/10.1007/978-3-030-86841-3_15
6. Ozgormusa, E., Smith, A.E.: A data-driven approach to grocery store block layout. Comput. Indust. Eng. **139**, 105562 (2020)
7. Filipe, S., Henriques Marques, S., Fátima Salgueiro. M. de: Customers' relationship with their grocery store: Direct and moderating effects from store format and loyalty programs. J. Retail. Consum. Serv. **37**, 78–88 (2017)
8. Chou, X., Ognibene Pietri, N., Loske, D., Klumpp, M., Montemanni, R.: Optimization strategies for in-store order picking in omnichannel retailing. In: Dolgui, A., Bernard, A., Lemoine, D., von Cieminski, G., Romero, D. (eds.) APMS 2021. IAICT, vol. 631, pp. 603–611. Springer, Cham (2021). https://doi.org/10.1007/978-3-030-85902-2_64
9. Montemanni, R., Smith, D.H., Rizzoli, A.E., Gambardella, L.M.: Sequential ordering problems for crane scheduling in port terminals. Int. J. Simul. Process Model. **5**(4), 348–361 (2009)
10. Gambardella, L.M., Montemanni, R., Weyland, D.: An Enhanced Ant Colony System for the Sequential Ordering Problem. In: Klatte, D., Lüthi, H.J., Schmedders, K. (eds.) Operations Research Proceedings 2011, pp. 355–360. Springer, Heidelberg (2012). https://doi.org/10.1007/978-3-642-29210-1_57
11. Bernard, S.: Cashiers work-time: between a productivity mentality and a service mentality. Sociologie du travail **49**, e129–e144 (2007)

Logistics and Freight Transportation

Multi-start Heuristics for Unit-Capacity Orienteering Problems

Alexander Beckmann(✉) and Jürgen Zimmermann

Clausthal University of Technology, Clausthal-Zellerfeld, Germany
{alexander.beckmann,juergen.zimmermann}@tu-clausthal.de
https://www.wiwi.tu-clausthal.de/ueber-uns/abteilungen/betriebswirtschaftslehre-und-unternehmensforschung

Abstract. We address a planning problem for the underground transit of goods with containers in mining companies. A problem instance contains tasks and capacities for their fulfillment. Tasks are transports of containers from their current locations to given destinations and deliveries and pickups of empty containers and materials. Besides, we address combinations of those tasks with precedence constraints. Storage locations provide a stock of empty containers and materials and are destinations for pickup containers and materials. Heterogeneous workers and a fleet of heterogeneous vehicles are available for performing the tasks. Each task has an assigned profit and can have a due date, which grants additional profit on observance. The objective is to maximize the benefit of fulfilled tasks and met due dates. We developed two heuristic solution approaches, creating tours either sequentially or in parallel. Furthermore, these constructive algorithms are extended to multi-start versions using randomization of vehicle and worker selections, container-type selection, and insertion criteria.

Keywords: Drayage problem · Orienteering problem · Unit-capacity

1 Introduction

The problem studied is motivated by a planning task in a German mine. In the application, roll-off containers are used for transport activities. The problem is based on the container drayage problem [4], which is also studied under different names by, for example, [5] and [1], and extends it with numerous task types to integrate activities of material delivery, pickup, and transposition. In addition, the integration of heterogeneous vehicles and workers provides further practical relevance. Finally, we assume that the available capacities cannot fully realize the task volume. There is, therefore, no commitment to fulfill tasks, which makes our problem an orienteering problem. Gunawan et al. [2] give an overview of variants of the orienteering problem. Our problem thus extends and combines problems from the literature to cover an orienteering problem in mines.

N. Trautmann and M. Gnägi (Eds.): OR 2021, LNOR, pp. 247–253, 2022.
https://doi.org/10.1007/978-3-031-08623-6_37

Section 2 contains a general description of the problem. In Sect. 3 the developed heuristic solution approaches are presented. Finally, Sect. 4 includes a performance analysis, in which the results of our heuristic approaches are compared to results obtained by Xpress using a mixed integer model.

2 Problem Description

In the orienteering problem under investigation the nodes represent logistic tasks. Each task has a profit value assigned. Additionally, a subset of the tasks has a due date that, if met, yields additional profit. The tasks comprise the delivery, collection, or transport of containers and materials. Heterogeneous workers and a heterogeneous fleet of vehicles are available to perform the tasks. The workers differ in their competence to handle materials and drive vehicles while vehicles have different speeds and can each use different containers. The number of tours is limited by either the number of workers or vehicles. We also consider the specification of a maximum number of tours, which can be relevant in practice if workers are not exclusively deployed for transport activities. We assume that the available capacities cannot fully realize the task volume. There is, therefore, no commitment to fulfill tasks, and a selection of tasks is necessary. Since there are generally several options for servicing a task, a decision must be made on the execution option. Finally, the execution options must be distributed among the tours and put in order.

Table 1 gives an overview of the considered task types and their required information, marked '+'. In the case of container and material deliveries, one or more container types can be used. The necessity for a decision about which container type to use is signed '⊕'. A material specification can be omitted for transports if and only if an empty container has to be moved. The task type material pickup requires the material to be already loaded into a container.

Table 1. Basic tasks

Identifier	Task	Origin	Destination	Container	Material
C^T	Transport	+	+	+	+
C^D	Container delivery		+	⊕	
C^P	Container pickup	+		+	
M^D	Material delivery		+	⊕	+
M^P	Material pickup	+		+	+

Other activities can be mapped by connecting the task types listed above. A sequence of tasks is referred to as a task chain. The tasks of a chain always concern a container, which may have to be selected. If a decision about a container type has to be made, this is done in the first task of a chain. In that case, the container type of a succeeding task is unknown in advance and determined by the choice made for the first task. The tasks of a chain have to be performed by one vehicle. Table 2 shows the considered activities, their required information, and the resulting task chains. For example, the transposition of loose material is mapped by a container delivery followed by a transport and an optional pickup of the used container.

Table 2. Task chains

Activity	Origin	Dest.	Container	Material	Task chain
Collection of loose material	+		⊕	+	$C^D \rightarrow M^P$
Delivery of material with container pickup		+	⊕	+	$M^D \rightarrow C^P$
Transposition of loose material (with container pickup)	+	+	⊕	+	$C^D \rightarrow C^T$ $(\rightarrow C^P)$
Transposition of material with container pickup	+	+	+	+	$C^T \rightarrow C^P$

In general, there are several execution options for tasks, except for transports. Container and material deliveries have a container demand, while container and material pickups have a container supply. Material deliveries and pickups additionally have a material demand and a supply, respectively. Container locations (CL) and material locations (ML) can provide a supply and take on the role of demand by receiving containers or material. The execution options result from matching supply and demand for containers and materials. They represent the movement of a container, possibly extended by loading activities for material. An overview of the considered execution options is given in Table 3.

Table 3. Execution options of basic tasks

Identifier	Providing	Sinks		Sources		Receiving
	Task	ML	CL	CL	ML	Task
CLC^D				+		+
C^PCL	+		+			
CLMLM^D				+	+	+
MLM^D					+	+
M^PML	+	+				
M^PMLCL	+	+	+			
$C^P C^D$	+					+
$M^P M^D$	+					+
C^PMLM^D	+				+	+
M^PMLC^D	+	+				+
M^PMLMLM^D	+	+			+	+

In the upper half of the table, execution options that only concern one single task are listed. The execution options CLC^D match the supply of a container location with the demand of a container delivery. Execution options $\text{CLML}M^D$ and $\text{ML}M^D$ for material deliveries differ in the usage of a newly or already loaded container. In one case, the execution option starts at a container location, in the other case at a material location. The execution options for container and material pickups are formed analogously in reverse order. The lower half of the table shows the execution options that link two tasks. Inserting one or more material locations between the tasks is necessary when material needs to be unloaded or loaded. In execution option $M^P\text{MLML}M^D$, the loaded material must first be unloaded at a material location before the requested material is loaded at another material location, if necessary. By linking two tasks, the compatible container types of the connecting execution option result from the intersection of the compatible container types of both tasks.

3 Constructive Heuristic

The heuristic solution approaches construct tours by composing execution options. They are based on the approach H2 by [3]. An artificial execution option represents the depot at the beginning and end of each tour. First, a sequential approach for the generation of tours is described. Unless the specified maximum number of tours is reached, a combination of a worker and a vehicle is chosen. The choice is made according to the profit potential, i.e., the maximum profit value that a combination can achieve, ignoring time and resource capacity. Tasks already scheduled in other tours are not considered. As long as execution options are available, the best insertion option, consisting of execution option and insertion position, is determined. The profitability serves as an evaluation criterion, as a quotient of benefit and insertion duration. The benefit includes tasks profit, compliance of their due dates, and violation of due dates of subsequent tasks. If the best insertion option has negative profitability, the construction of a tour terminates. Otherwise, the best insertion option is implemented, and available execution options and tasks are updated.

The determination of the best insertion option takes the direct predecessor tasks into account. Thus, tasks can only be considered at positions of a tour if their direct predecessor is scheduled at a previous position of the same tour. If a task has a successor, the task's execution options are also evaluated in combination with the execution options of the subsequent task. First, the best execution option for the considered task at the respective position is determined. If the task includes a container decision, the best insertion option is stored for every possible container type. In addition, for cross-task execution options with transport as a preceding task, we consider the previous execution of the transport as an option. This enables the use of containers that would otherwise not be considered for implementation due to the low profit of the corresponding transport task. Subsequently, the best execution option of the initial task is determined for all available execution options of the subsequent task, taking into account container compatibility. For each usable container type, the best execution option of the subsequent task is recorded, and finally, the combination of the best execution options of the task and its successor is evaluated. If the value of a combination exceeds the best value found so far, the insertion of the combination's first execution option is recorded as the best insertion option. Considering only the first execution option allows a better evaluation of the execution options of the successor task in subsequent iterations by taking into account a possible further use of a container. Otherwise, containers could be prematurely deposited at the container or material locations, and a possibly good re-use option would be excluded.

Parallel tour planning is considered as a variant of the procedure. In contrast to sequential trip planning, the parallel method determines vehicle-employee pairs for trips at first. The specified maximum number of tours gives the number of tours. The set of available tasks and execution options are to be managed for each tour. The best insertion option can be stored for each tour, and the best option of all tours is realized. When scheduling an execution option, the best insertion option of the tour in question must be re-determined. If the best insertion option of another tour is no longer available, it must also be updated. Finally, both constructive algorithms are extended to multi-start versions using randomization of vehicle and worker selections, container-type selection, and insertion criteria. A randomization factor r determines the degree of randomization of the insertion criteria. Each criteria value is modified by multiplication with a uniformly distributed random number from the interval $[1 - r, 1]$. The deterministic heuristics correspond to a randomization factor value $r = 0$.

4 Performance Analysis

In order to evaluate the performance of our two constructive solution approaches and their multi-start variants we conducted a computational study. The study was performed on an Intel Core i7-8700 CPU and 64 GB RAM under Windows 10. As a basis for comparison, a linear mixed-integer model was implemented and solved using Xpress 8.11. Our constructive heuristics have been implemented in C#. We used a self developed instance set consisting of 320 instances. The

Table 4. Results of the solver, deterministic heuristics and highly randomized heuristic approaches

	600 s	Deterministic		Multi start, r = 0.95, 600 s	
#tours	Xpress	Sequential	Parallel	Sequential	Parallel
1	37.1	11.7		0.2	0.4
2	49.2	13.1	20.8	0.3	1.8
4	59.7	9.8	22.2	0.2	2.3
8	85.5	6.6	21.7	0.3	1.5

instances differ in the composition of task types, number of tasks and the available resources. The smallest instances contain 12 jobs, the largest instances contain 96 jobs, with task chains considered as one job.

Table 4 shows the results for deterministic and multi-start test runs of the two heuristic approaches, as well as the Xpress solver. The multi-start runs were performed with randomization factor value $r = 0.95$ and a time limit of ten minutes. Indicated are the results grouped by instance size. The results are given as relative deviation between the objective function value found by the respective heuristic and the best objective function value found by any of the approaches, i.e. $\frac{f_{best}-f_H}{f_{best}}$. For example, the objective function values for instances with one tour obtained by Xpress deviate on average 37.1% from the best objective function values found. Further results are shown in Table 5 for the heuristic approaches with randomization factor values 0.15, 0.5 and 0.95 and time limits one and sixty seconds. The randomization factor values leading to the best results for a heuristic approach with a given time limit and instance size are highlighted. The results show that the Xpress solver delivers only comparatively poor results, even for small instances. The sequential approach dominates the parallel approach over all instance sizes in both the deterministic and multi-start versions. For a time limit of one second high randomization factors provide better results in the smaller instance sizes due to the higher number of solutions generated within the time limit compared to larger instances. Over sixty seconds run time the medium randomization factor provides the best results. Finally, after one second, the sequential approach provides solutions that on average deviate less than four percent from those obtained after sixty seconds.

Table 5. Results of the heuristic variants for 1 s and 60 s time limit

	1 s						60 s					
	Sequential			Parallel			Sequential			Parallel		
#tours	0.15	0.5	0.95	0.15	0.5	0.95	0.15	0.5	0.95	0.15	0.5	0.95
1	5.9	2.3	**1.3**	6.2	2.6	**1.7**	4.1	**0.1**	0.7	4.5	**0.4**	1.0
2	4.7	**3.4**	3.7	5.9	4.9	**4.7**	2.6	**0.2**	1.2	3.8	**1.7**	2.5
4	**3.2**	3.5	4.0	5.3	**5.2**	5.4	**0.6**	**0.6**	1.5	3.0	**2.7**	3.3
8	**2.6**	3.0	3.5	4.1	**3.9**	4.3	**0.5**	0.6	1.2	2.1	**1.9**	2.3

References

1. Braekers, K., Caris, A., Janssens, G.K.: Integrated planning of loaded and empty container movements. OR Spect. **35**(2), 457–478 (2013)
2. Gunawan, A., Lau, H.C., Vansteenwegen, P.: Orienteering problem: a survey of recent variants, solution approaches and applications. Eur. J. Oper. Res. **255**(2), 315–332 (2016)
3. Laporte, G., Martello, S.: The selective travelling salesman problem. Discr. Appl. Math. **26**(2–3), 193–207 (1990)

4. Xue, Z., Zhang, C., Lin, W.H., Miao, L., Yang, P.: A tabu search heuristic for the local container drayage problem under a new operation mode. Transp. Res. Part E: Logist. Transp. Rev. **62**, 136–150 (2014)
5. Zhang, R., Yun, W.Y., Kopfer, H.: Heuristic-based truck scheduling for inland container transportation. OR Spect. **32**(3), 787–808 (2010)

Check for updates

ULD Build-Up Scheduling with Dynamic Batching in an Air Freight Hub

Ricardo Euler[1(✉)], Ralf Borndörfer[1], Timo Strunk[2], and Tuomo Takkula[2]

[1] Zuse Institute, Takustraße 7, 14195 Berlin, Germany
euler@zib.de

[2] Ab Ovo Germany GmbH, Prinzenallee 9, 40549 Düsseldorf, Germany

Abstract. Air freight is usually shipped in standardized unit load devices (ULDs). The planning process for the consolidation of transit cargo from inbound flights or locally emerging shipments into ULDs for outbound flights is called build-up scheduling. More specifically, outbound ULDs must be assigned a time and a workstation subject to both workstation capacity constraints and the availability of shipments which in turn depends on break-down decisions for incoming ULDs. ULDs scheduled for the same outbound flight should be built up in temporal and spatial proximity. This serves both to minimize overhead in transportation times and to allow workers to move freight between ULDs. We propose to address this requirement by processing ULDs for the same outbound flight in batches.

For the above build-up scheduling problem, we introduce a multi-commodity network design model. Outbound flights are modeled as commodities; transit cargo is represented by cargo flow volume and unpack and batch decisions are represented as design variables. The model is solved with a standard MIP solver on a set of benchmark data. For instances with a limited number of resource conflicts, near-optimal solutions are found in under two hours for a whole week of operations.

Keywords: Logistics · Airline applications

1 Introduction

Air freight is usually shipped in standardized unit load devices (ULD). Often these ULDs are routed through a hub airport. As they frequently contain freight for multiple destinations, they need to be unpacked (break-down) and reconsolidated (build-up). An intricate scheduling problem thus arises at the hub airport: Outbound ULDs need to be scheduled for reconsolidation in time for their departure while respecting constraints imposed by the availability of workstations and workforce. The amount of available shipments in turn is a function of break-down decisions for inbound ULDs subject to similar resource constraints.

Since many shipments cannot be stacked arbitrarily and also often come in odd shapes, it is desirable to build up multiple ULDs destined for the same

N. Trautmann and M. Gnägi (Eds.): OR 2021, LNOR, pp. 254–260, 2022.
https://doi.org/10.1007/978-3-031-08623-6_38

flight simultaneously and in spatial proximity in order to facilitate better packing options. An easy model of proximity is a partition of the workstations into groups. We refer to a set of identical ULDs for the same flight scheduled at the same time in the same workstation group as a batch. In general, it is not allowed to keep shipments that do not fit into an ULD in the build-up area. Instead, they have to be moved back to the warehouse. Hence, a welcome side effect of build-up in batches is a reduction in the number of movements necessary between the warehouse and the build-up area. From a modeling perspective, considering batches instead of individual ULDs reduces the amount of variables to consider, since outgoing ULDs of the same type (e.g. a container or pallet) are indistinguishable and need no longer be represented individually. Inbound ULDs, however, cannot be aggregated as they already contain freight. Additionally, they do not benefit from being deconsolidated in batches and, hence, they are not treated as such. We call the resulting scheduling problem the build-up scheduling problem with dynamic batch building (BSP).

Build-up scheduling (without batches) is categorized as one step of the sequential air cargo load planning problem in [3], which also contains a comprehensive literature review of related problems. The authors survey three modeling approaches for the scheduling of personnel for ULD build-up [5–7]. Among these, [6] also schedules workers for break-down operations. However, build-up and break-down demand are parameters and not interdependent in their model. All of these models only consider personnel scheduling and do not take individual ULDs, batches or workstations into account. A variant of build-up scheduling without batches is studied in [2]. The same author also introduced the benchmark instances [8] on which we base our computational study. Recently, [4] studied the problem of scheduling both personnel and batch build-ups under constraints on the availability of workstations. Their model treats the creation of batches from incoming cargo as a preprocessing step such that batches appear as jobs with a definite release time, deadline and resource consumption. Here, workstations are not split into groups. Our approach differs from both [4] and [2] in several key aspects. First, we do not consider explicit personnel constraints but assume these to be implicitly given by the availability of workstations. Secondly, we also consider break-down processes and, thirdly, we aim to maximize the size of batches in workstation groups. To the best of our knowledge, this is the first work to incorporate dynamic batch building and interdependent build-up and break-down scheduling into a single model.

Despite the name, BSP is difficult to classify using classical scheduling notation (see e.g. [1]). BSP consists of two parallel processor scheduling problems connected by cargo flow constraints. Here, the value of a build-up job in the objective function depends on the amount of freight made available by finished break-down jobs. Note however, that this is not a precedence relationship. In fact, outbound ULDs might be constructed even if few or no relevant inbound ULDs are unpacked and maximizing the amount of cargo placed in an outbound ULD is part of the objective function.

2 A Multi-commodity Network Design Model with Edge Activity

BSP can be addressed using a network design approach. Let $T = (t_1, \dots, t_{|T|})$ be a discretized time horizon for which ULD build-ups and break-downs need to be scheduled. We are given a set of ULD types V with capacities $c_v \in \mathbb{N}$ and build-up times b_v for all $v \in V$. We denote the set of departing flights by K. Each departing flight $k \in K$ has a departure time δ_k, a freight demand d_k, a (financial) cost of one unit of unshipped cargo l_k and a number of pre-planned ULDs $p_{v,k}$ for all $v \in V$. Each ULD requires a workstation for consolidation. We aggregate workstations that are close to each other into disjoint workstation groups W. The number of workstations in a group $w \in W$ is its capacity c_w. Inbound ULDs I are already assigned a type and an inbound flight in the input data. Therefore, we directly assign each of them a break-down duration β_i, a freight volume $\lambda_{i,k}$ for all outgoing flights $k \in K$ and an arrival time α_i.

We define a candidate batch decision $b \in B$ to be a five-tuple $(t_b, n_b, k_b, v_b, w_b)$ where $n_b \in \mathbb{N}$ is the number of ULDs in the batch, $t_b \in [0, \delta_{k_b} - b_{v_b}] \cap T$ the starting time of build-up, $v_b \in V$ the ULD type used, $k_b \in K$ the outgoing flight and $w_b \in W$ the assigned workstation group. An ULD unpack decision $u \in U$ is a tuple (t, i) with $i \in I$ and $t \in [\alpha_i, t_{|T|}] \cap T$.

Now, BSP can be formalized as finding a set of decisions $D = \bar{B} \cup \bar{U}$ with $\bar{B} \subset B$ and $\bar{U} \subset U$ such that no flight k is assigned more ULDs of type v than $p_{k,v}$, no more than c_w ULDs are scheduled for any workstation group $w \in W$ at any given point in time, the general storage capacity is never exceeded and no inbound ULD i is unpacked more than once while both minimizing the amount of unshipped cargo and maximizing the average batch size. Minimizing freight losses is essential for customer satisfaction, while maximizing batch size has organizational benefits. Hence, we treat the BSP as a single-objective problem using a parametrization that prioritizes loss avoidance over batch building.

To solve BSP, we propose a time-expanded fixed-charge multi-commodity network design model in which the arcs represent unpack and batch decisions or the storage unit. Consider the network $\mathcal{N} = (I \cup S \cup K, A)$, where nodes in $S = \{s_{t_1}, \dots, s_{t_{|T|}}\}$ represent the storage facility at various time points. Then, we define the arc set A as follows. For each unpack decision $u = (t, i) \in U$ an arc $(i, s_{t+\beta_i})$ is created. For each batch decision $b = (t_b, n_b, k_b, v_b, w_b) \in B$, an arc (s_{t_b}, k_b) is created if $t_b + b_{v_b} \leq \delta_{k_b}$. Note that this introduces multi-arcs and that the multiarcs between s_t and k represent all possible batches one can start building for k at time point t. Arcs $(s_{t_j}, s_{t_{j+1}})$ are introduced for all $t_j \in \{t_1, \dots, t_{|T|-1}\}$ representing cargo kept in the storage facility during the time interval $[t_j, t_{j+1}]$. Finally, arcs (i, k) are added for all $i \in I$, $k \in K$. Cargo that is routed along these arcs is considered unscheduled and penalized with high weights.

We write $\mathcal{A} \subset A$ with $\mathcal{A} := B \cup U$. The set $\mathcal{A}$ are the design arcs, while arcs in $A \backslash \mathcal{A}$ are always active. An example of the resulting network can be found in Fig. 1. In our MIP formulation, all batches that differ only in their size correspond to columns that are multiples of each other. To mitigate this

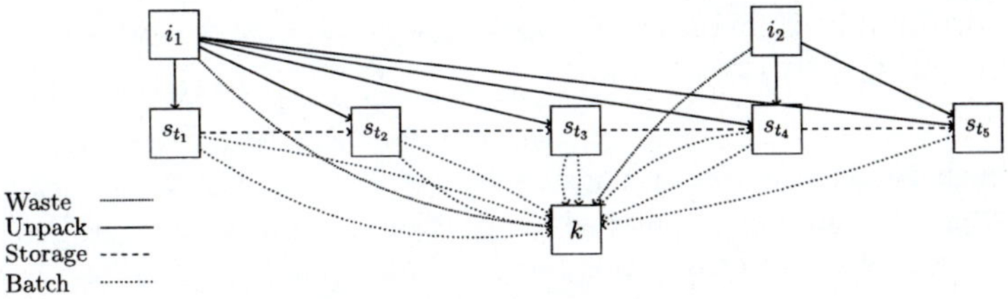

Fig. 1. The time-expanded cargo flow network of the build-up scheduling problem for two inbound ULDs and one departing flight. Here, i_1 and i_2 are sources and k is a sink. The time horizon consists of five time points. Build-up, break-down durations and departure and arrival times of flights are factored into the construction of the graph and determine the presence of edges.

dual degeneracy, we introduce *activity* variables and reduce batches $b \in B$ to four-tuples $b = (t_b, k_b, v_b, w_b) \in B$.

Then, the build-up scheduling problem can be formulated as the following MIP:

$$\min \quad \sum\nolimits_{a\in A} w_a x_a + \sum\nolimits_{k\in K} \sum\nolimits_{a\in A} w_a^k f_a^k \tag{1}$$

$$\text{s.t} \quad \sum\nolimits_{k\in K} d_k f_a^k \leq c_a \quad \forall a \in A \setminus \mathcal{A} \tag{2}$$

$$\sum\nolimits_{k\in K} d_k f_a^k \leq c_a y_a \quad \forall a \in \mathcal{A} \tag{3}$$

$$\sum\nolimits_{a\in\delta+(v)} f_a^k - \sum\nolimits_{a\in\delta-(v)} f_a^k = \gamma_v^k \quad \forall k \in K \, \forall v \in V \tag{4}$$

$$f_a^k \leq x_a \quad \forall k \in K \, \forall a \in \mathcal{A} \tag{5}$$

$$\sum\nolimits_{a\in\mathcal{A}} \alpha_a^r y_a \leq L^r \quad \forall r \in R \tag{6}$$

$$y_a \leq M_a x_a \quad \forall a \in \mathcal{A} \tag{7}$$

$$x_a \in \{0,1\} \quad \forall a \in \mathcal{A} \tag{8}$$

$$y_a \in [0, M_a] \cap \mathbb{Z} \quad \forall a \in \mathcal{A} \tag{9}$$

$$f_a^k \in [0,1] \quad \forall a \in A \, \forall k \in K. \tag{10}$$

Here, f_a^k is the amount of cargo for flight k passed along an arc $a \in A$, x_a indicates whether a is active and y_a represents the number of ULDs constructed on a. Depending on the arc type, constraints (2) and (3) impose bounds on storage or ULD capacity. Constraints (4) ensure flow conservation with $\gamma_v^k := \lambda_{i,k}/d_k$ if $v \in I$, $\gamma_v^k := 1$ if $v \in K$ and $\gamma_v^k := 0$ otherwise. Note that $\sum_{i\in I} \gamma_{i,k} = 1 \forall k \in K$. Constraints (5) ensure that flow only passes through active arcs. Finally, constraints (6) summarize resource limits on active arcs. These are: Ensuring that for each $i \in I$ only one unpack arc is active, that batch activity for flight $k \in K$ and ULD type $v \in V$ is smaller than $p_{k,v}$ and finally that for all $t \in T$ and $w \in W$ the workstation utilization is at most c_w. By introducing costs on the slack of the second type of resource constraints, penalties for planned but offloaded ULDs can be introduced. Constraints (7) limit the activity of active arcs. Here, for a batch arc $b \in B$ we have $M_b = \min\{c_{w_b}, p_{v_b,k_b}, \lceil \frac{d_{k_b}}{c_{v_b}} \rceil\}$. Note that $M_u = 1$ for all $u \in U$. In the objective function, we set $w_a^k = d_k l_k$ if $a = (i,k) \in I \times K$ and $w_a = 0$ otherwise. Also $w_a = 1$ if $a \in B$ and $w_a = 0$ otherwise. Hence, we aim to minimize the number of batches in order to maximize the average batch size. As a consequence, late build-up is incentivized.

3 Computational Study

We based our computational study on the data set provided in [2,8]. The data set consists of shipments extracted from anonymized real-world booking data that is randomly assigned to a real flight schedule of one week with 82 outbound flights. In cooperation with our industry partner Ab Ovo Germany GmbH, we augmented this data as follows. We created 28 time horizons from the week consisting of all possible combinations of between one and seven consecutive days. In the original data set each shipment has a release time at which it becomes available. We grouped shipments with similar arrival times together to form an inbound ULD using a randomly drawn ULD type. The ULD's arrival time is its earliest shipment's release time minus its break-down time. Workstation groups are not part of the data set. We modeled these around settings which appeared sensible to us and our industry partner. We created three different set-ups with 12, 24 and 48 workstations partitioned in groups of six. Hence, the testset consists of 84 instances in total. We did not apply any storage capacity or other restrictions to ULD break-down in this study. Thus, the problem is reduced to a variant of the problem studied in [2] without personnel constraints but using the additional batching objective and workstation groups. We investigated scenarios both with offload penalties for unscheduled outbound ULDs calculated following [2] and without. In line with [2], we assumed that 66% of nominal ULD capacity is available. As this resulted in severe overbooking of some flights, we additionally investigated scenarios with 90% capacity. We implemented the MIP model of Sect. 2 with (*Activ*) and without nonzero (*Free*) costs for batch variables. Additionally, we implemented a standard network design formulation without activity variables, where batch sizes are explicitly enumerated (*Enum*). All implementations were carried out in Python3 using Gurobi 9.1.2. All tests were conducted on a Dell PowerEdge M620v3. Results are reported in Table 1. We found that BSP becomes easier when more workstations are available. If 48 workstations are present, this effect diminishes, most likely due to the increased number of integer variables.

Table 1. Computational results. *Activ* and *Free* refer to the formulation as defined in Sect. 2 with and without batch costs, respectively. *Enum* refers to a standard network design formulation with batch costs. For all 12 scenarios defined by the number of workstations (*#WS*), offload penalties (*Off*) and usable ULD capacity (*Cap*), we report the number of instances solved to optimality (*Opt*), the average gap (*Gap*, %) and the average run time (t) in seconds. Run times (of the MIP solver) were capped at 7200 s.

#WS	Off	Cap	Free			Enum			Activ		
			Opt	Gap	t	Opt	Gap	t	Opt	Gap	t
12ws	•	66	1	2.16	6943.85	0	13.12	7205.23	0	4.78	7201.80
24ws	•	66	28	0.00	13.31	28	0.00	156.96	25	0.01	2229.53
48ws	•	66	28	0.00	22.97	28	0.00	152.62	28	0.00	1501.74
12ws	•	90	4	1.83	6174.67	0	13.17	7206.93	0	5.84	7202.44
24ws	•	90	28	0.00	58.55	28	0.00	151.91	28	0.00	137.02
48ws	•	90	28	0.00	202.59	28	0.00	129.86	28	0.00	117.86
12ws	◦	66	7	4.29	5405.67	3	12.65	6447.57	3	7.28	6511.30
24ws	◦	66	28	0.00	8.85	5	0.04	5926.12	8	0.03	5468.70
48ws	◦	66	28	0.00	15.27	5	0.07	6011.70	7	0.05	5644.11
12ws	◦	90	28	0.00	204.73	5	15.26	6128.94	4	8.72	6255.31
24ws	◦	90	28	0.00	44.97	6	1.55	5966.61	5	1.53	5959.77
48ws	◦	90	28	0.00	147.67	4	2.02	6444.38	4	1.87	6193.93

BSP is generally easier when minimizing the number of batches is not part of the objective. This effect is distinctly more pronounced in scenarios without offload penalties. In general, offload penalties result in more instances solved to optimality and lower run times. We believe this to be due to a reduction in dual degeneracy. Without penalties, scheduling additional empty ULDs has no effect on the objective function, which can lead to the presence of a high number of optimal solutions. Comparing *Enum* and *Activ*, we find that introducing activity variables gives mixed results. *Activ* reports a lower gap in all but one scenario. In scenarios without offload penalties, *Activ* solves three more instances to optimality and reports lower run times in four out of six scenarios. When offload penalties are applied, however, *Activ* reports significantly higher run times than *Enum* for ULD capacities of 66% while being slightly faster for ULD capacities of 90%. These scenarios differ mainly in the amount of offloaded freight. The reason for this variation in performance is not yet well understood.

In conclusion, the BSP proves challenging especially when workstation resources are scarce and offloads are not penalized. The uneven performance of the activity-based formulation warrants further investigation. Future work will also include the scheduling of break-down operations.

References

1. Blazewicz, J., Ecker, K.H., Pesch, E., Schmidt, G., Weglarz, J.: Handbook on scheduling: From Theory To Applications. Springer Science & Business Media, New York (2007). https://doi.org/10.1007/978-3-540-32220-7
2. Brandt, F.: The air cargo load planning problem. Ph.D. thesis, Karlsruher Institut für Technologie (KIT) (2017). https://doi.org/10.5445/IR/1000075507
3. Brandt, F., Nickel, S.: The air cargo load planning problem - a consolidated problem definition and literature review on related problems. Eur. J. Oper. Res. **275**(2), 399–410 (2019). https://doi.org/10.1016/j.ejor.2018.07.013
4. Emde, S., Abedinnia, H., Lange, A., Glock, C.H.: Scheduling personnel for the build-up of unit load devices at an air cargo terminal with limited space. OR Spect. **42**(2), 397–426 (2020). https://doi.org/10.1007/s00291-020-00580-2
5. Nobert, Y., Roy, J.: Freight handling personnel scheduling at air cargo terminals. Transp. Sci. **32**(3), 295–301 (1998). https://doi.org/10.1287/trsc.32.3.295
6. Rong, A., Grunow, M.: Shift designs for freight handling personnel at air cargo terminals. Transp. Res. Part E: Logist. Transp. Rev. **45**(5), 725–739 (2009). https://doi.org/10.1016/j.tre.2009.01.005
7. Yan, S., Chen, C.H., Chen, M.: Stochastic models for air cargo terminal manpower supply planning in long-term operations. Appl. Stoch. Models Bus. Ind. **24**(3), 261–275 (2008). https://doi.org/10.1002/asmb.710
8. ACLPP Instances. https://github.com/fbrandt/ACLPP. Commit: 3516c2b

Route Planning Under Uncertainty: A Case Study on Objectives Apart from Mean Travel Time

Jan Gertheiss[1](✉) and Florian Jaehn[2]

[1] Department of Mathematics and Statistics, School of Economics and Social Sciences, Helmut Schmidt University, Hamburg, Germany
jan.gertheiss@hsu-hh.de

[2] Institute for Management Science and Operations Research, School of Economics and Social Sciences, Helmut Schmidt University, Hamburg, Germany

Abstract. We take the perspective of an individual passenger and consider the problem of choosing a route from a start point S to a target T from a relatively small set of options. We assume that travel times are not deterministic but subject to some stochastic mechanism/uncertainty. For modeling and analyzing travel times, we use stochastic simulation based on mixtures of gamma distributions. Instead of focusing on mean travel times, we discuss multiple criteria for decision making. Our approach is illustrated by an example from public transportation: traveling from Göttingen to Cologne by ICE train. Furthermore, we discuss ways how to extend our approach; e.g., by inferring model parameters from historical data.

Keywords: Stochastic route planning · Public transport · Monte Carlo simulation

1 Introduction

Route planning often tries to minimize the expected travel time (especially in public transport). However, individuals might follow different objectives such as on-time arrival, minimization of the risk of missing connections in public transport, the price, or combinations of these. In a case study, we analyze how these objectives interact. We consider the following real world problem: traveling from Göttingen (train station) to Cologne Central by ICE train as shown in Fig. 1. We restrict ourselves to two (plausible) options: A, via Hanover Central, or B, via Frankfurt (Central and/or Airport). More generally speaking, however, we consider the problem of choosing a route from a start point S to a target T from a relatively small set of options. In case of railway systems, this decision can, for instance, be made by selecting the option with minimum travel time according to train schedules. Taking the perspective of an individual passenger who is able/going to depart from Göttingen at 9 am, on a Thursday in March 2021, the default connection offered by Deutsche Bahn (DB) is:

N. Trautmann and M. Gnägi (Eds.): OR 2021, LNOR, pp. 261–267, 2022.
https://doi.org/10.1007/978-3-031-08623-6_39

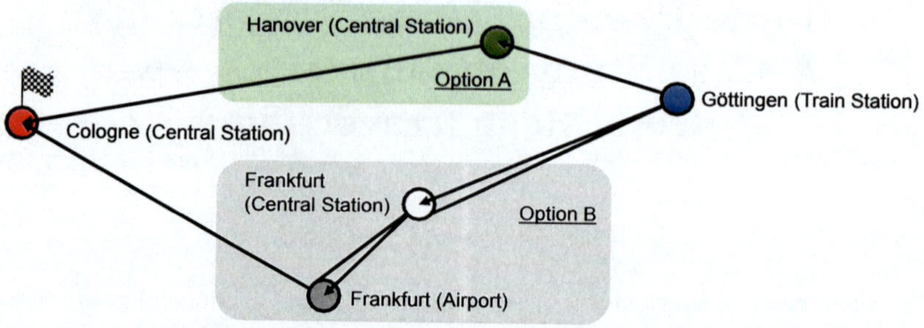

Fig. 1. The case study considered: traveling from Göttingen to Cologne (central station) by ICE train; note, when traveling via Frankfurt, you may change at Frankfurt Central, Airport, or both (e.g., if missing a direct connecting train at Central).

- Depart from Göttingen at 9:53, arrive at Frankfurt Airport 11:51
- Depart from Frankfurt Airport at 12:09, arrive at Cologne Central 13:05

Without any discounts, the price (2nd class) is 106.70 Euro. However, we may also choose to pay 83.90 Euro (full price, 2nd class) and:

- Depart from Göttingen at 9:40, arrive at Hanover Central 10:17
- Depart from Hanover Central at 10:31, arrive at Cologne Central 13:09

Furthermore, there are earlier trains from Göttingen to Frankfurt Airport or Hanover (e.g., at 9:16 or 9:18, respectively), which would result in longer transit times, according to schedule. In summary, there are a couple of options and it is by no means clear which one to choose. In particular, departure and arrival times are not deterministic but exhibit some stochastic behavior, compare [2]. For instance, if a passenger has to arrive at 2 pm (i.e., 14:00) the latest, it might be advisable to take an earlier train in order to decrease the risk of missing a connecting train. In general, we are not focusing on efficient algorithms here to select the best option according to some criteria; compare, e.g., [4], but propose a tool to provide passengers with additional information (in addition to schedules) when planning a journey. The basic approach as presented and illustrated in Sect. 2 is Monte Carlo simulation. Potential improvements and extensions using historical and real time data are discussed in Sect. 3.

2 Stochastic Simulation

For modeling/simulating (relevant) departure and arrival times D_{ij} and A_{ij} of train i from/at station j, respectively, we use

$$D_{ij} = \delta_{ij} + V_{ij}, \quad A_{ij} = \alpha_{ij} + W_{ij} + U_{ij},$$

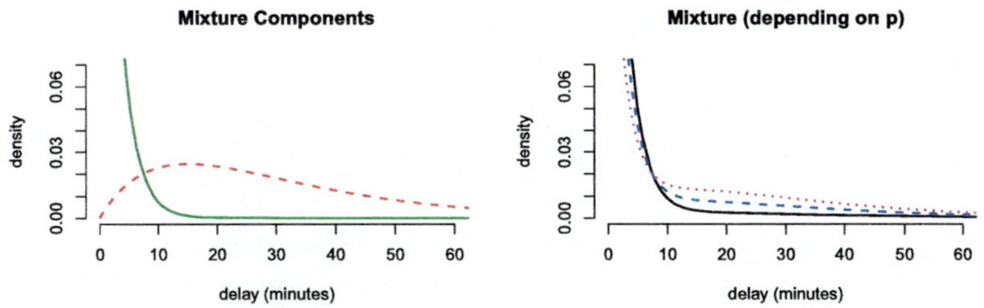

Fig. 2. Mixture components (left) and resulting mixtures (right) for $p = 0.5$ (dotted pink), $p = 0.7$ (dashed blue), and $p = 0.9$ (solid black).

where δ_{ij}, α_{ij} is the departure/arrival time according to schedule. $W_{ij} = D_{i,j-1} - \delta_{i,j-1} - \tau$, if $D_{i,j-1} > \delta_{i,j-1} + \tau$, and zero otherwise, is the delay at departure from station $j - 1$ with soft threshold $\tau > 0$. The latter takes into account that small delays typically do not jeopardize the train's schedule. V_{ij}, U_{ij} are iid random variables following a distribution with density f. For both V_{ij} and U_{ij} we assume a mixture of two gamma distributions, since the most simple, standard approach of using a purely exponential distribution for modeling headway times does not seem realistic; compare, e.g., [3]. That means,

$$f(x) = pg_1(x) + (1-p)g_2(x), \tag{1}$$

where g_1 is the density of a gamma distribution with shape a_1 and scale s_1, g_2 refers to a_2 and s_2; $p \in [0, 1]$ is a weight parameter, which can be interpreted such that an observation comes with probability p from a (latent) class with density g_1, and with probability $1 - p$ from a subpopulation with g_2. More specifically, for our case study, we assume that $a_1 = 1$, $s_1 = 2.5$, which gives an exponential distribution with mean (delay) of 2.5 min, see the green density in Fig. 2 (left). The interpretation is that a train comes with probability p from a 'population' without any major incidents (note, DB considers a train 'on time' if the delay is less than 6 min), where a standard model holds. With probability $1-p$, however, a more serious issue occurs, leading to a gamma distribution with $a_2 = 2$, $s_2 = 15$ (the red curve in Fig. 2, left). The resulting mixture distribution, i.e., density f from (1), for various p is shown in Fig. 2 (right). For our case study, we assume $p = 0.9$ (the black curve in Fig. 2, right), which leads to an overall probability of 82.5% for a delay of less than 6 min, and 92.7% for $<$ 16 min. This appears reasonable having the overall delay of DB long distance trains in 2021 (until April) in mind as given in Table 1. For the soft threshold, we use $\tau = 3$ minutes. Furthermore, we assume walking times of 3 min if changing in Hanover, 2 min at Frankfurt Airport, and 5 min at Frankfurt Central.

Table 1. Delay (overall) of DB long distance trains between January and April 2021 [6] and corresponding settings/numbers in the simulation study.

Delay	January	February	March	April	Simulation
<6 min	83.3%	74.9%	81.1%	81.4%	**82.5%**
<16 min	92.7%	86.7%	91.9%	92.1%	**92.7%**

Figure 3 shows the realized travel times across 10,000 runs of the simulation if starting at Göttingen train station at 9:00 (top left) or 9:30 (bottom left), and taking the first train headed to Frankfurt or Hanover, respectively. In addition, the relative frequency is given whether traveling via Frankfurt or Hanover would have been faster (right). Please note, for calculating travel times, we used 9:00/9:30 as the starting time point, since waiting at the station should be considered part of the trip. The dotted lines (Fig. 3, left) indicate travel times according to schedule/connection (via Frankfurt) proposed by the DB app (see Sect. 1). It is seen that the route via Frankfurt, which is more expensive, was faster about 75% of all trips. Overall, however, differences in travel times between option A and B are rather small (see Fig. 3, left). Furthermore, we see that there is some variation in travel times. In particular, if departing early and choosing to go via Frankfurt, there is some chance of catching an earlier/delayed train there which makes the passenger arrive at Cologne ahead of schedule. This is also seen from Fig. 4, where actual arrival times are given (note, arrival time according to schedule is 13:05/13:09, see Sect. 1). More importantly, however, the dashed line marks the latest arrival time acceptable due to some pretended appointment at 14:00. For instance, if a passenger arrives at Göttingen train station at 9:30 with a ticket via Hanover (compare the second connection given in Sect. 1), the probability/risk of missing the appointment is about 16% according to our simulation (Fig. 4, bottom left). As we can see, this risk can be reduced by taking an earlier train and/or choosing a connection via Frankfurt. Of course, one reason is that by taking an earlier train the chance of catching a connecting train is increased. For instance, when departing from Göttingen around 9:40 as suggested by the schedule given in Sect. 1, the risk of missing the connecting train in Hanover is around 14% according to our model. By taking an ICE train already at 9:00 (or somewhat later) in Göttingen, the risk is reduced to 4%. Generally speaking, our simulation based approach provides, for instance, the means for a passenger who wants to buy his/her ticket in advance to make an informed decision whether a higher price of a specific connection is justified by a shorter, expected travel time, lower risk of arriving late for an appointment, etc.

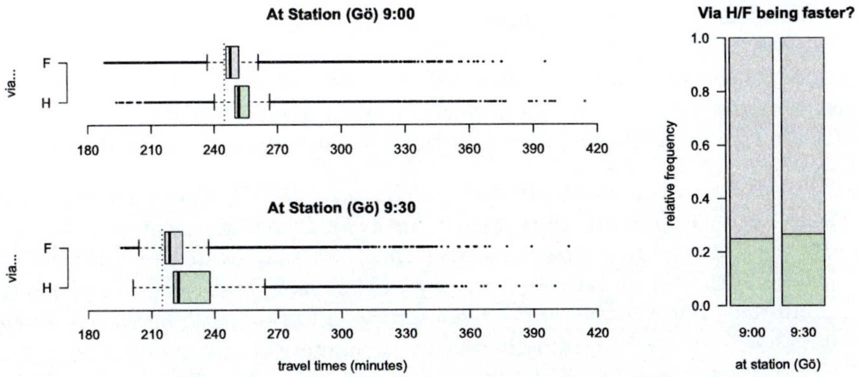

Fig. 3. Left: simulated travel times starting 9:00 am or 9:30 am in Göttingen (Gö); the travel time according to schedule for the connection proposed by DB Navigator is marked by the dotted line. Right: relative frequencies of traveling via Hanover (H, green/bottom) or Frankfurt (F, gray/top) being faster.

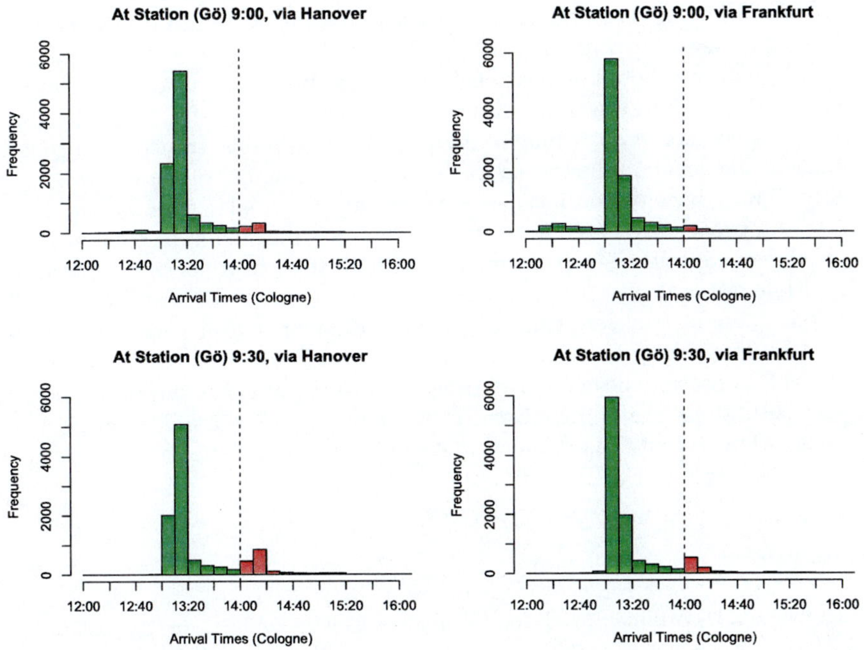

Fig. 4. Simulated arrival times when starting 9:00 am (top) or 9:30 am (bottom) in Göttingen; the pretended appointment at Cologne Central at 2 pm (14:00) is marked by the dashed line.

3 Discussion and Outlook

Our simulation study already produced some interesting results. Of course, however, it should only be seen as a nucleus for larger and extended follow-up studies and research. In particular, there are the following ways to go from here:

1. The parameters used for simulating departure/arrival times appear plausible from a global perspective (compare Table 1). In practice, however, headway distributions are not constant across time and space, but depend on various factors such as time of the year, day of the week, hour of the day, weather conditions, route characteristics, construction works, etc. Therefore historical data has to be used to estimate crucial parameters of probability distributions and to relate those parameters to covariates as mentioned above. Very importantly, this goes beyond simple parameters such as mean waiting time, but also includes variance and shape, for instance by using advanced statistical methods (distributional regression) such as GAMLSS [5].
2. Our framework should be *adaptive* in terms of using additional information that comes in while traveling. In particular, real time data on train delays etc. should be used to update (estimated) probability distributions and hence the model used for simulation.
3. The framework should be *open* such that it can be extended to include aspects such as the way to the train station, e.g., using public transportation, shared ride services, etc. Also, it may be adapted for local public transport in general.
4. In cases like local public transport, however, the number of options to go from S to T may become too large to simulate all potential routes. So the Monte Carlo approach presented here needs to be incorporated in or combined with intelligent algorithms for optimizing with respect to a specific criterion, or multiple criteria.
5. Often enough, travelers can only collect data on travel time distributions by travelling along these routes. Thus, when deciding on routes, there is a tradeoff between expected travel time and future data accuracy. In economics, such situations are called 'Bandit Problems' [1]. We are not aware of an application to route planning so that this denotes an interesting direction of future research.

References

1. Bergemann, D., Valimaki, J.: Bandit problems. Cowles Foundation Discussion Paper (2006)
2. Barbeito, G., Moll, M., Bein, W., Pickl, S.: Deterministic and stochastic simulation: a combined approach to passenger routing in railway systems. In: Neufeld, J.S., Buscher, U., Lasch, R., Möst, D., Schönberger, J. (eds.) Operations Research Proceedings 2019. ORP, pp. 659–665. Springer, Cham (2020). https://doi.org/10.1007/978-3-030-48439-2_80
3. Li, Q., Chen, P., Nie, Y.: Finding optimal hyperpaths in large transit networks with realistic headway distributions. Eur. J. Oper. Rese. **240**, 98–108 (2015)

4. Liu, Y., Blandin, S., Samaranayake, S.: Stochastic on-time arrival problem in transit networks. Transp. Res. Part B: Methodol. **119**, 122–138 (2019)
5. Stasinopoulos, M.D., Rigby, R.A., De Bastiani, F.: GAMLSS: a distributional regression approach. Statist. Model. **18**, 248–273 (2018)
6. Deutsche Bahn. https://www.deutschebahn.com/de/konzern/konzernprofil/zahlen_fakten/puenktlichkeitswerte-1187696. Accessed 19 May 2021

Robust Multistage Yard Crane Scheduling in Container Terminals

Tobias Marx and Michael Hartisch(✉)

University of Siegen, Unteres Schloss 3, 57072 Siegen, Germany
{tobias.marx,michael.hartisch}@uni-siegen.de

Abstract. We deal with robust yard crane scheduling inside a storage block of a container terminal. We consider scenarios with high utilization rates of the cranes, where uncertainties regarding the arrival of transport vehicles at the sea and landside lead to uncertain schedules for the yard cranes. If uncertainties are not incorporated in the scheduling process, the resulting plans can quickly become suboptimal, leading to extended overall completion times that result in delays and can incur penalty costs as a consequence. We present a multistage robust optimization model that allows us to minimize the worst-case objective value by taking into account the uncertain arrival times. Furthermore, we compare our approach with a deterministic model.

Keywords: Robust optimization · Uncertainty · Quantified programming · Scheduling · Yard cranes · Container terminals

1 Introduction

As part of the international transport chains, container terminals contribute to the global exchange of goods and must ensure efficient and on-time handling of high quantities of containers. Container terminals may differ in their design and the transport vehicles involved, but almost always have an area for temporary storage of containers. These storage blocks are a central component and the scheduling of the cranes within the blocks is particularly important, as their operations can have a direct impact on the efficiency and therefore the costs of the entire terminal. Due to the often relatively slow movements of the cranes within the storage blocks, these can become the bottleneck of the terminal [5]. Further information regarding the operations and components of a container terminal can be found in [6].

In our work we focus on the uncertain arrivals of containers in the transfer areas of the storage blocks at seaports and the scheduling of yard cranes (YC) to move the containers. The arrival of containers at the storage blocks depend on multiple uncertain factors [4]. All transport vehicles involved, such as trucks, trains, ships and horizontal transporters within the terminal, may experience delays due to, for example, technical defects or environmental influences. However, predetermined plans can quickly become suboptimal if these uncertain events are not taken into

N. Trautmann and M. Gnägi (Eds.): OR 2021, LNOR, pp. 268–273, 2022.
https://doi.org/10.1007/978-3-031-08623-6_40

account during scheduling. This leads to an increased overall completion time and further delays in the subsequent process, which can result in penalty costs.

We introduce a robust multistage scheduling model in Sect. 2. In Sect. 3, we compare our model with a deterministic model before summarizing and concluding in Sect. 4.

2 Robust Multistage Scheduling

For the robust multistage optimization problem, we use a Quantified IntegerLinear Programming (QIP) formulation. This is a formal extension of Linear Programming (IP), where variables are either existentially or universally quantified. The existentially quantified variables represent decisions made by the planner and the universally quantified variables represent uncertain events. The variables are explicitly ordered, resulting in a multistage optimization model where the solution is a strategy for assigning existentially quantified variables to react optimally to any realization of the universally quantified variables. The objective value of a strategy is given by the worst-case realization of the universally quantified variables. Further details can be found in [2]. In our approach we use a decision-dependent uncertainty set, i.e. the domain of the universally quantified variables depends on realizations of earlier existentially (and universally) quantified variables [3].

Our model aims at minimizing the job's tardiness by finding a robust adjustable job sequence, i.e. jobs are assigned and executed sequentially, incorporating new information gained about actual release dates. We consider a container storage block where container movements are handled by a single YC with two opposite transfer areas. The job specifications for the YC are provided using a test data generator [1] and include the starting and destination locations of the containers, release and due dates for the jobs, and values for a potential delay in the release date. In our case, the parameter values are subject to a uniform distribution. We suppose that the movement along the bays and rows can occur simultaneously and that during the lifting and lowering of the spreader for loading or unloading of the containers no further crane movements are performed for safety considerations.

Within the storage blocks, different job types are possible, such as storage and retrieval on the sea and landside or housekeeping operations. The determination of job j in the processing sequence at time stage t by the existential variables s_j^t and the possible delay of the release date of job j by the universal decision variables ξ_j^t in stage t are carried out alternately, whereby the first job in the sequence is not subject to any delay and only jobs in the subsequent decision stages that have not already been processed may be delayed. The overall number of delayed release dates is bounded by G. Therefore, the decision-dependent uncertainty set Ξ^t in stage t is given by

$$\Xi^t(s^0, ..., s^{t-1}, \xi^1, ..., \xi^{t-1}) = \left\{ \begin{array}{c} \xi^t \in \{0,1\}^J \mid \sum_{t \in T} \sum_{j \in J} \xi_j^t \leq G \\ \wedge\ \xi_j^t \leq 1 - \sum_{t'=0}^{t'<t} s_j^{t'}\ \forall j \in J \end{array} \right\}. \tag{1}$$

We will use Ξ^t and omit stating the dependencies when clear. In Tables 1 and 2 the remaining parameters and variables are introduced, respectively.

Table 1. Sets and Parameters

Symbol	Description
J	set of jobs $\{0,\dots,N\}$ and existential decision stages
T	set of universal decision stages $\{1,\dots,N\}$
$G \in \mathbb{N}$	overall limit for the number of release date delays
$M \in \mathbb{N}$	big M
$A \in \mathbb{N}$	$A_{j,k}$: crane travel time from destination of job k to starting slot of job j
$L \in \mathbb{N}$	L_j: crane transport time of container j from starting slot to destination
$R \in \mathbb{N}_0$	R_j: release date of job j
$D \in \mathbb{N}$	D_j: due date of job j
$P \in \mathbb{N}_0$	P_j: possible delay in release date

Table 2. Variables

Symbol	Quantification	Description
$s \in \{0,1\}^{T\cup\{0\}\times J}$	$(\exists)$	s_j^t: indicates whether job j is executed in stage t
$\xi \in \{0,1\}^{T\times J}$	$(\forall)$	ξ_j^t: indicates whether the release date of job j is delayed with announcement in stage t
$x \in \mathbb{N}_0^J$	$(\exists)$	x_j: tardiness of container j
$c \in \mathbb{N}^J$	$(\exists)$	c_j: completion time of job j

$$\min \sum_{j\in J} x_j \tag{2}$$

$$\text{s.t.} \exists s_j^0 \in \{0,1\}^J \tag{3}$$

$$\forall \xi_j^1 \in \Xi^1 \quad \exists s_j^1 \in \{0,1\}^J$$

$$\forall \xi_j^2 \in \Xi^2 \quad \exists s_j^2 \in \{0,1\}^J$$

$$\dots \forall \xi_j^N \in \Xi^N \quad \exists s_j^N \in \{0,1\}^J \; x \in \mathbb{N}_0^J \; c \in \mathbb{N}^J :$$

$$\sum_{j\in J} s_j^t = 1, \qquad \forall t \in T \cup \{0\} \tag{4}$$

$$\sum_{t\in T\cup\{0\}} s_j^t = 1, \qquad \forall j \in J \tag{5}$$

$$R_j + L_j + (s_j^0 - 1) * M \le c_j, \qquad \forall j \in J \tag{6}$$

$$c_k + A_{j,k} + L_j + (s_j^t + s_k^{t-1} - 2) * M \le c_j \tag{7}$$

$$\forall t \in T,\, j \in J,\, k \in J,\, k \neq j$$

$$R_j + \sum_{t' \geq 1}^{t} (\xi_j^{t'} * P) + A_{j,k} + L_j + (s_j^t + s_k^{t-1} - 2) * M \leq c_j, \tag{8}$$

$$\forall t \in T,\, j \in J,\, k \in J,\, k \neq j$$

$$c_j - D_j \leq x_j, \qquad \forall j \in J \tag{9}$$

The objective function (2) aims at minimizing the overall tardiness. The order and domains of the variables in the multistage model are defined in the Quantification Sequence (3). At each existential decision stage, a job is assigned to the job sequence (4) and each job may be executed only once across all stages (5). The very first job performed by the YC is not affected by uncertainty (6). There are two constraints for the subsequent jobs in the processing sequence, as job j cannot start until both the previous job k is completed (7) and the release date of job j (plus the possible delay) has occurred (8). Only tardiness is considered and early submission cannot be utilized as compensation for tardiness (9). We can solve such instances using a special solver for QIPs [3].

In Sect. 3 we will compare our robust multistage approach with an IP model, that does not take uncertainty into account and aims for a deterministic, non-adjustable job sequence. This IP is similar to the QIP to a major part, but in particular does not incorporate the universally quantified variables. We do not present the entire model, but rather discuss the components based on the model presented above. The IP model has the same objective function (2) and constraints (4), (5), (6), (7) and (9). Constraint (8) is adapted for the IP model by neglecting the possible delays:

$$R_j + A_{j,k} + L_j + (s_j^t + s_k^{t-1} - 2) * M \leq c_j \quad \forall t \in T,\, j \in J,\, k \in J,\, k \neq j$$

Thus, the job sequence is determined regardless of delays of the release dates. However, the objective function also minimizes the tardiness relating to the completion times of the jobs. In order to compare the two models, the job sequence from the IP model should be tested for scenarios.

3 Computational Experiments

We first discuss an illustrative example with six jobs, for which the job specifications are given in Table 3. Additionally the job sequence from the IP solution and the worst-case realization of the optimal strategy from the QIP model are shown. This example includes three storage operations each from the sea and the landside of the storage block and is visualized in Fig. 1, with the blue lines representing the crane's routes with loaded container which are invariant for each job sequence. The routes of the unloaded crane between the individual jobs, which are shown in red for the IP solution and green for the QIP solution, result from the assignment of the jobs to the particular stage and thus the position in the job sequence. The optimal sequence from the deterministic model starts with job

4 in bay 21 and the optimal worst-case job sequence of the robust model starts with job 1 in bay 0. Since we include release and due dates in our considerations and, in the case of the QIP model, also potential delays of the release date, the job sequence does not only result from the shortest paths between the individual jobs. For our example, we set $G = 1$ so that only one release date is delayed. Also note that in our model we only consider the required crane operating time and not the exact path between two points. Hence, the exact travel paths in the figure are suggestions and can also be represented differently.

Table 3. Job specifications and the optimal deterministic job sequence from the IP as well as the optimal worst-case sequence from the QIP.

	Start		Destination		Dates		Stage	
j	bay	row	bay	row	R_j	D_j	ip	qip
0	0	5	9	7	45	130	4	5
1	0	6	17	3	27	126	5	0
2	0	4	19	6	49	107	2	2
3	21	5	18	6	32	93	1	3
4	21	1	14	5	15	104	0	1
5	21	4	5	3	52	124	3	4

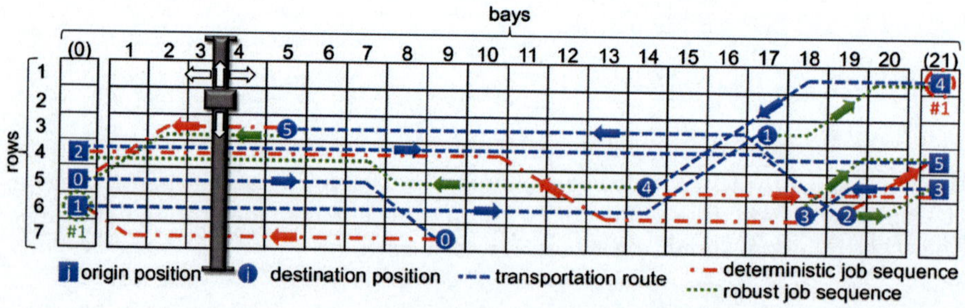

Fig. 1. Crane movements within the storage blocks.

We also conducted experiments on 100 instances consisting of 6 to 8 jobs and similar job types as in the example above with $G = 1$. In order to be able to compare the two models described, we applied the job sequences from the solution of the IP model in a QIP model for each instance and computed the solution for the worst-case scenario. Figure 2 displays the number of instances with respect to their improvement in the objective value resulting from considering the uncertainty in a multistage manner. For 90% of all instances, the QIP solution was at least 50% better in the given comparison.

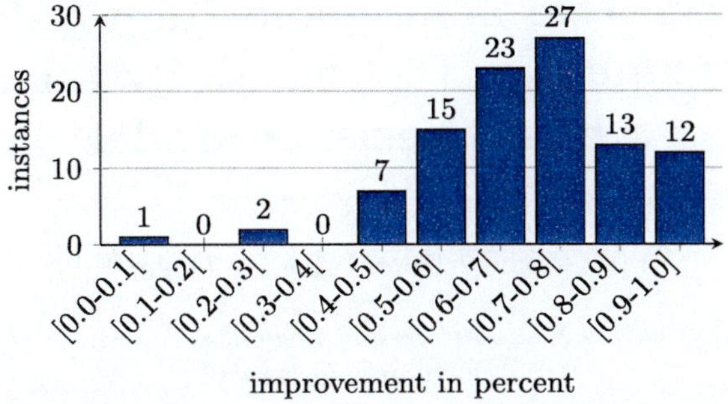

Fig. 2. Improvement in the objective value of the robust QIP solution compared to the worst-case realization for the corresponding deterministic job sequence from the IP.

4 Conclusion

We introduced a quantified programming model for robust multistage scheduling of a yard crane within a container storage block. We illustrated our model by visualizing the crane movements of an instance with six storage operations split between both sides of the storage block. By conducting experiments on 100 instances, we demonstrated the benefits of robust multistage scheduling. Future work may focus, for example, on the integration of additional cranes in the storage block or on special considerations of important scenarios in the robust multistage approach.

References

1. Briskorn, D., Jaehn, F., Wiehl, A.: A generator for test instances of scheduling problems concerning cranes in transshipment terminals. OR Spectrum **41**(1), 45–69 (2018). https://doi.org/10.1007/s00291-018-0529-z
2. Goerigk, M., Hartisch, M.: Multistage robust discrete optimization via quantified integer programming. Comput. Oper. Res. **135**, 105434 (2021)
3. Hartisch, M.: Quantified integer programming with polyhedral and decision-dependent uncertainty. Ph.D. thesis, Universität Siegen (2020)
4. He, J., Tan, C., Zhang, Y.: Yard crane scheduling problem in a container terminal considering risk caused by uncertainty. Adv. Eng. Inform. **39**, 14–24 (2019)
5. Ng, W., Mak, K.: Yard crane scheduling in port container terminals. Appl. Math. Model. **29**(3), 263–276 (2005)
6. Stahlbock, R., Voß, S.: Operations research at container terminals: a literature update. OR Spectrum **30**, 1–52 (2008)

Integrative Zoning and Item-to-Zone Assignment in Pick&Pass Systems – A Basic Decision Model

Ralf Gössinger and Regina Thelen(✉)

Chair of Production Management and Logistics, University of Dortmund, Dortmund, Germany
regina.thelen@tu-dortmund.de

Abstract. Order picking is the warehouse process of consolidating stored items according to customer orders. When designing a zoned order picking system, two problems need to be solved in the light of performance objectives: zoning and item-to-zone assignment. Usually, these problems are tackled in such a way that assignment is done after the zones were formed. Since both problems are interrelated, we propose an integrative decision model and solve it by means of optimization software. Based on a numerical study the suitability of the model and weaknesses of the solution approach are pointed out.

Keywords: Warehouse design · Mixed-integer programming

1 Introduction

Zoning (ZG) and storage location assignment (SLA) are two important tasks of warehouse operations management [11,12]. ZG divides the warehouse into several sub-areas, such that in each sub-area a specific group of workers is responsible for picking the items according to orders. SLA can refer to different aggregation levels: bin location assignment [6] and assignment to zones [9]. The latter case is relevant for this paper.

With respect to the objective of makespan minimization, the ZG and SLA decisions are *interdependent* [12]. The path length within a zone increases with the number of bin locations (zone size), since the dimensions of the zone increase and the locations contained in a picking route are distributed over a larger area. The strength of this effect is influenced by the item-to-zone assignment, which is usually based on the items' demand frequencies and cross-correlations. Both reduces the inter-zone path length, but ZG unfolds an ambivalent impact: While the effectiveness of frequency-based SLA decreases, that of correlation-based SLA increases in the zone size. Performing ZG and SLA in an isolated way, induces longer lead times than necessary and thus competitive disadvantages.

Most of the existing approaches solve the sub-problem of SLA in an isolated way [4]. It is usually suggested to perform ZG at first and SLA afterwards [8,10].

N. Trautmann and M. Gnägi (Eds.): OR 2021, LNOR, pp. 274–279, 2022.
https://doi.org/10.1007/978-3-031-08623-6_41

This decomposition saves computational effort, but may not lead to optimal solutions as soon as the decisions are interdependent. A perfect coordination of such decisions can be achieved by considering them in an integrative model [1,2]. However, additional decision variables increase the complexity of the model, so that it is questionable whether it can be solved exactly with acceptable computational effort. There are few suggestions for integrative solution approaches to the problem at hand [1,5]. The approach [5] is based on the assumption of an evenly divided warehouse, which severely limits the scope for ZG. Hence, only a few configurations are possible and can be compared in a simulation study. In [1] a three-step solution approach is developed. At first, the best SLA is determined using simulation, second, the zone locations are optimized, and third, the impact of uncertainty is analyzed by means of simulation. In contrast to existing approaches, to the best of our knowledge, we are the first who derive an optimization model that treats both decisions simultaneously without using simulation studies. In this connection, we focus on pick&pass systems. These are zoned systems in which each picker only works at one zone and each zone only comprises a sub-set of items. Order-related containers are routed along a sequence of zones to be filled up consecutively with the items needed for order fulfillment [3].

In the remainder of this paper, we continue these considerations and propose a decision model that simultaneously decides on ZG and SLA (Sect. 2). In order to evaluate the proposed model in terms of solution quality and computational effort and to identify regularities of the solutions found, we conduct a numerical study and analyze the results by statistical means (Sect. 3). Finally, in Sect. 4, the main findings and the next steps of our future research are pointed out.

2 Model

The warehouse comprises C^L shelves of identical size at which the items $k(k = 1, \ldots, K)$ are stored. The inventory of each item occupies exactly one shelf. Hence, $K \leq C^L$ holds. A maximum number of J zones can be formed. Each zone $j(j = 1, \ldots, J)$ consists of at least $\underline{C}$ and at maximum $\overline{C}$ neighboring shelves to which items can be assigned to. An order $n(n = 1, \ldots, N)$ is a bundle of items with composition $y_{nk} \in \{0, 1\}$ and occurrence probability p_n. Order fulfillment is performed sequentially. Starting at the I-point ($j = 0$), the order-related container is forwarded to the first relevant zone. After the container has arrived in a zone, items relevant for both, order and zone, are taken out from the shelves and placed in the container. Subsequently, the container is passed to the next relevant zone. This process is continued up until all required items are picked. Finally, the filled up container is forwarded to the O-point ($j = J + 1$). Assuming that times for taking an item out of the shelf are identical for each item and each shelf, decision-relevant *makespancomponents* are the inter-zone time $t^w_{jj'}$ and the intra-zone time t^z_j. Both are dependent on warehouse design as well as *basic decisions* on the activation of zones u_j and the assignment of items to zones x_{kj}. In addition, *decisions on details* refer to the routing of orders. Whether zone j is (not) relevant for order n is captured by the variable b_{nj}. The route of order n has to be determined by combining direct connections $w_{jj'n}$ between the zones j and j'. As I- and O-point are relevant for each order, their activation and relevance for routes are parameters ($u_0, u_{J+1} = 1; b_{n0}, b_{nJ+1} = 1 \forall n$). Based on these assumptions, a MILP model results:

$$\text{Min} \sum_{j,n,k} p_n \cdot y_{nk} \cdot t_j^z \cdot x_{kj} + \sum_{j=0}^{J} \sum_{j'=j+1}^{J+1} \sum_{n} p_n \cdot t_{jj'}^w \cdot w_{jj'n} \quad (1)$$

$$\sum_j x_{kj} = 1 \quad \forall k \quad (2)$$

$$\sum_k x_{kj} \leq u_j \cdot \overline{C} \quad \forall j = 1, \ldots, J \quad (3)$$

$$\sum_k x_{kj} \geq u_j \cdot \underline{C} \quad \forall j = 1, \ldots, J \quad (4)$$

$$b_{nj} \cdot M \geq \sum_k y_{nk} \cdot x_{kj} \quad with \quad M > K \cdot C^L \quad \forall j = 1, \ldots, J \quad (5)$$

$$w_{jj'n} = 0 \quad \forall j \geq j', j' = 1, \ldots, J, n \quad (6)$$

$$w_{jj'n} \leq u_j \quad \forall j < j', j' = 1, \ldots, J, n \quad (7)$$

$$w_{jj'n} \leq u_{j'} \quad \forall j < j', j' = 1, \ldots, J, n \quad (8)$$

$$\sum_{j'=0}^{j-1} w_{j'jn} = b_{nj} \quad \forall n, j = 1 \ldots, J \quad (9)$$

$$\sum_{j'=j+1}^{J+1} w_{jj'n} = b_{nj} \quad \forall n, j = 1, \ldots, J \quad (10)$$

$$b_{nj} \in \{0, 1\} \quad \forall j = 1, \ldots, J, n \quad (11)$$

$$x_{kj} \in \{0, 1\} \quad \forall j = 1, \ldots, J, k \quad (12)$$

$$u_j \in \{0, 1\} \quad \forall j = 1, \ldots, J \quad (13)$$

$$w_{jj'n} \in \{0, 1\} \quad \forall j = 1, \ldots, J, j' = 1, \ldots, J, n \quad (14)$$

The $objective function$ (1) minimizes the expected makespan. The decisions have to be made subject to *restrictions*: (2) Each item is assigned to exactly one zone and articles can only be assigned to activated zones. The size of an activated zone is restricted from above and from below (3, 4). During fulfillment, an order has to visit each zone, which at least stores one of the ordered items (5).

The calculation of distances covered during order fulfillment is based on the routes: (6) defines that an order visits the zones according to ascending zone index. The route of an order is composed of connections between activated zones only (7, 8). Furthermore, each zone contained in the route has exactly one predecessor (9) and one successor (10).

The initial analyses are based on a simple warehouse design: Shelves of identical length l are arranged in a single row. I- and O-point are in the same location, which is situated between two zones in the middle of the shelf row. In the left and right halves of the shelf row, the zone index increases as the distance from the I/O point increases. The flow of containers from I-point via zones to O-point is bidirectional with constant speed v^w. In each zone, the container stays at the centered base as long as all relevant items are picked. Required items are picked sequentially. Distances between base and shelves of a zone are covered with constant speed v^z. Based on these assumptions and calculating the zone size with $s_j = \sum_k x_{kj}$, $t^w_{jj'}$ and t^z_j are:

$$t^w_{0j} = t^w_{jJ+1} = \frac{l}{v_w} \cdot \begin{cases} \sum_{j'=1}^{j-1} s_{j'} + \frac{s_j}{2} & : j \leq \frac{J}{2} \\ \sum_{j'=\frac{J}{2}+1}^{j-1} s_{j'} + \frac{s_j}{2} & : j \geq \frac{J}{2} + 1 \end{cases} \quad \forall j = 1, \ldots, J \tag{15}$$

$$t^w_{jj'} = \frac{l}{v^w} \cdot$$
$$\begin{cases} \sum_{j''=j+1}^{j'-1} s_{j''} + \frac{s_j + s_{j'}}{2} & : j+1 \leq j' \wedge (j,\ j' \leq \frac{J}{2} \vee j,\ j' \geq \frac{J}{2} + 1) \\ \sum_{j'=1}^{j-1} s_{j'} + \sum_{j''=\frac{J}{2}+1}^{j'-1} s_{j'} + \frac{s_j + s_{j'}}{2} & : j+1 \leq j' \wedge j \leq \frac{J}{2} \wedge j' \geq \frac{J}{2} + 1 \end{cases}$$
$$\forall j = 1, \ldots, J-1, j+1 \leq j' \leq J \tag{16}$$

$$t^z_j = \frac{l}{v^z} \cdot \frac{s_j}{2} \quad \forall j = 1, \ldots, J \tag{17}$$

In order to include these calculations in the model, the time components are defined as variables ($t^w_{jj'} \in R_0^+ \forall j, j'; t^z_j \in R_0^+ \forall j$) and restricted by the time calculation from below. The products $t^z_j \cdot x_{kj}$ and $t^w_{jj'} \cdot w_{jj'n}$ are replaced by aggregate continuous variables using standard linearization techniques [7].

3 Numerical Study

The purpose of the full-factorial numerical study is to provide insights into the model behavior in terms of solution time, quality and structure. Varied warehouse parameters are the number of articles $K \in \{12; 20; 28\}$ and the maximum relative zone size $\overline{c} \in \{0.25; 0.5; 0.75\}$, which is the ratio of $\overline{C}$ and K. The minimum zone size is fixed to $\underline{C} = 1$. Varied order parameters are the number $N \in \{20; 40; 60\}$ and the heterogeneity $\eta \in \{0.2; 0.3; 0.4\}$ of orders. The coefficient $\eta \in [0, 1]$ expresses the average dissimilarity of the orders regarding the articles they contain: $\eta = \sum_{n=1}^{N-1} \sum_{n'=n+1}^{N} \sum_{k=1}^{K} |y_{nk} - y_{n'k}| / (N \cdot (N-1)/2 \cdot K)$.

In the case of four parameters, a systematic variation of three specifications per parameter leads to 81 combinations. Due to the randomly generated order information, we consider 3 runs with different order information for each combination, so that 243 instances result, which are sufficient for meaningful statistical analyses ($R^2 > 0.5; p < 0.01$). We used CPLEX 20.1 for solving the instances on a Windows PC (2.70 GHz Intel Core i7 CPU, 16 GB RAM) with a solution time limit of 30 min. 177 instances are solved to optimality. For the other 66 instances, intermediate results and integrality gaps (GAP) are recorded. To receive the detailed statistical results, please send an email to pl.wiwi@tu-dortmund.de.

For analyzing the *solution time* (ST) an exponential regression model is estimated that explains 65% of variations by the varied factors with a high significance. ST increases exponentially in K, N and η and decreases exponentially in $\overline{c}$, whereby K has the strongest impact. This underlines the necessity of alternative solution approaches when bigger instances need to be solved.

Regarding the *solution quality*, the frequency of solution process interruptions (IN) and the variation of GAP can be explained by second-order polynomial linear regressions with two-factor interactions. With a high significance, more than 50% of the variations can be traced back to the examined factors. IN indicates that the probability of reaching the solution time limit most strongly depends on K, the impact of which is strengthened by N and η. The GAP analysis reveals $\overline{c}$ as the main driver. Its impact is strengthened by both, K and N, and weakened by η. Both results allow the conclusion that although the model can be solved with a standard solver, a good solution quality can only be achieved for small instances in acceptable time. Accordingly, heuristic approaches should be used to solve instances of realistic size.

The analyses of the *solution structure* concentrate on the makespan Z and the number of zones U. The procedure is analogous to the analyses of the solution quality. At least 66% of variations are explained by the varied factors with a high significance. The analyses reveal non-linear influences for all observed variables. With respect to Z, the most important influencing factors are K and N. Their extending impact is mainly weakened by $\overline{c}$. The main factor influencing U is $\overline{c}$, while the relevance of other factors is neglectable. In contrast to K and N, $\overline{c}$ can primarily be determined by the warehouse management. Hence, it is the central parameter to control the solution structure. Due to the non-linear conditional impact of $\overline{c}$ and its interactions with K and N, it needs to be carefully adjusted.

4 Conclusions

In this paper, a new optimization model is proposed that integratively decides on zones and storage locations in a pick&pass system. It is structured in such a way that the calculation of inter-zone and intra-zone times can be adapted to specific warehouse designs without changing the main parts of the model. Exact linearizations result in a MILP that can be solved by means of standard solvers. Based on a numerical study, the model behavior is statistically evaluated in terms of solution time, solution quality and solution structure. The analyses reveal that the model can be exactly solved for small instances in acceptable time. Bigger instances often cannot be solved within a reasonable time limit and the quality of "unfinished" solutions worsens with increasing instance dimensions. In addition, the maximum relative zone size has been identified as a parameter that unfolds significant impact on the solution quality and structure. These results motivate four directions of subsequent research: (1) Heuristic approaches are to be tested in order to achieve near-optimal solutions for bigger instances. (2) It has to be explained how the "maximum relative zone size" parameter can be advantageously set. (3) Both time components should be formally described for other warehouse designs. (4) The drivers of interdependencies between ZG and SLA need to be analyzed depending on warehouse design.

References

1. Amorim-Lopes, M., Guimarães, L., Alves, J.: Improving picking performance at a large retailer warehouse by combining probabilistic simulation optimization, and discrete-event simulation. Int. Trans. Oper. Res. **28**, 687–715 (2021)
2. Bottani, E., Volpi, A., Montanari, R.: Design and optimization of order picking systems: an integrated procedure and two case studies. Comput. Ind. Eng. **137**, 106035 (2019)
3. Choy, K.L., et al.: Assess the effects of different operations policies on warehousing reliability. Int. J. Prod. Res. **52**, 662–678 (2014)
4. de Koster, R.B.M., Le-Duc, T., Roodbergen, K.J.: Design and control of warehouse order picking: a literature review. Eur. J. Oper. Res. **182**, 481–501 (2007)
5. de Koster, R.B.M., Le-Duc, T., Zaerpur, N.: Determining the number of zones in a pick-and-sort order picking system. Int. J. Prod. Res. **50**, 757–771 (2012)
6. Glock, C.H., Grosse, E.H.: Storage policies and order picking strategies in U-shaped order-picking systems with a movable base. Int. J. Oper. Res. **16**, 4344–4357 (2012)
7. Gribkovskaia, I., et al.: Optimization model for a livestock collection problem. Int. J. Phys. Distrib. Logist. Manag. **36**, 136–152 (2006)
8. Jane, C.C., Laih, Y.W.: A clustering algorithm for item assignment in a synchronized zone order picking system. Eur. J. Oper. Res. **166**, 489–496 (2005)
9. Kress, D., Boysen, N., Pesch, E.: Which items should be stored together? A basic partition problem to assign storage space in group-based storage systems. IISE Trans. **49**, 13–30 (2017)
10. Petersen, C.G.: Considerations in order picking zone configuration. Int. J. Oper. Prod. Manag. **22**, 793–805 (2002)
11. van Gils, T., Ramaekers, K., Caris, A.: Designing efficient order picking systems by combining planning problems: state of the art classification and review. Eur. J. Oper. Res. **267**, 1–15 (2018)
12. van Gils, T., Ramaekers, K., Braekers, K.: Increasing order picking efficiency by integrating storage, batching, zone picking, and routing policy decisions. Int. J. Prod. Econ. **197**, 243–261 (2018)

Mobility and Traffic

New Optimization Guidance for Dynamic Dial-a-Ride Problems

Christian Ackermann(✉) and Julia Rieck

Institute for Business Administration and Information Systems, Operations Research Group, University of Hildesheim, Universitätsplatz 1, 31141 Hildesheim, Germany
ackermann@bwl.uni-hildesheim.de

Abstract. In the dial-a-ride problem, customers have to be transported from different pickup to drop-off locations. Various constraints such as time windows and a maximum ride time per passenger need to be considered. In the dynamic version of the problem, not all customer requests are known in advance, but arrive during the operation time. Therefore, the maximization of the number of served customers is usually set as the optimization goal. Nevertheless, in the vast majority of known heuristics, the total distance is used to guide the optimization. In this paper, we present different metrics that should enable the evaluation of the insertion potential of future customers and lead to a higher acceptance rate through their use in solution procedures. We show that even a single metric can provide better results than the distances and present a *Markov decision process*-based approach to enable an agent trained by *reinforcement learning* to perform even more anticipatory decision making by considering multiple metrics simultaneously.

Keywords: Dynamic dial-a-ride problem · Metrics · Reinforcement learning

1 Introduction

As a compromise between expensive cabs and unflexible public transport, on-demand ridesharing services continue to gain in popularity. In order to model sharing concepts, the dial-a-ride problem (DARP) as a variant of the vehicle routing problem with time windows can be used. In this problem, a set of requests is served by a fleet of vehicles. Each request consists of taking a customer from a pickup to a drop-off point. The customers specify time windows for pickup or drop-off as well as a maximum ride time (often a linear function depending on the direct travel time).

The DARP combines a *routing* and a *scheduling* part. First, the routing has to determine which vehicle serves which customer in what order. The subsequent scheduling determines the exact timing so that time windows and the maximum ride time are respected. The goal of the static problem, in which all requests are known in advance, is usually to minimize the total travel distance. In a

N. Trautmann and M. Gnägi (Eds.): OR 2021, LNOR, pp. 283–288, 2022.
https://doi.org/10.1007/978-3-031-08623-6_42

dynamic DARP, only some of the requests are known at the beginning of the planning horizon, the remainder are received during the operation. How many of the requests are dynamic is specified by the *degree of dynamism*. Due to the consideration of requests arriving at short notice, it may not be possible to serve all requests. Hence, the maximization of the number of requests that can be accepted is typically chosen as the optimization goal.

There are well known approaches that adapt the scheduling part to be able to respond to the dynamic aspect of the DARP (see, e.g., [1]). The idea is to generate a schedule that facilitates the insertion of future requests. However, the routing part is usually disregarded. Instead, the corresponding approaches focus on insertion heuristics that may be followed by local search procedures. These are executed for a certain number of iterations or until a new event (e.g., arrival of a new request) occurs (see, e.g., [2]). Nevertheless, the respective insertion and local search methods only minimize distances, which leads to an efficient use of vehicles but does not explicitly maximize the insertion potential for future requests.

In this contribution, we seek to find a metric whose optimization in the routing part of a solution procedure leads to a higher acceptance rate than just using the distances. To this end, we evaluate different metrics for estimating the insertion potential of future requests (cf. Sect. 2). In Sect. 3, we show the superiority of one of the metrics over the previously used distances. In Sect. 4, we additionally present a *reinforcement learning*-based approach that could enable even more anticipatory decision making. Section 5 contains a summary and provides an outlook on future research.

2 Metrics for Estimating the Insertion Potential

The metrics used to guide the optimization are intended to reflect the potential of a solution to allow future request insertions. According to the literature and own preliminary experiments, primarily the time windows and secondarily the maximum ride time are the critical constraints. Consequently, we incorporate the subsequent four metrics (it is preferable if the metrics have high values):

Mean Waiting Time: For each arc $\langle A, B\rangle$ in the current solution (A and B are customer nodes), it is determined how long the respective vehicle waits on that arc. Thereby, the waiting time between end of service and departure at A as well as between arrival and start of service at B are taken into account. To evaluate the entire solution, the average waiting time is calculated over all arcs that are still modifiable, i.e., have not yet been fully or partially traversed.

Ellipse Area: Let $\langle A, B\rangle$ again be an arc in the solution. In order to now maximize the probability of being able to insert a random next request between A and B, we calculate where the node to be inserted (pickup or drop-off) could be located (assuming suitable time windows). We specify the earliest departure time at node A (ED_A) and the latest arrival time at node B (LA_B). The difference of LA_B and ED_A is the total available time (at_{AB}) for the trip from A to B. Determining all points C that can be reached within this time via the route

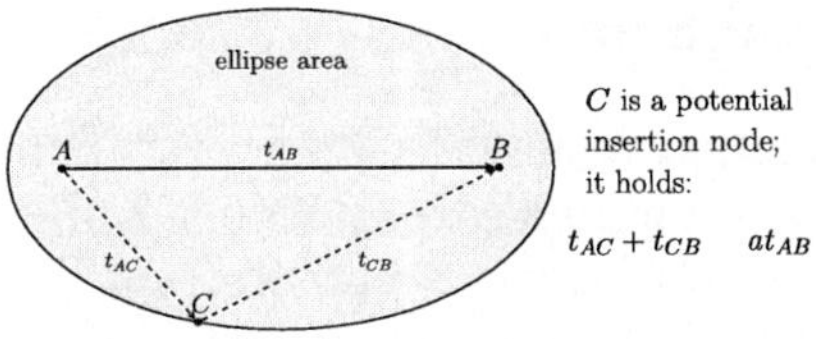

Fig. 1. Example of the ellipse area metric

$A \rightarrow C \rightarrow B$ results in an ellipse with focal points A and B (see Fig. 1). We take the area of the ellipse as an indicator of the probability of inserting a random request on $\langle A, B\rangle$. With (1), we obtain the area according to:

$$\text{ellipse area} = \frac{\pi}{4} \cdot at_{AB} \cdot \sqrt{(at_{AB})^2 - (t_{AB})^2} \tag{1}$$

with t_{AB} being the direct travel time from A to B. To evaluate the entire solution, a vehicle score S_k is determined by averaging the area of the ellipse over all arcs still to be traversed by vehicle k. Formula (2) is then used to combine the scores of each of the K existing vehicles into a joint score S:

$$S = \left(\frac{1}{K}\sum_{k=1}^{K}\sqrt{S_k}\right)^2 \tag{2}$$

Mean Ride Time Slack: We identify the average ride time slack (difference between actual and maximum ride time per customer) as a measure of how strict the current transportation plan is. The larger the average ride time slack of a customer not yet visited, the larger detours are possible without violating the customer's maximum ride time.

Mean Time Window Slacks: In addition, we integrate the *forward*, *central*, and *backward slack* times as proposed in [3]. These slack times consider the average difference between the planned start of service at a node and the opening (forward), center (central) or closing (backward) of the associated time window.

To evaluate how well these metrics can predict the insertion potential, we collected different information from temporary solutions during the application of a traditional distance-based solution procedure and noted whether a new request could be inserted. We trained a *random forest classifier* by *supervised learning* based on about 60,000 solutions to estimate whether the next request can be added to the current transportation plan or not. Input parameters were the current timestamp (as a measure of progress within the planning horizon), the number of requests accepted so far, and one of the collected metrics. The quality of the classification was measured by the area under the ROC curve (AUC). All metrics except the *mean ride time slack* were able to predict the insertion potential better than the pure distances. Particularly, the *ellipse area* achieved the best AUC with 0.9456 (in contrast to the distances with 0.9284).

3 Computational Study

Always taking the solution with the highest probability of inserting the next request is a very short-term and greedy strategy. In order to check whether this nevertheless leads to a higher acceptance rate in total, we have changed the acceptance criterion at all points in our optimization procedure. Instead of taking the solution with the shorter total distance, we choose the solution with the higher value of the metric under consideration.

The solution procedure used for the routing part is designed as follows: The initial insertion follows the ejection chain approach presented in [4]. All feasible insertion positions for the new request r are examined and the best one (according to the considered metric) is selected. If all insertion positions are infeasible, a request whose time windows overlap with r is removed from the current solution, the insertion procedure is repeated for r, and the removed request is reinserted again. This is iterated for all eligible requests. In case that no feasible solution is found, the request is rejected. Otherwise the best resulting solution is accepted and improved by a local search phase. This process essentially follows the approach in [5]. Neighborhood operators considered are the exchange of two requests, the relocation of one request, a 2-opt operator applied to route segments that do not have a customer in the vehicle at the beginning and end respectively, and a restricted 4-opt operator, where four consecutive arcs are eliminated. Once an improved solution is found, a simulated annealing criterion is applied to check whether it should be accepted or not. For the scheduling part and the feasibility check, the approach from [1] was used.

In the evaluation, we used 72 instances, which were generated based on realistic characteristics. The number of customer requests varied from 50 to 200, the time windows were 10 or 20 min each, the degree of dynamism ranged from 20% to 80%, and dynamic customers were known either 30 or 90 min before their pickup time window opened. The number of vehicles varied from 2 to 6 depending on the number of customers, the service area was 15 × 15 distance units, the duration of the planning horizon was 10 h, and the vehicles had a fixed capacity of 4 requests and a uniform speed of one distance unit per minute. The procedure was run 10 times with each metric and different durations of the local search for all instances, considering the average value over all 10 runs.

The *ellipse area* was the only metric that produced better results than the total distance. Table 1 shows the detailed results. While the first two columns display the properties of the instances, columns 3–5 and 6–8 compare the acceptance rate and the number of accepted customers for the *distance* (D) and the *ellipse area* (EA) metrics respectively, and provide the respective percentage differences. Column 9 presents a comparison of the number of instances in which the distance or the ellipse area led to a better solution as well as the number of instances in which the same results were achieved.

It is clearly evident that, on average, the ellipse area metric gives better results than the distances (regardless of the instance properties) and also solves most instances better with respect to the objective function. Larger time windows seem to amplify the difference due to the higher degree of freedom. How short-term the

Table 1. Comparison of distance and ellipse area metrics for optimization guidance

Instance property		Acceptance rate [%]			Accepted customers			InstComp
		D	EA	Incr [%]	D	EA	Incr[%]	D/EA/=
Time windows	10 min	90.88	91.79	1.00	129.30	130.46	0.90	32/70/6
	20 min	93.27	94.93	1.78	133.13	135.10	1.48	19/89/0
Booking in advance	30 min	91.97	93.26	1.40	130.83	132.47	1.25	29/78/1
	90 min	92.19	93.46	1.38	131.60	133.09	1.13	22/81/5
# Requests	50	90.26	92.28	2.24	45.13	46.14	2.24	7/26/3
	100	89.57	91.36	2.00	89.57	91.36	2.00	8/28/0
	150	92.70	93.85	1.24	139.05	140.78	1.24	16/54/2
	200	93.62	94.41	0.84	187.24	188.81	0.84	20/51/1
Degree of dynamism	20%	91.38	93.14	1.93	130.16	132.35	1.68	21/51/0
	50%	91.62	92.38	0.83	130.91	131.54	0.48	21/49/2
	80%	93.24	94.56	1.42	132.58	134.45	1.41	9/59/4
Local search duration	500 ms	91.92	93.03	1.21	130.94	132.24	0.99	21/50/1
	1000 ms	92.14	93.42	1.39	131.32	132.89	1.20	15/54/3
	2000 ms	92.18	93.63	1.57	131.38	133.20	1.39	15/55/2

requests become known does not seem to have a decisive influence. For smaller instances, the advantage of the ellipse area is stronger than for larger instances. Regarding the proportion of dynamic requests, the worst results are obtained when the ratio of static and dynamic requests is balanced. When the degree of dynamism is high, the optimization potential is greater and this obviously affects the results. Please note that the advantage of the ellipse area metric becomes more salient when increasing the duration of the local search.

4 MDP-Based Approach

As a first step towards an approach that can also account for combinations of multiple metrics, we modeled the scenario as a *Markov decision process* (MDP). The MDP serves as a basis for training a *reinforcement learning* (RL) agent. A *decision point* occurs when a decision must be made whether to choose a new solution over the previous one. The *actions* indicate which of the two solutions is chosen. The *states* include for both solutions the current timestamp, the number of accepted customers as well as the values of the metrics to be used. A unit *reward* is given if the solution with more customers than the other is selected. In case that both solutions have the same number of customers, no reward is given. The *objective* is to maximize the sum of future rewards, i.e., the sum of requests to be accepted. This is to enable the model to learn to make decisions that lead to states in which a request can be integrated. The *transition function* is non-deterministic in that the next state is formed from the chosen solution and the next solution found.

In a proof-of-concept, we used Q-learning and a classical Q-table to validate the MDP-based approach for general functionality. Due to the use of Q-tables, the state space had to be kept very small. We strongly discretized the time and only used $\{-1, 0, 1\}$ to indicate whether the first solution was worse, as good, or better than the second solution with respect to the considered metric. The results show that in the case of different numbers of requests, the solution with more requests should be chosen and otherwise the solution with the higher *ellipse area metric*. This confirms the observations from Sect. 3. To exploit the full potential of the approach, *deep reinforcement learning* should be used by replacing the Q-table with a neural network so that the magnitude of the differences in the metrics can also be taken into account. Alternatively, *approximate dynamic programming* techniques can be used (see, e.g., [6]).

Please note that a complex MDP brings the following advantage: The common greedy approach of accepting a request whenever possible can be suboptimal. As long as the optimization goal is to maximize the number of accepted requests and not, e.g., the revenue generated, short or "easy" requests are more lucrative. An anticipatory RL-trained agent could accordingly identify a non-lucrative request and reject it despite an insertion opportunity, thus maintaining the possibility of accepting several more requests in the future instead.

5 Conclusion and Outlook

We developed different metrics and analyzed their suitability for predicting the insertion potential of future requests in dynamic dial-a-ride problems. The computational study showed that the presented *ellipse area*, which considers the number of reachable points between two nodes, was able to increase the customer acceptance rate by over 1.5% in the used method. In addition, we presented an MDP-based approach that could enable an RL agent to learn a combination of metrics and thus be even better to decide which solution actually leads to a higher acceptance rate. Further research will address the difficulties of the MDP-based approach (e.g., very similar Q-values for the states due to sparse reward and non-deterministic transition function).

References

1. Berbeglia, G., Cordeau, J.F., Laporte, G.: A hybrid tabu search and constraint programming algorithm for the dynamic dial-a-ride problem. Informs J. Comput. **24**(3), 343–355 (2012)
2. Vallee, S., Oulamara, A., Cherif-Khettaf, W.R.: New online reinsertion approaches for a dynamic dial-a-ride problem. J. Comput. Sci. **47**, 101199 (2020)
3. Chassaing, M., Giboulot, V., Lacomme, P., Quilliot, A., Ren, L.: Determination of robust solutions for the dynamic dial-a-ride problem. In: CIE45 Proceedings (2015)
4. Luo, Y., Schonfeld, P.: Online rejected-reinsertion heuristics for dynamic multivehicle dial-a-ride problem. Transp. Res. Rec. **2218**(1), 59–67 (2011)
5. Braekers, K., Caris, A., Janssens, G.K.: Exact and meta-heuristic approach for a general heterogeneous dial-a-ride problem with multiple depots. Transp. Res. Part B **67**, 166–186 (2014)
6. Ulmer, M.W.: Anticipation vs. reactive reoptimization for dynamic vehicle routing with stochastic requests. Networks **73**(3), 277–291 (2018)

Benders Decomposition for the Periodic Event Scheduling Problem

Niels Lindner[1(✉)] and Rolf van Lieshout[2]

[1] Zuse Institute Berlin, Takustr. 7, 14195 Berlin, Germany
lindner@zib.de

[2] Department of Operations, Planning, Accounting, and Control, Eindhoven University of Technology, PO Box 513, 5600, MB Eindhoven, The Netherlands
r.n.v.lieshout@tue.nl

Abstract. The Periodic Event Scheduling Problem (PESP) is the central mathematical model behind the optimization of periodic timetables in public transport. We apply Benders decomposition to the incidence-based MIP formulation of PESP. The resulting formulation exhibits particularly nice features: The subproblem is a minimum cost network flow problem, and feasibility cuts are equivalent to the well-known cycle inequalities by Odijk. We integrate the Benders approach into a branch-and-cut framework, and assess the performance of this method on instances derived from the benchmarking library PESPlib.

Keywords: Periodic timetabling · Periodic event scheduling problem · Benders decomposition · Mixed integer programming

1 Introduction

Public transport is an important pillar of everyday mobility, and its expansion is indispensable in order to increase the share of climate-friendly traffic. A large part of public transportation networks is operated in a periodic manner, and this creates the need for periodic timetable optimization by mathematical methods. The standard model for this purpose is the Periodic Event Scheduling Problem (PESP), which is difficult to solve, both in theory and practice. We investigate a Benders decomposition approach to PESP, providing a new mixed integer programming formulation. We formally define our setting in Sect. 2. The Benders reformulation is presented and analyzed in Sect. 3. We evaluate the method computationally in Sect. 4.

2 The Periodic Event Scheduling Problem

2.1 Problem Definition

The input to the Periodic Event Scheduling Problem is given by a 5-tuple (G, T, ℓ, u, w), where

N. Trautmann and M. Gnägi (Eds.): OR 2021, LNOR, pp. 289–294, 2022.
https://doi.org/10.1007/978-3-031-08623-6_43

- $G = (V, E)$ is a directed graph,
- $T \in \mathbb{N}$ is a period time,
- $\ell \in \mathbb{R}^E_{\geq 0}$ is a vector of lower bounds,
- $u \in \mathbb{R}^E_{\geq 0}$ is a vector of upper bounds, $u \geq \ell$,
- $w \in \mathbb{R}^E_{\geq 0}$ is a vector of weights.

A *periodic timetable* is a vector $\pi \in [0, T)^V$ such that there exists a *periodic tension* $x \in \mathbb{R}^E$ such that

$$\forall (i,j) \in E: \ \ell_{ij} \leq x_{ij} \leq u_{ij} \quad \text{and} \quad \pi_j - \pi_i \equiv x_{ij} \mod T. \tag{1}$$

A periodic timetable π assigns times in $[0, T)$ to the vertices in V, a periodic tension x fixes arc durations within the bounds, and constraint (1) ensures the compatibility of π and x modulo the period time T.

Definition 1 [12]. *Given* (G, T, ℓ, u, w) *as above, the* Periodic Event Scheduling Problem (PESP) *is to find a periodic timetable* π *along with a periodic tension* x *such that* $w^\top x$ *is minimum.*

Equivalently, one may minimize the weighted periodic slack $w^\top (x - \ell)$. If π is a periodic timetable, then a periodic tension x with minimum $w^\top x$ compatible to π can be computed by

$$x_{ij} := [\pi_j - \pi_i - \ell_{ij}]_T + \ell_{ij}, \qquad \text{for all } (i,j) \in E,$$

where $[\cdot]_T$ denotes the modulo T operator taking values in $[0, T)$.

In the context of periodic timetabling in public transport, vertices often model departure or arrival events of vehicles at stations. Arcs represent, e.g., driving or dwelling of vehicles, transfers for passengers, and safety conditions [6]. The weights typically reflect the number of passengers making use of a vehicle or a transfer, so that the PESP objective amounts to minimizing the total travel time of all passengers.

2.2 Incidence-Based MIP Formulation

Let $A \in \{-1, 0, 1\}^{V \times E}$ denote the incidence matrix of G, i.e., the matrix whose columns are the unit vector differences $e_j - e_i$ for $(i,j) \in E$. By (1), a vector x is a periodic tension for a periodic timetable π if and only if $\ell \leq x \leq u$ and $A^\top \pi \equiv x \mod T$. In particular, we can write $x = A^\top \pi + Tp$ for some integer vector $p \in \mathbb{Z}^E$. This allows to express PESP as the following mixed-integer linear program (MIP), cf. [5,7,10]:

$$\begin{aligned} \text{Minimize} \quad & (Aw)^\top \pi + T w^\top p \\ \text{s.t.} \quad & \ell \leq A^\top \pi + Tp \leq u \\ & \pi \in \mathbb{R}^V \\ & p \in \mathbb{Z}^E. \end{aligned} \tag{2}$$

The domain of π can be extended beyond $[0,T)$: for each feasible solution (π, p) to (2), the vector $[\pi]_T$ is a periodic timetable in $[0,T)^V$, $\pi - [\pi]_T \equiv 0 \bmod T$, and $([\pi]_T, p + A^\top(\pi - [\pi]_T)/T)$ has the same objective value as (π, p).

To make (2) even more compact, consider the digraph $\overline{G} = (V, \overline{E})$, where $\overline{E}$ contains all arcs in E and additionally a reverse copy $\overline{e}$ for each arc $e \in E$. Define $c(p)_e := u_e - Tp_e$ and $c(p)_{\overline{e}} := -\ell_e + Tp_e$ for all $e \in E$. Then

$$\ell \leq A^\top \pi + Tp \leq u \quad \Leftrightarrow \quad \overline{A}^\top \pi \leq c(p), \tag{3}$$

where $\overline{A}$ denotes the incidence matrix of $\overline{G}$.

3 Benders Decomposition

We apply a classical Benders decomposition [1] to the MIP formulation (2), considering as Benders subproblem the dual of the linear program (LP) that arises for a fixed vector $p \in \mathbb{Z}^E$.

3.1 Analysis of the Subproblem

Using (3), the Benders subproblem reads

$$\begin{aligned} \text{Maximize} \quad & -c(p)^\top f \\ \text{s.t.} \quad & \overline{A} f = -Aw \\ & f \geq 0. \end{aligned} \tag{4}$$

In this form, (4) is equivalent to an uncapacitated minimum cost flow problem in the network $\overline{G}$ with balance $-Aw$ and cost $c(p)$. This has also been observed in [8] in a different context. In particular, minimum cost flow algorithms can be applied to solve the Benders subproblem rather than general-purpose LP solvers.

Lemma 1. *The Benders subproblem (4) is always feasible.*

Proof. By Gale's theorem [3], (4) is feasible if and only if for every subset $S \subseteq V$ the sum of balances $\sum_{v \in S}(-Aw)_v$ is at most the capacity of all arcs leaving S. As capacities are infinite, we only need to consider such S that do not admit any leaving arc. However, as $\overline{G}$ contains for each arc a reverse copy, S can only be a union of connected components of $\overline{G}$ and hence of G. But then the rows of A corresponding to the vertices in S add to 0, so that $\sum_{v \in S}(-Aw)_v = 0$.

We now turn to boundedness of (4). An oriented cycle in G is a vector $\gamma \in \{-1, 0, 1\}^E$ such that $\{e \in E \mid \gamma_e \neq 0\}$ becomes a cycle when undirecting G. Any oriented cycle can be decomposed as $\gamma = \gamma_+ - \gamma_-$, where $\gamma_+ := \max(\gamma, 0)$ is the forward part, and $\gamma_- := \max(-\gamma, 0)$ is the backward part of γ.

Lemma 2. *For $p \in \mathbb{Z}^E$, the following are equivalent:*

a) *The Benders subproblem (4) is bounded for p.*
b) *There is no directed cycle in $\overline{G}$ of negative cost w.r.t. $c(p)$.*

c) *For all oriented cycles γ in G holds*

$$\gamma^\top p \leq \left\lfloor \frac{\gamma_+^\top u - \gamma_-^\top \ell}{T} \right\rfloor.$$

Proof. The equivalence of a) and b) is well-known for network flow problems. By LP duality and Lemma 1, a) is equivalent to the feasibility of the LP arising from (2) when fixing p. That the latter is in turn equivalent to b) resp. c) is indicated in [9] and is explained in detail in the proof of Theorem 4.3 in [11].

Remark 1. In the PESP literature, the inequalities in Lemma 2c are known as Odijk's cycle inequalities [9], which are the base for a cutting plane algorithm to construct a feasible, but not necessarily optimal, periodic timetable [10].

3.2 Master Problem

Having discussed the subproblem, we now turn to the master problem:

Theorem 1. *The following mixed integer program solves PESP:*

$$\begin{array}{llll} \text{Minimize} & z & & \\ \text{s.t.} & z - Tw^\top p \geq -c(p)^\top f, & f \text{ feasible for (4)}, & \\ & c(p)(C) \geq 0, & C \text{ directed cycle in } \overline{G}, & (5) \\ & z \in \mathbb{R}, & & \\ & p \in \mathbb{Z}^E. & & \end{array}$$

An optimal periodic timetable π^* can be recovered from an optimal solution (z^*, p^*) by solving the LP that arises from (2) by fixing p to p^*.

The proof of Theorem 1 is straightforward using the standard Benders decomposition technique [1]. The second line of constraints in (5) corresponds to the Benders feasibility cuts, which, by Lemma 2, are equivalent to Odijk's cycle inequalities for each oriented cycle. The first line of constraints correspond to the Benders optimality cuts. It is sufficient to consider these cuts only for vertices of the polyhedron $\{f \geq 0 \mid \overline{A}f = -Aw\}$, i.e., extremal flows given by spanning tree structures, i.e., flows $f \geq 0$ that can be positive only on the arcs of a spanning tree of $\overline{G}$. This way, the MIP (5) is endowed with a finite description, however, there will in general be exponentially many spanning trees, and exponentially many cycles. When solving the Benders master problem with a MIP solver in practice, it is therefore necessary to generate the constraints dynamically.

Remark 2. Any PESP instance can be preprocessed so that it is no restriction to assume $p \in \{0, 1, 2\}^E$ [5]. This is useful for breaking symmetries in (5).

Remark 3. Spanning trees in $\overline{G}$ provide lower bounds on the optimal objective value by means of the Benders optimality cuts in (5). On the other hand, spanning trees in $\overline{G}$ correspond to spanning trees in G with an additional marking of the tree arcs as either original or reversed. The latter is the combinatorial structure behind the vertices of the periodic tension polytope [7]. In particular, the value $w^\top x$ of any such vertex x is an upper bound on the optimal value.

4 Computational Results

We implemented a branch-and-cut algorithm to solve formulation (5) using the generic callback framework of CPLEX 12.10. The subproblem is solved using the network simplex implementation available in CPLEX. To stabilize and accelerate the solution process, we use the method for cut loop stabilization described in [2]. The computations were carried out on a Dell Precision 7520 running Windows 10 with an Intel Core i7-7820HQ processor at 2.9 GHz and with 16 GB of RAM.

We test the Benders approach on sub-instances of `R1L1`, one of the instances from the benchmark library PESPlib [4]. Table 1 presents the objective (weighted slack), optimality gap and number of cuts, obtained with a computation time of 20 min. We find that the Benders approach terminates with a large optimality gap for all instances. The resulting solution for instance `R1L1-0.8` is known to be optimal, but it appears that the Benders optimality cuts are not strong enough to close the optimality gap. For the other instances, the found solutions are worse than best known solutions, hence the Benders approach is not competitive with other approaches, neither on the primal nor on the dual side. The number of generated cuts for all instances is very large, indicating that the cuts are relatively weak.

Table 1. Results of the branch-and-cut algorithm.

Instance	Objective	Optimality gap (%)	Cuts
`R1L1-0.8`	1 032 021	44.1	3 743
`R1L1-0.7`	3 568 074	81.4	30 828
`R1L1-0.6`	9 080 015	82.8	23 925

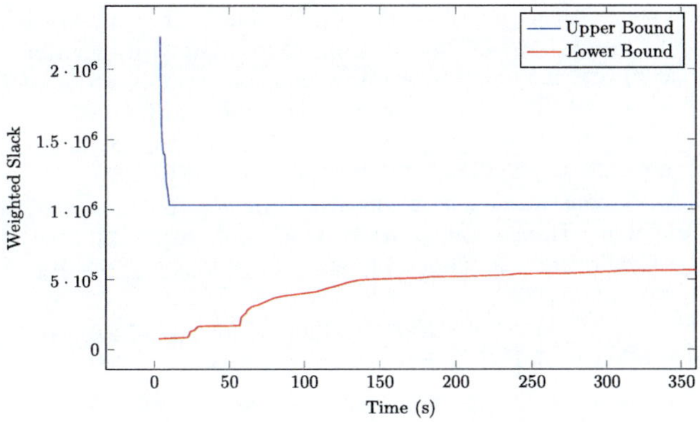

Fig. 1. Evolution of the lower and upper bound for instance `R1L1-0.8`.

The typical behavior of the Benders approach is illustrated in Fig. 1, visualizing the evolution of the lower and upper bounds on the instance `R1L1-0.8`. We observe that the optimal solution is already found within 10 s. On the other hand, there is a large gap between lower and upper bound, with the lower bound increasing at a diminishing rate. This is rather disappointing, as the underlying digraph G of `R1L1-0.8` has only 23 nodes and 36 arcs, and the incidence-based MIP formulation (2) can be solved to optimality by CPLEX within less than a second.

We therefore conclude that Benders decomposition, despite its attractive theoretical properties, does not seem to be beneficial compared to the well-established solution methods for the PESP.

References

1. Benders, J.F.: Partitioning procedures for solving mixed-variables programming problems. Numer. Math. **4**(1), 238–252 (1962)
2. Fischetti, M., Ljubić, I., Sinnl, M.: Redesigning Benders decomposition for large-scale facility location. Manag. Sci. **63**(7), 2146–2162 (2017)
3. Gale, D.: A theorem on flows in networks. Pacific J. Math. **7**(2), 1073–1082 (1957)
4. Goerigk, M.: PESPlib – a benchmark library for periodic event scheduling (2012). http://num.math.uni-goettingen.de/~m.goerigk/pesplib/
5. Liebchen, C.: Periodic timetable optimization in public transport. Ph.D. thesis, Technische Universität Berlin (2006)
6. Liebchen, C., Möhring, R.H.: The modeling power of the periodic event scheduling problem: railway timetables - and beyond. In: Geraets, F., Kroon, L., Schöbel, A., Wagner, D., Zaroliagis, C.D. (eds.) Algorithmic Methods for Railway Optimization. Lecture Notes in Computer Science, pp. 3–40. Springer, Berlin, Heidelberg (2007)
7. Nachtigall, K.: Periodic Network Optimization and Fixed Interval Timetables. Habilitation thesis, Universität Hildesheim (1998)
8. Nachtigall, K., Opitz, J.: Solving periodic timetable optimisation problems by modulo simplex calculations. In: Fischetti, M., Widmayer, P., (eds.), 8th Workshop on Algorithmic Approaches for Transportation Modeling, Optimization, and Systems (ATMOS 2008), vol. 9, OpenAccess Series in Informatics (OASIcs), Dagstuhl, Germany, Schloss Dagstuhl–Leibniz-Zentrum fuer Informatik (2008)
9. Odijk, M.A.: Construction of periodic timetables, part 1: a cutting plane algorithm. Technical report 94-61, TU Delft (1994)
10. Odijk, M.A.: A constraint generation algorithm for the construction of periodic railway timetables. Transp. Res. Part B: Methodol. **30**(6), 455–464 (1996)
11. Peeters, L.: Cyclic Railway Timetable Optimization. Ph.D. thesis, Erasmus Universiteit Rotterdam (2003)
12. Serafini, P., Ukovich, W.: A mathematical model for periodic scheduling problems. SIAM J. Disc. Math. **2**(4), 550–581 (1989)

Optimal Line Plans in the Parametric City and the Impact of In-Motion Costs

Berenike Masing(✉), Niels Lindner, and Ralf Borndörfer

Zuse Institute Berlin, Berlin, Germany
{masing,lindner,borndoerfer}@zib.de

Abstract. Line planning in public transport involves determining vehicle routes and assigning frequencies of service such that travel demands are satisfied. We evaluate how line plans, which are optimal with respect to in-motion costs (IMC), the objective function depending purely on arc-lengths for both user and operator costs, performs with respect to the value of resources consumed (VRC). The latter is an elaborate, socio-economic cost function which includes discomfort caused by delay, boarding and alighting times, and transfers. Even though discomfort is a large contributing factor to VRC and is entirely disregarded in IMC, we observe that the two cost functions are qualitatively comparable.

Keywords: Line planning · Public transport · Mixed-integer programming

1 Introduction

The goal of line planning in public transport is to find the best line plan, i.e., to determine vehicle routes and assigning frequencies of service in order to cover travel demands. However, what constitutes the optimal solution depends on the point of perspective: Operators' main interest is in cost reduction, while good solutions for users include short travel times, few transfers, and low waiting times. We will consider two objective functions which are comprised of both operator and user costs to address both viewpoints: The value of the resources consumed (VRC), a socio-economic cost function, which puts a price to travelers' waiting times and transfers, and includes delays along the vehicle routes due to passengers boarding and disembarking, etc. In contrast, the second cost function considers only the in-motion costs (IMC) and disregards delay factors. As the former is highly nonlinear, it cannot be applied to our mixed integer programming formulation of the line planning problem. Instead, we solve the problem with IMC as objective and analyze how the results obtained with the much simpler version perform with respect to VRC.

2 Line Planning in the Parametric City

2.1 Line Planning

To describe the line planning problem, we choose the mixed integer linear programming formulation ($MILP$) introduced by Borndörfer et al. [1] with a few

N. Trautmann and M. Gnägi (Eds.): OR 2021, LNOR, pp. 295–301, 2022.
https://doi.org/10.1007/978-3-031-08623-6_44

adjustments. A solution (f, y) consists of an integral line plan $f = (f_l)_{l\in L}$ indicating with which frequency line $l \in L$ is used, and a non-negative passenger flow $y = (y_p)_{p\in P}$. A path p is used if and only if $y_p > 0$, the analogue holds for the lines. The set L of potential line candidates is given by all circulations in the directed graph $G = (V, A)$ representing the city. The set of admissible passenger paths P consists of all simple paths in said graph. Let us further define the sets L_a of all lines and P_a of all paths using arc $a \in A$. Let further $P_{s\to t}$ be the set of all $s-t$-paths. With the parameters $K > 0$ as vehicle and $\Lambda > 0$ as street capacity, the model formulation is then

$$(MILP) \quad \min \overbrace{\sum_{l\in L}(c_v + c_s K)\tau_l f_l}^{c_o^I} + \overbrace{\sum_{p\in P} p_v \tau_p y_p}^{c_u^I} \qquad =: IMC \tag{1}$$

$$\text{s.t.} \sum_{p\in P_{s\to t}} y_p = d_{st} \qquad \forall (s,t) \in V \times V \tag{2}$$

$$\sum_{p\in P_a} y_p - \sum_{l\in L_a} f_l K \leq 0 \qquad \forall a \in A \tag{3}$$

$$\sum_{l\in L_a} f_l \leq \Lambda \qquad \forall a \in A \tag{4}$$

$$f_l \in \mathbb{N} \quad \forall l \in L \qquad \forall l \in L \tag{5}$$

$$y_p \geq 0 \qquad \forall p \in P. \tag{6}$$

We refer to [4,5] for an explanation of all constraints and focus only on the objective function here: IMC (1) is a sum of operator costs c_o^I and user costs c_u^I. Operator costs depend on the total arc length of line l, i.e., $\tau_l = \sum_{a\in l} \tau_a$ and the costs per vehicle c_v and per seat c_s. The user costs only depend on the path-lengths $\tau_p = \sum_{a\in p} \tau_a$ and a price for in-vehicle time p_v. This means that both user satisfaction and operator costs are based on pure travel times; the number of transfers or delay is not taken into account. In particular, line activation or turnaround costs are not included. This allows us to restrict the line pool L to the set of simple cycles in the directed graph, because the costs for line l with frequency f_l are simply $\sum_{a\in l}(c_v + c_s K)\tau_a f_l$ which means that we can truncate l into cycles and assign frequency f_l to each of them. The costs of the sum of these sub-cycles then correspond to the costs of l.

2.2 Parametric City

We choose the *Parametric City* [3] as city representative, since this model balances generality and simplicity: It is comprised of a *helm* graph $\mathcal{G} = (V, A)$ of $2n + 1$ nodes, as depicted in Fig. 1 for $n = 6$. Its geometric shape and the associated demand $d_{s,t}, (s,t) \in V \times V$ (cf. Table 1) represent the most prominent features of the city, e.g., size, degree of mono- or polycentricity, economical importance of districts. This can be controlled by a choice of parameters, an overview is given in Table 2, a more detailed explanation can be found in [3,4].

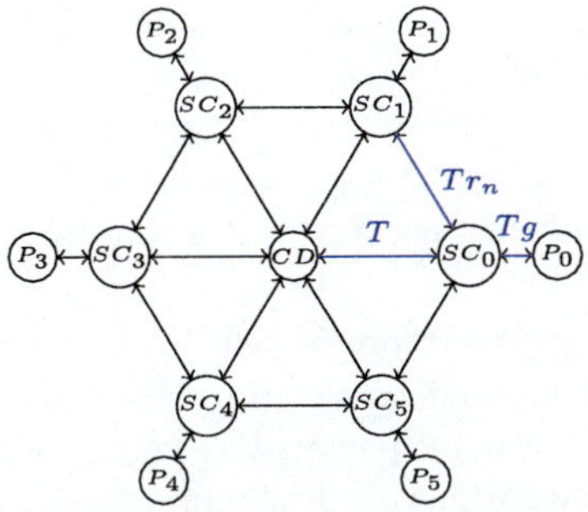

Fig. 1. Graph $\mathcal{G}$ with $n = 6$

Table 1. Demand $d_{s,t}$ (not listed vertex-pairs correspond to $d_{s,t} = 0$)

s,t	SC_i	$SC_j, j \neq i$	CD
P_i	$\frac{aY}{n}\beta$	$\frac{aY}{n(n-1)}\gamma$	$\frac{aY}{n}\alpha$
SC_i	0	$\frac{(1-a)Y}{n(n-1)}\tilde{\gamma}$	$\frac{(1-a)Y}{n}\tilde{\alpha}$

Table 2. Parameters in the Parametric City, f.o.t. = fraction of travelers $\in [0, 1]$

n	no. of subcenters/peripheries
T	arc length (SC_i, CD)
g, r_n	factors for arc length $(SC_i, P_i), (SC_i, SC_{i\pm1})$
Y	total patronage
a	f.o.t. from P_i
$\alpha\,(\tilde{\alpha})$	f.o.t. from P_i (SC_i) to CD
β	f.o.t. from P_i to SC_i
$\gamma\,(\tilde{\gamma})$	f.o.t. from P_i (SC_i) to SC_j, $i \neq j$
	$\alpha + \beta + \gamma = 1$, $\tilde{\alpha} + \tilde{\gamma} = 1$, $\alpha/\gamma = \tilde{\alpha}/\tilde{\gamma}$

3 Value of the Resources Consumed

The value of resources consumed is the socioeconomic cost function introduced by Fielbaum et al. [2]. Just like IMC, it combines operator costs C_o^{v} and user costs C_u^{v}. However its aim is to include delay and other comfort-factors:

$$VRC = \overbrace{\sum_{l \in L} (c_v + K c_s) t_l^{\circ} f_l}^{C_o^{\mathrm{v}}} + \overbrace{Y(p_{\mathrm{v}} \bar{t}_{\mathrm{v}} + p_{\mathrm{w}} \bar{t}_{\mathrm{w}}) + p_r R}^{C_u^{\mathrm{v}}}.$$

Here, t_l° denotes the cycle time of line l, which in addition to in-motion times, includes the times for passengers to get on and off the vehicles, i.e.,

$$t_l^{\circ} = \sum_{a \in l} \tau_a + (x_a^+ + x_a^-)t,$$

where t is the time needed for one person to board or alight and $x_{(v,w)}^+$ resp. $x_{(v,w)}^-$ is the expected number of people in a single vehicle on arc (v, w) that board at v resp. leave at w. The values of $x_{(v,w)}^+$ and $x_{(v,w)}^-$ are computed with the following assumptions in mind as stated by Fielbaum et al. [2, p. 302]: 1) "Buses operate with a regular headway", 2) "Passengers arrive at a constant rate", 3) "In the case of common lines passengers will be assigned proportional to frequency". By taking an arc-wise perspective, and considering the aggregated frequencies per arc $F_a := \sum_{l \in L_a} f_l$, we can compute

$$X^-_{(v,w)} := \sum_{l\in L_{(v,w)}} \sum_{\substack{p\in P_{(v,w)}\setminus P_{(w,x)} \\ \text{for } (w,x)\in l}} \frac{y_p f_l}{F_{(v,w)}}, \quad X^+_{(v,w)} := \sum_{l\in L_{(v,w)}} \sum_{\substack{p\in P_{(v,w)}\setminus P_{(u,v)} \\ \text{for } (u,v)\in l}} \frac{y_p f_l}{F_{(v,w)}}$$

which corresponds to the total number of alighting passengers at w and boarding at v along arc (v, w) respectively. This leads to $x^-_{(v,w)} := X^-_{(v,w)}/F_{(v,w)}$ and $x^+_{(v,w)} := X^+_{(v,w)}/F_{(v,w)}$. For a more detailed explanation we refer to [4].

The user costs include costs for time spent in vehicles, waiting for an arriving vehicle, as well as transfer penalties, each priced by parameters $p_{\mathrm{v}}, p_{\mathrm{w}}$ and p_r respectively. Again, we make the same assumptions as Fielbaum et al., who describe that in-vehicle "passengers are delayed by the boarding and alighting process; the starting node, where this time is already incorporated as waiting time; and the last node, where passengers alight such that the first alighting has no delay and the last has full delay, making the average half of the total alighting time." [2, p. 303]. Since we only have a passenger-to-path description without an explicit passenger-to-line assignment, we introduce $\mathbb{P}^p_w$ as the probability of path p transferring lines at node w :

$$\mathbb{P}^p_w := \begin{cases} 0 & \text{if } w \text{ is the first or last node of } p \\ \displaystyle\sum_{l\in L_{(v,w)}\setminus L_{(w,x)}} \frac{f_l}{F_{(v,w)}} & \text{if } (v,w,x) \text{ are three consecutive nodes on } p. \end{cases}$$

The average travel of one path $p = (v_0, v_1, \ldots, v_q)$ can then be described by

$$t_{\mathrm{v}}(p) = \sum_{\substack{(v,w)\in p \\ (v,w)\neq(v_{q-1},v_q)}} \left((1-\mathbb{P}^p_w)\underbrace{t(x^-_{(v,w)} + x^+_{(w,x)})}_{\substack{\text{others boarding/} \\ \text{alighting on line} \\ \text{at } w}} + \mathbb{P}^p_w \underbrace{\frac{t}{2}\frac{x^-_{(v,w)}}{F_{(v,w)}}}_{\substack{\text{own alighting} \\ \text{at } w}} \right)$$
$$+ \underbrace{\frac{t}{2}\frac{x^-_{(v_{q-1},v_q)}}{F_{(v_{q-1},v_q)}}}_{\substack{\text{own alighting} \\ \text{at endnode } v_q}} + \underbrace{\tau_p}_{\substack{\text{total time in} \\ \text{motion}}} .$$

The average in-vehicle time per passenger is then $\bar{t}_{\mathrm{v}} = \left(\sum_{p\in P} y_p t_{\mathrm{v}}(p)\right)/Y$.

The "average waiting time [is] half of the headway" [2, p. 303] which allows for the following description of average waiting times per path $t_{\mathrm{w}}(p)$ and average waiting time in total $\bar{t}_{\mathrm{w}}$ as

$$t_{\mathrm{w}}(p) = \sum_{(v,w)\in p} \frac{\mathbb{P}^p_v}{2F_{(v,w)}}, \quad \text{and } \bar{t}_{\mathrm{w}} = \sum_{p\in P} t_{\mathrm{w}}(p)/Y.$$

To stay consistent, we also compute the total number of transfers via the probabilities by

$$R = \sum_{p \in P} \sum_{(v,w) \in p} \mathbb{P}^p_w.$$

4 Computational Results

As mentioned, we solved *MILP* with respect to in motion costs and evaluated what *VRC* these solutions have. As input we fixed all parameters in the Parametric City as summarized in Table 3 and varied the demand: For each parameter α, β, γ we chose all values between 0.025 and 0.0975 with a step size of 0.025 and set the other two demand parameters equal. Figure 2 shows the components of both cost functions, as well as how they performed in comparison. As expected, the value of resources consumed is considerably larger than *IMC* as it incorporates *IMC* in addition to delay factors. We denote this by discomfort costs $disc. = VRC - IMC$. It turns out, that discomfort costs make up 44% of the *VRC* on average. What is surprising however, is that two cost functions behave comparably: If *IMC* increases, so does *VRC*, and the same holds for both the respective user and operator costs. In the figures on the right, this becomes more evident: *IMC* relative to *VRC* are nearly constant. Even if user costs contribute more to *VRC* than to *IMC*, the trend is comparable. This is a rather surprising result, since *VRC* differs from *IMC* in such a fundamental way, and which is evident in the large fraction of discomfort costs; particularly in face of our restriction of the line pool to cycle-like lines. The latter means that lines are rather short – any bidirectional line uses only one arc in two directions, which leads to many transfers. This does not contribute to in-motion costs, however when considering discomfort costs as well, this line pool can be regarded as a worst-case scenario. Despite this particularly 'uncomfortable' choice of lines, both cost functions have a similar behavior. The great advantage of using *IMC* as objective in *MILP* is that the problem can be solved to optimality, whose results are in turn qualitatively comparable to *VRC*.

Table 3. Parameter choices of computational experiments (* in [$/h])

$n = 8$	$T = 0.5$ [h]	$g = 1/3$	$a = 0.8$	$K = 100$	$Y = 24000$
$c_v = 10.65^*$	$c_s = 0.204^*$	$p_v = 1.48^*$	$p_w = 4.44*$	$p_r = 0.59^*$	$t = 2.5$ [s]

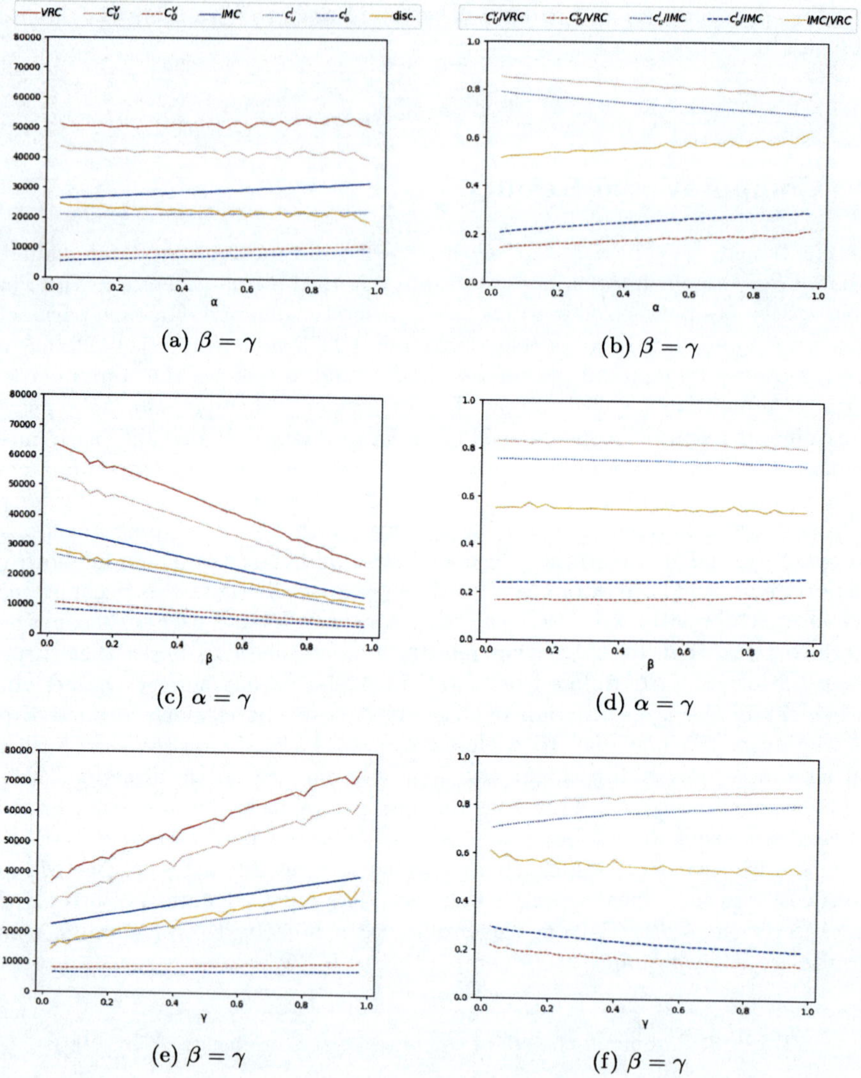

(a) $\beta = \gamma$ (b) $\beta = \gamma$

(c) $\alpha = \gamma$ (d) $\alpha = \gamma$

(e) $\beta = \gamma$ (f) $\beta = \gamma$

Fig. 2. VRC and IMC in absolute (left) and relative (right) values

References

1. Borndörfer, R., Grötschel, M., Pfetsch, M.E.: A column-generation approach to line planning in public transport. Transp. Sci. **41**(1), 123–132 (2007). https://doi.org/10.1287/trsc.1060.0161
2. Fielbaum, A., Jara-Díaz, S., Gschwender, A.: Optimal public transport networks for a general urban structure. Transp. Res. Part B: Methodol. **94**, 298–313 (2016). https://doi.org/10.1016/j.trb.2016.10.003

3. Fielbaum, A., Jara-Diaz, S., Gschwender, A.: A parametric description of cities for the normative analysis of transport systems. Netw. Spatial Econ. **17**(2), 343–365 (2016). https://doi.org/10.1007/s11067-016-9329-7
4. Masing, B.: Optimal line planning in the parametric city. Master's thesis, Technische Universität Berlin (2020)
5. Masing, B., Lindner, N., Borndörfer, R.: The price of symmetric line plans in the parametric city (2022). https://doi.org/10.48550/ARXIV.2201.09756

Strategic Road Safety Dashboard: Visualizing Results of Accident Data Mining

Katherina Meißner(✉) and Julia Rieck

University of Hildesheim, Hildesheim, Germany
{meissner,rieck}@bwl.uni-hildesheim.de

Abstract. Road safety is a major concern, as accidents kill on average 3,600 people per day. In order to reduce the number of road accidents, the police or local authorities jointly implement actions and measures to increase road safety. Therefore, it is necessary to analyze and predict the different circumstances of accidents comprehensively. Only with the knowledge, e.g., about the temporal pattern, locations, or road conditions, meaningful actions can be derived and implemented. A framework to support strategic planning of road safety measures is designed that consists of several consecutive data mining stages, i.e., frequent itemset mining, time series clustering, forecasting, and scoring. An informative and comprehensible presentation of the results is necessary to make them usable for the planning of measures. With a strategic road safety dashboard, we enable police managers to identify accident blackspots and especially their temporal pattern for different feature combinations.

Keywords: Descriptive accident analytics · Data mining · Road safety dashbord

1 Introduction

For the planning and implementation of actions and measures to improve road safety, police managers or local authorities must be equipped with adequate tools (e.g., a dashboard). They need to know the *circumstances* that lead to accidents in different locations and at different times in order to properly schedule measures such as speed reductions, new stop signs or patrol routes. Accident circumstances include many aspects that are recorded as data sets of *attributes* (e.g., weather) and corresponding *values* (e.g., rain, snow etc.). A large number of combinations of attributes and values (i. e., *features*) are possible and must be considered in the strategic planning of police actions. By observing the temporal pattern of the various accident circumstances, changes in certain *feature combinations* can be detected. For example, if there is a steep increase in the frequency of a combination or a change in the seasonality, then appropriate countermeasures must be initiated urgently. The actions typically cannot be implemented ad hoc, but must be planned months in advance. Based on many years of experience,

N. Trautmann and M. Gnägi (Eds.): OR 2021, LNOR, pp. 302–308, 2022.
https://doi.org/10.1007/978-3-031-08623-6_45

police managers might tend to investigate already known relationships between accident features. This could easily leave unfamiliar feature combinations undetected and thus unaddressed by road safety measures. In order to also detect unknown relationships, we have developed a complex *data mining framework*. By applying *unsupervised*, descriptive methods (e.g., frequent itemset mining), it is possible to identify "interesting" feature combinations with, e.g., critical changes in frequency in the corresponding time series. Forecasting methods may be used to anticipate behavior in the future and initialize preventive measures as soon as possible. Although advanced data mining methods can certainly provide good results, it is important, especially for road safety actions, that the results of the methods are presented in a comprehensible way and can be easily operationalized by police managers. For this purpose, we propose a *web-based interface* providing an appropriate presentation for the expert audience.

2 Related Work on Road Safety Dashboards

In this section, we outline possible solutions for designing a road safety dashboard depicting results from data mining studies. While *dashboards* typically aim to visualize key performance indicators at an aggregate level in real time, *decision support systems* are more often used for long-term planning and often provide more in-depth data exploration. Road safety dashboards are usually situated between the two poles, as effective planning rarely demands real-time data while requiring *descriptive visualizations*.

Feng et al. [2] use a dashboard to display some predefined statistical reports of certain attributes and a graph of daily accident counts over a selected time period. In addition, they present an interactive map that provides a rough overview of the accident situation in the UK with some filtering options for a more detailed insight into the type of accidents. In their dashboard, Ait-Mlouk et al. [1] combine different data mining approaches for analyzing accidents. By integrating multiple criteria analysis within association rule mining, it is possible to present only the most interesting rules according to the decision maker's preferences. The tool also displays a time series of predicted accident numbers and casualties. Jiang et al. [3] use association rules to determine the key factors of fatal and non-fatal run-off-road accidents and present a dashboard for *spatial visualization* of accidents that belong to a certain rule. By highlighting the *blackspots* on the map, the user's attention is particularly drawn to high-risk locations for the particular combination of accident features. The European road safety decision support system [4] provides existing knowledge about traffic risks and possible countermeasures. A literature repository of studies and synopses allows for an overview of the estimated effects for each road accident risk factor and corresponding measures. In addition, a tool to evaluate the economic efficiency of actions in terms of cost and extent of road safety increase is implemented.

The results of unsupervised methods are particularly interesting for visualization in a dashboard, as they potentially point out new knowledge on circumstances and locations. The presented dashboards lack a method to properly select the "interesting" information from a large amount of data, because they do not take the temporal patterns of the circumstances into account, as the prediction of accident circumstances is not included in the selection process.

3 Strategic Planning Support Data Mining Framework

Here, we give a short overview of the underlying data mining methods in our framework. The parameters for the methods were chosen by comparing the results on three different datasets (i. e., London, Wales, and Scotland in the UK). The primary objective of our planning support framework (see Fig. 1) is to identify correlations within accident data that are either unknown to police managers or have a particular temporal pattern that warrants a closer inspection. In [5], four consecutive data mining stages (a)–(d) are defined, the results of which are displayed in the convenient dashboard introduced in this paper.

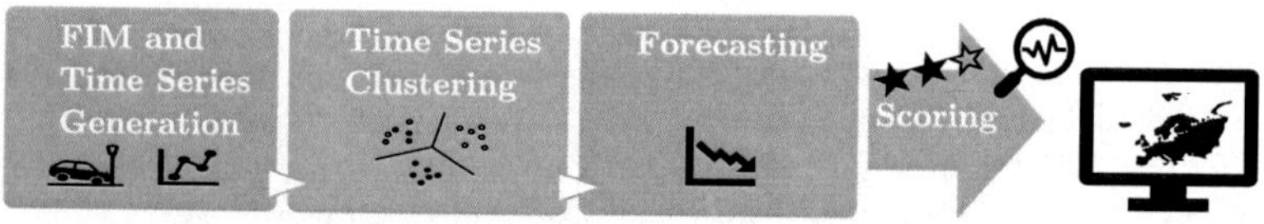

Fig. 1. Strategic planning support framework

(a) Frequent Itemset Mining and Time Series Generation: Frequent itemset mining (FIM) enumerates all possible *itemsets* I (i. e., sets of accident features) and determines their support values, reflecting the (relative) frequency of each itemset. With this unsupervised approach, we are able to detect unknown relationships within the data. In order to evaluate the temporal patterns of the itemsets, we separate the accident data set D into T monthly sets D_t, such that $D = \cup_t D_t, \forall t = 1, \ldots, T$. For each itemset I, we then determine the relative support x_t^I for each data set D_t and thereby generate a monthly *time series* X_I.

(b) Time Series Clustering: Even with small data sets, step (a) generates many itemsets and thus time series. To identify itemsets with an "interesting" temporal pattern that might occur even if it has low overall frequency, we apply time series clustering to group similar time series. This allows to identify one suitable forecasting method for each cluster, rather than finding one for each of the time series. Since clustering approaches depend on the underlying data, it is crucial to find an appropriate parameter configuration consisting of a scaling method, a distance measure, a clustering method, and a number of clusters.

(c) Time Series Forecasting: When all time series are grouped, we identify the most suitable forecasting method for each cluster by taking the corresponding centroid time series into account. We consider simple (e.g., random walk, naïve) and statistical methods (e.g., ARIMA, exp. smoothing) as well as sophisticated approaches like neural networks. Well-known error measures like root-mean-squared error help to identify a good approach per cluster.

(d) Scoring: In order to identify the most interesting accident feature combinations, we employ a predictive scoring procedure based on three different aspects of time series. A time series is interesting, e.g., if it shows an increasing trend (1) compared to the mean of the previous period's values. Consequently, the feature combination occurs in a growing percentage of accidents and should therefore be investigated in more detail. Another time series might be worth looking at because it shows significant fluctuations (2) in the current period, i. e., deviations from the mean, which are not caused by regular seasonal fluctuations. The information about seasonality results from the ratio of the variance of the time series without and with seasonality. A third aspect is the root-mean-squared error (3) introduced by the centroid-based forecasting method. If the prediction is not reliable, the time series should be submitted to a user for review, as it is possibly assigned to an inappropriate cluster. For the final score the three individual aspects (1), (2), and (3) are summed while applying individual weights $\gamma_1, \gamma_2, \gamma_3$ to modify their influence on the selection of interesting feature combinations. The time series with the highest scores are presented in the strategic road safety dashboard, where the default weights $(\gamma_1, \gamma_2, \gamma_3) = (0.4, 0.4, 0.2)$ can be adjusted based on the analyst's interests. Please note that we only weakly incorporate the forecasting error (3) into the resulting score. Unlike the other two aspects, this one does not primarily assess "interestingness" in terms of road accidents, but rather evaluates the goodness of the framework itself.

4 Strategic Road Safety Dashboard

Police managers need information on "interesting" accident feature combinations in order to strategically plan road safety measures. Hence, the results are visualized in a user-friendly and convenient dashboard, where we focus on descriptive analytics and employ predictions only in the scoring process. On a landing page, the region of interest can be selected and the weights of the score calculation are adjusted according to individual needs. For example, one user wants to see those itemsets that are very poorly predicted to ensure that the time series clustering approach does not "hide" incorrectly grouped time series (increase γ_3). Another user might be interested in accident feature combinations that show high non-seasonal fluctuation (increase γ_2) or feature combinations that provide a high increase in the relative number of accidents in the last period (increase γ_1). The ten most interesting itemsets are selected and displayed with their score values.

For a case study, we consider London as the region of interest and select the itemset $I = \{$2 vehicles, Urban, Junction controlled, Road class 'A'$\}$, as it exhibits one of the highest score values. The spatial distribution of accidents for this specific feature combination can be examined on two detail pages. The page "Map and Time Series" (see Fig. 2a) displays the accidents for a pre-defined period colored by their severity. The respective time series of relative support values is displayed as well. If some irregularities within the time series are detected, it might be advisable to compare the accidents of different seasons and years. In Fig. 2a, we can identify an unusually high support value in summer 2018 compared to the same months in 2017. In order to investigate the corresponding

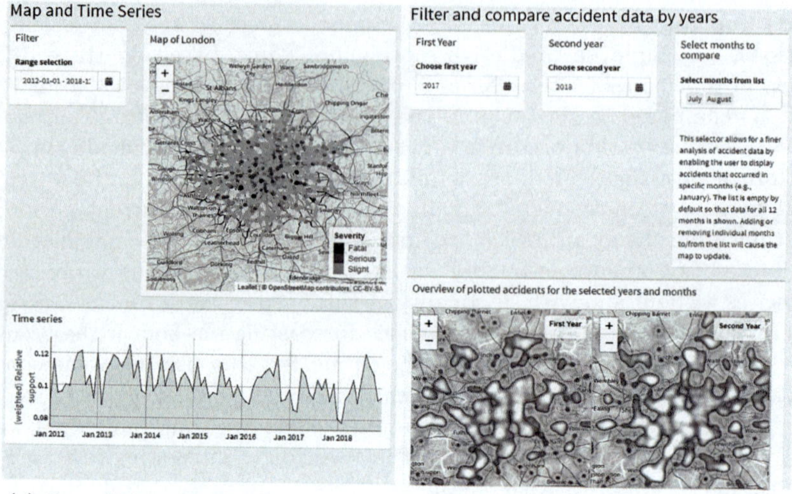

(a) Page "Map and Time Series" (b) Page "Comparing two Seasons"

Fig. 2. Strategic road safety dashboard detail pages http://applications.wirtschaftsinformatik-hildesheim.de:443/

accident locations, we provide two ways to drill down into the data. On the page "Comparing two Seasons", the user can define the years to be taken into account and also select several months to compare. For both years, the accidents of the chosen months are displayed on separate maps, as shown in Fig. 2b. Here, it is also possible to observe accident blackspots instead of accident points. Particularly for small geographical areas with many accidents, it is useful to consider accident blackspots for the analysis to identify risky locations more easily. The accidents in our selected itemset for London happened at controlled junctions on major roads. Hence, the blackspots reflect the most important routes coming out of London and we can clearly detect a shift in the blackspot locations towards the south-east between 2017 and 2018. At these spots, controlled junctions should be closely inspected and accident prevention measures implemented, such as changes in the control of the intersections or monitoring of red light violations via safety cameras. In order to support the choice of appropriate safety measures, we embed a link to the SafetyCube decision support system. The repository of studies on safety measures can be searched with the accident features of the itemset under consideration in order to plan actions accordingly.

The exemplary dashboard is not yet subject to a regular update routine, as exemplified in Fig. 3. The dashboard data requires monthly updates to display the accidents of the previous month on the map. For the new dataset D_t, the relative support values for all existing itemsets are calculated and forecasting and scoring are applied to the now extended time series. After a few minutes, the police manager can then analyze the updated results and begin planning appropriate road safety measures. In order to detect structural changes in the data

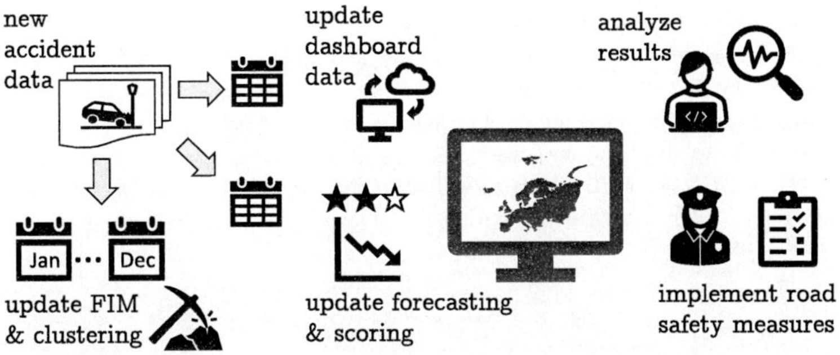

Fig. 3. Update procedure for the strategic road safety dashboard

(e.g., new frequent itemsets or changes in time series clusters), the entire underlying framework from Fig. 1 should be updated about once a year, which should also include re-evaluating and possibly adjusting the configuration parameters.

5 Conclusion and Future Work

We presented a strategic road safety dashboard based on accident data mining. The underlying framework, consisting of unsupervised methods for detecting unknown feature combinations and forecasting their temporal patterns, provides results that can help police managers to plan road safety measures. In the dashboard, these results are presented in a comprehensible way. Thus, a spatial analysis of accident circumstances and the accident numbers over time in a given accident situation is facilitated. Thereby, it is possible to individually adjust the weighting of the different aspects of "interestingness" and determine temporal changes in accident blackspot locations.

So far, we have only analyzed accident circumstances. The next step it to include vehicle and accident data to enrich the results with substantial insights. Moreover, the forecasting accuracy could be enhanced by including external data (e.g., weather forecasts or surrounding conditions like sport events). To establish a productive system, the settings page should be extended to support updates of the framework.

References

1. Ait-Mlouk, A., Agouti, T.: DM-MCDA: a web-based platform for data mining and multiple criteria decision analysis: a case study on road accident. SoftwareX **10**, 100323 (2019)
2. Feng, M., Zheng, J., Ren, J., Liu, Y.: Towards big data analytics and mining for UK traffic accident analysis, visualization and prediction. In: Proceedings of 12th International Conference on Machine Learning and Computing, pp. 225–229. ACM, New York (2020)

3. Jiang, F., Yuen, K., Lee, E., Ma, J.: Analysis of run-off-road accidents by association rule mining and geographic information system techniques on imbalanced datasets. Sustainability **12**(12), 4882 (2020)
4. Martensen, H., Diependaele, K., Daniels, S., et al.: The European road safety decision support system on risks and measures. Accid. Anal. Prev. **125**, 344–351 (2019)
5. Meißner, K., Rieck, J.: Data mining framework to derive measures for road safety. In: Perner, P. (ed.) Machine Learning and Data Mining in Pattern Recognition, pp. 625–639. ibai-publishing, Leipzig (2019)

OR in Engineering

Pooling of Contracts for Outsourcing Problems with Two-Dimensional Asymmetric Information

Alexander Herbst(✉)

University of Siegen, Chair of Technology Management,
Unteres Schloss 3, 57072 Siegen, Germany
alexander.herbst@uni-siegen.de
https://www.wiwi.uni-siegen.de/technologiemanagement/

Abstract. In this paper, we model an outsourcing problem as a specific principal-agent relationship in which two-dimensional hidden characteristics describe the agent's type. Assuming that the principal knows the joint probability distribution on the continuous type space, a standard solution technique for the resulting contracting problem is stochastic optimization on the set of incentive compatible menus of contracts from which the agent can choose a single contract according to the take-it-or-leave-it principle, respectively. In practice, however, the menu which maximizes the expected utility for the principal generally consists of infinitely many single contracts and cannot be determined analytically for all kinds of probability distributions.

To address this issue, we present a novel two-step approach which, in a first step, partitions the rectangular type space into a predefined number of subsets and, in a second step, computes an optimal incentive compatible menu of contracts containing a mutual contract for each subset of pooled types by using quadratic programming. Within our computational study we finally show that our method not solely bypasses the above described solution difficulties but also guarantees small optimality gaps by using only few contracts.

Keywords: Agent systems · Stochastic programming · Transportation

1 Introduction and Problem Description

Principal-agent models have various applications in the wide field of supply-chain-management where they appear in various shapes [1]. Within this framework the design of contracts between two parties is a common thread between many approaches, where—from a practical standpoint—the amount of potential contract constellations should be finite, even if the set of possible scenarios is infinite [2]. Let us consider the following problem.

We want to process $Q \in \mathbb{R}^+$ units of some good and have the opportunity to outsource a partial quantity $q \in [0; Q]$ to an external service provider while the remaining quantity $Q - q$ has to be handled by own resources. First of all,

N. Trautmann and M. Gnägi (Eds.): OR 2021, LNOR, pp. 311–317, 2022.
https://doi.org/10.1007/978-3-031-08623-6_46

we give a general definition to express the performance of an outsourcing policy. Based on this, we can define the utility of the principal in a general manner.

Definition 1 (Performance function). *For an outsourced quantity* $q \in [0; Q]$ *we define the performance function*

$$t(q; x) := \min\{t \in \mathbb{R}^+ : \text{ at time } t \text{ at least } x \text{ units are available}\}.$$

Definition 2 (General utility function of the principal). *Let* $\phi : \mathbb{R}_0^+ \to \mathbb{R}$, $\psi : [0; Q] \to \mathbb{R}$ *and* $m : [0; Q] \to \mathbb{R}_0^+$ *with* $\int_0^Q m(x)\, dx = 1$. *We define the general utility function of the principal as*

$$\mathcal{U}(q) := \phi(\tau(q)) + \psi(q) := \phi\left(\int_0^Q m(x) t(q; x)\, dx\right) + \psi(q).$$

Claim. Under the assumption of rational behavior, the principal will always outsource the quantity $q^* := \arg\max_{q \in [0;Q]} \mathcal{U}(q)$.

Example 1. The term $\psi(q)$ can be used to depict costs. Considering linear cost functions for both principal and agent where θ_p and θ_a monetary units arise to process one quantity unit, respectively, allows us to write $\psi(q) = -(\theta_p(Q - q) + \theta_a q + \mathcal{U}^a(q))$, with $\theta_a q + \mathcal{U}^a(q)$ monetary units being payed to the agent.

Example 2. The term $\tau(q)$ evaluates the outsourcing policy q by mapping the performance function $t(q; \cdot)$ to a real non-negative value. If principal and agent have linear processing times where they need δ_p and δ_a time units to process one quantity unit, respectively, one can show that choosing $m = 1/Q \cdot \mathbb{1}_{[0;Q]}$ reveals

$$\tau(q) = \frac{1}{Q} \int_0^Q t(q; x)\, dx = \frac{\delta_p + \delta_a}{2Q} q^2 - \delta_p q + \frac{\delta_p}{2} Q.$$

Example 3. Linear reward function $\phi(t) = R - c \cdot t$ for some $R, c > 0$.

2 Incentive Compatibility on Continuous Type Spaces

In what follows let ψ be as in Example 1, τ as in Example 2 and ϕ as in Example 3. As the principal knows his own type, (δ_p, θ_p) is assumed to be fixed. On the other hand, the agent's true type (δ_a, θ_a) is generally not assumed to be known to the principal but can be restricted to a compact set. Summarizing, we consider

$$\mathcal{U}_{\delta,\theta}(q) = R - c \cdot \left(\frac{\delta_p + \delta}{2Q} q^2 - \delta_p q + \frac{\delta_p}{2} Q\right) - (\theta_p(Q - q) + \theta q + \mathcal{U}^a_{\delta,\theta}(q)) \quad (1)$$

for some fixed $(\delta, \theta) := (\delta_a, \theta_a) \in \mathcal{T} := \Delta \times \Theta := \left[\underline{\delta}, \overline{\delta}\right] \times \left[\underline{\theta}, \overline{\theta}\right]$. Here, the term $\mathcal{U}_{\delta,\theta}(q)$ is directly related to the agent's utility as defined below.

We claim that a cooperation only happens on the basis of a contract $\mathcal{C} := (q, d, p)$ with some quantity $q \in [0; Q]$, deadline $d \in \mathbb{R}^+$ and payment of $p \in \mathbb{R}^+$ monetary units. For (δ, θ) fixed we want to define the agent's utility $\mathcal{U}^a_{\delta,\theta}(\mathcal{C})$ for accepting contract $\mathcal{C}$ as a quasi-linear function of quantity q and payment p, but only if deadline d can be satisfied. Otherwise, the contract should never be chosen.

Definition 3 (Utility of the agent). *For a contract* $\mathcal{C} := (q, d, p)$ *we define*

$$\mathcal{U}^a_{\delta,\theta}(\mathcal{C}) := \begin{cases} p - \theta q, & \delta q \leq d, \\ -1, & \text{otherwise.} \end{cases} \tag{2}$$

In order to deal with uncertainty about the true type (δ, θ) it is an essential approach to design contract alternatives for different possible scenarios and equip them with specific incentives that rationally affect the agent in our interest [3].

Definition 4 (Incentive compatibility). *A set of contracts* $\{\mathcal{C}(\delta, \theta),\, (\delta, \theta) \in \mathcal{T}\}$ *is called incentive compatible if the following expression holds:*

$$\mathcal{U}^a_{\delta,\theta}(\mathcal{C}(\delta, \theta)) \geq 0 \quad \wedge \quad (\delta, \theta) \in \underset{(\delta',\theta') \in \mathcal{T}}{\arg\max}\, \mathcal{U}^a_{\delta,\theta}(\mathcal{C}(\delta', \theta')) \qquad \forall (\delta, \theta) \in \mathcal{T}. \tag{3}$$

Loosely speaking, only incentive compatible menus of contracts are controllable for the principal. We are thus interested in the contract allocation which maximizes the principal's utility subject to incentive compatibility—usually resulting in a separate contract for each $(\delta, \theta) \in \mathcal{T}$. Nevertheless, such menus of infinitely many contracts are generally not only hard to compute [3] but also impractical for real-live applications [2]. For these reasons, we will only state an upper bound for the respective maximal utility in the following theorem.

Theorem 1 (Upper bound for the principal's utilty). *We assume that the distribution of* (δ, θ) *is given by a density* $f : \mathcal{T} \to \mathbb{R}_0^+$ *with* $\int\int f(\delta, \theta)\, d\theta\, d\delta = 1$. *By using* $f(\theta|\delta) := f(\delta, \theta) / \int_{\underline{\theta}}^{\overline{\theta}} f(\delta, \theta)\, d\theta$, $F(\theta|\delta) := \int_{\underline{\theta}}^{\theta} f(\delta, \xi)\, d\xi / \int_{\underline{\theta}}^{\overline{\theta}} f(\delta, \theta)\, d\theta$ *and the principal's utility (1), an upper bound for the utility of the optimal incentive compatible menu of contracts is given by* $\int\int \mathcal{U}_{\delta,\theta}(q^*(\delta, \theta))\, d\theta d\delta$, *where*

$$q^*(\delta, \theta) = \max\left\{\min\left\{\frac{Q}{c(\delta_p + \delta)}\left[c\delta_p + \theta_p - \left(\theta + \frac{F(\theta|\delta)}{f(\theta|\delta)}\right)\right]; Q\right\}; 0\right\} \tag{4}$$

is chosen as outsourcing policy and $\mathcal{U}^a_{\delta,\theta}(q(\delta, \theta)) = \int_{\theta}^{\overline{\theta}} q(\delta, \xi)\, d\xi$ *as agent's utility.*

Proof. The choice $\mathcal{U}^a_{\delta,\theta}(q(\delta, \theta)) = \int_{\theta}^{\overline{\theta}} q(\delta, \xi)\, d\xi$ is necessary for incentive compatibility due to the investigation of the one-dimensional model (only considering costs) in [3]. We are interested in the function $q(\cdot, \cdot)$ which maximizes the respective expected utility without involving further incentive compatibility constraints. Assuming $R - (c\delta_p/2 + \theta_p)Q = 0$ w.l.o.g. the expected utility writes

$$\begin{aligned} &\int_{\underline{\delta}}^{\overline{\delta}} \int_{\underline{\theta}}^{\overline{\theta}} \left(-c\tfrac{\delta_p+\delta}{2Q} q(\delta, \theta)^2 + (c\delta_p + \theta_p - \theta) q(\delta, \theta) - \int_{\theta}^{\overline{\theta}} q(\delta, \xi)\, d\xi\right) f(\delta, \theta)\, d\theta\, d\delta \\ = &\int_{\underline{\delta}}^{\overline{\delta}} \int_{\underline{\theta}}^{\overline{\theta}} \left(-c\tfrac{\delta_p+\delta}{2Q} q(\delta, \theta)^2 + \left(c\delta_p + \theta_p - \left(\theta + \tfrac{F(\theta|\delta)}{f(\theta|\delta)}\right)\right) q(\delta, \theta)\right) f(\delta, \theta)\, d\theta\, d\delta \end{aligned}$$

where the transformation follows from partial integration (compare [3]). Pointwise optimization regarding $0 \leq q \leq Q$ finally leads to $q^*(\delta, \theta)$ as in (4). We remark that all deadlines are implicitly set sharp, i.e. $d^*(\delta, \theta) = \delta q^*(\delta, \theta)$. □

3 Pooling of Contracts

In this section we want to present an alternative approach which propagates menus of finitely many contracts. As a natural feature of this method whole subsets of types (δ, θ) are depicted by only one single contract, in other words: they are pooled [2]. In the last section we will show that we can often exploit much of the optimal utility (estimated by the upper bound from Theorem 1) with few contracts, but for a theoretical basis we will start with two definitions.

Definition 5 (Grid-wise partition). *For $N_\delta, N_\theta \in \mathbb{N}$ define the grid points $\delta_i := \underline{\delta} + i/N_\delta(\overline{\delta} - \underline{\delta})$, $i = 0, ..., N_\delta$, and $\theta_j := \underline{\theta} + j/N_\theta(\overline{\theta} - \underline{\theta})$, $j = 0, ..., N_\theta$. We call a partition $P = \{P_1, ..., P_K\}$ of $\mathcal{T}$ grid-wise (with respect to N_δ and N_θ) if*

- *$P_k = \cup_{l=1}^{N_k} S_{i_l, j_l} := \cup_{l=1}^{N_k} ([\delta_{i_l - 1}; \delta_{i_l}] \times [\theta_{j_l - 1}; \theta_{j_l}])$ for all $k = 1, ..., K$,*
- *P_k is connected for all $k = 1, ..., K$.*

We want to denote $k(i, j)$ as that (unique) $k \in \{1, ..., K\}$ with $S_{i,j} \in P_k$.

Definition 6 (Well-shaped partition). *A cell $P_k \in P$, where P is grid-wise, is called well-shaped if there is a $(\delta^{(k)}, \theta^{(k)}) \in P_k$ with $\delta \leq \delta^{(k)}$ and $\theta \leq \theta^{(k)}$ for all $(\delta, \theta) \in P_k$. A partition P is called well-shaped if all cells P_k are well-shaped.*

Table 1. Well-shaped partitions with probabilities for uniform distribution on $\mathcal{T}$

0.04	0.04	0.04	0.04	0.04
0.04	0.04	0.04	0.04	0.04
0.04	0.04	0.04	0.04	0.04
0.04	0.04	0.04	0.04	0.04
0.04	0.04	0.04	0.04	0.04

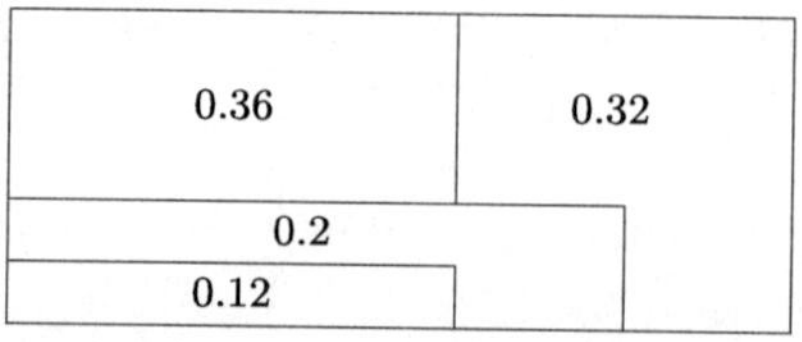

Examples for well-shaped partitions can be seen in Table 1. By computing the probabilities $\mathfrak{p}_k$ of each cell P_k we get a discretisation of the density f. Based on this, we want to determine the optimal menu of contracts that assigns a single contract (q_k, d_k, p_k) to each cell P_k of a given well-shaped partition. Dropping the constant $R - (c\delta_p/2 + \theta_p)Q$ in (1), setting $p_k = \theta^{(k)} q_k + \mathcal{U}_{\delta^{(k)}, \theta^{(k)}}(q_k)$ and incorporating $d_k = \delta^{(k)} q_k$ implicitly we consider the following quadratic program.

Problem 1 (Optimal incentive compatible menu of contracts on given partition).

$$\max_{\substack{q_1,...,q_K \\ p_1,...,p_K}} \sum_{k=1}^{K} \mathfrak{p}_k \left(-\frac{c(\delta_p + \delta^{(k)})}{2Q} q_k^2 + (c\delta_p + \theta_p) q_k - p_k \right) \tag{5}$$

$$
\begin{aligned}
\text{s.t. } & p_{k_{(i,N_\theta)}} - \theta_{N_\theta} q_{k_{(i,N_\theta)}} \geq 0 && \forall i \in \{1,...,N_\delta\} && (6)\\
& p_{k_{(i,j)}} - \theta_j q_{k_{(i,j)}} = p_{k_{(i,j+1)}} - \theta_j q_{k_{(i,j+1)}} && \forall k(i,j) \neq k(i,j+1) && (7)\\
& q_{k_{(i,j)}} \geq q_{k_{(i,j+1)}} && \forall k(i,j) \neq k(i,j+1) && (8)\\
& p_{k_{(i,j)}} - \theta_{j-1} q_{k_{(i,j)}} \geq p_{k_{(i+1,j)}} - \theta_{j-1} q_{k_{(i+1,j)}} && \forall k(i,j) \neq k(i+1,j) && (9)\\
& 0 \leq q_k \leq Q, \quad p_k \geq 0 && \forall k \in \{1,...,K\} && (10)
\end{aligned}
$$

Theorem 2. *Problem 1 computes the optimal incentive compatible menu of contracts on a given well-shaped partition $P = \{P_1, ..., P_K\}$.*

Proof. As an agent with type $(\delta, \theta) \in S_{i,j} \subseteq P_k$ never chooses a contract $k' \neq k$ with $\delta > \delta^{(k')}$ ((2) and (3)) we only have to consider the case $\delta \leq \delta_i \leq \delta^{(k')}$. For $k' = k(i', j')$, $i' \geq i$, fixed, $l = k(i', j)$ and $\lambda = (\theta - \theta_{j-1})/(\theta_j - \theta_{j-1})$ we get

$$
\begin{aligned}
p_k - \theta q_k &= \lambda(p_k - \theta_j q_k) + (1-\lambda)(p_k - \theta_{j-1} q_k)\\
&\geq \lambda(p_l - \theta_j q_l) + (1-\lambda)(p_l - \theta_{j-1} q_l) = p_l - \theta q_l
\end{aligned}
$$

due to (7) and (9). Furthermore the combination of (7) and (8) yields

$$
p_l - \theta q_l = p_l - \theta^{(l)} q_l + (\theta^{(l)} - \theta) q_l \geq p_{k'} - \theta^{(l)} q_{k'} + (\theta^{(l)} - \theta) q_{k'} = p_{k'} - \theta q_{k'}
$$

if $j' > j$ (the case $j' < j$ works similarly). All in all we get $p_k - \theta q_k \geq p_{k'} - \theta q_{k'}$ for all $k' \neq k$ and $p_k - \theta q_k \geq 0$, where the latter holds due to (6) and (7). This means that incentive compatibility is ensured.

While the necessity of (6) and (9) for incentive compatibility is obvious, the investigation of the one-dimensional setting in [3] directly enables the derivation that (7) and (8) are essential as well. This proves our theorem. □

As Problem 1 demands a predefined well-shaped partition P as input, the question about a clever choice of P arises on a higher level. To address this issue we propagate a heuristic approach given in Algorithm 1. The algorithm starts with the optimal menu of contracts on a grid partition of given size and gradually merges adjacent cells which promise the lowest utility loss. Relaxation of Problem 1 means that the monotonicity constraints (8) and (9) are deactivated. In the end we compute the optimal menu of contracts on the final partition P.

Algorithm 1. Pooling of Contracts

1: **Inputs:**
    Parameters K, Q, c, δ_p, θ_p, type-space $\mathcal{T}$, density f
2: **Initialize:**
    Grid partition P with $N := N_\delta \cdot N_\theta > K$ cells, probabilities $\mathfrak{p}_1, ..., \mathfrak{p}_N$
    Optimal incentive compatible menu of contracts on P (solve Problem 1)
3: **while** $N > K$ **do**
4:     Compute list L of cell pairs $p := (k_1, k_2)$ such that $k_1 \cup k_2$ is well-shaped
5:     **for** $p \in L$ **do**
6:         Compute optimal hypothetic contract on $P_{new} := P_{k_1} \cup P_{k_2}$ by optimizing the relaxed Problem 1 on the remaining $N-1$ cells in which the variables for all other cells $k \neq k_1, k_2$ are fixed to their current values
7:         Compute the hypothetic utility loss $\mathfrak{p}_{k_1}\mathcal{U}_{k_1} + \mathfrak{p}_{k_2}\mathcal{U}_{k_2} - (\mathfrak{p}_{k_1} + \mathfrak{p}_{k_2})\mathcal{U}_{new}$
8:     **end for**
9:     Merge cells $P_{k_1^*}$ and $P_{k_2^*}$ with minimal hypothetic utility loss computed in line 7 to $P^*_{new} := P_{k_1^*} \cup P_{k_2^*}$ and implement the respective contract from line 6
10:     Update payments for all other cells P_k with $P_k \cap P^*_{new} = \emptyset$ due to (7)
11:     Set $N \leftarrow N - 1$
12: **end while**
13: Compute optimal incentive compatible menu of contracts on P (solve Problem 1)

4 Computational Study and Conclusion

We test the performance of Algorithm 1 for different sizes K of the final menu of contracts with parameters $Q = 1000$, $c = 2$, $\theta_p = \delta_p = 40$, $\mathcal{T} = [20; 60]^2$ and two different densities f. Both densities are truncated bivariate normal distributions on $\mathcal{T}$ with parameters as in Table 2, the first containing a strongly negative correlation ($\varrho = -0.8$) and the second a strongly positive correlation between δ and θ ($\varrho = 0.8$). We consider three different initial grid sizes for which we state the relative gaps to the upper bound from Theorem 1 (%UB) and the total computation times in seconds (SEC). All tests were made on a MacBook Pro (3.1 GHz, 16 GB RAM) with Python 3.7.5 and CPLEX 12.7.1 as QP-Solver.

Table 2. $\mu_\delta = \mu_\theta = \sigma_\delta = 40$; $\sigma_\theta = 10$; LEFT: $\varrho = -0.8$, RIGHT: $\varrho = 0.8$

K	Init: $5 \cdot 5$		Init: $10 \cdot 10$		Init: $20 \cdot 20$	
	%UB	Sec	%UB	Sec	%UB	Sec
3	79.5	0.0	84.6	0.6	69.7	15.7
5	88.0	0.0	89.6	0.6	79.5	15.7
10	92.5	0.0	93.8	0.6	92.8	15.6
20	93.1	0.0	96.0	0.6	96.1	15.7
50	-	-	96.9	0.5	97.9	15.5
100	-	-	97.0	0.1	98.4	15.3
200	-	-	-	-	98.6	13.4
400	-	-	-	-	98.6	1.2

K	Init: $5 \cdot 5$		Init: $10 \cdot 10$		Init: $20 \cdot 20$	
	%UB	Sec	%UB	Sec	%UB	Sec
3	81.9	0.0	68.9	0.6	70.7	14.9
5	87.9	0.0	84.4	0.6	76.5	14.9
10	91.6	0.0	92.0	0.6	89.1	14.9
20	92.2	0.0	94.7	0.6	94.5	14.8
50	-	-	95.6	0.5	96.6	14.7
100	-	-	96.0	0.1	97.1	14.2
200	-	-	-	-	97.4	12.6
400	-	-	-	-	97.7	1.3

Taking a look at Table 2 we observe a remarkable exploitation of the upper bound with 50 contracts or more. However, the best trade-off between the amount of contracts, solution quality and computation time seems to be a combination of medium menu size (i.e. $K \approx 10$) and medium initial grid size (i.e. $\approx 10 \cdot 10$)—a choice that keeps down two important aspects: firstly, the size of the quadratic program in the initialization step, and secondly, the number of heuristical (and therefore error-prone) merging operations.

Summarizing, our pooling approach can be seen as an extension to the one-dimensional case in [2] by enabling the computation of arbitrarily large incentive compatible menus of contracts for the described two-dimensional setting.

References

1. Fayezi, S., O'Loughlin, A., Zutshi, A.: Agency theory and supply chain management: a structured literature review. Supply Chain Manag. **17**(5), 556–570 (2012). https://doi.org/10.1108/13598541211258618
2. Kerkkamp, R.B.O., van den Heuvel, W., Wagelmans, A.P.M.: Robust pooling for contracting models with asymmetric information. Eur. J. Oper. Res. **273**(3), 1036–1051. ISSN:0377-2217. https://doi.org/10.1016/j.ejor.2018.08.041
3. Laffont, J.J., Martimort, D.: The Theory of Incentives: The Principal-Agent Model. Princeton University Press, Princeton (2002). https://doi.org/10.2307/j.ctv7h0rwr

Check for updates

From Design to Operation: Mixed-Integer Model Predictive Control Applied to a Pumping System

Tim Moritz Müller, Christoph Knoche, and Peter Franz Pelz(✉)

Chair of Fluid Systems, Technische Universität Darmstadt, Darmstadt, Germany
{tim.mueller,peter.pelz}@fst.tu-darmstadt.de

Abstract. The two most significant life cycle phases of products or systems are the design and operation phase. Both share their incredibly high level of complexity due to the available diversity in components, operating settings and interconnection variants. In the design process, two-stage stochastic optimisation problems have proven to be suitable, in which the operation is anticipated by considering various demand scenarios. Since the operation is characterised by uncertainty and fluctuation, it has to be ensured that the operation is also realised in an optimal way. In this contribution we show how the original planning problem is transformed into a control problem using the example of a pumping system.

Keywords: Mixed-integer model predictive control · Optimal control · Experimental validation · Pumping system · Water distribution network

The task of engineers is to design and operate technical systems. Designing a system is about finding a solution which ensures the function, fulfills all design specifications while having the highest possible quality (e.g. long service life or low costs). This corresponds to a constraint optimization problem [4]. The solution is the system design, i.e. system properties that can not be adjusted in the future (e.g. pump type). During operation, the system responds to uncertainty or fluctuations by adjusting the operating settings (e.g. pump speeds), aiming at a fast and accurate fulfillment of the function and low operational costs.

In order to ensure function fulfillment for all operation points and to estimate the influence of operation on quality (e.g. energy costs), the operation must be anticipated during the design. This leads to a two-stage stochastic optimization problem with recourse [5]. The first-stage problem, i.e. the design task, is defined in problem (1) in which $F^{\text{second}}(x, W)$ describes the optimal value of the second-stage problem, i.e. the operation problem (2):

$$\begin{aligned} \min_{x} \quad & f^{\text{first}}(x) + E_W\left[F^{\text{second}}(x, W)\right] \\ \text{subject to} \quad & g^{\text{first}}(x) \leq 0. \end{aligned} \qquad (1)$$

$$\begin{aligned} \min_{y} \quad & f^{\text{second}}(y, W) \\ \text{subject to} \quad & g^{\text{second}}(x, y, W) \leq 0. \end{aligned} \qquad (2)$$

Here x are the first-stage variables, y the second-stage variables and W uncertain parameters of the problem. The manifestation of the uncertainty can only

N. Trautmann and M. Gnägi (Eds.): OR 2021, LNOR, pp. 318–324, 2022.
https://doi.org/10.1007/978-3-031-08623-6_47

be determined in operation - i.e. after the values of x have been fixed. The two problems are coupled: the possible operation depends on the chosen design, whereas the design anticipates the operation. Thus, for the operation, optimization problem (2) is solved, where x becomes a previously determined parameter. Moreover, W has a specific value, which is measured or estimated.

The outline of the paper is as follows: We first present the technical application and corresponding optimization problems for the operation. Then, we present the developed control approach and the obtained experimental results.

1 Application and Model

Pump systems, so-called booster stations, are used to provide sufficient water pressure in high-rise buildings. Figure 1a) shows the down scaled test rig. Booster stations use multiple pumps, usually speed controllable, to respond to fluctuating demands. The crucial question for the design is which type of pumps to choose and how to interconnect them in order to have the lowest possible life cycle costs and sufficient functional performance. This has been investigated in previous work [2,3]. State of the art are parallel pumps of the same type. More cost and energy efficient are different pump types and topologies as well as decentrally placed pumps to reduce throttling losses [3].

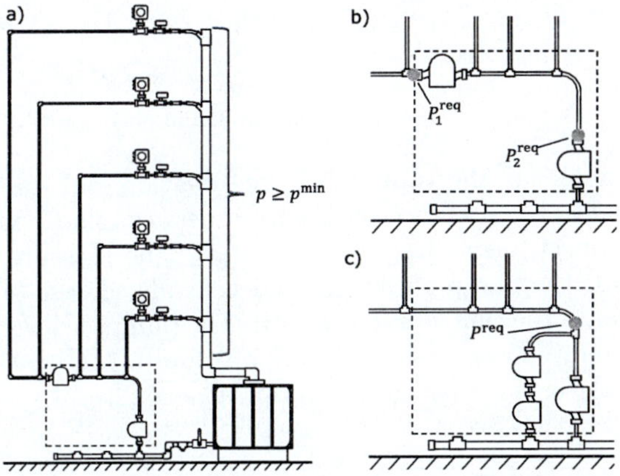

Fig. 1. In test rig a), the water flows from the tank via the booster station to 5 floors. In each floor the volumeflow can be measured and a valve can be used to control it. Subsequently, the water flows back into the tank via a drainpipe. Any topologies of the pumps can be realized. Pressure difference over the pumps and power consumption are measured. All pumps can be turned on and off and the speed can be controlled. Figure b) shows the decentralized and c) the centralized system variant which is investigated.

For operation, the question of the on/off states as well as the speeds of the pumps arises in order to achieve a low energy consumption. At the same time a minimum pressure must be ensured in the different floors, independent of the volumeflow demand. By a simple hydraulic model of the building, the pressure-volumeflow requirement at the pump or booster station outlet is calculated.

The planning problem considered in [2] corresponds to the Deterministic Equivalent Program of problem (1) for the pump system design. This is transformed into the control problem outlined above by dropping all constraints that only refer to the first stage. In addition, the first stage variables are fixed, i.e. the pump selection and topology become parameters and the investment costs can be removed from the objective function. This results in the optimization problem for the system adaptation (3). Overall, the optimization problem for the system adaption evolves from the original second stage problem by solely changing variable and parameter assignments.

The objective of the system adaptation is to minimize the power consumption (5a) by adjusting the speeds $n_i \in [0,1]$ and on/off states $x_i \in \{0,1\}$ of all pumps $i \in \mathcal{P}$. Lower and upper bounds must be respected (3b). Moreover, the minimum required power can be estimated to speed up the optimization (3c), cf. [2]. In order to evaluate the power consumption and ensure functional performance, the physical behavior must be taken into account (3d). A further description of the variables, parameter, sets and constraints can be found in [2].

$$\min_{n,x,po,p,\Delta p,q} \quad \sum\nolimits_{i\in\mathcal{P}} po_i \tag{3a}$$

subject to

$$\Delta p_i \leq \overline{\Delta P}_i x_i,\ p_i^{\text{out}} \leq \overline{P},\ q_i \leq \overline{Q} x_i,\ \underline{N}_i x_i \leq n_i \leq \overline{N}_i x_i,\ po_i \leq \overline{\Delta Po}_i x_i \quad \forall i \in \mathcal{P} \tag{3b}$$

$$\sum\nolimits_{i\in\mathcal{P}} po_i \geq Po^{\min} \tag{3c}$$

$$\text{system model (4a) - (4n)} \tag{3d}$$

In the system model (4) the volumeflow conservation and -fulfillment (4a)–(4c) is ensured. Besides, the pressure propagation between nodes including pressure losses of the connections is described (4d)–(4j). The pressure requirements are defined directly at the pump outlet in the decentralized case (4k) and at the booster station outlet in the centralized case (4l). The pumps are described by the quadratic pressure-volumeflow characteristic (4m) and cubic power-volumeflow characteristic (4n). For a more detailed description see [2].

$$q_i = \sum\nolimits_{j\in\mathcal{P}} q_{j,i} + q_i^{\text{source}} = \sum\nolimits_{j\in\mathcal{P}} q_{i,j} + q_i^{\text{sink}} \quad \forall i \in \mathcal{P} \tag{4a}$$

$$q_{i,j} \leq \overline{Q}T_{i,j},\ q_i^{\text{source}} \leq \overline{Q}T_i^{\text{source}},\ q_i^{\text{sink}} \leq \overline{Q}T_i^{\text{sink}} \quad \forall i,j \in \mathcal{P} \tag{4b}$$

$$\sum\nolimits_{i\in\mathcal{P}} q_i^{\text{sink}} = Q^{\text{bound}} \tag{4c}$$

$$p_i^{\text{in}} + \Delta p_i - p_i^{\text{out}} \overset{\leq +}{\geq -} (1-x_i)\overline{P} \quad \forall i \in \mathcal{P} \tag{4d}$$

$$p_i^{\text{out}} - \Delta p_{i,j}^{\text{frict,con}} - \Delta P^{\text{geo,con}} - p_j^{\text{in}} \overset{\leq +}{\geq -} \overline{P}(1-T_{i,j}) \quad \forall i,j \in \mathcal{P} \tag{4e}$$

$$p_i^{\text{in}} - \Delta p_{i,j}^{\text{frict,in}} - \Delta P^{\text{geo,in}} - P_i^{\text{source}} \overset{\leq +}{\geq -} \overline{P}(1-T_{i,j}^{\text{source}}) \quad \forall i \in \mathcal{P} \tag{4f}$$

$$p_i^{\text{out}} - \Delta p_{i,j}^{\text{frict,out}} - \Delta P^{\text{geo,out}} - p_i^{\text{sink}} \overset{\leq +}{\geq -} \overline{P}(1-T_{i,j}^{\text{sink}} x_i) \quad \forall i \in \mathcal{P} \tag{4g}$$

$$\Delta p_{i,j}^{\text{frict,con}} = 0.5\varrho\zeta_{i,j}^{\text{con}}(q_{i,j}/A)^2 \quad \forall i,j \in \mathcal{P} \tag{4h}$$

$$\Delta p_i^{\text{frict,in}} = 0.5\varrho\zeta_i^{\text{in}}(q_i^{\text{source}}/A)^2 \quad \forall i \in \mathcal{P} \tag{4i}$$

$$\Delta p_i^{\text{frict,out}} = 0.5\varrho\zeta_i^{\text{out}}(q_i^{\text{sink}}/A)^2 \quad \forall i \in \mathcal{P} \tag{4j}$$

$$p_i^{\text{out}} \geq P_i^{\text{req}} \quad \forall i \in \mathcal{P} \tag{4k}$$

$$p^{\text{sink}} \geq P^{\text{req}} \quad \forall i \in \mathcal{P} \tag{4l}$$

$$\Delta p = (\alpha_{i,0} - \zeta^{\text{inst}})\, {q_i}^2 + \sum\nolimits_{m=1}^{2} \alpha_{i,m}\, {q_i}^{2-m}\, {n_i}^m \quad \forall i \in \mathcal{P} \tag{4m}$$

$$po_i = \beta_{i,4} + \sum\nolimits_{m=0}^{3} \beta_{i,m}\, {q_i}^{3-m}\, {n_i}^m \quad \forall i \in \mathcal{P} \tag{4n}$$

Some of the system quantities are uncertain and the system requirements are constantly changing due to the fluctuating demand. However, these quantities cannot be measured directly for the most part, but must be estimated. For this purpose, a similar optimization problem is used, where the control quantities of the system n and x, which are variables of the system adaptation (3), become fixed parameters. In turn, the uncertain model parameters α, β, ζ are now variables. These are optimized so that the model reflects as good as possible the measured real values, which results in an NLP for the optimal system calibration (5). To do so, the sum of the squared, normalized deviation ε of measured (upper case) and modeled (lower case) pressures and pump power is minimized (5b). To stabilize the control and to prevent too aggressive optimization, the correction of the system parameters between two time steps is integrated. Thus, the character of the model shall be preserved and prevent that, for example, cubic or quadratic components are set to zero. The correction is calculated from the sum of the squared, normalized changes of the system parameters δ (5c). The two parts are weighted by λ_ε and λ_δ (5a). The system model (4), with the described modifications of variables and parameters, is considered as well.

$$\min_{\zeta,\alpha,\beta,\varepsilon,\delta,po,p,\Delta p,q} \quad \lambda_\varepsilon \varepsilon + \lambda_\delta \delta \tag{5a}$$

subject to

$$\varepsilon = \sum_{i\in\mathcal{P}} \left(\frac{p_i^{\text{out}} - P_i^{\text{out,meas}}}{\overline{P}}\right)^2 + \left(\frac{p_i^{\text{sink}} - P_i^{\text{sink,meas}}}{\overline{P}}\right)^2 + \left(\frac{\Delta p_i - P_i^{\text{meas}}}{\overline{\Delta P_i}}\right)^2 + \left(\frac{po_i - Po_i^{\text{meas}}}{\overline{\Delta Po_i}}\right)^2 \tag{5b}$$

$$\delta = \sum_{i\in\mathcal{P}} \left(\zeta_i^{\text{in,past}} - \zeta_i^{\text{in}}\right)^2 + \left(\zeta_i^{\text{inst,past}} - \zeta_i^{\text{inst}}\right)^2 + \sum_{i,j\in\mathcal{P}} \left[\left(\zeta_{i,j}^{\text{con,past}} - \zeta_{i,j}^{\text{con}}\right)^2 + \sum_{m\in 0,1,2} \left(\frac{\alpha_{i,m}^{\text{past}} - \alpha_{i,m}}{\alpha^{\text{lit}}}\right)^2 + \sum_{m\in 0,1,2,3} \left(\frac{\beta_{i,m}^{\text{past}} - \beta_{i,m}}{\beta^{\text{lit}}}\right)^2\right] \tag{5c}$$

system model (4a) - (4n) (5d)

2 Control Approach

In the following, we describe how the two optimization problems interact to control the pumping system, cf. Fig. 2.

The minimum requested pressure in the floors, the system design, and the water demand in the form of valve positions are inputs for the approach. The system design (pump types and topology) origins from the design problem [2]. This is used to create the optimization models for operation.

The control algorithm is highlighted in gray. Based on the pressure demand $p^{\min}$ and the measured flow demand of the building Q_s, the pressure and flow requirements of the booster system, P^{req}, Q^{bound}, are calculated in the preprocessing (hydraulic model of the building, not considered in this paper). These are used in the system adaptation together with the estimated model parameters of the previous iteration to calculate the optimal operating states, n, x, which are then applied at the test rig. At the same time, the system calibration is executed and the model parameters α, β, ζ are fitted based on the measured quantities and the operating states of the previous iteration.

The algorithm runs continuously. This means that as soon as one control iteration has been completed, the next one starts immediately. System adaptation and calibration run in parallel, so that the system adaptation uses the adjusted system parameters of the last iteration and vice versa.

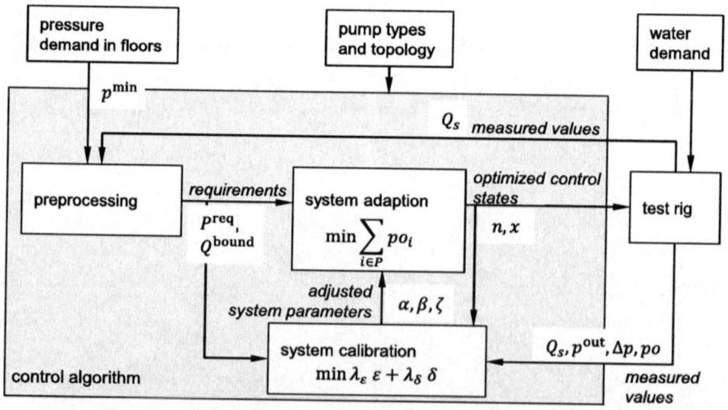

Fig. 2. Control approach

Thus, in each time step, a mixed-integer nonlinear program (MINLP) for the system adaptation (*optimize the operation to achieve high efficiency*) and a continuous nonlinear program (NLP, i.e. without binary decision variables) for system calibration (*optimize the model to fit the real behavior*) is solved. This corresponds to a mixed-integer model predictive control, where one time step is predicted and optimization and model are combined in one MINLP. In addition, the model parameters are adaptively fine-tuned via an NLP.

The approach is implemented using Labview for the test rig control, with a link to Matlab used for optimization. The problems are modeled with YALMIP [1] and solved with SCIP 7 [6].

3 Results

In addition to the variants shown in Fig. 1 (decentral (b) and series-parallel (c)) a reference system with parallel pumps of the same type with conventional control (ref) and the approach shown here (opt) has been investigated. The energy consumption for a test case of 30 min, which is based on the random water demand during the course of a day in a real building, is shown in Table 1.

Table 1. Experimentally obtained results for four different system variants

		Parallel		Series-parallel	Decentral
		Ref	Opt		
Energy consumption	in Wh	42.5	38.25	35.69	28.44
	in %	100	90	83.98	66.92
Median pressure deviation in mbar		15.6	3.1	5.8	7.9

It can be clearly seen that the energy consumption decreases due to the optimization in the parallel system. The other topologies also show lower energy consumption. In [3] a energy saving of 19% have been predicted for the decentralized system variant in the design phase based on five load demands. Those are even exceeded. It is shown that energy-efficient control is possible even outside of the load demands anticipated in the design, provided the complexity is managed with optimization techniques. The median pressure deviations are also consistently small, so that it can be assumed that the minimum pressures are sufficiently fulfilled.

4 Conclusion

We have shown how the approach of a two-stage stochastic optimization problem for systems design can be adopted to realize a controller for the operation by simply changing parameter and variable assignments. This shows the versatility of optimization models and how they can be adapted for various purposes. Here, two optimization programs interact to determine the optimal operating settings on the one hand and to fit the model parameters to determine the uncertain parameters on the other hand. The results are applied to a real-world pump system showing that the functional requirements can be met and the power consumption is reduced compared to a conventional approach. However, it should be noted that the initial implementation and modeling effort is high and that a central controller is required.

Acknowledgment. Results were obtained in project No. 17482 N/1, funded by the German Federal Ministry of Economic Affairs and Energy (BMWi) approved by the Arbeitsgemeinschaft industrieller Forschungsvereinigungen "Otto von Guericke" e.V. (AiF).

References

1. Löfberg, J.: YALMIP: a toolbox for modeling and optimization in MATLAB. In: In Proceedings of the CACSD Conference. Taipei, Taiwan (2004)
2. Müller, T.M., Altherr, L.C., Leise, P., Pelz, P.F.: Optimization of pumping systems for buildings: experimental validation of different degrees of model detail on a modular test rig. In: Neufeld, J.S., Buscher, U., Lasch, R., Möst, D., Schönberger, J. (eds.) Operations Research Proceedings 2019. ORP, pp. 481–488. Springer, Cham (2020). https://doi.org/10.1007/978-3-030-48439-2_58
3. Müller, T.M., Leise, P., Lorenz, I.-S., Altherr, L.C., Pelz, P.F.: Optimization and validation of pumping system design and operation for water supply in high-rise buildings. Optim. Eng. **22**(2), 643–686 (2020). https://doi.org/10.1007/s11081-020-09553-4
4. Pelz, P.F., Groche, P., Pfetsch, M., Schäffner, M. (eds.): Mastering Uncertainty in Mechanical Engineering. Springer, Cham (2021). https://doi.org/10.1007/978-3-030-78354-9
5. Shapiro, A., Philpott, A.: A Tutorial on Stochastic Programming (2007)
6. Vigerske, S., Gleixner, A.: SCIP: global optimization of mixed-integer nonlinear programs in a branch-and-cut framework. Optim. Methods Softw. **12**, 1–31 (2017). https://doi.org/10.1080/10556788.2017.1335312

A Finite Element Approach for Trajectory Optimization in Wire-Arc Additive Manufacturing

Johannes Schmidt(✉), Johannes Buhl, and Armin Fügenschuh

Brandenburg University of Technology Cottbus-Senftenberg,
Platz der Deutschen Einheit 1, 03046 Cottbus, Germany
{johannes.schmidt,johannes.buhl,fuegenschuh}@b-tu.de

Abstract. In wire-arc additive manufacturing (WAAM), the desired workpiece is built layer-wise by a moving heat source depositing droplets of molten wire on a substrate plate. To reduce material accumulations, the trajectory of the weld source should be continuous, but transit moves without welding, called deadheading, are possible. The enormous heat of the weld source causes large temperature gradients, leading to a strain distribution in the welded material which can lead even to cracks. In summary, it can be concluded that the temperature gradient reduce the quality of the workpiece. We consider the problem of finding a trajectory of the weld source with minimal temperature deviation from a given target temperature for one layer of a workpiece with welding segments broader than the width of the weld pool. The temperature distribution is modeled using the finite element method. We formulate this problem as a mixed-integer linear programming model and demonstrate its solvability by a standard mixed-integer solver.

Keywords: Additive manufacturing · Mixed-integer linear programming · Finite element method · Heat equation · Path optimization

1 Introduction

In the last decades, the field of additive manufacturing (AM) developed into an advantageous alternative to common metal cutting manufacturing processes, due to its ability to produce complex workpieces without substantial material removal. One of these processes is the so-called wire-arc additive manufacturing (WAAM). In this process, a wire is molten by an electrical arc or laser and deposited in droplets on an underlying substrate plate building the workpiece layer-wise. The weld source moves around freely and is capable of transiting without welding, called deadheading. A crucial factor to the quality of the manufactured workpiece is the trajectory of the weld source. Due to its enormous heat, high thermal gradients can occur, leading to stress inside the material. Possible results are distortions of the workpiece or even cracks. Controlling the

N. Trautmann and M. Gnägi (Eds.): OR 2021, LNOR, pp. 325–330, 2022.
https://doi.org/10.1007/978-3-031-08623-6_48

temperature gradients and aiming for a homogeneous temperature distribution within the workpiece reduces these effects. Therefore, detailed planning of the welding trajectory is advantageous. A review about the effect of the chosen weld strategy to the process time and workpiece quality can be found in [3]. We tackle the problem of finding a path of the weld source with minimal absolute deviation to a given target temperature for a single layer of the workpiece. In [4], the strain simulation for a given welding trajectory using a finite element method is presented. For workpieces with wall strength as broad as the width of the weld pool, the problem has been studied in [1]. Thus, we consider in this work wall strengths broader than the width of the weld pool, motivated in [5]. A mixed-integer linear problem is set up to compute the trajectory and track the temperature during the process at the same time. The temperature distribution within the workpiece is affected by the heat input of the weld source, heat conduction, and thermal radiation. These three aspects are combined in the heat equation with a Robin boundary condition. It is discretized using the finite element method and incorporated into the model. Its solvability by a standard mixed-integer solver is demonstrated on a test instance.

2 Mathematical Model

Given the structure of the desired workpiece, the two-dimensional layers are obtained by slicing it vertically, see Fig. 1. Each layer consists of segments that have to be welded and intersection points between them. Thus, every layer can be considered as an undirected graph $G = (\mathcal{V}, \mathcal{W})$, with nodes at the intersection points and the segments between them as edges. This graph has not to be connected and can contain several components $G_i = (\mathcal{V}_i, \mathcal{W}_i)$ $(i = 1, \ldots, n)$. Let $\mathcal{V}^{odd}$ and $\mathcal{V}_i^{odd}$ denote the set of nodes with odd degree in G or its component G_i, respectively.

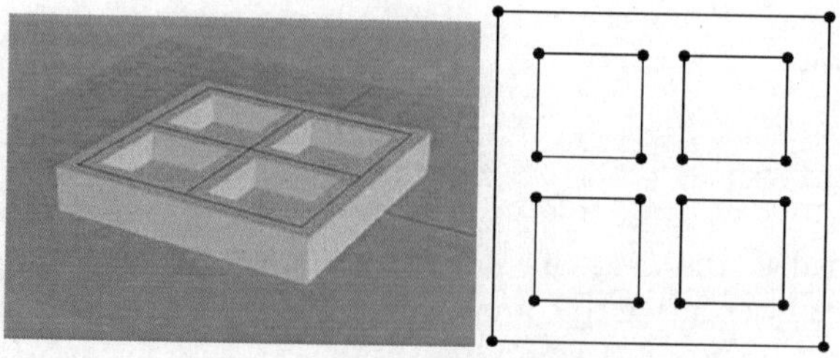

Fig. 1. A prototypical workpiece *(left)* and a single layer of it *(right)*

In practical application, the heat source is much slower while welding than during deadheading. Thus, we assume that transition moves are done immediately without consuming time. Since the movement speed v_w of the heat source while welding and the length $l_{i,j}$ of each segment $(i, j) \in \mathcal{W}$ is known apriori, the time to weld the complete layer is given by $T = \sum_{(i,j)\in\mathcal{W}} \frac{l_{i,j}}{v_w}$. Introducing the time step length Δt, the time horizon $[0, T]$ is discretized as a set of discrete time steps $\mathcal{T}_0 = \{0, 1, \ldots, T^{max}\}$, with $T^{max} = \sum_{(i,j)\in\mathcal{W}} \tau_{i,j} = \sum_{(i,j)\in\mathcal{W}} \left\lceil \frac{l_{i,j}}{v_w \Delta t} \right\rceil$ and $\tau_{i,j}$ describing the time to weld segment $(i,j) \in \mathcal{W}$. As abbreviations we use $\mathcal{T} = \mathcal{T}_0 \setminus \{0\}$, $\mathcal{T}^- = \mathcal{T} \setminus \{T^{max}\}$, and $\mathcal{T}_0^- = \mathcal{T}_0 \setminus \{T^{max}\}$.

2.1 Path Generation

Modeling each layer as an undirected graph, the search for a welding trajectory of the heat source becomes the problem of computing a continuous path for the graph, a problem closely related to the Chinese Postman problem [2], where the artificial edges are now the deadheading moves. The number of the necessary artificial edges ω in the solution of the Chinese Postman problem for a connected graph G is $\omega = \frac{|\mathcal{V}^{odd}|}{2}$. In our setting, the start and the endpoint of the path can be different. Thus, the number of necessary deadheading moves within each component can be calculated by $\omega_i = \frac{|\mathcal{V}_i^{odd}|}{2} - 1$. Furthermore, there are $n-1$ moves necessary to navigate between the components. In total, the number of necessary deadheading moves for the whole graph G is given by $\sum_{i=1}^{n} \omega_i + n - 1$.

Let $\mathcal{U} \subseteq \mathcal{V} \times \mathcal{V}$ denote the set of all possible deadheading moves. All segments can be welded in arbitrary direction, thus the set $\mathcal{W}$ is extended to $\overline{\mathcal{W}} = \{(i,j) \in \mathcal{W} \mid (i,j) \in \mathcal{W} \vee (j,i) \in \mathcal{W}\}$ and we introduce further sets $\mathcal{W}^* = \{(i, t_i, j, t_j) \in \mathcal{V} \times \mathcal{T}_0 \times \mathcal{V} \times \mathcal{T} \mid (i,j) \in \overline{\mathcal{W}},\ t_j = t_i + \tau_{i,j}\}$ and $\mathcal{U}^* = \{(i,j,t) \in \mathcal{U} \times \mathcal{T}^-\}$ to relate all welding and deadheading moves to the time, respectively. Let $w_{i,t_i,j,t_j} \in \mathcal{W}^*$ be a binary variable indicating if segment $(i,j) \in \overline{\mathcal{W}}$ is processed from time step $t_i \in \mathcal{T}_0$ to time step $t_j \in \mathcal{T}$ and $u_{i,j,t} \in \mathcal{U}^*$ a binary variable indicating if connection $(i,j) \in \mathcal{U}$ is used for deadheading at time step $t \in \mathcal{T}^-$. Then, the problem of finding a welding path can be stated as

$$\sum_{i,j,t_j:(i,0,j,t_j)\in\mathcal{W}^*} w_{i,0,j,t_j} = 1, \tag{1}$$

$$\sum_{i,t_i,j:(i,t_i,j,T^{max})\in\mathcal{W}^*} w_{i,t_i,j,T^{max}} = 1, \tag{2}$$

$$\sum_{t_i,t_j:(i,t_i,j,t_j)\in\mathcal{W}^*} w_{i,t_i,j,t_j} + \sum_{t_j,t_i:(j,t_j,i,t_i)\in\mathcal{W}^*} w_{j,t_j,i,t_i} = 1 \quad \forall (i,j) \in \mathcal{W}, \tag{3}$$

$$\begin{aligned} &\sum_{h,t_h:(h,t_h,i,t)\in\mathcal{W}^*} w_{h,t_h,i,t} + \sum_{h:(h,i,t)\in\mathcal{U}^*} u_{h,i,t} \\ &= \sum_{j,t_j:(i,t,j,t_j)\in\mathcal{W}^*} w_{i,t,j,t_j} + \sum_{j:(i,j,t)\in\mathcal{U}^*} u_{i,j,t} \qquad \forall i \in \mathcal{V},\ t \in \mathcal{T}, \end{aligned} \tag{4}$$

$$\sum_{(i,j,t)\in\mathcal{U}^*} u_{i,j,t} = \omega + n - 1, \tag{5}$$

$$\sum_{i,j:(i,j,t)\in\mathcal{U}^*} u_{i,j,t} \leq 1 \qquad \forall t \in \mathcal{T}. \tag{6}$$

The weld source has to start and end its path somewhere (1), (2), while every segment must be welded (3). The computed path has to be continuous (4) and the number of deadheading moves is limited (5). Equation (6) is not necessary, but a valid inequality since every time there are two consecutive deadheading moves $(i,j,t),(j,k,t) \in \mathcal{U}^*$ in one time step, they can be merged to $(i,k,t) \in \mathcal{U}^*$ reducing the traveled distance.

2.2 Temperature Distribution

To model the temperature distribution within one layer of the workpiece during the welding process, the heat input of the weld source, heat conduction, and thermal radiation have to be taken into account. The two dimensional heat equation

$$\frac{\partial \theta}{\partial t}(x,y,t) = \alpha \left(\frac{\partial^2 \theta}{(\partial x)^2}(x,y,t) + \frac{\partial^2 \theta}{(\partial y)^2}(x,y,t) \right) + q(x,y,t) \qquad \forall (x,y) \in \Omega,\ t \in (0,T], \tag{7.1}$$

$$\frac{\partial \theta}{\partial n}(x,y,t) = \kappa^e \left(\theta^{amb}(t) - \theta(x,y,t)\right) \qquad \forall (x,y) \in \partial\Omega,\ \forall t \in [0,T], \tag{7.2}$$

$$\theta(x,y,0) = \theta^{init}(x,y) \qquad \forall (x,y) \in \Omega, \tag{7.3}$$

describes the heat conduction within an area Ω with initial temperature distribution θ^{init} and ambient temperature θ^{amb}. The Robin boundary condition (7.2) allows a linear approximation of the thermal radiation [1].

To discretize the heat Eq. (7), $\tau_{i,j} - 1$ nodes are added equidistantly on every segment $(i,j) \in \mathcal{W}$ and collected in the set $\mathcal{V}^{int}$. To relate these interior nodes to their corresponding segment, a function $\xi : \mathcal{V}^{int} \mapsto \mathcal{W} \times \{1,\ldots,\tau_{i,j} - 1\}$ is introduced, reporting the respective segment and the nodes position along it for every interior node. Furthermore, let the variable $\theta_{i,t}$ describe the temperature of node $i \in \overline{\mathcal{V}} := \mathcal{V} \cup \mathcal{V}^{int}$ at time step $t \in \mathcal{T}_0$. We apply the finite element method according to [6] with node set $\overline{\mathcal{V}}$, the segments $(i,j) \in \mathcal{W}$ as boundary of the area Ω, and linear triangle elements. This yields the linear equation system

$$(M + \Delta t K)\vec{\theta}_{t+1} = \Delta t(\vec{q}_{t+1} \vec{f^H} + \vec{f^R}) + M\vec{\theta}_t, \tag{8}$$

with mass matrix $M = (m_{i,j})_{i,j}$, stiffness matrix $K = (k_{i,j})_{i,j}$, load vectors $\vec{f^H}$ and $\vec{f^R}$, and $\vec{\theta}_t$, $\vec{q}_t$ describing the vectors of the temperature and the heat input of all nodes at time step $t \in \mathcal{T}_0^-$, respectively.

The weld source is described by the piece-wise constant approximation of the Goldak heat source model from [1] with coefficients κ_k^w, $k = 1,\ldots,K^w$ and intervals $\mathcal{P}_k$. The temperature gain of a node at the center of the weld pool is given by the parameter φ^w. To simplify notations, the binary variable $w_{i,t}$ with

$$w_{i,t} = \begin{cases} \sum_{h,t_h:(h,t_h,i,t)\in\mathcal{W}^*} w_{h,t_h,i,t} + \sum_{h:(h,i,t)\in\mathcal{U}^*} u_{h,i,t} & ,i \in \mathcal{V}, t \in \mathcal{T}_0, \\ \sum_{\substack{(h,t_h,j,t_j)\in\mathcal{W}^* \\ \xi(i)=(h,j,k) \\ t=t_h+k}} w_{h,t_h,j,t_j} + \sum_{\substack{(j,t_j,h,t_h)\in\mathcal{W}^* \\ \xi(i)=(h,j,k) \\ t=t_h-k}} w_{j,t_j,h,t_h} & ,i \in \mathcal{V}^{int},\ t \in \mathcal{T}, \\ 0 & ,i \in \mathcal{V}^{int},\ t = 0, \end{cases} \tag{9}$$

is introduced, indicating above which node $i \in \overline{\mathcal{V}}$ the weld source is positioned at time step $t \in \mathcal{T}_0$. Note that the third case is necessary since the index sets of the sums in (9) are empty for $i \in \mathcal{V}^{int}$ at $t = 0$. The temperature gain of the nodes from the weld source is then given by

$$\varphi_{i,t} = \sum_{k=1}^{K^w} \sum_{\substack{j \in \mathcal{V} \cup \mathcal{V}^{int} \\ d^e_{i,j} \in \mathcal{P}_k}} \kappa^w_k \varphi^w w_{i,t} \qquad \forall i \in \overline{\mathcal{V}},\ \forall t \in \mathcal{T}_0, \tag{10}$$

where $d^e_{i,j}$ is the Euclidean distance between nodes $i, j \in \overline{\mathcal{V}}$. Let θ^{init}_i denote the initial temperature of node $i \in \overline{\mathcal{V}}$. Incorporating (10) into (8), the temperature distribution within the workpiece is modeled by

$$\theta_{i,0} = \theta^{init}_i, \qquad \forall i \in \overline{\mathcal{V}} \tag{11}$$

$$\sum_{j \in \overline{\mathcal{V}}} (m_{i,j} + \Delta t k_{i,j})\theta_{j,t} = \sum_{j \in \overline{\mathcal{V}}} m_{i,j}\theta_{j,t-1} + \Delta t \left(\varphi_{i,t} f^H_i + f^R_i\right) \quad \forall i \in \overline{\mathcal{V}},\ t \in \mathcal{T}. \tag{12}$$

The initial temperature of each node is set in (11) and then the linear equation system (12) is solved for every time step.

To achieve a preferably homogeneous temperature distribution, we minimize the absolute deviation of the temperature of the workpiece to a given target temperature θ^{tar}. Introducing further variables $\theta^+_{i,t}, \theta^-_{i,t} \in \mathbb{R}_+$ to describe the positive and negative portion of the absolute value function, the objective function is given by

$$\min \qquad \sum_{i \in \overline{\mathcal{V}}, t \in \mathcal{T}_0} \theta^+_{i,t} + \theta^-_{i,t} \tag{13}$$

with the additional constraint

$$\theta^{tar} - \theta_{i,t} = \theta^+_{i,t} - \theta^-_{i,t} \qquad \forall i \in \overline{\mathcal{V}},\ t \in \mathcal{T}_0. \tag{14}$$

3 Computational Results and Conclusions

The mixed-integer model consisting of constraints (1)–(6), (11), (12), (14), and objective (13) is implemented in AMPL using CPLEX 12.10 (default settings) for its solution on a Mac Pro with an Intel Xeon W running 32 threads parallel at 3.2 GHz clockspeed and 768 GB RAM. As an example instance, the layer displayed in Fig. 1 is used with a side length of 16 mm and inner squares of 5×5 mm. The parameter values are chosen according to [1], except for $\kappa^e = 0.5$ and $\theta^{tar} = \theta^{amb}(t) = 200\,°\mathrm{C}$, which are estimated. After 172,670 s, the optimal solution was obtained. It is displayed in Fig. 2, together with the temperature progression of node 12, where all aspects of heat transmission are observable. In the beginning, its temperature increases due to heat conduction through the

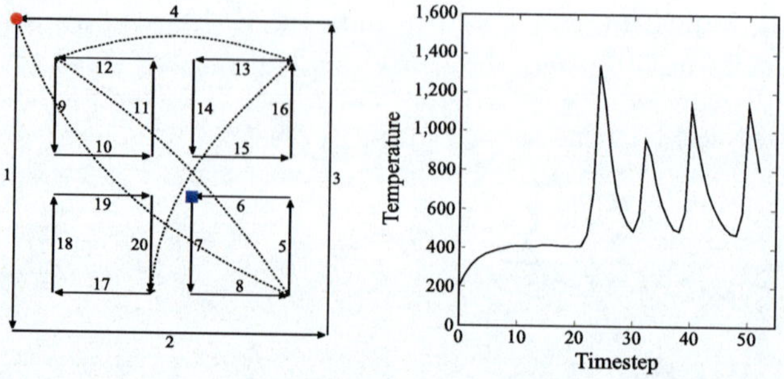

Fig. 2. Optimal welding path for the considered layer *(left)* and temperature progression of node 12 *(right)*. The weld trajectory is given by the numbers next to each segment. Dashed lines describe deadheading moves, the starting point of the path is marked by a red circle, and node 12 by a blue square

material. The weld source reaches this node at time step 24, leading to a first peak in its temperature progression. With this high temperature, the effect of heat conduction is outperformed by the heat loss due to radiation. Later, there is a significant temperature gain every time the node is within the area of effect of the weld source.

In our future work, we formulate new objective functions providing better solutions of the LP relaxation during the solution process. Furthermore, we consider workpieces where the wall strength of every segment can be arbitrary and extend the problem to several consecutive layers, where the chosen paths for each layer should differ to increase the stability of the workpiece and avoid joints.

References

1. Bähr, M., Buhl, J., Radow, G., Schmidt, J., Bambach, M., Breuß, M., Fügenschuh, A.: Stable honeycomb structures and temperature based trajectory optimization for wire-arc additive manufacturing. Optim. Eng. **22**(2), 913–974 (2020). https://doi.org/10.1007/s11081-020-09552-5
2. Edmonds, J., Johnson, E.L.: Matching, Euler tours and the Chinese postman. Math. Program. **5**, 88–124 (1973)
3. Jiang, J., Ma, Y.: Path planning strategies to optimize accuracy, quality, build time and material use in additive manufacturing: a review. Micromachines **11**(7), 633 (2020)
4. Montevecchi, F., Venturini, G., Grossi, N., Scippa, A., Campatelli, G.: Finite element mesh coarsening for effective distortion prediction in wire arc additive manufacturing. Addit. Manuf. **18**, 145–155 (2017)
5. Nguyen, L., Buhl, J., Bambach, M.: Continuous eulerian tool path strategies for wire-arc additive manufacturing of rib-web structures with machine-learning-based adaptive void filling. Addit. Manuf. **35**, 101, 265 (2020)
6. Taler, J., Ocłoń, P.: Finite element method in steady-state and transient heat conduction. Encycl. Thermal Stresses **4**, 1604–1633 (2014)

Optimization of Wear Related Material Costs of a Hydrostatic Transmission System via MINLP

Lena Charlotte Altherr[1], Philipp Leise[2], Marc Emanuel Pfetsch[3], and Andreas Schmitt[3](✉)

[1] Department of Electrical Engineering and Information Technology, Aachen University of Applied Sciences, Aachen, Germany
altherr@fh-aachen.de

[2] Department of Mechanical Engineering, TU Darmstadt, Darmstadt, Germany
philipp.leise@fst.tu-darmstadt.de

[3] Department of Mathematics, TU Darmstadt, Darmstadt, Germany
{pfetsch,aschmitt}@mathematik.tu-darmstadt.de

Abstract. This contribution presents a method to find an optimal topology and control for a hydrostatic transmission system that is equipped with a hydraulic accumulator. The goal is to minimize wear in the system while fulfilling a predefined load cycle given by speed and force requirements during the retraction and extension of a piston. The degrees of freedom of the design are the selection and the connection of valves with the system's piston, pressure source and tank as well as the sizing of the accumulator. We derive a mixed-integer nonlinear program, which contains continuous variables for the quasi-stationary flow, pressure and valve conditions, as well as binary variables to include selection decisions and valve circuits. Pressure and wear conditions are modeled by nonconvex nonlinear functions. To solve the problem, we use a reformulation which approximates the valve wear by a quadratic polynomial depending on volume flow and pressure difference and use a technique based on perspective cuts. Our optimization results show that the inclusion of the accumulator reduces the wear related material costs by one third.

Keywords: MINLP · Perspective cuts · Engineering optimization

1 Introduction

Hydraulic systems are required in a multitude of technical systems, e.g., in construction and agricultural machinery. An important factor influencing the availability of such systems is component wear. This motivates the search for systematic design methods within the engineering design process that lead to a reduced component wear. We present a system synthesis approach of a hydrostatic transmission system equipped with a hydraulic accumulator using a mixed-integer nonlinear program (MINLP). In more detail, we optimize the connection of valves, the sizing of an accumulator and the control of the valves such that

N. Trautmann and M. Gnägi (Eds.): OR 2021, LNOR, pp. 331–337, 2022.
https://doi.org/10.1007/978-3-031-08623-6_49

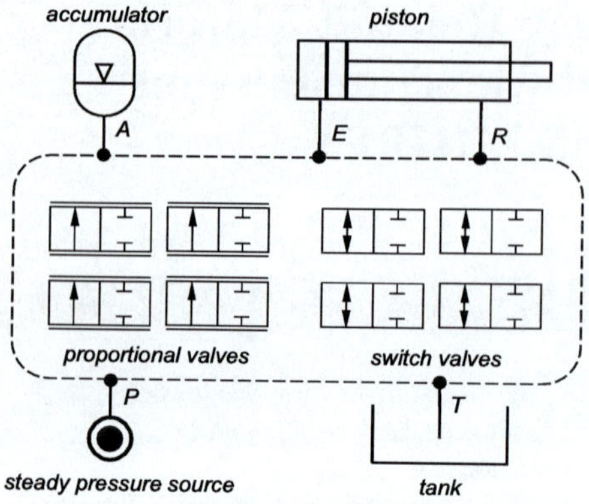

(a) Construction kit of the design problem to find an optimal connection of proportional and switch valves.

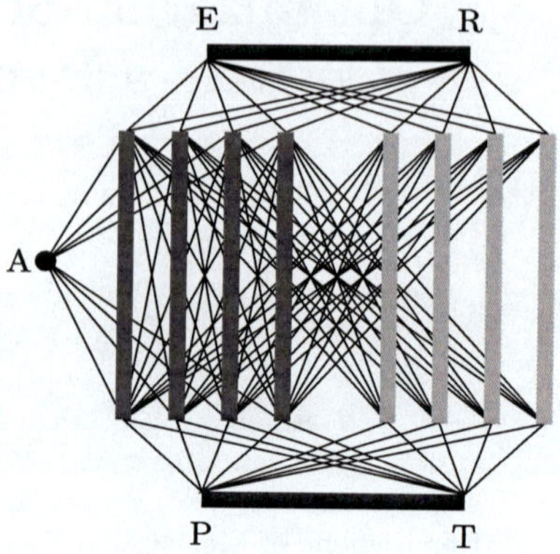

(b) Graph of the problem. Dark and light gray bars depict potential proportional and switch valves, respectively.

Fig. 1. Illustration of the problem.

a given movement pattern of a piston can be realized under minimal material wear, see Fig. 1a for a schematic depiction of the technical system.

Our model is an extension of the work [1], which considers this system without an accumulator. The presented system design approach can also be used to derive optimized digital hydraulic [6] systems. The integration of an accumulator has multiple benefits for the system: for instance, as shown by [8, p. 249], the volume flow demand can still be satisfied while having fluctuating requirements and pressure peaks can be reduced.

In previous work [1], the complex nonlinear physics and wear constraints necessitated the usage of linearization. In contrast, here we propose a reformulation of these nonlinearities, which allows the solving of a more precise MINLP-formulation. In the following we first introduce an optimization model for the problem. Afterwards we present a reformulation and apply perspective cuts. We then conclude with a presentation of the optimal solution for some test data.

2 Optimization Model

The function of the system is given by a load cycle $\mathcal{L} := \{\text{in}, \text{out}\}$ which controls the velocity $v^{\text{in}}/v^{\text{out}}$, force $F^{\text{in}}/F^{\text{out}}$ and time $t^{\text{in}}/t^{\text{out}}$ of retracting and extending the piston given by the points E and R in Fig. 1a. The needed pressure difference for this movement can be generated by connecting these two points with the pressure source P and/or the accumulator A via switch and proportional valves as well as by the adjustment of the proportional valves' lift.

The possible designs of the hydrostatic transmission system are represented by the directed graph $\mathcal{G} = (\mathcal{V}, \mathcal{A})$ depicted in Fig. 1b. The nodes $\mathcal{V}$ include the nodes for the extension (E) and the retraction (R) input of the piston, the accumulator (A), the pump (P) and the tank (T). Furthermore, there are two nodes for each of the eight valves. The set of arcs $\mathcal{A}$ contains an arc for each valve, where the arcs for proportional and switch valves are collected in the subsets $\mathcal{A}^{\mathrm{p}}$ and $\mathcal{A}^{\mathrm{s}}$, respectively. There are two arcs for the piston and the connection of the pressure source and the tank, respectively. Lastly, there are connections between all the components with some restrictions, e.g., the accumulator can only be connected to a proportional valve and switch valves can not be connected in series. These arcs are collected in the set $\mathcal{A}^{\mathrm{c}} \subset \mathcal{A}$.

The MINLP uses the following variables: Binary variables x_a for $a \in \mathcal{A}$ signify whether an arc/connection is used in the solution. The binary variables y_a^ℓ specify whether a valve on the arc $a \in \mathcal{A}^{\mathrm{p}} \cup \mathcal{A}^{\mathrm{s}}$ is open or closed in load case $\ell \in \mathcal{L}$. To model quasi-stationary physical conditions for each load case $\ell \in \mathcal{L}$, the volume flow on each arc $a \in \mathcal{A}$ is given by q_a^ℓ. Furthermore, the pressure at a given node $v \in \mathcal{V}$ is given by p_v^ℓ. For each proportional valve $a \in \mathcal{A}^{\mathrm{p}}$ we have its valve lift u_a^ℓ, its accumulated wear w_a^ℓ and the pressure loss Δp_a^ℓ. Lastly, we model the hydraulic accumulator using preloading pressure p^0 and volume V^0 as well as the fluid volume V^ℓ.

$$
\begin{aligned}
\min \quad & \sum_{a \in \mathcal{A}^{\mathrm{s}}} C^{\mathrm{s}} x_a + \sum_{a \in \mathcal{A}^{\mathrm{p}}} C^{\mathrm{p}} x_a + \sum_{a \in \mathcal{A}^{\mathrm{p}}} C^{\mathrm{p}} \left\lfloor N / \lfloor \overline{w} / (w_a^{\mathrm{in}} + w_a^{\mathrm{out}}) \rfloor \right\rfloor && && \text{(1a)}\\
\text{s.t.} \quad & \sum_{a \in \delta^-(v)} q_a^\ell - \sum_{a \in \delta^+(v)} q_a^\ell = 0, && v \in \mathcal{V} \setminus \{T, A\}, \ell \in \mathcal{L}, && \text{(1b)}\\
& q_a^\ell \, (1 - x_a) = 0, && a \in \mathcal{A}^{\mathrm{c}}, \ell \in \mathcal{L}, && \text{(1c)}\\
& q_a^\ell \, (1 - y_a^\ell) = 0, && a \in \mathcal{A}^{\mathrm{p}} \cup \mathcal{A}^{\mathrm{s}}, \ell \in \mathcal{L}, && \text{(1d)}\\
& (p_u^\ell - p_v^\ell) \, x_a = 0, && a = (u, v) \in \mathcal{A}^{\mathrm{c}}, \ell \in \mathcal{L}, && \text{(1e)}\\
& (p_u^\ell - p_v^\ell) \, y_a^\ell = 0, && a = (u, v) \in \mathcal{A}^{\mathrm{s}}, \ell \in \mathcal{L}, && \text{(1f)}\\
& (p_u^\ell - p_v^\ell) \, y_a^\ell = \Delta p_a^\ell, && a = (u, v) \in \mathcal{A}^{\mathrm{p}}, \ell \in \mathcal{L}, && \text{(1g)}\\
& \rho \, q_a^\ell |q_a^\ell| = 2 \Delta p_a^\ell \, (\zeta \, d \, u_a^\ell)^2, && a \in \mathcal{A}^{\mathrm{p}}, \ell \in \mathcal{L}, && \text{(1h)}\\
& w_a^\ell \geq K(u_a^\ell, q_a^\ell), && a \in \mathcal{A}^{\mathrm{p}}, \ell \in \mathcal{L}, && \text{(1i)}\\
& w_a^{\mathrm{in}} + w_a^{\mathrm{out}} \leq \overline{w}, && a \in \mathcal{A}^{\mathrm{p}}, \ell \in \mathcal{L}, && \text{(1j)}\\
& \underline{u} \, y_a^\ell \leq u_a^\ell \leq y_a^\ell, && a \in \mathcal{A}^{\mathrm{p}}, \ell \in \mathcal{L}, && \text{(1k)}\\
& 0 \leq V^\ell \leq V^0, && \ell \in \mathcal{L}, && \text{(1l)}\\
& V^\ell = V^{\ell-1} + \sum_{a \in \delta^-(v)} q_a^{\ell-1} - \sum_{a \in \delta^+(v)} q_a^{\ell-1}, && \ell \in \mathcal{L}, && \text{(1m)}\\
& p_{A,\ell} = p^0 \left(\frac{V^0}{V^\ell} \right)^n, && \ell \in \mathcal{L}, && \text{(1n)}\\
& (p_E^\ell - p_R^\ell) = \frac{F^\ell}{A^{\mathrm{pist}}}, \; q_{(E,R)}^\ell = v^\ell A^{\mathrm{pist}}, && \ell \in \mathcal{L}, && \text{(1o)}\\
& p_T^\ell = 1, \; p_P^\ell = \Delta P + 1, && \ell \in \mathcal{L}, && \text{(1p)}\\
& q \in \mathbb{R}^{\mathcal{A} \times \mathcal{L}}, \; p \in \mathbb{R}^{\mathcal{V} \times \mathcal{L}}, \; \Delta p, w, u \in \mathbb{R}^{\mathcal{A}^{\mathrm{p}} \times \mathcal{L}}, \\
& p^0, V^0, V^{\mathrm{in}}, V^{\mathrm{out}} \in \mathbb{R}_+, \; x \in \{0,1\}^{\mathcal{A}}, \; y \in \{0,1\}^{(\mathcal{A}^{\mathrm{p}} \cup \mathcal{A}^{\mathrm{s}}) \times \mathcal{L}}.
\end{aligned}
$$

The whole model is given by Eqs. (1a)–(1p). The objective is to minimize the material costs given by the usage of the valves, weighted by $C^{\mathrm{s}}/C^{\mathrm{p}}$ for the respective type and the replacement of worn out proportional valves. We use N maintenance intervals in which valves can be replaced. A valve is worn out, if the wear exceeds the upper wear bound $\overline{w}$. The number of replacements is given by $\lfloor \overline{w}/(w_a^{\mathrm{in}}+w_a^{\mathrm{out}}) \rfloor$. The accumulator is not considered within the objective. Constraints (1b)–(1g) enforce volume flow balance and pressure propagation between built/active components. Here, $\delta^-(v)$ and $\delta^+(v)$ denote the incoming and outgoing arcs of node v, respectively. Pressure loss only occurs for proportional valves and is approximated by Constraints (1h) following [7]. Parameters are the diameter d, the oil density ρ and the pressure loss coefficient ζ. The wear of a used proportional valve is modeled by (1i) and (1j). Here, $K(u_a^\ell, q_a^\ell)$ is a nonlinear function depending on the lift and the volume flow of the considered valve and parameters like the maximum valve lift, the movement time and further aspects. This wear model was derived by a dimensional analysis using experimental data, see [7] and is depicted in Fig. 2a. We also bound the valve lift in (1k). Constraints (1l)–(1n) cover boundary conditions, volume flow balance and the pressure of the accumulator. The expression $\ell - 1$ is an abbreviation for the opposite load case of ℓ. The final constraints give boundary conditions on several nodes and arcs in the graph. This includes the necessary pressure difference and volume flow for the piston movement (A^{pist} is the area of the piston) as well as the ambient pressure of 1 bar at the tank and the constant pressure increase due to the pump of ΔP.

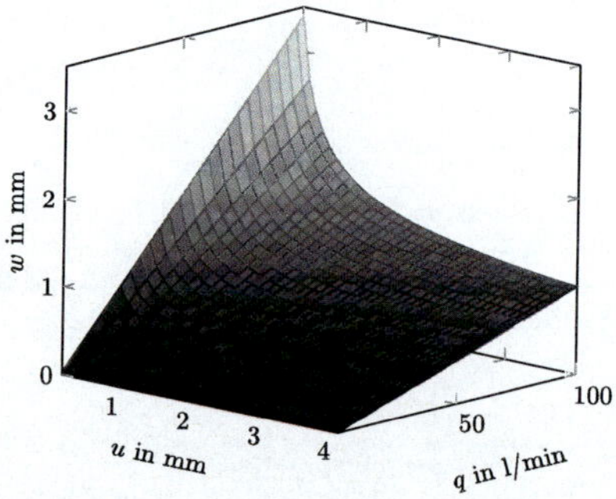

(a) According to the function K and [7].

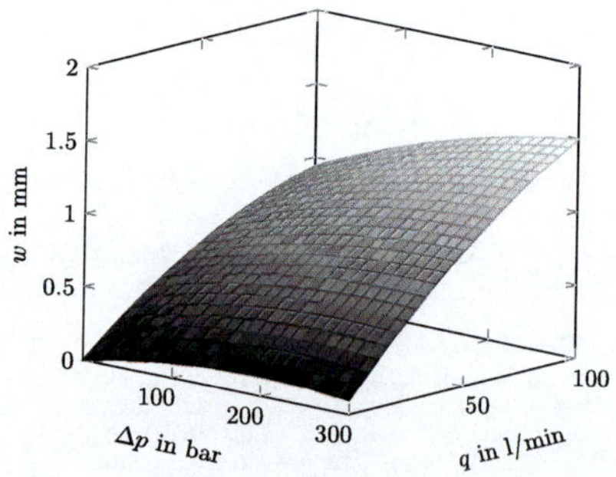

(b) Using the quadratic polynomial f.

Fig. 2. Approximation of the wear w in terms of the valve lift u and volume flow q or pressure difference Δp and volume flow q.

3 Solution Approach

To optimize the above MINLP using the solver SCIP [5], we use a reformulation and separate tight valid inequalities based on perspective cuts, see [4].

We first reformulate the wear part in the objective function, which forms a piecewise linear function with arguments $w_a^{\text{in}} + w_a^{\text{out}}$, as an aggregation of binary variables. Furthermore, we compute bounds on the volume flow and pressure variables to reformulate the bilinear constraints (1c)–(1g) with binary variables as big-M constraints. To simplify the absolute value in Constraint (1h) we split the volume flow and pressure increase variables into positive and negative parts and add binary activation variables.

Preliminary tests showed computational difficulties with the optimization of the interaction between volume flow, pressure difference, lift and wear of the proportional valves. To handle this, we neglect the valve lift variable and approximate the wear of a valve only in terms of volume flow and pressure difference using a two-dimensional quadratic polynomial f. This function is depicted in Fig. 2b. Thus, for nonnegative q_a^ℓ and Δp_a^ℓ we replace Constraints (1h)–(1k) by the system

$$2\Delta p_a^\ell\, (\zeta\, d\, \underline{u})^2 \le \rho\, (q_a^\ell)^2 \le 2\Delta p_a^\ell\, (\zeta\, d)^2, \qquad a \in \mathcal{A}^{\mathrm{P}}, \ell \in \mathcal{L}, \tag{2a}$$

$$w_a^\ell \ge f(q_a^\ell, \Delta p_a^\ell), \qquad a \in \mathcal{A}^{\mathrm{P}}, \ell \in \mathcal{L}. \tag{2b}$$

To further speed up the solution process, an adaptation of perspective cuts is used. These cuts use the model structure that a binary variable switches a continuous variable on/off and that there exists a convex nonlinear relationship between the continuous variables:

$$\{(\alpha, \beta, \gamma) \in \{0,1\} \times \mathbb{R}^n \times \mathbb{R} \,:\, \gamma \ge f(\beta), \underline{\beta}\,\alpha \le \beta \le \overline{\beta}\,\alpha\}.$$

This structure is also given in our problem when linking wear (γ) with volume flow and pressure loss (β) depending on activation of the valve (α). The quadratic approximation f is non-convex. However, the fixed convexity behavior of quadratic functions makes it possible to generate linear underestimators $f(\beta) \ge a^\top \beta + b$ following [2]. In [3] it is shown, that the inequalities $\gamma \ge a^\top \beta + b\, \alpha$ are valid for the above set. We dynamically separate these valid cutting planes within our optimization algorithm and significantly reduce the solution time.

4 Example Design

Exemplified results for system designs with and without an accumulator and with different valves and connections to ensure a given load-scenario are presented in Fig. 3. The pictures show the valve configurations for the extension phase. The retraction scenario is obtained by opening closed valves and closing open valves. Here, proportional valves are considered to be twice as expensive as switching valves. The load cases are determined by $v^{\text{in}} = 0.15\,\text{m/s}$, $F^{\text{in}} = 20\,\text{kN}$, $t^{\text{in}} = 20\,\text{s}$, $v^{\text{out}} = 0.6\,\text{m/s}$, $F^{\text{out}} = 5\,\text{kN}$, and $t^{\text{out}} = 5\,\text{s}$.

The use of the hydraulic accumulator reduces the wear and thus the material costs based on the underlying model assumptions by one third. The optimized solution configures the accumulator in such a way that it can handle the retraction cycle without the pump.

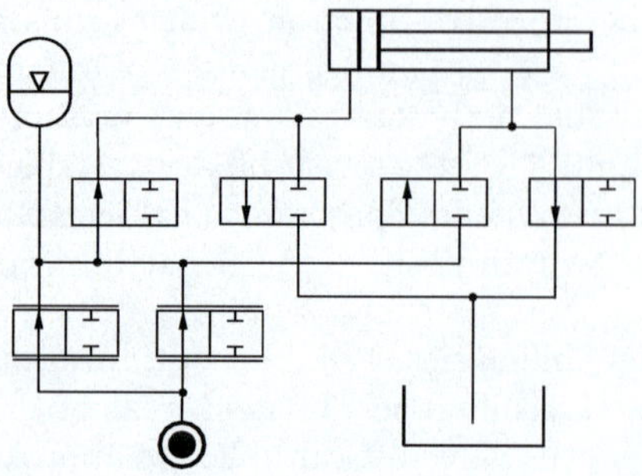
(a) With accumulator.

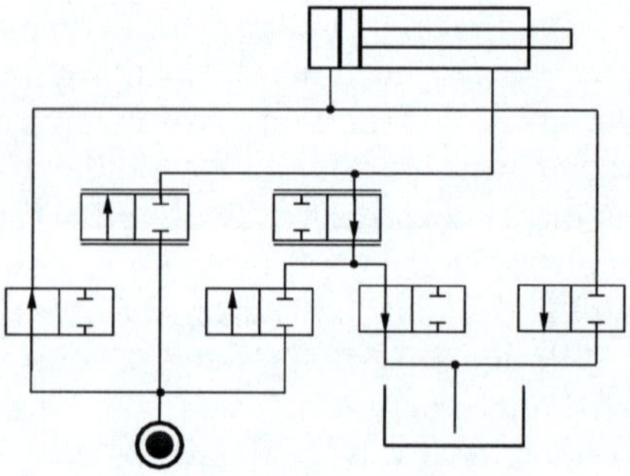
(b) Without accumulator.

Fig. 3. Optimal solution configuration for the extension scenario.

5 Conclusion and Outlook

We presented a more accurate nonlinear model of a hydrostatic transmission system, which we solved using refined solution strategies. Further research could focus on the combinatorial structure of the problem in order to solve models involving more valves. Furthermore, the investigation of a cost model for the accumulator sizing would be interesting.

Funding. This research was funded by Deutsche Forschungsgemeinschaft (DFG, German Research Foundation) – project A9 within SFB 805 (Project Number 57157498).

References

1. Altherr, L.C., Ederer, T., Farnetane, L.S., Pöttgen, P., Vergé, A., Pelz, P.F.: Multicriterial design of a hydrostatic transmission system via mixed-integer programming. In: Doerner, K.F., Ljubic, I., Pflug, G., Tragler, G. (eds.) Operations Research Proceedings 2015. ORP, pp. 301–307. Springer, Cham (2017). https://doi.org/10.1007/978-3-319-42902-1_41
2. Ballerstein, M.: Convex relaxations for mixed-integer nonlinear programs. Ph.D. thesis, ETH Zürich (2013)
3. Bestuzheva, K., Gleixner, A., Vigerske, S.: A computational study of perspective cuts. arXiv preprint arXiv:2103.09573 (2021)
4. Frangioni, A., Gentile, C.: Perspective cuts for a class of convex 0–1 mixed integer programs. Math. Program. **106**(2), 225–236 (2006). https://doi.org/10.1007/s10107-005-0594-3
5. Gamrath, G., et al.: The SCIP optimization suite 7.0. Technical report, Optimization Online, March 2020
6. Pan, M., Plummer, A.: Digital switched hydraulics. Front. Mech. Eng. **13**(2), 225–231 (2018). https://doi.org/10.1007/s11465-018-0509-7

7. Vergé, A., Pöttgen, P., Altherr, L.C., Ederer, T., Pelz, P.F.: Lebensdauer als Optimierungsziel – Algorithmische Struktursynthese am Beispiel eines hydrostatischen Getriebes. O+P Ölhydraulik und Pneumatik **60**(1–2), 114–121 (2016)
8. Will, D., Gebhardt, N., Ströhl, H.: Hydraulik - Grundlagen, Komponenten, Schaltungen, 3rd edn. Springer, Heidelberg (2007). https://doi.org/10.1007/978-3-540-79535-3

Project Management and Scheduling

Check for updates

Scheduling Projects with Converging and Diverging Material Flows Using IBM ILOG CP Optimizer—An Experimental Performance Analysis

Marco Gehring(✉), Rebekka Volk, Niklas Braun, and Frank Schultmann

Karlsruhe Institute of Technology (KIT), Institute for Industrial Production (IIP), Hertzstr. 16, 76187 Karlsruhe, Germany
Marco.Gehring@kit.edu

Abstract. We study the problem of scheduling construction or deconstruction projects subject to temporal and renewable resource constraints to minimize the makespan. Moreover, we take into account material flows between activities of the project. Material flows can cause delays due to the limited capacity of intermediate storage facilities. In construction projects, material flows follow a convergent structure as materials are supplied, pre-treated, and mounted throughout the project. In deconstruction projects, materials are released, treated and removed from the site, resulting in a divergent structure. An experimental performance analysis with IBM ILOG CP Optimizer on generated test instances reveals that solving the problem with converging material flows requires a multiple of the computation time for solving the problem with diverging material flows.

Keywords: Project scheduling · Material flows · Storage facilities

1 Introduction

There is a wide range of possible applications for the extensively studied *Resource-Constrained Project Scheduling Problem (RCPSP)*. The RCPSP deals with determining the start times for project activities so that temporal relations between activities and scarce renewable resources are taken into account, and the makespan is minimized. Our work focuses on applications involving material flows between activities, which can impose significant restrictions on the project execution due to storage or processing bottlenecks. Construction and deconstruction projects are typical examples of such applications. In construction projects, the material processing (e.g., supply, pre-treatment, pre-assembly, on-site transport) occurs upstream of the assembly activities. In deconstruction projects, however, the material processing (e.g., segmentation, conditioning, removal from site) occurs downstream of the disassembly activities. In both cases, the storage space on the (de-)construction site is often limited, e.g., if located in congested

N. Trautmann and M. Gnägi (Eds.): OR 2021, LNOR, pp. 341–346, 2022.
https://doi.org/10.1007/978-3-031-08623-6_50

urban areas. The problem of scheduling a project with material flows at minimal makespan can be considered as an extension of the RCPSP. It consists of determining the start times for project activities subject to temporal relations, scarce renewable resources and scarce storage resources.

Most of the scheduling literature on material flows and storage facilities addresses production planning settings where materials are usually regarded as a homogeneous resource that any activity can consume as soon as it is available (e.g., [2–4]). In contrast, material flows in (de-)construction projects implicitly prescribe a temporal order on the activities involved. For example, consider a case where two different walls need to be deconstructed, but only one wall requires special treatment due to a contamination. For modeling this case correctly, it is imperative to maintain the link between the respective dismantling activity and the material flow (i.e., the residual materials from the wall) released by it. This can be achieved by introducing temporal relations with minimum time lags between the dismantling activity and the subsequent activities responsible for processing the material flow. An overlap is allowed as long as no activity must be interrupted to wait for its materials. The case of constructing two different walls, where one wall requires special pre-treatment, is modeled analogously. However, the structure of the temporal relations differs depending on whether the wall is constructed or deconstructed. In construction (deconstruction) projects where material flows are located upstream (downstream), these temporal relations typically form a converging (diverging) structure. For convenience, we speak of *converging (diverging) material flows*. However, it is essential to note that we observe the material flows from a temporal perspective. Regardless of whether we refer to converging or diverging material flows, the physical material units can always flow together or apart when traversing storage facilities.

In project scheduling, there exist several well-known parameters which have an effect on the solvability of instances, such as the resource strength or the network complexity. When evaluating solution procedures, the observed performance measures are usually presented specifically for these instance parameters. In this work, we examine whether the material flow structure should be considered as an additional instance parameter. For this purpose, we experimentally investigate how the off-the-shelf constraint programming solver IBM ILOG CP Optimizer in IBM ILOG CPLEX Optimization Studio 12.9.0 performs at solving generated test instances with converging or diverging material flows. Our findings will lead to a better understanding and assessment of solution procedures for scheduling problems with storage resources.

2 Problem Statement

We consider a project composed of $i = 0, \ldots, n+1$ activities with durations $d_0, \ldots, d_{n+1}$. Activities 0 and $n+1$ are fictitious (i.e., $d_0 = d_{n+1} := 0$) and represent the start and the end of the project, respectively. The project scheduling problem (P) consists in determining a vector of start times $S := (S_0, \ldots, S_{n+1})$, briefly referred to as *schedule*, so that the project makespan S_{n+1} is minimized and constraints (3) to (7) are satisfied:

$$
\begin{array}{llll}
\text{(P)} & & & (1)\\
\min & S_{n+1} & & (2)\\
\text{s. t.} & S_j \geq S_i + d_{ij}^{min} & ((i,j) \in E); & (3)\\
& \sum_{i \in \mathcal{A}^{\rho}(S,t)} r_{ik}^{\rho} \leq R_k^{\rho} & (k \in \mathcal{R}^{\rho}, t \geq 0); & (4)\\
& 0 \leq \sum_{i \in \mathcal{A}_k^{\gamma+}(S,t)} r_{ik}^{\gamma} + \sum_{i \in \mathcal{A}_k^{\gamma-}(S,t)} r_{ik}^{\gamma} \leq R_k^{\gamma} & (k \in \mathcal{R}^{\gamma}, t \geq 0); & (5)\\
& S_0 = 0; & & (6)\\
& S_i \geq 0 & (i = 1, \ldots, n+1). & (7)
\end{array}
$$

Constraints (3) are *temporal constraints*, which are prescribed by a set of temporal relations $E \subset \{0, \ldots, n+1\}^2$ and a matrix of minimum time lags $(d_{ij}^{min})_{i,j=0,\ldots,n+1}$. If for two activities i, j, there exists a temporal relation $(i, j) \in E$, activity j must not start earlier than $d_{ij}^{min} \in \mathbb{Z}_{\geq 0}$ periods after the start of activity i. Note that, for simplification, we restrict ourselves to minimum time lags. The temporal constraints could be extended by maximum time lags straight-forwardly. Constraints (4) are *renewable resource constraints*, which are prescribed by a set of renewable resources $\mathcal{R}^{\rho}$, a maximum availability $R_k^{\rho} \in \mathbb{Z}_{\geq 0}$ for each renewable resource $k \in \mathcal{R}^{\rho}$ and a required amount $r_{ik}^{\rho} \in \mathbb{Z}_{\geq 0}$ of each renewable resource $k \in \mathcal{R}^{\rho}$ by each activity i. The active set $\mathcal{A}^{\rho}(S,t) := \{i \mid S_i \leq t < S_i + d_i\}$ comprises all activities executed at a time t, given a schedule S. Constraints (5) are *storage constraints* (cf. [2–4]), which are prescribed by a set of storage resources $\mathcal{R}^{\gamma}$, a maximum inventory $R_k^{\gamma} \in \mathbb{Z}_{\geq 0}$ for each storage resource $k \in \mathcal{R}^{\gamma}$ and a required amount $r_{ik}^{\gamma} \in \mathbb{Z}$ of each storage resource $k \in \mathcal{R}^{\gamma}$ by each activity i. If $r_{ik}^{\gamma} > 0$ ($r_{ik}^{\gamma} < 0$), activity i *replenishes* (*depletes*) r_{ik}^{γ} material units into (from) storage resource k. We assume that replenishments (depletions) take place at the start (end) of each activity, blocking the required upstream and downstream storage space during its entire execution time. Hence, the active set comprising all activities that have replenished (depleted) material into (from) k until a time t is defined as $\mathcal{A}_k^{\gamma+}(S,t) := \{i \mid r_{ik}^{\gamma} > 0 \wedge S_i \leq t\}$ ($\mathcal{A}_k^{\gamma-}(S,t) := \{i \mid r_{ik}^{\gamma} < 0 \wedge S_i + d_i \leq t\}$).

To describe material flows, we establish the following definitions:
An activity i is said to be *replenishing* (*depleting*) if and only if there exists at least one storage resource $k \in \mathcal{R}^{\gamma}$ with $r_{ik}^{\gamma} > 0$ ($r_{ik}^{\gamma} < 0$).
The material flows are said to be *converging* (*diverging*) if and only if for each replenishing (depleting) activity i there exists exactly one activity $j \neq i$ with

1. $(i, j) \in E$ ($(j, i) \in E$), i.e., activity i has exactly one successor (predecessor) j, and
2. $r_{jk}^{\gamma} + \sum_{(lj) \in E} r_{lk}^{\gamma} = 0$ ($r_{jk}^{\gamma} + \sum_{(jl) \in E} r_{lk}^{\gamma} = 0$) for each $k \in \mathcal{R}^{\gamma}$ with $r_{jk}^{\gamma} < 0$ ($r_{jk}^{\gamma} > 0$), i.e., each replenished material unit is also depleted and vice versa.

Hence, in-trees (out-trees) in the activity-on-node network associated with an instance of (P) indicate that we are dealing with converging (diverging) material flows. On the left side of Fig. 1, an exemplary activity-on-node network for an instance with converging material flows is depicted. We can mirror this network to obtain an instance with diverging material flows, as shown on the right side.

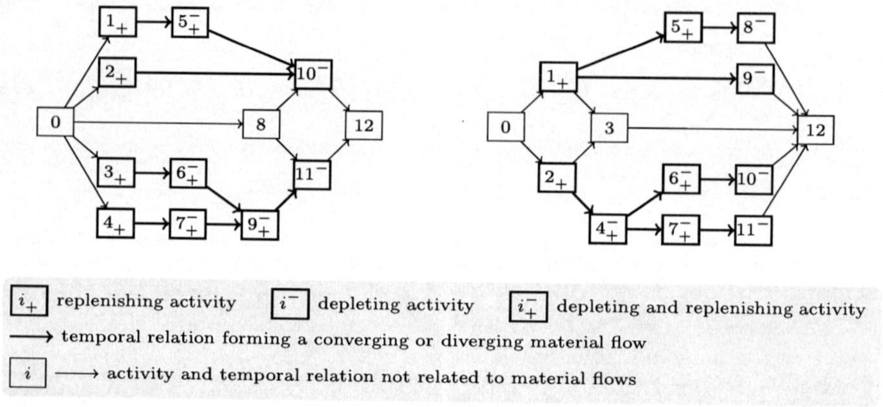

Fig. 1. Exemplary activity-on-node network for an instance with converging (left) and an instance with diverging (right) material flows

3 Structural Properties

As a generalization of the RCPSP, solving (P) to optimality is NP-hard. Schwindt and Trautmann [4] propose the following branch-and-bound algorithm: Initially, a polynomial longest path algorithm solves the resource relaxation of (P) (i.e., relaxation of constraints (4) and (5)). If the maximum inventory $\overline{R}_k^{\gamma}$ of a storage resource $k \in \mathcal{R}^{\gamma}$ is exceeded at a time t, replenishing activities are postponed sufficiently so that the inventory falls below $\overline{R}_k^{\gamma}$. For each possible combination of replenishing activities, a new enumeration node is created.

Imagine an inventory excess at a time t before the end of activity 9 from Fig. 1. Let us focus on the material flow processed in activity 9 and resolve this inventory excess by postponing preceding activities of activity 9. In case of diverging material flows, each depleting activity has exactly one predecessor. Thus, we postpone preceding activity 1 to resolve our inventory excess (cf. right network in Fig. 1). In case of converging material flows, each depleting activity can have multiple predecessors. Thus, we can postpone preceding activity 6 or 7 or both to resolve our inventory excess (cf. left network in Fig. 1). Due to this structural difference, we expect more enumeration nodes with converging material flows. Therefore, we assume that solving instances with converging material flows requires more computational effort than instances with diverging material flows. In the following section, we experimentelly verify this assumption using IBM ILOG CP Optimizer. We chose CP Optimizer since it is easily accessible and achieved good results on similar problems.

4 Experimental Performance Analysis

4.1 Instance Generation

To generate test instances of (P), we use instances with $n_{\text{PSPLIB}} = \{30, 120\}$ activities from the PSPLIB [1] as a starting point. We randomly select a portion $PREL \in \{0.25, 1\}$ of all PSPLIB-activities and associate these activities with material flows. To this end, we randomly simulate $NREL \in \{50, 200\}$ directed material flow paths through a flow network of five storage resources and eight processing steps for each selected activity. Each material flow path describes the material flow of one material unit. Next, we aggregate the material flow paths for each selected activity so that the respective material units are described by a common path as long as they pass through the same processing steps and storage facilities. The aggregation results in a diverging material flow, where we derive an activity for each of its processing steps. We link the activities by temporal relations according to the structure of the simulated material flow. Since the processing of materials may require renewable resources, we define a resource factor $RF \in \{0, 0.5\}$ which prescribes the average portion of PSPLIB-resources for which the PSPLIB-activities and the activities derived from processing steps compete. We set the maximum inventory $R_k^\gamma := INV \in \{200, 1000\}$ for each $k \in \mathcal{R}^\gamma$. For each possible combination $(n_{\text{PSPLIB}}, PREL, NREL, RF, INV)$ we generate three instances with varying durations of the activities derived from processing steps. Finally, we mirror the material flow structure of each generated instance so that we obtain a corresponding instance with converging material flows (such as depicted in Fig. 1). In total, we get $2^5 \cdot 3 = 96$ instance pairs, where each instance comprises 106 to 2348 activities. Instances of one pair have the same characteristics apart from the direction of the material flows.

4.2 Results and Evaluation

We modeled (P) using CP Optimizer's "interval variables" for representing activities and "cumul functions" for representing renewable resource and storage requirements. We carried out all experiments on an AMD Ryzen 9 (4.0 GHz, 12 cores) with 128 GB RAM using the automatic search and the default settings. Table 1 shows the results for the different instance generation parameter levels.

The percentages and the average computation times confirm our assumption that solving (P) requires more computational effort with converging material flows than diverging material flows. For ten instances with converging flows, CP Optimizer could not find a feasible solution within a time limit of one hour per instance. These are mainly the instances with $INV = 200$ and $NREL = 200$, i.e., with tight storage constraints. CP Optimizer could not prove infeasibility. In contrast, CP Optimizer found a feasible solution for all instances with diverging material flows within less than one minute per instance. Concerning the computation times, solving instances with converging material flows requires a multiple of the time for solving instances with diverging material flows. Parameters n_{PSPLIB} and $PREL$ have the strongest impact on the computation times.

Table 1. Percentage of feasible/optimal solutions and average computation times until a feasible/optimal solution was found for converging (con.) and diverging (div.) material flows within a time limit of one hour per instance

Parameter	Level	Feasible (%)		Optimal (%)		Feasible (s)[a]		Optimal (s)[a]	
		con.	div.	con.	div.	con.	div.	con.	div.
Overall	-	89.58	100.00	86.46	96.88	25.16	4.12	128.19	27.69
n_{PSPLIB}	30	95.83	100.00	91.67	95.83	4.23	0.40	49.72	6.97
	120	83.33	100.00	81.25	97.92	49.23	8.41	214.71	50.53
PREL	0.25	95.83	100.00	93.75	100.00	4.96	1.13	22.17	11.07
	1	83.33	100.00	79.17	93.75	48.38	7.56	257.14	47.90
NREL	50	100.00	100.00	93.75	95.83	23.18	3.83	71.10	41.73
	200	79.17	100.00	79.17	97.92	27.66	4.50	194.30	11.42
RF	0	89.58	100.00	89.58	100.00	23.84	3.76	83.81	5.41
	0.5	89.58	100.00	83.33	93.75	26.47	4.49	177.12	52.25
INV	200	79.17	100.00	77.08	93.75	21.77	4.30	137.89	42.06
	1000	100.00	100.00	95.83	100.00	27.84	3.99	120.60	16.44

[a] Values averaged over the levels of all parameters except the one under consideration; instance pairs were excluded if no feasible/optimal solution was found for one of both instances.

5 Conclusion

We dealt with an extension of the RCPSP, which takes into account material flows between activities and constraints on storage resources. Solving generated instances with CP Optimizer showed that the computational effort significantly depends on whether we deal with converging or diverging material flows. Although the compared instances have the same characteristics apart from the direction of the material flows, optimizing instances with converging flows required 128 s on average compared to 28 s for optimizing instances with diverging flows. This suggests that the structure of material flows should be considered when designing and evaluating problem-specific models and solution procedures. We suspect that CP Optimizer performs faster with diverging material flows since this structure results in a smaller number of enumeration nodes.

References

1. Kolisch, R., Sprecher, A.: PSPLIB - a project scheduling problem library. Eur. J. Oper. Res. **96**, 205–216 (1996)
2. Neumann, K., Schwindt, C.: Project scheduling with inventory constraints. Math. Methods Oper. Res. **56**, 513–533 (2002)
3. Neumann, K., Schwindt, C., Trautmann, N.: Scheduling of continuous and discontinuous material flows with intermediate storage restrictions. Eur. J. Oper. Res. **165**, 495–509 (2005)
4. Schwindt, C., Trautmann, N.: Batch scheduling in process industries: an application of resource-constrained project scheduling. OR Spectr. **22**, 501–524 (2000)

Scheduling Heating Tasks on Parallel Furnaces with Setup Times and Conflicts

Julia Lange[1(✉)], Philipp Fath[2], and David Sayah[2]

[1] Chair of Logistics, TU Kaiserslautern, 67663 Kaiserslautern, Germany
julia.lange@wiwi.uni-kl.de

[2] FZI Research Center for Information Technology, 76131 Karlsruhe, Germany
{fath,sayah}@fzi.de

Abstract. Heating processes constitute costly manufacturing steps in metalworking industries, where planning problems are challenging due to heterogeneous furnaces and various product properties. A mathematical formulation for a real-world furnace scheduling problem described by unrelated parallel machines with job families, sequence-dependent setup times and job conflicts is presented. Model enhancements are discussed and a computational study comparing model variants is reported. The results give insights into applicability and promising model enhancements.

Keywords: Scheduling · Parallel unrelated machines · Mathematical optimization · Metalworking industry

1 Introduction

Heating activities in metalworking industries are usually operated continuously and in parallel which leads to challenging scheduling tasks. Here, a real-world setting of a hardening shop with doubled-lined pusher furnaces and chamber ovens is considered. Typically, processing a job in a chamber furnace takes long processing and short setup time, while it is vice versa in a pusher furnace. The main goal is to determine an efficient heating schedule for a given set of jobs with individual and furnace-dependent resident times.

Jobs requiring the same heating procedure form job families. However, the jobs of one family may have rather different processing times. Further, setup times occur between jobs of different families. Due to disproportionately long heating times, certain job families are incompatible with the pusher furnaces. Additionally, certain heating procedures are not allowed to be operated at the same time by one pusher furnace. Thus, if certain job pairs of distinct families are assigned to adjacent pusher furnace lines their processing may not overlap in time. This is referred to as a job conflict. Generally, job family-dependent characteristics are transferred to every single job. The resulting furnace scheduling problem is modeled as scheduling tasks on unrelated parallel machines with

N. Trautmann and M. Gnägi (Eds.): OR 2021, LNOR, pp. 347–352, 2022.
https://doi.org/10.1007/978-3-031-08623-6_51

sequence- and machine-dependent setup times, job conflicts and incompatibilities. Desired is the minimization of the total completion time of all jobs.

Similar scheduling problems are investigated in application- as well as theory-driven research areas. In [4], a furnace scheduling problem appearing in steel production is discussed, while the authors point out that situations with different furnace types are barely studied. The process of reheating and consecutive hot rolling is studied in [8], whereby energy-efficient schedules are obtained for parallel pusher furnace lines. Parallel machine scheduling typically distinguishes between identical and unrelated machines. For scheduling jobs on unrelated machines with sequence-dependent setup times, a comprehensive study on mixed-integer programming (MIP) formulations is reported in [1] and an exact solution method for problems with additional machine availability is proposed in [6]. The total completion time objective is heuristically approached in a small computational study on randomly generated instances in [9]. A problem involving unrelated machines and family-sequence-dependent setup times is tackled using different heuristic methods in [7]. Scheduling job families on identical machines with setup times is studied by applying an exact solution approach to different MIP formulations with total weighted completion time in [5]. In [3], no overlap job conflicts occurring on all identical machines are discussed and the performance of a commercial solver on three different MIP models is evaluated. These models are not fully applicable here, since conflicts are also machine-dependent in our case. This work extends existing models by considering (i) furnaces with different working modes and (ii) job- and machine-dependent conflicts.

Our contribution is a mathematical formulation of this scheduling problem along with basic model enhancements (Sect. 2). A computational study based on real-world data is conducted to evaluate schedules obtained by an out-of-the-box MIP solver. The results (Sect. 3) indicate the appropriateness of our model. We conclude with promising research opportunities in Sect. 4.

2 Mathematical Problem Formulation

2.1 Problem Description and Notation

The given furnace scheduling problem is modeled as scheduling a set of jobs $J = \{1, \ldots, n\}$ on a set of machines $M = \{1, \ldots, m\}$. For every job j, a set $M_j \subseteq M$ of compatible machines is given together with processing times p_{jk} for all machines $k \in M_j$. Setup times s_{ijk} are given for every job pair i, j processed on a compatible machine k. To describe the job- and machine-dependent conflicts, a set $J^{\text{con}} \subseteq \{(j,h) \mid j,h \in J, j \neq h\}$ of ordered pairs of conflicting jobs is defined. If the jobs h and j are conflicting, two ordered pairs (h,j) and (j,h) appear in J^{con}. Further, a set $M^{\text{con}} \subseteq \{(k,l) \mid k,l \in M, k < l\}$ of ordered pairs of machines on which conflicts may occur is defined.

The goal is to find a feasible schedule which respects incompatibility and conflicts and minimizes the sum of the completion times C_j of all jobs $j \in J$.

2.2 Mixed-Integer Programming Model

In accordance with the literature (see e.g. [3]), we use binary sequencing variables. A variable x_{ijk} equals 1, if job i directly precedes job j on a machine k (0, otherwise). Further, a variable $z_{ij} = 1$ forces job j to start after job i is completed if i and j are conflicting jobs assigned to a machine pair $(k, l) \in M^{\text{con}}$. Note that a separation of assignment and sequencing as given in [3] is not reasonable here, since setup times are job- and machine-related. We introduced a dummy job 0 to describe the start of the job sequence on each machine. Let $J_0 = J \cup \{0\}$ and $M_0 = M$.

$$\sum_{j \in J} C_j \rightarrow \text{min!} \tag{a}$$

s.t.

$$\sum_{i \in J_0, i \neq j} \sum_{k \in M_i \cap M_j} x_{ijk} = 1 \qquad j \in J \tag{b}$$

$$\sum_{j \in J, i \neq j} \sum_{k \in M_i \cap M_j} x_{ijk} \leq 1 \qquad i \in J \tag{c}$$

$$\sum_{j \in J, k \in M_j} x_{0jk} \leq 1 \qquad k \in M \tag{d}$$

$$\sum_{h \in J_0 \setminus \{i,j\}, k \in M_h} x_{hik} \geq x_{ijk} \qquad i, j \in J, i \neq j; k \in M_i \cap M_j \tag{e}$$

$$C_i + s_{ijk} + p_{jk} - V^{\text{f}}_{ijk}(1 - x_{ijk}) \leq C_j \qquad i \in J_0, j \in J, i \neq j; \; k \in M_i \cap M_j \tag{f}$$

$$\sum_{\substack{i \in J_0, i \neq j, \\ k \in M_i}} x_{ijk} + \sum_{\substack{i \in J_0, i \neq h, \\ l \in M_i}} x_{ihl} - 1 \leq z_{hj} + z_{jh} \qquad (j, h) \in J^{\text{con}}; (k, l) \in M^{\text{con}}, \; k \in M_j, l \in M_h \tag{g}$$

$$C_h + p_{jk} - V^{\text{h}}_{jk}\Big(2 - z_{hj} - \sum_{\substack{i \in J_0, i \neq j, \\ k \in M_i}} x_{ijk}\Big) \leq C_j \qquad (j, h) \in J^{\text{con}}; k \in M_j : \; \exists (l_1, l_2) \in M^{\text{con}} : \; (l_1 = k \wedge l_2 \in M_h) \vee (l_1 \in M_h \wedge l_2 = k) \tag{h}$$

$$C_0 = 0 \tag{i}$$

$$x_{ijk} \in \{0, 1\} \qquad i \in J_0, j \in J, i \neq j; \; k \in M_i \cap M_j \tag{j}$$

$$z_{ij} \in \{0, 1\} \qquad (i, j) \in J^{\text{con}} \tag{k}$$

The minimization of the total completion time is defined in (a). Constraints (b) and (c) form appropriate job sequences, while constraints (d) assure that each machine is assigned at most one job sequence. Note that it is feasible to leave machines empty if reasonable. Constraints (e) assure the connectedness of job sequences. Constraints (f) implement correct job completions times and avoid cyclic orderings. If two conflicting jobs are assigned to a machine pair in M^{con}, these jobs are non-overlapping by constraints (g) and (h). Note that it is sufficient to explicitly account for job j being processed on machine k, while the assignment of job h is only indirectly involved through the sequencing variable z_{hj}. The decision variables are defined as binary in (j) and (k).

The constants V^{f}_{ijk} and V^{h}_{jk} are determined by approximating the worst-case makespan of any job sequence on any machine. Assuming that all jobs are processed on one machine realizing their longest setup and processing times, a basic estimate is $V = \sum_{j\in J} \max_{i\in J_0, i\neq j} \left(\max_{k\in M_i\cap M_j} (s_{ijk} + p_{jk})\right)$. With this, we define $V^{\text{f}}_{ijk} = V + (s_{ijk} + p_{jk})$ and $V^{\text{h}}_{jk} = V + p_{jk}$.

2.3 Model Enhancements

We consider the following additional constraints:

$$z_{jh} + z_{hj} \leq 1 \qquad (j,h) \in J^{\text{con}} \tag{I}$$

$$\sum_{i\in J_0, i\neq j} \sum_{k\in M_i\cap M_j} (s_{ijk} + p_{jk})\, x_{ijk} \leq C_j \qquad j \in J \tag{II}$$

$$C_i + \min_{k\in M_i\cap M_j} (s_{ijk} + p_{jk}) - V^{\text{III}}_{ij}\Big(1 - \sum_{k\in M_i\cap M_j} x_{ijk}\Big) \leq C_j \qquad i \in J_0, j \in J: i \neq j, M_i \cap M_j \neq \emptyset \tag{III}$$

$$C_h + \min_{k\in M_j} p_{jk} - V^{\text{IV}}_{j}\Big(2 - z_{hj} - \sum_{i\in J_0, i\neq j} \sum_{k\in M_i\cap M_j} x_{ijk}\Big) \leq C_j \qquad (j,h) \in J^{\text{con}} \tag{IV}$$

While constraints (I) strengthen the sequencing of conflicting jobs by imposing valid upper bounds on z_{jh}-variables, inequalities (II) implement lower bounds on the completion times. Constraints (III) and (IV) reinforce completion time differences between ordered pairs of jobs. The constants V^{III}_{ij} and V^{IV}_{j} are calculated according to the scheme described above.

Symmetry breaking may be reasonable for problems with unrelated machines that can be classified into homogeneous groups. A set of ordered pairs of identical machines $M^{\text{sym}} \subseteq \{(k,l) \mid k,l \in M, k < l\}$, on which symmetric schedules may occur, is introduced. Given that jobs are indexed by positive integers, restrictions on the indices of the first jobs on machines k and l can be raised, while an empty sum is taking the value of 0.

$$\sum_{\substack{j\in\{1,\dots,h-1\},\\ k\in M_j}} x_{0jk} \geq x_{0hl} \qquad h \in J, (k,l) \in M^{\text{sym}}, l \in M_h \tag{V}$$

Note that concerning the other characteristics of the jobs, it needs to hold that incompatibility is equivalent for identical machines.

3 Computational Study

3.1 Experimental Setting

The instance data is randomly taken from a one-week production of more than 300 jobs in the hardening shop. There exist two pusher furnaces with two parallel furnace lines each ($k = 1, 2, 3, 4$) and four chamber furnaces ($k = 5, 6, 7, 8$). Job conflicts may occur in the pusher furnaces, so that $M^{\text{con}} = \{(1, 2), (3, 4)\}$. The lines in each pusher furnace and furnaces of the same type behave identically, so that $M^{\text{sym}} = \{(1, 2), (3, 4), (1, 3), (5, 6), (6, 7), (7, 8)\}$. For each instance, Table 1 shows the number of jobs n, the conflict ratio (CR) $\frac{1}{2n} \cdot |J^{\text{con}}|$ and the incompatibility ratio (IR) $\frac{1}{n} \cdot$ (#jobs with incompatibility). Note that due to the family-related job characteristics, the instances feature some specialties. Only job types that are compatible to all furnaces may be conflicting. Thus, CR increases over-proportionately to the number of jobs, when IR is constant. Contrarily, processing and setup times vary greatly and their behavior is only furnace- and job- but not family-dependent. In addition, processing times are shorter for pusher furnaces so that the majority of the compatible jobs are scheduled on them while chamber furnaces may be left empty.

The model is implemented in Java using Google OR-Tools and SCIP solver. The code including instances can be found at [2]. With a time limit of 600 s, the basic model (M) proposed in Sect. 2.2 and the enhanced versions using combinations of (I)–(V) are solved on an Intel Core i7-8565U with 1.80 GHz and 16 GB RAM.

3.2 Computational Results

Table 1 summarizes the best total completion time values obtained by solving the model variants. Optimal results are indicated by *, the minimal value per instance by bold face printing and runs without feasible solution by 'none'. Note that the indicator constraints (f) and (h) lead to extremely weak lower bounds. In our preliminary tests, the largest impact on lower bounds was found when

Table 1. Best objective function values obtained with different model variants

n	CR	IR	M	M+II	MII+I	MII+III	MII+IV	MII+V	M+(I–IV)	M+(I–V)
10	2,6	0,10	5704*	5704*	5704*	5704*	5704*	5704*	5704*	5704*
12	4,2	0	5359	4280	**4212**	4280	**4212**	**4212**	**4212**	**4212**
14	4,9	0,07	7651	**2785**	**2785**	2795	2805	**2785**	2825	**2785**
16	3,6	0	28467	9267	9146	**8566**	9940	10951	8822	10095
18	1,3	0	41514	**20277**	21947	21479	21687	24382	23676	22448
20	4,7	0	36365	**16484**	18248	16734	19998	16706	20022	18638
25	8,2	0,04	53055	**18319**	18611	22971	19768	21080	20120	21601
30	10,5	0	79180	48904	38122	56238	**23633**	None	29521	42309
40	5,9	0	335508	**147642**	182283	184708	None	None	672253	None
60	14,4	0,17	488762	**365597**	None	408558	None	462772	607143	2407113

adding constraints (II) to M. Thus, we report experiments focusing on the other enhancing constraints based on M with (II), referred to as (MII). Adding constraints (I), (III), (IV) and (V) to MII only slightly enhances the best schedules found for some instances. M+II shows an overall good performance. In line with the findings in [3], it can be stated that instances with more than 20 jobs are very difficult to solve. Surprisingly, symmetry breaking constraints (V) do not show a significant improvement.

4 Conclusions

This paper examines a real-world furnace scheduling problem. The problem consists of heterogeneous furnaces, setup times, incompatibility, and job- and machine-dependent job conflicts. We present a compact mathematical formulation. A computational study based on real-world instances evaluates the model and enhanced variants. The results generally support the appropriateness of the formulation for smaller instances. The performance improvement caused by additional completion time bounding is to be highlighted. However, to solve instances of practically relevant size, stronger enhancements of the model should be considered together with matheuristic approaches and the integration of human expert knowledge. Also, broader studies with different machine settings, job characteristics and real-world objective functions are necessary.

References

1. Fanjul-Peyro, L., Ruiz, R., Perea, F.: Reformulations and an exact algorithm for unrelated parallel machine scheduling problems with setup times. Comput. Oper. Res. **101**, 173–182 (2019)
2. Fath, P., Sayah, D., Lange, J.: Furnace scheduling, July 2021. https://doi.org/10.5281/zenodo.5146130
3. Há, M.H., Ta, D.Q., Nguyen, T.T.: Exact algorithms for scheduling problems on parallel identical machines with conflict jobs. Tech. rep. arXiv:2102.06043 (2021)
4. Ilmer, Q., Haeussler, S., Missbauer, H.: Optimal synchronization of the hot rolling stage in steel production. IFAC-PapersOnLine **52**(13), 1615–1619 (2019)
5. Kramer, A., Iori, M., Lacomme, P.: Mathematical formulations for scheduling jobs on identical parallel machines with family setup times and total weighted completion time minimization. Euro. J. Oper. Res. **289**(3), 825–840 (2021)
6. Lopes, M.J.P., de Carvalho, J.V.: A branch-and-price algorithm for scheduling parallel machines with sequence dependent setup times. Euro. J. Oper. Res. **176**(3), 1508–1527 (2007)
7. Pinheiro, J.C.S.N., Arroyo, J.E.C., Fialho, L.B.: Scheduling unrelated parallel machines with family setups and resource constraints to minimize total tardiness. In: Proceedings of GECCO 2020. ACM (2020)
8. Tang, L., Ren, H., Yang, Y.: Reheat furnace scheduling with energy consideration. Int. J. Prod. Res. **53**(6), 1642–1660 (2014)
9. Tavakkoli-Moghaddam, R., Taheri, F., Bazzazi, M., Izadi, M., Sassani, F.: Design of a genetic algorithm for bi-objective unrelated parallel machines scheduling with sequence-dependent setup times and precedence constraints. Comput. Oper. Res. **36**(12), 3224–3230 (2009)

Scheduling a Two Stage Proportionate Flexible Flow Shop with Dedicated Machines and No Buffers

Heiner Ackermann, Lena Schwehm, and Christian Weiß(✉)

Fraunhofer Institute for Industrial Mathematics ITWM,
Fraunhofer-Platz 1, 67663 Kaiserslautern, Germany
{heiner.ackermann,lena.schwehm,christian.weiss}@itwm.fraunhofer.de

Abstract. In this paper, we study a scheduling problem derived from an application in the chemical processing industry. At the chemical plant we consider, a single base reactor prepares different starting products for several distinct production lines, each consisting of a number of further, specialized reactors. Importantly, there are no buffers, so after completion a job may block a reactor from processing further jobs, if no successor reactor is ready to start it. First, we show how to model the scheduling problem as a special version of the well-known flexible flow shop problem. Then we prove that in the general version of the problem it is strongly NP-hard to minimize the makespan. Towards a solution, we propose and compare several different construction heuristics.

Keywords: Scheduling · Flexible flow shop · No-buffer

1 Introduction

Derived from an actual chemical plant, consider a production process, where a single base machine preprocesses jobs for m distinct production lines, consisting of several machines each (note that machines are also called reactors in our application). Each production line processes one type of job. Jobs of the same type are identical, with processing times depending only on the machines, not on the individual jobs. Jobs of different types may have different processing times, even on the common, single base machine.

For the purpose of this paper, each production line is only represented by its bottleneck machine. This is not an oversimplification, since, as processing times are deterministic and only dependent on the machines, preceding and succeeding machines can always be scheduled in such a way, that processing in the production line happens continuously for each job. Thus, formally, we consider a two stage *flexible flow shop*, with a single, common base machine at the first stage and a set of dedicated machines at the second stage. All jobs first are processed by the base machine on the first stage and then transferred onto the second stage machine representing their target production line.

N. Trautmann and M. Gnägi (Eds.): OR 2021, LNOR, pp. 353–358, 2022.
https://doi.org/10.1007/978-3-031-08623-6_52

Denote by M_B the base machine and by $M_i, i = 1, \ldots, m$, the machines representing the production lines. The job set partitions into m job types $\mathcal{J} = \mathcal{J}_1 \cup \ldots \cup \mathcal{J}_m$, one for each second stage machine. Furthermore, denote by n_i the number of jobs of type $\mathcal{J}_i$, $i = 1, \ldots, m$, and by $n = \sum_{i=1}^{m} n_i$ the number of all jobs. All jobs of the same type $\mathcal{J}_i$, $i = 1, \ldots, m$, are identical, with processing times p_i^B on the base machine and p_i on their dedicated second stage machine. Scheduling models where processing times depend only on the machines and not on the jobs are sometimes called *proportionate* [7]. We number the jobs of each type and denote by J_{ij}, $i = 1, \ldots, m$, $j = 1, \ldots, n_i$, the j-th job of type $\mathcal{J}_i$.

Crucially, there are no buffers between the first and second stage. Thus, if second stage machine M_i is not ready to start a new job $J_{i,j}$, $i = 1, \ldots, m, j = 1, \ldots, n_i$, once it is finished on the base machine M_B, then job $J_{i,j}$ stays on machine M_B blocking it from processing further jobs until machine M_i is ready.

For a given schedule S, the *makespan* is defined as the maximum completion time of any job. Our goal is to find a schedule of minimum makespan. For brevity, we denote this problem by PFFDNB(m) (proportionate flexible flow shop with dedicated machines and no buffer).

2 Related Work

To the best of our knowledge, problem PFFDNB(m) has not been studied in the literature before. However, there exists research considering the closely related problem where buffers are unlimited and processing times are job-dependent by, e.g., [5,6,8,11].

In particular, for the problem with two dedicated machines at the second stage, denoted in the literature by FF2(1, 2), minimizing the makespan is strongly NP-hard [5,6,11]. Interestingly, this remains true if buffers are removed. The proof we present below for our setting can be adjusted to need only two job types, if jobs from one job type are allowed to have different processing times.

Exact solution algorithms making use of dominance rules and branch and bound are discussed in [3,4,10]. See also [6] for a survey. Towards approximation, there is a $(2 - \frac{1}{m})$-approximation algorithm to minimize the makespan in the FF2(1, m) setting. The performance bound is tight [5,6].

The results presented in this paper are part of a wider master thesis, see [9]. The practical application mentioned above is studied more closely in [1].

3 NP-Completeness of Problem PFFDNB(m)

In this section we prove that problem PFFDNB(m) is strongly NP-hard if the number of machines m is part of the input. This can be shown via a polynomial reduction from the well-known problem 3-PARTITION [2].

Theorem 1. *There exists a polynomial reduction from Problem 3-PARTITION to problem* PFFDNB(m).

Proof. Given an instance of 3-PARTITION with $3k$ integers a_i and target value b, construct the following instance of PFFDNB(m) with $m = 3k + 1$. For each integer a_i in the instance of 3-PARTITION add a dedicated machine M_i, $i = 1, \ldots, 3k$. Machine M_i process a job type with exactly one job $J_{i,1}$, with processing times $p_i^B = a_i$ on the base machine and $p_i = 1$ on the dedicated M_i. Furthermore, add a job type $\mathcal{J}_0$ consisting of k jobs $J_{0,1}, \ldots, J_{0,k}$, with processing times $p_0^B = 1$ on the base machine and $p_0 = b$ on the dedicated machine M_0. Using this construction, we show that a solution exists for the original instance of 3-PARTITION, if and only if the constructed instance of PFFDNB(m) has a schedule of makespan $kb + k + 1$. This can be seen by observing that the base machine has a total load of $kb + k$ and has to process continually, if a makespan of $kb + k + 1$ is to be achieved. Details can be found in [9].

The reduction immediately yields the desired result.

Corollary 1. *Problem* PFFDNB(m) *is strongly NP-hard if the number of job types m is part of the input.*

4 Heuristics

We now consider construction heuristics for problem PFFDNB(m). Note, that since there is no gain in delaying jobs to be finished later, a schedule is fully defined by the sequence of jobs on the base machine M_B. Thus, at any decision point, i.e., whenever the base machine becomes empty, the only decision is which job to schedule next. In what follows, we propose three different heuristics and then empirically compare their performances.

Note that, apart from knowledge of the number of jobs to be produced of each type, all three algorithms proposed below can be used in an online setting, with near instantaneous decision computation to decide the next job to process. This makes them useful in a practical setting, where uncertainties in the job processing times can lead to frequent re-scheduling. Indeed, in a practical study, we successfully applied the ideas below (adjusted for special requirements of the industrial setting) to support online shop floor dispatching [1].

In order to define the construction heuristics, we need the following additional notation. Let $\bar{\mathcal{J}} \subset \mathcal{J}$ be a subset of the job set and let $\bar{S}$ be a schedule for $\bar{\mathcal{J}}$. Denote by $C_i(\bar{S})$ the last finish time of any job $J_{i,j} \in \bar{\mathcal{J}}$ on machine M_i. Denote by $C_B(\bar{S})$ the last start time of any job on a second stage machine, i.e., the time when the base machine is ready to start the next chosen job $J_{i,j} \in \mathcal{J} \setminus \bar{\mathcal{J}}$. Furthermore denote by $\bar{n}_i(\bar{S})$ the number of jobs of type $\mathcal{J}_i$ which are already scheduled in $\bar{S}$, i.e., $\bar{n}_i(\bar{S}) = |\mathcal{J}_i \cap \bar{\mathcal{J}}|$. Also, denote by $R(\bar{S})$ the set of indices $i = 1, \ldots, m$ such that jobs still remain to be scheduled for job type $\mathcal{J}_i$, i.e., $R(\bar{S}) = \{i = 1, \ldots, m \mid n_i - \bar{n}_i > 0\}$.

Define the gap time $G_i(\bar{S})$ of job type $\mathcal{J}_i$ as the idle time caused on the base machine (due to blocking) by scheduling a job of type $\mathcal{J}_i$ at time $C_B(\bar{S})$, i.e.,

$$G_i(\bar{S}) = \max\left\{0, C_B(\bar{S}) + p_i^B - C_i(\bar{S})\right\}.$$

Finally, define the work load $W_i(\bar{S})$ of job type $\mathcal{J}_i$ as the remaining processing time needed for job type $\mathcal{J}_i$ on its dedicated second stage machine M_i, i.e.,

$$W_i(\bar{S}) = (n_i - \bar{n}_i) \times p_i.$$

4.1 Heuristic 1: MinGapMaxJobs

For our first heuristic, at any decision point we pick a job from job type $\mathcal{J}_i$ which fulfills the following two properties:

1. there are still unscheduled jobs of job type $\mathcal{J}_i$, i.e. $i \in R(\bar{S})$, and
2. job type $\mathcal{J}_i$ causes the least amount of blocking on the base machine, i.e., job type $\mathcal{J}_i$ minimizes $G_i(\bar{S})$.

If there is a tie, in particular when several job types cause no blocking, then we choose a job from the tied job type which still has the most jobs left to schedule, i.e. which maximizes $n_i - \bar{n}_i$ amongst all tied job types. If there is still a tie, we pick a job arbitrarily from any of the job types remaining tied.

4.2 Heuristic 2: MinGapMaxWorkLoad

The second heuristic is similar to the first, only with a different tie breaker. In case of a tie, pick a job of the type which has the largest work load still to schedule, i.e., the job type $\mathcal{J}_i$ which maximizes $W_i(\bar{S}) = (n_i - \bar{n}_i(\bar{S})) \times p_i$. If there is still a tie, pick a job arbitrarily from any of the job types remaining tied.

4.3 Heuristic 3: Weighted MinGapMaxWorkLoad

The third heuristic is actually a class of heuristics, depending on a weight. It is similar to the second heuristic, but instead of minimizing the gap first and using the work load to break ties, we compare a weighted sum of both criteria. In order to combine the two criteria, we first norm them onto the [0, 1] interval. Let

$$r_{gap,i}(\bar{S}) = \frac{G_i(\bar{S}) - \min_{i \in R(\bar{S})} \{G_i(\bar{S})\}}{\max_{i \in R(\bar{S})} \{G_i(\bar{S})\} - \min_{i \in R(\bar{S})} \{G_i(\bar{S})\}},$$

such that $r_{gap,i}(\bar{S}) = 1$ if job type $\mathcal{J}_i$ maximizes $G_i(\bar{S})$ and $r_{gap,i}(\bar{S}) = 0$ if job type $\mathcal{J}_i$ minimizes $G_i(\bar{S})$. If all job types produce the same gap, then set $r_{gap,i}(\bar{S}) = 0$ for all job types. Similar, let

$$r_{wl,i}(\bar{S}) = 1 - \frac{W_i(\bar{S}) - \min_{i \in R(\bar{S})} \{W_i(\bar{S})\}}{\max_{i \in R(\bar{S})} \{W_i(\bar{S})\} - \min_{i \in R(\bar{S})} \{W_i(\bar{S})\}},$$

such that $r_{wl,i}(\bar{S}) = 0$ if job type $\mathcal{J}_i$ maximizes $W_i(\bar{S})$ and $r_{wl,i}(\bar{S}) = 1$ if job type $\mathcal{J}_i$ minimizes $W_i(\bar{S})$. If all job types have the same remaining work load, then set $r_{wl,i}(\bar{S}) = 0$ for all job types. Note, that since we want to choose a job which produces a gap as small as possible but has as large as possible remaining work load, it is desirable to minimize both values $r_{gap,i}(\bar{S})$ and $r_{wl,i}(\bar{S})$. For a given weight ω, $0 \leq \omega \leq 1$ then define

$$r^{\omega}_{gap+wl,i}(\bar{S}) = \omega r_{gap,i}(\bar{S}) + (1 - \omega) r_{wl,i}(\bar{S}).$$

Then, for the weighted version of MinGapMaxWorkLoad, at any decision point we choose next a job from the job type which still has jobs left to schedule and which minimizes $r^{\omega}_{gap+wl,i}(\bar{S})$. In case of a tie, we pick a job arbitrarily from one of the tied job types. We denote this heuristic by MinGapMaxWorkLoad(ω).

5 Numerical Results

We performed computational experiments in order to test and compare the quality of our proposed heuristics. For this purpose, we randomly generated 720 problem instances, 240 each for the number of second-stage machines $m = 2, 3$, and 5. For each number of machines, the 240 instances can be further divided into three sets of 80 instances, with smaller to larger number of jobs. Again, each set of 80 instances can be further divided into four sets of 20 instances with shorter to longer processing times. For details see [9]. For each instance, an optimal solution was pre-computed via an MIP.

First, we tested which weights for algorithm MinGapMaxWorkLoad(ω) produced the best results. It turns out that $\omega = 0.65$ yields the best mean approximation over all instances as well as the best mean approximation for each individual number of dedicated machines $m = 2, 3, 5$. Note that larger weights ω in general seem to perform better for larger numbers of dedicated machines m.

Then, we compared MinGapMaxWorkLoad(0.65) to the other heuristics. Maybe surprisingly, in terms of approximation over all instances, the MinGapMaxWorkLoad heuristic performs nearly as good as any of its weighted versions MinGapMaxWorkLoad(ω). In fact, while it performs on average slightly worse than MinGapMaxWorkLoad(0.65), (with an approximation factor of 1.027 to 1.026 over all instances), it matches the mean quality of the weighted version for $m = 2$ and also produces a smaller spread of approximation factors for that setting. When the number of machines increases, MinGapMaxWorkLoad(ω) with $\omega = 0.65$ starts to outperform its unweighted counter part.

The heuristic MinGapMaxJobs is outperformed by both other heuristics, although it is close in performance for instances with $m = 2$ dedicated machines.

6 Conclusions and Future Work

In this paper we studied a proportionate, two-stage, flexible flow shop with a common base machine at the first stage and m dedicated machines at the second stage, one for each type of job. The problem is derived from an industrial application. Crucially, there are no buffers between the first and the second stage of the flow shop.

We proved that minimizing the makespan for an instance of PFFDNB(m) is NP-hard in the strong sense, if the number of job types m is part of the input. Then we proposed several heuristics and compared them computationally. Additional details and results, including polynomially solvable special cases, exact methods (MIPs), and additional heuristics can be found in [9].

The main open question for future work is to resolve the complexity status of our problem when the number of dedicated machines m is fixed. To the best of our knowledge, this is currently still open even for the special case of $m = 2$.

References

1. Ackermann, H., et al.: Dispatching for batch chemical processes using Monte-Carlo simulations - a practical approach to scheduling in operations. In: Bortz, M., Asprion, N. (eds.) Simulation and Optimization in Process Engineering, pp. 339–364. Elsevier (2022)
2. Garey, M.R., Johnson, D.S.: Computers and Intractability; A Guide to the Theory of NP-Completeness. W. H. Freeman & Co., New York (1990)
3. Hadda, H., Dridi, N., Hajri-Gabouj, S.: A note on the two-stage hybrid flow shop problem with dedicated machines. Optim. Lett. **6**, 1731–1736 (2012)
4. Hadda, H., Dridi, N., Hajri-Gabouj. S.: Exact resolution of the two stage hybrid flow shop with dedicated machines. Optim. Lett. **8**, 2329–2339 (2014)
5. Herrmann, J.W., Lee, C.-Y.: Three-machine look-ahead scheduling problems. Department of Industrial and Systems Engineering, University of Florida (1992)
6. Hwang, F.J., Lin, B.M.T.: Survey and extensions of manufacturing models in two-stage flexible flow shops with dedicated machines. Comput. Oper. Res. **98**, 103–112 (2018)
7. Panwalkar, S.S., Smith, M.L., Koulamas, C.: Review of the ordered and proportionate flow shop scheduling research. Naval Res. Logist. **60**(1), 46–55 (2013)
8. Riane, F., Artiba, A., Elmaghraby, S.: Sequencing hybrid two-stage flowshops with dedicated machines. Int. J. Prod. Res. **40**, 4353–4380 (2002)
9. Schwehm, L.: Two Staged Job-Shops with Shared Machines on the First and Dedicated Machines on the Second Stage. Master thesis, Dpt. of Mathematics, TU Kaiserslautern (2020)
10. Wang, S., Liu, M.: A heuristic method for two-stage hybrid flow shop with dedicated machines. Comput. Oper. Res. **40**(1), 438–450 (2013)
11. Yang, J.: A new complexity proof for the two-stage hybrid flow shop scheduling problem with dedicated machines. Int. J. Prod. Res. **48** (2010)

Revenue Management

Towards Transfer Learning for Revenue and Pricing Management

Alexander Kastius and Rainer Schlosser(✉)

Hasso Plattner Institute, University of Potsdam, Potsdam, Germany
{alexander.kastius,rainer.schlosser}@hpi.de

Abstract. Reinforcement Learning (RL) has proven itself as a powerful tool to optimize pricing processes. With the support of deep non-linear function approximation tools, it can handle complex and continuous state and action spaces. This ability can leverage the utility of pricing algorithms in markets with a vast number of participants or in use cases where additional product features should be considered in the pricing system. One problem with those tools is their apparent demand for training data, which might not be available for a single market. We propose to use techniques instead, that leverage the knowledge of different problems. Several similar algorithms have been proposed in the past years to allow RL algorithms to operate efficiently on various processes simultaneously. DISTRAL continuously merges information from different decision processes towards a distilled policy and uses the joint policy to update the market-specific source policies. We will discuss the influence of such regularization mechanisms. Multi-market pricing problems are used to illustrate their impact.

1 Introduction

Dynamic pricing serves the purpose of finding optimal pricing policies given a specified market situation. In many cases, the market under assessment can be specified by static and dynamic features. A pricing algorithm then has to discover the relationship between market parameters, price choice, and revenue.

Usually, the information about the relationship between those three components is sparse, only consisting of past experience and possibly human domain knowledge. Pricing algorithms aim at estimating this function based on experience.

The three major components of this challenge, parameters, price, and revenue, closely resemble the three fundamental components of a reinforcement learning problem. RL consists of algorithms and methods that aim to solve a decision process formulated in three dimensions: state, action, and reward. The state determines the current parameters of the process, the actions are the possible choices available at a specific state, and the reward signal provides feedback to the algorithm, whether the chosen action was well suited given the state. The agent faces a state, chooses an action, and receives a reward, which leads to a trajectory starting from the initial state s_0 to the current state s_n:

N. Trautmann and M. Gnägi (Eds.): OR 2021, LNOR, pp. 361–366, 2022.
https://doi.org/10.1007/978-3-031-08623-6_53

$$(s_0, a_0, r_0, s_1, a_1, r_1, ..., s_n, a_n, r_n)$$

Those three components can be mapped to a pricing problem:

- State: Consists of static and dynamic market parameters. Static parameters specify the properties of the market under assessment. Dynamic market parameters resemble the current state of the market, for example, competitors' prices.
- Action: Consists of the agent's price choice in the allowed price-span.
- Reward: Consists of the revenue achieved since the last pricing decision for this market.

Many RL algorithms use function approximation with artificial neural networks (ANNs) to represent what a good action choice consists of [6]. This setup demands noticeable amounts of data to find the optimal network configuration. In many situations, real-world traders do not have those amounts of data available for a single product in a single market. As both the product and the market determine the relationship between price choice and revenue, this becomes a challenge.

To overcome this challenge, we propose to rely on transfer learning. The idea of transfer learning is to use the information available for one problem and re-use it for a related problem. In practice, this means that data collected either for similar products or similar markets could be re-used to estimate the price-revenue relationship. The influence of external information for a given problem has to be balanced carefully, as the discrepancies between two problems are unknown, and fitting to an unrelated problem can diminish an agent's performance.

For this purpose, algorithms like DISTRAL have been proposed, which make use of a set of trained agents that influence each other by exchanging information via a central policy [7]. In the following sections, we introduce the fundamentals of DISTRAL, explain the mapping of a multi-market pricing problem to the core concepts of RL, and display possibilities to use tools like DISTRAL for revenue management (RM). Some problems that arise when implementing such a mapping will be displayed later on.

This paper is organized as follows. In Sect. 2, we discuss related work. In Sect. 3, we provide a detailed introduction to RL and DISTRAL. Section 4 provides a detailed description of how we map a RM problem on multiple markets to the RL setup. Section 5 concludes.

2 Related Work

While considering DISTRAL as an example in this paper, it is not the only available algorithm designed explicitly for multi-task RL. In the near past, a noticeable amount of alternative mechanisms have been presented. An overview is available in [8]. DISTRAL is exemplary for a group of algorithms that the authors group as policy transfer methods, in which a policy is copied from one task to another either directly or indirectly. Instead of mapping concepts between several tasks to comparable domains manually, as described later, such mappings can be learned automatically, see, for instance, the algorithm described in [1].

A related area of research in RM is omnichannel pricing. In omnichannel pricing, the same product is sold over multiple channels, e.g., online and in-store. While being more specific than the generic setup considered in this paper, the omnichannel problem can be mapped to this setup by considering each channel a different market. For example, [4] studies dynamic pricing for retailers which do sell the same stock of products both in-store and online. Their model considers cross-channel interactions. As their model considers inter-market dependencies, this offers an advantage over the setup described here, which considers each market independently.

3 Reinforcement Learning

The challenge in RL consists of finding the optimal policy to solve a Markov decision process [6]. A policy $\pi(a_t|s_t)$ determines the probability that the agent chooses action a_t in state s_t. A policy is considered optimal if it yields the maximum discounted reward of all policies. Solving an RL problem consists of two tasks: policy evaluation and policy improvement. Policy evaluation determines the expected future reward of a specific policy or action given a specified state. This information can then be used to improve the policy. If an action that is different from the policy's action choice in this state has a higher expected future reward, the policy is changed to the action with the highest expected value in this state.

This concept can be further extended to policy gradient algorithms, which keep a parametric representation of the policy. The policy can then be implemented using ANNs to map states to actions. The parameters of the policy are adjusted by following the policy gradient theorem, which allows deriving a gradient with regard to the parameters based on past experience and an additional value estimator. There are several examples of this available, one of those algorithms is Soft Actor-Critic (SAC) [3]. SAC was used by us for pricing problems in the past and has shown exceptional performance but suffers from the disadvantage that it requires tremendous amounts of data to work successfully. A detailed analysis of SAC's results in duopolies and oligopolies is available in [5].

To overcome the demand for data, we consider using data from multiple sources. This approach is equivalent to solving multiple RL problems at the same time. Given the information from multiple markets, it is necessary to find the optimal policy for each one. A well-performing algorithm is then able to exchange information between all problems under consideration to improve learning performance.

Training a single policy to work well on all problems under consideration is difficult, as the policy has to encode differences of each problem. DISTRAL chooses a different approach. Each problem under consideration has a specified policy, which is influenced by two mechanisms. First, all problem-specific policies are distilled towards a unified policy. Then, every problem-specific policy is adjusted towards the unified policy. This way, information from one problem can be stored in the unified policy if it is considered relevant for every problem. Every problem can then re-use this by taking the unified policy into account.

While a detailed description of DISTRAL's internals is available in [7], a brief description will be given here. For a set of n problems, a set of $n+1$ policies is kept: $\pi_0, ..., \pi_n$. π_0 is considered the distilled policy, while π_1 to π_n correspond to a single one of the n problems. For each problem, a different reward function $R_i(a_t, s_t)$ and a state transition distribution are given, but can only be learned by observation. All policies are optimized according to the following loss function (via tuning parameters $c_{KL}, c_{Ent} \geq 0$ and discount $\gamma \in (0, 1)$):

$$J(\pi_0, \{\pi_i\}_{i=1}^n) = \sum_{i=1}^{n} \mathbf{E}_{\pi_i}[\sum_{t \geq 0} \gamma^t R_i(a_t, s_t) - c_{KL}\gamma^t \log \frac{\pi_i(a_t, s_t)}{\pi_0(a_t, s_t)} - c_{Ent}\gamma^t \log \pi_i(a_t, s_t)].$$

This loss function formulation achieves three goals:

- It aims to maximize the expected discounted reward of each task-specific policy.
- It aims to minimize the difference between each task-specific policy and the distilled policy π_0 using the Kullback-Leibler divergence.
- Each policy incorporates an additional entropy-regularization term. Entropy regularization aims at avoiding premature convergence in RL [3].

The solution algorithm is derived from this by reapplying the concept of policy gradient algorithms. Given a parametric representation of the policies, a gradient of the loss function can be computed with regard to those parameters. This mechanism is implemented by using a combination of two ANNs. h represents the distilled policy, f represents the task-specific policy. For f, a different set of parameters ϕ_i is kept for each task. The actual policies are then derived from both networks:

$$\pi_0(a_t|s_t) = \frac{\exp(h_{\phi_0}(a_t|s_t))}{\sum_{a' \in A} \exp(h_{\phi_0}(a'|s_t))}$$

$$\pi_i(a_t|s_t) = \frac{\exp(\alpha h_{\phi_0}(a_t|s_t) + \beta f_{\phi_i}(a_t|s_t))}{\sum_{a' \in A} \exp(\alpha h_{\phi_0}(a'|s_t) + \beta f_{\phi_i}(a'|s_t))}$$

with $\alpha = c_{KL}/(c_{KL} + c_{Ent})$ and $\beta = 1/(c_{KL} + c_{Ent})$ and $i = 1, ..., n$. This architecture immediately applies changes in the distilled policy to all tasks without requiring additional updates. The combination of this setup, the loss function, and the policy gradient theorem allow simultaneous learning of several tasks.

4 Pricing on Multiple Markets with RL

DISTRAL can be used to solve multi-market pricing problems. Each market forms a task. We consider a task as a unit of an action space, a state space, reward, and a state transition distribution. The latter maps a state and an action to the respective following state and the reward achieved in that transition.

For multi-task RL with DISTRAL, we must consider some limitations regarding both the action and the state space. The unified action space of all tasks has to be chosen so that all price spans of all markets are covered. As we consider different products with possibly different price ranges, the unified action space might diverge from a reasonable price range for a single problem. To solve this issue, we propose to transform all price ranges to the same bandwidth and apply this transition whenever an action choice needs to be applied. When learning occurs, this change has to be inverted. This means that, for each market i of n markets, we define lower and upper limits for the market, $p_l^{(i)}$ and $p_u^{(i)}$. As DISTRAL in its natural form can only be applied to discrete action spaces due to the necessary computation of the sums in the respective equations, the theoretically continuous price band has to be discretized, which then requires a step size $p_s^{(i)}$ as well. The step size has to be chosen in a way that $(p_u^{(i)} - p_l^{(i)})/p_s^{(i)}$ is constant for all tasks i. This number then defines the number of actions in the unified action space A. In practice, this might lead to a problem that requires attention from the user: If the borders are not chosen well, the market-specific adjustment from the task-specific policies has to be very strong, which hinders information exchange between tasks.

This effect can be overcome by ensuring that similar actions do have similar semantics in both markets. To achieve this goal, all problems must have similar semantics in their respective state descriptions. To normalize the input accordingly, a stringent data model is required. In previous evaluations by the authors, the market features only consisted of competitor prices. It can be expected that the amount of competitors in the market is different for each product. As the ANNs used for previous experiments require equally shaped inputs, this requires either a tool to process sequences or a unification of the input. Both alternatives are possible, as there are specialized ANN structures available that allow processing sequences efficiently, for example, recurrent neural networks, which parse elements one by one and can remember the outputs of previous iterations [2]. We propose a different encoding for simplicity: Each market participant is represented by his price and a flag that indicates his presence in this market.

A unified setup, as suggested in the two-column setup of DISTRAL, might struggle with this. Every policy needs to learn to incorporate the correct features and ignore others explicitly. Specialized network setups that also incorporate L1 regularization might be of use in this case [2].

As previously indicated, even with this setup, further normalization of inputs is required. The price range of the competitors might be different between all markets. To allow easier input recognition, those have to be normalized as well. A normalization with zero mean and a standard deviation of one is considered desirable for applications using ANNs [2]. To achieve this goal, past market data could be used. When past data is not available, the discretization parameters can be used instead. Every competitor's price of market i will be normalized by $p_u^{(i)}$.

The different markets will then be observed, and the initially randomized policies will be applied. The learning can occur as specified by the algorithm, including distillation of the centralized policy π_0.

5 Conclusions

We have proposed a setup that allows DISTRAL to be used in multi-market pricing problems to overcome data limitations when such tools are used with deep ANNs. It consists of mapping concepts of a pricing problem to the fundamental components of a Markov decision process and further additions to simplify the problem. Furthermore, the problem has to be adjusted to take the limitations of DISTRAL regarding the action and state space into account.

A detailed analysis of the issues raised by the adjustments mentioned in the previous sections has to be performed. This analysis includes robustness regarding different market setups, which can occur if either the demand model or the competitor behavior does differ fundamentally between regions, trading places, or products of a single company. Our model assumes independence of all markets under consideration. As this does not hold in practice, further adjustments have to be taken into account.

References

1. Ammar, H.B., Taylor, M.E.: Reinforcement learning transfer via common subspaces. In: Vrancx, P., Knudson, M., Grześ, M. (eds.) ALA 2011. LNCS (LNAI), vol. 7113, pp. 21–36. Springer, Heidelberg (2012). https://doi.org/10.1007/978-3-642-28499-1_2
2. Goodfellow, I.J., Bengio, Y., Courville, A.C.: Deep Learning. Adaptive Computation and Machine Learning, MIT Press, Cambridge (2016)
3. Haarnoja, T., Zhou, A., Abbeel, P., Levine, S.: Soft actor-critic: off-policy maximum entropy deep reinforcement learning with a stochastic actor. In: ICML 2018, 10-15, 2018. Proceedings of Machine Learning Research, vol. 80, pp. 1856–1865. PMLR (2018)
4. Harsha, P., Subramanian, S., Uichanco, J.: Dynamic pricing of omnichannel inventories. Manuf. Serv. Oper. Manag. **21**(1), 47–65 (2019)
5. Kastius, A., Schlosser, R.: Dynamic pricing under competition using reinforcement learning. J. Revenue Pricing Manag. **21**, 50–63 (2022)
6. Sutton, R.S., Barto, A.G.: Reinforcement Learning - An Introduction. Adaptive Computation and Machine Learning, MIT Press, Cambridge (1998)
7. Teh, Y.W., et al.: Distral: robust multitask reinforcement learning. In: Advances in Neural Information Processing Systems 30: Annual Conference on Neural Information Processing Systems 2017, pp. 4496–4506 (2017)
8. Zhu, Z., Lin, K., Zhou, J.: Transfer learning in deep reinforcement learning: a survey. CoRR abs/2009.07888 (2020)

Stochastic Dynamic Pricing Under Duopoly Competition with Mutual Strategy Adjustments

Rainer Schlosser(✉) and Alexander Kastius

Hasso Plattner Institute, University of Potsdam, Potsdam, Germany
{rainer.schlosser,alexander.kastius}@hpi.de

Abstract. In practical applications, firms use data-driven dynamic pricing strategies to increase their rewards in the presence of competition. Merchants are forced to steadily adjust their strategies in response to changing market environments caused by competitors that update their pricing strategies over time. In this paper, we study mutual updates of dynamic pricing strategies in an infinite horizon duopoly model with stochastic demand. We use dynamic programming techniques to compute strategies that take anticipated price reactions of the competitor into account. We consider cases in which (i) strategies are mutually observable and (ii) have to be identified from single price observations over time. For both scenarios, we analyze and compare the long-term interaction of competing self-adaptive strategies.

1 Introduction

Firms offering goods on online marketplaces have to face increasing competition and steadily changing conditions. One reason for the increasing competition is the rising application of automated repricing algorithms and the resulting shortening of time spans between price updates. The time pressure and demand stochasticity make it challenging for firms to determine prices fast and efficiently (often for a large number of products) while still ensuring to employ prices that maximize expected revenues. But at the same time, online marketplaces also provide numerous advantages. Sellers are now able to observe the market situation at any given point in time and set prices accordingly. Having historical market data at hand also enables sellers to learn the demand over time and better understand the consumers' decision making. More interestingly for the context of this paper, firms can learn the competitors' strategies. Pricing strategies that use demand knowledge and anticipate competitors' actions will thus be of increasing interest.

Nevertheless, revealing a competitor's strategy is a highly challenging task as price exploration can be costly and a competitor's strategy may change.

The main contributions of this paper are the following:

- We analyze mutual strategy updates under duopoly competition.
- We identify examples of symmetric equilibrium feedback pricing strategies.

N. Trautmann and M. Gnägi (Eds.): OR 2021, LNOR, pp. 367–372, 2022.
https://doi.org/10.1007/978-3-031-08623-6_54

- We compare settings with full and partial knowledge of the competitor's strategy.

This paper is organized as follows. In Sect. 2, we discuss related work. In Sect. 3, we describe the stochastic dynamic duopoly model and show how dynamic programming techniques can be used to determine price response strategies. In Sect. 4, we study the long-term interaction when (i) strategies are mutually observable and (ii) have to be identified from single price observations over time. Concluding remarks and ideas for future research are given in the final Sect. 5.

2 Related Work

Dynamically selling products on online marketplaces is a classical application of revenue management theory. The problem is closely related to the field of dynamic pricing, which is summarized in the books by, e.g., [10]. The surveys [2] and [1] provide an excellent overview of recent pricing models under competition.

In [7], different demand learning techniques are applied to estimate demand in competitive real-life markets. However, in their model, neither price anticipations nor equilibrium strategies are considered. [8] study adaptive learning strategies of two competing competitors under reference price effects. [3] study similar problems using reinforcement learning techniques (Deep Q-learning Networks and Soft Actor Critic) abstaining from explicit demand and state dynamics.

The combined problem of (i) updating prices, (ii) learning demand probabilities, and (iii) identifying strategy equilibria in competitive markets is challenging as information is incomplete. For analyzing and evaluating the complex interplay and long-term behavior of mutual self-adaptive pricing strategies usually simulations have to be used [4,9]. Note, since dynamics are constantly changing in multi-agent markets stability issues arise making it hard to identify equilibrium strategies.

3 Model Description

We consider the scenario, in which two competing merchants sell goods on an online marketplace. Motivated by practical applications, in which firms adjust their prices subsequently at distinct points in time, in our model, time is split into periods of fixed length, which start with firm 1's price update a (out of the set of prices A). After a certain share h of the period, $h \in (0, 1)$, the competitor (firm 2) reacts to firm 1's action a and updates its old price p to p' (cp. the setup used in [8]).

In general, a firm's strategy can be characterized by a probability distribution of how to respond to a certain competitor price. In this context, in our model, the probability that firm 2 reacts to firm 1's price $a \in A$ (after delay h) with the price $p' \in A$ is denoted by

$$P_{react}(a, p') \in [0, 1] \quad with \quad \sum\nolimits_{p' \in A} P_{react}(a, p') = 1 \quad \forall a \in A. \tag{1}$$

Further, we assume that arriving customers base their buying decision on the two current competitors' prices. As demand learning is not in focus, we assume that the customer's behavior is known or has already been estimated. In this context, we use a logistic demand model to describe an example behavior of customer's demand, which is based on a real-world dataset, cf. [7]. It returns the probability of having a sale within one unit of time (for a given price pair) according to several features, including price rank or difference to the competitor's price. The overall customer behavior also reflects that lower price levels increase sales. An overview of all features used is given in Table 1. Note, for the price rank the indicator function $1_{\{.\}}$ is used.

From firm 1's perspective the probability that a customer buys a product (within one period) is denoted by $P_{buy}(a,p)$ whereas firm 2's sales probability is $P_{buy}(p,a)$, $a,p \in A$. Note that in our model the sales probability (e.g., of firm 1) depends on (i) the current competitor price (p) and (ii) the price a chosen for one period. Further, it is also affected by (iii) the competitor's price reaction p' and (iv) the reaction delay h (of firm 2). Hence, based on (1), i.e., if p' and h can be *anticipated*, for the first time span $(t,t+h)$ for firm 1 we obtain $\tilde{P}_{buy}^{(h)}(a,p) := h \cdot P_{buy}(a,p)$, $a,p \in A$, and for the remaining share of the period, i.e., $(t+h,t+1)$, we have the sales probability $\tilde{P}_{buy}^{(1-h)}(a,p') := (1-h)\cdot P_{buy}(a,p')$, $a,p' \in A$.

Finally, for each firm the objective is to maximize its expected discounted total future rewards G (over an infinite horizon with discount factor $\delta < 1$ and unit costs $c \geq 0$). Given a current competitor price, firm 1 looks for a response strategy (a_t) that (in the presence of firm 2's strategy (p_t)) maximizes

$$E(G) = \sum_{t=0}^{\infty} \delta^t \cdot (a_t - c) \cdot \left(\tilde{P}_{buy}^{(h)}(a_t, p_{t-1+h}) + \sum_{p_{t+h} \in A} P_{react}\,(a_t, p_{t+h}) \cdot \tilde{P}_{buy}^{(1-h)}(a_t, p_{t+h}) \right) \tag{2}$$

Taking firm 1's perspective and using (i) the assumed buying probabilities based on P_{buy}, (ii) the reaction time h, and (iii) *given* price reaction probabilities P_{react} for the competitor (firm 2), the associated value function $V(p)$, $p \in A$, of

Table 1. Features for a logistic demand model (cp. [8]) to calibrate demand probabilities for a period of length 1; a is the own price (firm 1), p is the price of the competitor (firm 2).

Features	Regressors $x(a,p)$	Coefficients β
Constant/Intercept	1	−3.82
Price rank	$1 + 0.5 \cdot 1_{\{a=p\}} + 1_{\{a>p\}}$	−0.56
Price difference	$a - p$	−0.01
Average market price	$(p+a)/2$	−0.02

the considered duopoly problem, cf. (2), is determined by the Bellman equation, $p \in A$,

$$V(p) = \max_{a \in A} \Big\{ \sum_{p' \in A} P_{react}(a, p') \cdot \Big((a - c) \cdot \Big(\tilde{P}_{buy}^{(h)}(a, p) + \tilde{P}_{buy}^{(1-h)}(a, p') \Big) + \delta V(p') \Big) \Big\} \tag{3}$$

The system (3) can be solved using standard dynamic programming methods (e.g., using value iteration or linear programming). The associated price reaction policy $a^*(p)$ (i.e., how to respond to price p) is determined by the arg max of (3), $p \in A$.

4 Mutual Strategy Adjustments

In this section, we evaluate mutual strategy adjustments of both firms based on (3) under full knowledge (Sect. 4.1) and partial knowledge from single price observations (Sect. 4.2).

4.1 Iterating Mutual Observable Response Strategies

In the following, we let both firms optimally adjust their strategies (using (3)) in order to identify equilibrium strategies. We use the framework described above and consider the following example.

Example 1. Let $c = 3$, $\delta = 0.995$, $h = 0.5$, and $A := \{1, 2, ..., 80\}$. We assume demand $P_{buy}(a, p) = h \cdot e^{x(a,p)'\beta}/(1 + e^{x(a,p)'\beta})$, cf. Table 1, to define $\tilde{P}_{buy}^{(h)}$ and $\tilde{P}_{buy}^{(1-h)}$.

To iterate (non-randomized) optimal mutual strategy adjustments between both (symmetric) firms we use an initial starting strategy denoted by $S^{(0)}(p)$, $p \in A$. Recall, in general, optimal pure response strategies do not have to converge to mutual optimal strategies. Instead, we may obtain repeating cycles of strategy adjustments [6]. However, if the initial strategy is not "too" aggressive, i.e., composed by comparably high prices, the approach can lead to the identification of equilibria. In the context of Example 1, for $S^{(0)}(p) \equiv s_0$, $s_0 \geq 45$, we observe that after 15 iterations the optimal response strategies converge to a pure (symmetric) equilibrium strategy S^*, which is such that no firm has an incentive to deviate from S^*. Figure 1a depicts the iteration process of strategy adjustments. While early iterations are depicted in lighter shades of grey the final equilibrium strategy S^* is shown in green.

The equilibrium strategy is of the following structure. If the competitor's price is either below a certain low price p_{min} or a above a certain large price p_{max}, it is best to react with the upper price level p_{max}. If the competitor's price is slightly under p_{max}, it is optimal to undercut that price by one price unit as long as the competitor's price is above a certain medium price p_{med}. If the competitor's price is below p_{med}, the strategy uses a strong price drop to p_{min}.

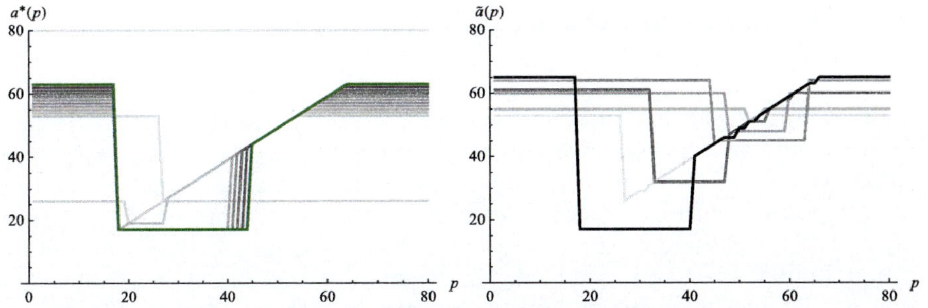

(a) Mutual iterated strategy adaptions (via full knowledge) towards the equilibrium S^* (green)

(b) Response strategies over time (from no initial knowledge in t=0 to 1500 observed periods)

Fig. 1. Comparison of the evolution of self-adapting strategies under full and partial information. Taking firm 1's perspective, the strategies $a^*(p)$ (left window) and $\tilde{a}(p)$ (right window) show how to respond to a given competitor price p, $p \in A$ (lighter shades of grey show earlier results).

Instead of restoring the price level oneself, the price drop forces the competitor to raise the price again (cf. avoiding a race to the bottom). This shifts the price range – in which the undercutting price battle takes place – to a higher level, which is advantageous for both competitors.

4.2 Self-adjusting Strategies Based on Mutual Price Reactions

In this section, we assume that the competitors' strategies are mutually *not* observable and have to be revealed. We let both competitors play an adaptive learning strategy that is based on mutually observed price reactions. We follow the approach described in [8] refraining from reference price effects. The approach balances price exploration and exploitation over time in an incentive-driven framework using an optimistic initiation based on artificially added observations of high price reactions of the competitor. Each firm regularly computes response strategy updates according to (3) taking into account new realized price observations of both firms, i.e., own actions and observed competitor responses, which characterize P_{react}, cf. (1).

To evaluate this approach, we again use the setup as described in Example 1 above. Figure 1b depicts the evolution of strategy adjustments of firm 1 over 1500 periods of time (due to the symmetric structure of the setup the results for firm 2 are similar). Again, early iterations are depicted in lighter shades of grey; the most mature strategy associated with time $t = 1500$ is shown in black.

Comparing Fig. 1b with 1a, we observe that while the evolution of adaptively optimized response strategies is different in the beginning they have clear similarities when looking at later iterations. We find that if the firms have gathered a certain amount of observations, in the second model with less information, cf. Fig. 1b, strategies occur that are of the same structure as the equilibrium strategy S^* of the first model with full information exchange, cf. Fig. 1a.

5 Extensions and Future Work

We have studied subsequent strategy adaptions in a discrete time duopoly model under different information schemes. In case of full knowledge, i.e., if the competitors' pricing strategies are mutually observable, we find that equilibrium strategies can be identified using iterated optimal strategies adaptions. In case the competitor's current strategy is not observable, we showed that single mutual price observations can be used to reveal and steadily update estimations of the competitor's current response behavior. We observe that based on this information regular mutual strategy adjustments can lead to strategy evolutions that are similar to equilibrium strategies under full knowledge, which are in line with classic research for simpler models with deterministic demand, cf. [5].

In future research, we will look for equilibria in mixed strategies and analyze whether it pays off for both firms to exchange information about their strategies in order to identify equilibria earlier. Another direction to gain stability is to extend the state space by including the observed strategy adaptions of the competitor in response to certain own strategy amendments. This may allow to foresee whether and in which way a competitor will change his/her behavior making it possible to directly include these consequences in the firm's own decision-making.

References

1. den Boer, A.V.: Dynamic pricing and learning: historical origins, current research, and new directions. Surv. Oper. Res. Manag. Sci. **20**, 1–18 (2015)
2. Chen, M., Chen, Z.L.: Recent developments in dynamic pricing research: multiple products, competition, and limited demand information. Prod. Oper. Manag. **24**, 704–731 (2015)
3. Kastius, A., Schlosser, R.: Dynamic pricing under competition using reinforcement learning. J. Revenue Pricing Manag. **21**, 50–63 (2022)
4. Kephart, J.O., Hanson, J.E., Greenwald, A.: Dynamic pricing by software agents. Comput. Netw. **32**(6), 731–752 (2000)
5. Maskin, E., Tirole, J.: A theory of dynamic oligopoly, II: price competition, kinked demand curves, and edgeworth cycles. Econometrica **56**(3), 571–599 (1988)
6. Schlosser, R., Boissier, M.: Optimal price reaction strategies in the presence of active and passive competitors. In: ICORES 2017, pp. 47–56 (2017)
7. Schlosser, R., Boissier, M.: Dynamic pricing under competition on online marketplaces: a data-driven approach. In: Proceedings of the 24th ACM SIGKDD International Conference on Knowledge Discovery & Data Mining, pp. 705–714 (2018)
8. Schlosser, R., Richly, K.: Dynamic pricing under competition with data-driven price anticipations and endogenous reference price effects. J. Revenue Pricing Manag. **18**, 451–464 (2019)
9. Serth, S., et al.: An interactive platform to simulate dynamic pricing competition on online marketplaces. In: EDOC 2017, pp. 61–66 (2017)
10. Talluri, K.T., Van Ryzin, G.J.: The Theory and Practice of Revenue Management. Springer, New York (2006). https://doi.org/10.1007/b139000

Supply Chain and Production Management

Data-Driven Decision Making for Strategic Production Planning in a Brewing Company

Markus Mickein(✉), Matthes Koch, and Knut Haase

Institut für Verkehr, Logistik und Produktion, Universität Hamburg, Hamburg, Germany
Markus.Mickein@uni-hamburg.de

Abstract. Changing consumer preferences present a difficult challenge to the brewing industry and hence necessitate adaptations of the production system due to changed requirements. Beer manufacturing is subject to restrictive dependencies, such as lead times, shelf life, multilevel processes, and storage tank restrictions. Therefore, we propose a multilevel capacitated lot-sizing problem (MLCLSP) that covers brewery-specific constraints. The investigated brewing company produces 220 finished and 100 semifinished products on 13 production and 8 storage resources within 3 production levels. Since the brewery-specific MLCLSP cannot be solved within reasonable computing time in the case at issue, we introduce a priority-based fix-and-relax-and-optimize heuristic. We present a computational study and managerial insights regarding changes in consumer preferences (i.e., additional products with low demand). We prove that the priority-based strategy dominates the standard strategy in solution quality and computation time. Furthermore, we demonstrate the potential of the MLCLSP-based planning approach for strategic decision making by revealing the impact on the entire production system.

Keywords: Strategic planning and management · Production and inventory systems · Decision support systems

1 Introduction

Strategic planning of complex production systems is difficult but of great importance for the brewing industry because of fast-changing consumer preferences. Recently, consumers demand individual (i.e., innovative and unique) beers which result in additional products with low demand. However, large breweries are designed for high production volumes and hence require adaptations. Most material requirement planning (MRP) tools within an enterprise resource planning (ERP) system do not support strategic decision making. Therefore, we propose a multilevel capacitated lot-sizing problem (MLCLSP) that covers brewery-specific constraints. Since the proposed mathematical program cannot be solved within reasonable computing time in the case at issue (i.e., 220 finished products, 100

N. Trautmann and M. Gnägi (Eds.): OR 2021, LNOR, pp. 375–381, 2022.
https://doi.org/10.1007/978-3-031-08623-6_55

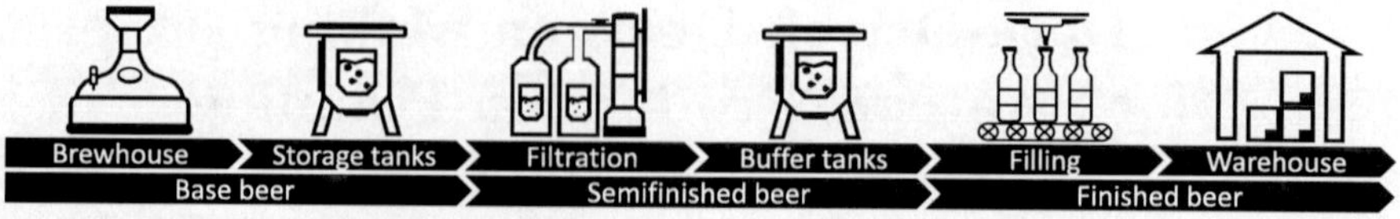

Fig. 1. Production process of a brewery

semifinished products, 13 production resources, 8 storage resources, 3 production levels, and 52 weeks within the planning horizon), we introduce a priority-based fix-and-relax-and-optimize (FRO) heuristic. The priority-based approach first determines a basic production schedule for priority products and then supplements it with secondary products. In this study, we investigate necessary adaptations of the production system due to changed consumer preferences. Furthermore, we demonstrate computational benefits of the priority-based solution strategy for the FRO heuristic.

Figure 1 shows the brewery production process from base beer to finished beer. The brewhouse brews the base beer. The fermentation and maturation processes stay at least two weeks in the storage tanks. Next, the filtration removes yeast and other particles from the base beer. Different recipes (e.g., mixing ratios) result in several types of beer. Before filling, buffer tanks temporarily store the semifinished beer. Finally, the warehouse stores the finished beer.

The MLCLSP has been extensively studied in the literature. However, only a few studies address the brewery production planning problem. Förster et al. [1] propose a capacitated lot-sizing problem that considers availability constraints of reusable bottles in a brewing company. Baldo et al. [2] adopt the synchronized and integrated two-level lot-sizing and scheduling problem (SITLSP) to the brewing industry. Both studies focus on operational and tactical issues. In contrast, our model approach supports strategic decision making, hence the periods are weeks instead of days or hours.

2 Brewery-Specific MLCLSP

Problem Statement

The proposed MLCLSP is based on the general formulation by [3] and the brewery-specific extension by [4]. We take into account limited production and storage capacity but allow the use of overtime and external warehousing. Each product is allocated to one production and one storage resource. We aggregate identical storage tanks to tank groups but assign each liquid to an individual tank. The model takes into account lead times for fermentation and maturation. The maximum permissible duration of advance production of semi-finished beer is one week. We consider setup losses and assume setup times to be sequence-independent. The model considers an initial and final inventory to guarantee the production capability at the beginning of and beyond the planning horizon. Base beer requires production in discrete batch sizes. Furthermore, demand fluctuations necessitate to hold a safety stock to meet service level agreements. To avoid spoilage of perishable goods, we consider the shelf life of finished beers.

Table 1. Set and parameter notation used for the brewery-specific MLCLSP

Indices and index sets		**Parameters**	
$\mathcal{I}, \mathcal{J}$	Set of products with index i, j	b_i	Batch size of product i, for $i \in \mathcal{B}$
$\mathcal{B}$	Subset of base beer products, $\mathcal{B} \subset \mathcal{I}$	$c_{t,r}$	Capacity of resource r in period t
$\mathcal{S}$	Subset of semifinished beer products, $\mathcal{S} \subset \mathcal{I}$	$d_{t,i}$	Demand of product i in period t
$\mathcal{R}$	Set of resources with index r	f_i	Production time for product i
$\mathcal{M}$	Subset of production resources, $\mathcal{M} \subset \mathcal{R}$	g_i	Setup time for product i
$\mathcal{N}$	Subset of storage resources, $\mathcal{N} \subset \mathcal{R}$	h_i	Holding cost of product i
$\mathcal{H}$	Subset of tank groups, $\mathcal{H} \subset \mathcal{N}$	$k_{t,i}$	Given safety stock level of product i in period t
$\mathcal{T}$	Set of periods with index t	$l_{t,i}$	Permitted minimum production quantity of product i in period t
$\mathcal{I}_r$	Set of products i assigned to resource r	$m_{t,i}$	Given demand within shelf life of product i in period t
$\mathcal{J}_i$	Set of products j that are direct successors of product i	n_r	Number of tanks within tank group r, for $r \in \mathcal{H}$
		o_r	Overapacity cost of resource r
		$p_{i,j}$	Quantity of product i required to produce product j
		s_i	Setup cost of product i
		$u_{t,i}$	Permitted maximum production quantity of product i in period t
		v_i	Specific tank volume of product i (depending on assigned tank group)
		w_i	Setup loss of product i

Model Formulation

The mathematical program optimizes inventories $I_{t,i}$, lot-sizes $Q_{t,i}$, and setups $X_{t,i}$ for each period t and product i. In addition, the model determines discrete batch sizes $M_{t,i}$ and the number of used storage tanks $N_{t,i}$ for each period t and base beer product $i \in \mathcal{B}$. Besides, the model calculates the overcapacity $O_{t,r}$ for each period t and resource r. Table 1 lists the used set and parameter notation.

$$\min F = \sum_{t\in\mathcal{T}} \sum_{i\in\mathcal{I}} h_i \cdot I_{t,i} + \sum_{t\in\mathcal{T}} \sum_{i\in\mathcal{I}} s_i \cdot X_{t,i} + \sum_{t\in\mathcal{T}} \sum_{r\in\mathcal{R}} o_r \cdot O_{t,r} \tag{1}$$

s.t.

$$I_{t-1,i} + Q_{t-\lambda_i,i} - \sum_{j\in\mathcal{J}_i} (p_{i,j} \cdot Q_{t,j} + w_j \cdot X_{t,j}) - d_{t,i} = I_{t,i} \qquad \forall i,t \tag{2}$$

$$I_{0,i} \leq I_{T,i} \qquad \forall i \tag{3}$$

$$I_{t,i} \leq Q_{t,i} \qquad \forall i \in \mathcal{S}, t \tag{4}$$

$$Q_{t,i} = b_i \cdot M_{t,i} \qquad \forall i \in \mathcal{B}, t \tag{5}$$

$$Q_{t,i} \leq u_{t,i} \cdot X_{t,i} \qquad \forall i, t \qquad (6)$$

$$Q_{t,i} \geq l_{t,i} \cdot X_{t,i} \qquad \forall i, t \qquad (7)$$

$$I_{t,i} \leq v_i \cdot N_{t,i} \qquad \forall i \in \mathcal{B}, t \qquad (8)$$

$$\sum_{i \in \mathcal{I}_r} N_{t,i} \leq n_r \qquad \forall r \in \mathcal{H}, t \qquad (9)$$

$$\sum_{i \in \mathcal{I}_r} I_{t,i} \leq c_{t,r} + O_{t,r} \qquad \forall r \in \mathcal{N}, t \qquad (10)$$

$$\sum_{i \in \mathcal{I}_r} (f_i \cdot Q_{t,i} + g_i \cdot X_{t,i}) \leq c_{t,r} + O_{t,r} \qquad \forall r \in \mathcal{M}, t \qquad (11)$$

$$I_{t,i} \in [k_{t,i}, m_{t,i}] \qquad \forall i, t \qquad (12)$$

$$X_{t,i} \in \{0, 1\} \qquad \forall i, t \qquad (13)$$

$$O_{t,r}, Q_{t,i} \geq 0 \qquad \forall r, i, t \qquad (14)$$

$$M_{t,i}, N_{t,i} \in \mathbb{N} \qquad \forall i \in \mathcal{B}, t \qquad (15)$$

The objective function (1) minimizes the inventory, setup, and overcapacity costs. The inventory balance (2) requires the fulfillment of the demand in each period from stock or actual production. The lead time λ_i takes into account the fermentation and maturation for base beer. The cycle condition (3) requires to produce the quantity demanded within the planning horizon. Equation (4) ensures a maximum permitted advance production quantity of semifinished products. Equation (5) considers the production of base beer in discrete batch sizes. Equation (6) ensures the setup condition. Equation (7) respects minimum lot sizes. Equations (8) and (9) take the storage tank restrictions into account; (8) determines the required number of tanks, and (9) respects the available number of tanks. The storage capacity constraint (10) considers the limited storage capacity and allows external warehousing. The production capacity constraint (11) includes the production and setup time and allows the use of worker overtime. The variable declaration (12) defines a minimum and maximum inventory level by a given safety stock and demand within shelf life.

3 Priority-Based Fix-and-Relax-and-Optimize Heuristic

To solve the brewery-specific MLCLSP within reasonable computing time, we introduce the priority-based FRO heuristic. First, the fix-and-relax (FR) heuristic generates an initial solution and, second, the fix-and-optimize (FO) heuristic improves this solution [5]. Each subproblem of the FR heuristic considers the integer condition for a small part of integer variables while neglecting the integer condition for the remaining variables [6]. The FO approach decomposes the main problem into smaller subproblems with fewer integer variables to be optimized while fixing the previous solution of the remaining variables [7].

The priority-based solution strategy contains two phases. The first phase determines a basic production schedule for prioritized products. The second

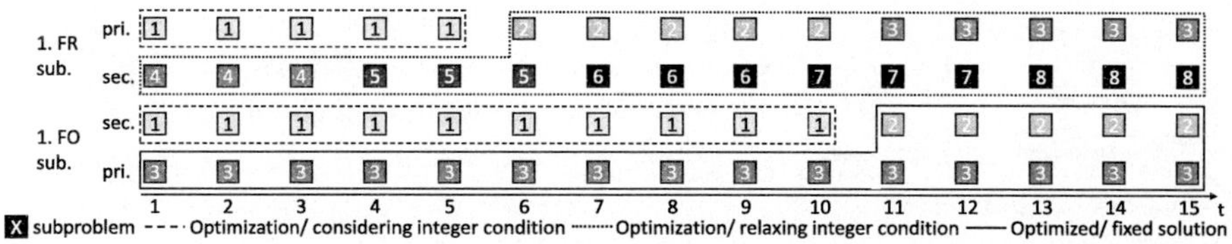

Fig. 2. Procedure of the priority-based FRO heuristic

phase adds secondary products to this basic production schedule. Thereby, we define products with relatively high demand as priority products. Figure 2 shows the procedure of the priority-based FRO heuristic. The first FR subproblem includes the periods one to five of the priority products. The variables within this subproblem consider the integer condition while the remaining variables are relaxed. The first FO subproblem includes the periods one to ten of the secondary products. The variables within this subproblem are optimized again while the remaining variables are fixed. As soon as each subproblem is optimized once, the procedure ends. The given numerical example contains eight FR and three FO subproblems. The priority-based solution strategy also applies to related problems in which attributes can be prioritized.

4 Computational Study and Managerial Insights

We implement the mathematical program and solution approach in GAMS/ CPLEX (v32.1/ v12.10) running on a CPU with 64 cores (4.4 GHz) and 256 GB of RAM.

We investigate the expected consumer preferences represented by four portfolio scenarios. The scenarios differ in additional products (AP) and relative share of additional demand (AD), e.g., in the first scenario, ten additional products lead to an increase in total demand of one percent. Table 2 displays the objective values and computing times depending on the portfolio scenario and solution strategy of the FRO heuristic (i.e., priority-based and standard). The

Table 2. Objective values and computing times depending on the portfolio scenario and solution strategy

Scenario	Parameter	Results (priority strategy)		Results (standard strategy)	
	AP/AD	Obj. v. (T€)	Com. t. (s.)	Obj. v. (T€)	Com. t. (s.)
#0		2,871	975	3,108	932
#1	10/1%	3,495	1,180	3,387	1,236
#2	20/1%	3,726	1,544	3,941	1,690
#3	10/5%	4,125	1,437	4,173	1,740
#4	20/5%	5,978	1,600	5,918	2,154

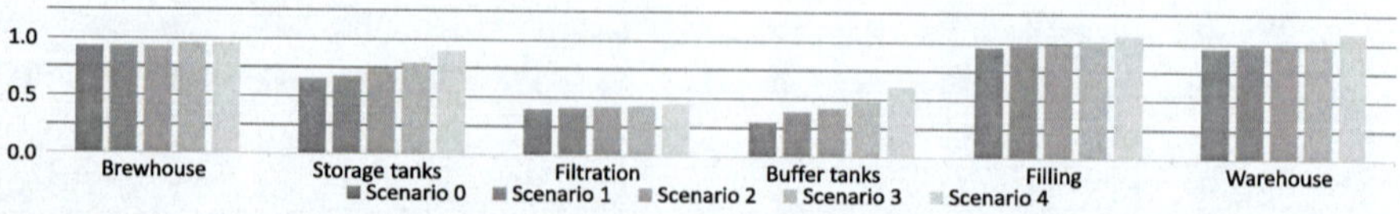

Fig. 3. Capacity utilization for each resource within the multilevel production system

priority-based solution strategy achieved lower costs and computing times compared to the standard solution strategy (i.e., time and product decomposition). Thereby, greater resource scarcity increases the computing time. However, additional demand has a greater impact than additional products. Moreover, the portfolio scenario 4 exhibits the highest costs due to an intensive use of overcapacity.

Figure 3 shows the capacity utilization for each resource within the multilevel production system. The expected consumer preferences affect the entire production system, especially the tank resources due to specific restrictions, e.g., single occupancy. Portfolio scenarios 1 to 3 do not necessitate adaptations of the production system. Therefore, we recommend to introduce those portfolios under consideration of the additional operating costs. However, portfolio scenario 4 exceeds the available capacity of the filling line and warehouse. Nevertheless, this scenario is feasible with acceptance of worker overtime and external warehousing.

5 Conclusion

In this study, we propose a brewery-specific MLCLSP and a priority-based FRO heuristic. We demonstrate that the priority-based strategy dominates the standard strategy in solution quality and computation time. Furthermore, we show that the MLCLSP-based planning approach supports strategic decision making by revealing the impact on the entire production system. The results highlight the importance of multilevel planning to avoid inefficient investments.

References

1. Förster, A., Haase, K., Tönnies, M.: Ein modellgestützter Ansatz zur mittelfristigen Produktions- und Ablaufplanung für eine Brauerei. J. Bus. Econ. **76**(12), 1255–1274 (2006). https://doi.org/10.1007/s11573-006-0061-5
2. Baldo, T.A., Santos, M.O., Almada-Lobo, B., Morabito, R.: An optimization approach for the lot sizing and scheduling problem in the brewery industry. Comput. Ind. Eng. **72**, 58–71 (2014). https://doi.org/10.1016/j.cie.2014.02.0080360-8352
3. Billington, P.J., McClain, J.O., Thomas, L.J.: Mathematical programming approaches to capacity-constrained MRP systems: review. Formul. Probl. Reduct. Manag. Sci. **29**, 1126–1141 (1983). https://doi.org/10.1287/mnsc.29.10.1126

4. Mickein, M., Koch, M., Haase, K.: A decision support system for brewery production planning at Feldschlösschen. J. App. Anal. (2021). https://doi.org/10.1287/inte.2021.1101
5. Stadtler, S., Sahling, F.: A lot-sizing and scheduling model for multi-stage flow lines with zero lead times. Eur. J. Oper. Res. **225**, 404–419 (2013). https://doi.org/10.1016/j.ejor.2012.10.011
6. Stadtler, H.: Multilevel lot sizing with setup times and multiple constrained resources: internally rolling schedules with lot-sizing windows. Oper. Res. **51**, 487–502 (2003). https://doi.org/10.1287/opre.51.3.487.14949
7. Helber, S., Sahling, F.: A fix-and-optimize approach for the multi-level capacitated lot sizing problem. Int. J. Prod. Econ. **123**, 247–256 (2010). https://doi.org/10.1287/opre.51.3.487.14949

The Repair Kit Problem with Fixed Delivery Costs

Christoph Rippe(✉)

Otto von Guericke University Magdeburg, Magdeburg, Germany
christoph.rippe@ovgu.de

Abstract. The repair kit problem is the problem of managing the spare parts inventory of a field service technician. Contrary to most previous contributions we acknowledge that fixed costs per delivery to the technician make up an important share of the total field repair costs. Thus, we treat the replenishment frequency as a decision variable and suggest to manage the content of the repair kit using individual (R, s, S)-policies for each spare part with common review periods R for all parts sourced from the same supplier. We derive a closed-form expression for the job-fill-rate service level and suggest a heuristic to determine the length of the review period(s) as well as the reorder and order-up-to-levels for spare parts carried in the repair kit. Using a numerical experiment we show that lowering the replenishment frequency can lead to a substantial cost reduction. That is because delivery cost reductions outweigh the costs for additional safety stock.

Keywords: Repair kit problem · Inventory management · Stochastic models

1 Introduction

Manufacturers who offer on-site repair services to their customers must provide their service technicians with a set of spare parts called a repair kit. The multiple-job repair kit problem (RKP) is the problem of managing the content of a repair kit that can only be restocked after several customers have been visited in a tour. This problem was first studied by [1–3], who trade off holding costs for spare parts stocked in the repair kit against the service level that can be achieved or against penalty costs incurred for failed repair attempts. [4] consider additional part-specific fixed order costs that are associated with material handling activities in a warehouse. Thus they apply individual (s, S)-policies rather than base stock policies to manage the repair kit. Even though papers on the RKP agree that repair kits are restocked from only one or very few suppliers (e.g. a regional and a national warehouse), the opportunity to reduce delivery costs for shipments from the supplier to the service technicians by means of coordinated replenishment has been largely overlooked. So far the only contribution that considers fixed costs per delivery is a deterministic-demand RKP studied

N. Trautmann and M. Gnägi (Eds.): OR 2021, LNOR, pp. 382–388, 2022.
https://doi.org/10.1007/978-3-031-08623-6_56

by [5]. In this paper, our contribution is to integrate fixed delivery costs into the more realistic stochastic-demand RKP studied by [1–4]. Thus, our contribution is at the intersection of RKP and stochastic joint replenishment problem (SJRP). What distinguishes our problem from papers on the SJRP is that we consider a job-fill-rate service level that requires a more complex analysis than the part-fill-rates used for backorder cost calculations in the SJRP literature (for a review see [8]). We suggest to apply (R, s, S)-policies as introduced by [7] for our RKP and derive closed-form expressions for expected costs and the job-fill-rate service constraint for given policy parameters. Further, we introduce a heuristic solution procedure for our RKP and present a numerical experiment that demonstrates the cost savings potential of longer replenishment cycles. Our work can be seen as an extension to [4], that allows for cycle times different from one. For this reason, we follow their notation wherever possible.

2 Problem Description

There are N different parts in the repair kit. The number of repair jobs per repair tour and the number of units of each part $i = 1, \ldots, N$ required for a single job are stochastic. Let $p_c(l), c_{\min} \leq l \leq c_{\max}$ denote the probability for l customers in one tour and $p_i(k), d_{\min} \leq k \leq d_{\max}$ define the probability that k units of part i are required for a job. Demands for different parts are independent. By $P_c^t(l)$ and $P_i^{J|j}(k)$ we define the probability for l customers in t tours and the probability for a demand of k items of part i in j jobs. We obtain

$$P_c^t(l) = \sum_{\substack{l_1,\ldots,l_t \\ l_1+\cdots+l_t=l}} \prod_{r=1}^{t} p_c(l_r) \quad (1) \qquad P_i^{J|j}(k) = \sum_{\substack{k_1,\ldots,k_j \\ k_1+\cdots+k_j=k}} \prod_{m=1}^{j} p_i(k_m). \quad (1)$$

The repair kit is replenished from G suppliers. Each part $i = 1, \ldots, N$ is sourced from exactly one warehouse with $w(i) = g \in \{1, \ldots, G\}$ defining this warehouse. To manage the content of the repair kit we suggest periodic (s, S)-policies. That means the inventory positions (IPs) of all parts sourced from one supplier g are reviewed every R_g tours and if the IP of a part i is at or below a reorder level s_i at review time it is raised to an order-up-to level S_i. All orders

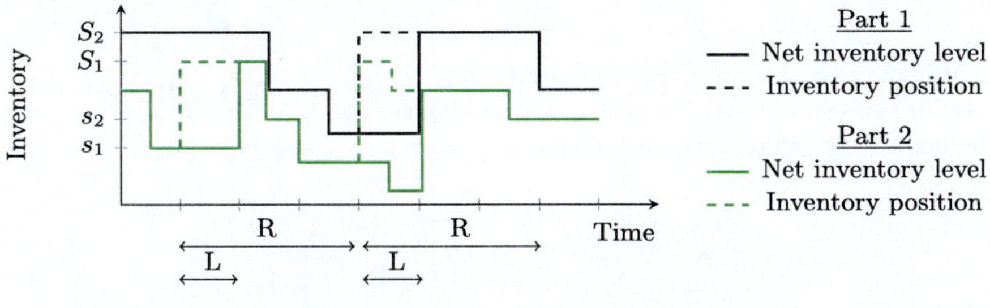

Fig. 1. Exemplary inventory development of two parts from the same supplier

from the supplier are delivered together after a lead time of L_g tours. See Fig. 1. for an example.

We define this policy by (R, s, S) with $R = (R_1, \ldots, R_G)$, $s = (s_1, \ldots, s_N)$ and $S = (S_1, \ldots, S_N)$. Let $\pi_i^{IP}(k)$ define the steady state distribution of part i's IP after order placement at review time. For a detailed derivation of $\pi_i^{IP}(k)$ we refer to [4]. Let us refer to the point in time $L_{w(i)}$ tours after review time as potential delivery time. The net inventory level (IL) at potential delivery time corresponds to the IP at review time minus demand during the lead time. With a review period longer than one tour we need to explicitly consider the IL after each tour within one review period. For a part i with $w(i) = g$ let us denote the steady state distribution of the IL $r = 0, \ldots, R_g - 1$ tours after the last potential delivery time by $\pi_i^{IL|r}(k)$. We obtain

$$\pi_i^{IL|r}(k) = \sum_{l=\max(s_i+1,k)}^{S_i} \pi_i^{IP}(l) \times \sum_{j=(L_g+r)\cdot c_{\min}}^{(L_g+r)c_{\max}} P_i^{J|j}(k)\ P_c^{L_g+r}(j). \tag{2}$$

With h_i defining the unit holding costs for part i per tour we can derive the expected holding costs per tour as

$$EHC = \sum_{i=1}^{N} \frac{h_i \sum_{r=0}^{R_{w(i)}-1} \sum_{k=1}^{S_i} k\ \pi_i^{IL|r}(k)}{R_{w(i)}}. \tag{3}$$

We incur fixed order costs f_i for every order of part i and fixed delivery costs of F_g for every shipment from supplier g. Let P_i^o and $P_i^{o|j}$ denote the probability that an order for part i is placed at review time, regardless of the number of customers during the last review period and given that j customers have been visited respectively.

$$P_i^{o|j} = \sum_{l=s_i+1}^{S_i} \pi_i^{IP}(l) \times \sum_{k=l-s_i}^{j\ d_{\max}} P_i^{J|j}(k) \tag{4}$$

$$P_i^o = \sum_{j=R_g c_{\min}}^{R_g c_{\max}} P_i^{o|j}\ p_c^t(j) \tag{5}$$

Whenever at least one part sourced from supplier g needs to be replenished at review time a delivery from that supplier to the service technician is initiated. The chance P_g^D that a delivery from supplier g is triggered at review time is

$$P_g^D = \sum_{j=R_g c_{\min}}^{R_g c_{\max}} \left(1 - \prod_{i|w(i)=g} \left(1 - P_i^{o|j}\right)\right) P_c^{R_g}(j). \tag{6}$$

Using (5)–(6) we can determine the expected fixed order costs and the expected delivery costs per tour as follows

$$EOC = \sum_{i=1}^{N} \frac{f_i P_i^o}{R_{w(i)}} \tag{7}$$

$$EDC = \sum_{g=1}^{G} \frac{F_g P_g^D}{R_g}. \tag{8}$$

We aim to minimize the sum of holding, order, and delivery costs per tour subject to a job-fill-rate (JFR) constraint. The JFR is defined as the average probability that a repair job can be completed with the spare parts available in the repair kit. As shown earlier the availability of different parts varies significantly across tours within a review period. Let us again use the potential delivery time as a reference point. Then the joint availability of multiple parts depends on the combination of the number of tours that elapsed since the last potential delivery times for each part. The number of possible combinations is given by the least common multiple of the different review periods $R_g, g = 1, \ldots, G$. Thus, we need to consider the average JFR across an observation period of $\text{lcm}(R_1, \ldots, R_G)$ tours to cover all possible cases. Let us assume synchronized potential deliveries at the start of the observation period. Then the chance that job $j = 1, \ldots, c_{\max}$ in tour $T = 0, \ldots \text{lcm}(R_1, \ldots, R_G) - 1$ is completed is

$$p_j^{c,T} = \begin{cases} \prod_{i=1}^{N} \left[p_i(0) + \sum_{l=1}^{S_i} \pi_i^{IL|T \bmod R_{w(i)}}(l) \sum_{k=0}^{l} p_i(k) \right], & j = 1 \\ \prod_{i=1}^{N} \left[p_i(0) + \sum_{l=1}^{S_i} \sum_{n=0}^{S_i - l} \sum_{m=l+n}^{S_i} p_i(l) P_i^{J|j-1}(n) \pi_i^{IL|T \bmod R_{w(i)}}(m) \right], & j > 1. \end{cases} \tag{9}$$

Dividing the expected number of completed jobs by the expected number of jobs during the observation period we obtain the JFR.

$$JFR = \frac{\sum_{T=0}^{\text{lcm}(R_1,\ldots,R_G)-1} \sum_{j=1}^{J} p_j^{c,T} \sum_{l=\max(j,c_{\min})}^{c_{\max}} p_c(l)}{\text{lcm}(R_1, \ldots, R_G) \sum_{l=c_{\min}}^{c_{\max}} l \, p_c(l)}. \tag{10}$$

3 Heuristic Solution Approach

The multi-job RKP presented in Sect. 2 is an integer optimization problem with $2 \cdot N + G$ decision variables. Just like other multi-job problems [1–4] our problem cannot be solved to optimality for real-life-sized repair kits. Even for a fixed combination of review periods $(R_1, \ldots, R_G)$ optimal reorder and order-up-to-levels can only be determined for very small numbers of parts N. For larger problems, the solution must be determined heuristically. To this end, we propose to adapt [4]'s job heuristic (JH) to account for delivery costs and review periods different from one. Their JH calculates reorder and order-up-to-levels jointly. That means in a first step the differences $Q_i := S_i - s_i, i = 1, \ldots, N$ are determined. Given these fixed differences a greedy algorithm is applied to iteratively increase stock levels until the target JFR is reached. Differing from [4] we suggest to determine the quantities Q_i in the following way: Let us first assume constant and continuous demand for each spare part. For a given review period of R_g the time between to consecutive orders for each part with $w(i) = g$ must be $m_i \cdot R_g$, where m_i is an integer multiplier. Under these conditions [6] derives the optimal integer multiplier as

$$m_i^* = \left\lceil \frac{1}{2}\sqrt{1 + \frac{8 f_i}{h_i D_i R_g^2}} - \frac{1}{2} \right\rceil, \tag{11}$$

where D_i is the demand for part i per time unit. Because demand is stochastic in our case we consider $\mathrm{E}[D_i]$. We set the quantity Q_i equal to the expected demand during $m_i^* \cdot R_g$ tours, but at least 1 unit.

$$Q_i = \max(\lfloor m_i^* R_g \mathrm{E}[D_i] \rceil, 1) \tag{12}$$

With the quantities Q_i calculated as described above we apply [4]'s greedy algorithm to determine s and S, obviously using the formulas for the JFR and the expected holding costs derived in Sect. 2. That way we obtain a heuristic solution for a given combination of review periods. For small numbers of suppliers, we can repeat this procedure for all reasonable combinations of review periods, to identify the best one. For instances with 500 parts and 3 suppliers we were able to run this procedure in less than 30 min on a Mac Pro 7.1 with an Intel 24-Core Xeon W-processor.

4 Numerical Experiment

The point of this experiment is to compare a situation in which orders are placed after every tour (as considered by [4]) to situations in which the review period is longer than just one day. We assume there is only one supplier from which all spare parts are sourced. Orders are delivered after a lead time of 2 tours. We consider a repair kit that consists of 500 parts. At most one unit of each part may be required by one customer. The unit demand probabilities for each part are drawn from a continuous uniform distribution on $[0, 0.01408]$. The number of customers per tour ranges from 1 to 3 with the following probabilities: 1: 25%, 2: 70%, 3: 5%. The value of each part is drawn from a log-normal distribution with a mean of 55€ and a standard deviation of 154€. Both, demand and value scenario have been designed to resemble the characteristics of the dataset described by [4]. The fixed order costs are set to 1€ for each part and each order. Using the algorithm described in Sect. 3, we determine reorder and order-up-to levels for all review periods from 1 (daily) to 5 (weekly) given different combinations of fixed delivery cost, holding cost rate, and target job fill rate.

Table 1 shows the review period lengths that yield the least total costs (in brackets) and the corresponding relative cost decreases compared to situations with daily reviews. For most combinations, weekly reviews perform best. Only in case holding cost rates are high and delivery costs are low, it can be beneficial to review the inventory more frequently. In these cases, longer review periods would lead to an increase in holding costs that outweighs any further savings on delivery costs. Across all scenarios, the average possible cost savings are 31.55%, which shows that considering the review period length can lead to substantial benefits. These savings can be attributed to lower delivery costs with longer review periods. This is offset by only a slight increase in holding costs. We find that it takes surprisingly little additional safety stock to counteract the increased demand uncertainty caused by an extended review period. The average total number and the average total value of all units in the repair kit increase by less than 20%, when reviews are conducted weekly rather than daily in all cases.

Table 1. Relative cost savings in % comparing the best review period length (in brackets) to the daily review option

	Fixed delivery costs								
	5			10			20		
	Holding cost rate (in % per year)								
	5	10	20	5	10	20	5	10	20
JFR = 0.8	33 (5)	26 (5)	16 (5)	47 (5)	40 (5)	29 (5)	59 (5)	53 (5)	44 (5)
JFR = 0.95	26 (5)	17 (5)	9 (4)	40 (5)	31 (5)	20 (5)	54 (5)	45 (5)	34 (5)
JFR = 0.99	22 (5)	12 (4)	5 (3)	36 (5)	24 (5)	14 (4)	50 (5)	40 (5)	27 (5)

5 Conclusions

We presented an extension to the multi-job RKP that integrates fixed costs per delivery and treats the review period length as a decision variable. With a numerical experiment, we could demonstrate that in many cases weekly replenishments should be favored over daily replenishments of a service technician. That way delivery costs can be reduced significantly against only a small increase in safety stocks. The latter increase is so moderate that most likely a van suitable for a daily restocked repair kit will also fit the weekly replenished kit.

References

1. Heeremans, D., Gelders, L.: Multiple period repair kit problem with a job completion criterion: a case study. Eur. J. Oper. Res. **81**, 239–248 (1995)

2. Teunter, R.H.: The multiple-job repair kit problem. Eur. J. Oper. Res. **175**, 1103–1116 (2006)
3. Bijvank, M., Koole, G., Vis, I.F.A.: Optimising a general repair kit problem with a service constraint. Eur. J. Oper. Res. **204**, 76–85 (2010)
4. Prak, D., Saccani, N., Syntetos, A., Teunter, R., Visintin, F.: The repair kit problem with positive replenishment lead times and fixed ordering costs. Eur. J. Oper. Res. **261**, 893–902 (2017)
5. Saccani, N., Visintin, F., Mansini, R., Colombo, M.: Improving spare parts management for field services: a model and a case study for the repair kit problem. IMA J. Manage. Math. **28**, 185–204 (2016)
6. Wildeman, R.E., Frenk, J.B.G., Dekker, R.: An efficient optimal solution method for the joint replenishment problem. Eur. J. Oper. Res. **99**, 433–44 (1997)
7. Viswanathan, S.: Note. Periodic review (s, S) policies for joint replenishment inventory systems. Manage. Sci. **43**(10), 1447–1454 (1997)
8. Khouja, M., Goyal, S.: A review of the joint replenishment problem literature: 1989–2005. Eur. J. Oper. Res. **186**, 1–16 (2008)

Disassembly Line Balancing with Collaborative Robots

Christian Weckenborg(✉)

Institute of Automotive Management and Industrial Production, Technische Universität Braunschweig, Mühlenpfordtstr. 23, 38106 Braunschweig, Germany
c.weckenborg@tu-braunschweig.de
https://www.tu-braunschweig.de/aip/pl

Abstract. Against the background of scarce natural resources, the recovery of components and materials from discarded products is increasingly relevant. The disassembly of products, however, must be profitable for the involved actors. Therefore, disassembly is frequently conducted efficiently using disassembly lines. The balancing of such disassembly lines and decisions on the disassembly depth of the affected products are simultaneously considered to maximize profitability. Today, partial disassembly dominates this industry. While this is economically advantageous, a high ecological potential may remain unused. In recent years, collaborative robots can support workers in the manual disassembly of products to further enhance the disassembly processes' efficiency and profitability. However, the increase in efficiency due to the partial automation of disassembly tasks may also result in a higher realized disassembly depth and thus account for ecological advantages. In our contribution, we investigate this effect for an illustrative example using a model-based approach.

Keywords: Disassembly line balancing · Collaborative robots · Cobots · Profit-oriented

1 Introduction

In today's societies, economic growth and wealth on a large scale remain based on the physical production and distribution of goods. The resources required in the production processes are predominantly non-renewable and finite. Therefore, the establishment of a circular economy has become increasingly important in the past years. The recycling of products, their remanufacturing, or alternative end-of-life options may ensure industries' future supply of (secondary) raw material and components with significantly lower environmental impact than their primary equivalents. The disassembly of the products is frequently a prerequisite for further end-of-life options.

Corporations pursuing disassembly activities have to tradeoff the costs of their business (e.g., purchasing the goods to be recycled, employment of workers, and procurement of machines and facilities) and the associated revenues

N. Trautmann and M. Gnägi (Eds.): OR 2021, LNOR, pp. 389–394, 2022.
https://doi.org/10.1007/978-3-031-08623-6_57

(e.g., selling the recovered materials or components) to achieve a profitable business setting. If similar products are disassembled in a high quantity, the use of disassembly lines is efficient. The recyclers simultaneously have to decide on the disassembly depth, i.e., which materials or components to recover from the considered products, and the balancing of the disassembly line, i.e., the allocation of the required disassembly tasks according to the considered disassembly depth among the workers or machines in the available stations. In such settings, partial disassembly might result as economically advantageous for certain products, whereas the complete disassembly allows for recovery of all components and may be ecologically beneficial.

In recent years, novel technologies have been developed to complement or substitute human workers for particular tasks at relatively low costs. Therefore, both economic and ecological criteria of disassembly lines may simultaneously be enhanced. In this contribution, we investigate this effect on an illustrative example considering collaborative robots (cobots) in manual disassembly lines. Cobots can operate next to human workers without additional safety equipment and thus provide a low-threshold means of partial automation. To this end, we discuss the main problem characteristics and refer to the associated literature in Sect. 2. Section 3 discusses our modeling approach. We provide insights by an illustrative example in Sect. 4 and derive the cobots' capabilities to enhance economic and ecological indicators of the disassembly processes. The paper closes with an outlook in Sect. 5.

2 Disassembly Line Balancing with Collaborative Robots

In this section, we describe the problem characteristics of balancing disassembly lines with collaborative robots. Generally, the decisions of the disassembly line balancing problem (DLBP) are similar to those of the assembly line balancing problem (ALBP) [3]. Therefore, the required (dis-)assembly tasks have to be allocated among the opened stations and assigned to deployed resources (e.g., workers or cobots) considering precedence relations restricting the tasks' potential sequence and a given cycle time. The effect of cobots was already evaluated for ALBP [9,10]. However, significant differences among ALBP and DLBP exist, e.g., through an alternative representation of the precedence relations [6] and the option of only partial disassembly [2]. Three main attributes further specify the problem considered in the article at hand.

The first attribute refers to the problem of equipment selection. Generally, contributions assume stations to be homogeneous, i.e., stations are equipped with identical resources. This assumption has to be overcome to allow for alternative resources with different and potentially limited capabilities to be deployed among the stations. In a recent literature review on DLBP, the authors postulate the necessity to increasingly consider opportunities of disassembly automation and identify a lack of research for this attribute [8]. Subsequently, researchers increasingly consider this attribute [11].

The second attribute addresses the necessity of task scheduling within the stations. As cobots can be deployed next to human workers, both worker and cobot can work at the same workpiece simultaneously, i.e., they conduct different disassembly tasks in parallel. Therefore, precedence relations also have to be respected within the stations, which requires scheduling the tasks. The scheduling is generally required for approaches that consider more than one arbitrary resource within a station. In the aforementioned literature review the authors barely find two-sided or multi-manned stations in DLBP approaches [8]. Recent approaches address this attribute [5].

The third attribute refers to the economic character of the problem suggesting a profit-oriented objective [1,2]. To the best of author's knowledge, however, no previous contribution on DLBP investigates the link between economic and ecological indicators and their interrelations with the availability of cobots.

3 Model

This section discusses the required sets, indices, decision variables, the objective function, and the constraints of our mathematical model. The considered goods comprise a set of available disassembly tasks $i, l \in I$. Three types of precedence relations between tasks may restrict the tasks' executability and feasible sequences. If a task has multiple AND-predecessors, each of those needs to be finished before the considered task can start. Among the OR-predecessors of a certain task, at least one has to be finished earlier. Among the OR-successors of a task, only one can be conducted at all. A set of stations $k, h \in K$ may be opened. Each station offers $e \in E$ workplaces, to each of which one resource type $r \in R$ can be assigned. There is a set of components $j \in J$ which disassembly activities can recover. The binary variables x_{ikre} are set to one, if task i is assigned to station k and resource type r in workplace e. Analogously, the binary variables y_{kre} determine the resources allocated to stations and workplaces. The number of opened stations is represented by continuous variable u. The number of recovered components j is evaluated by continuous variables q_j. Binary variables f_{ilk} are used to ensure consideration of precedence relations within stations. f_{ilk} is set to one if task i has to precede task j in station k. To this end, the start time of task i in its respective station is depicted in continuous variables s_i.

The objective function (1) maximizes the achieved profit P^{total}, consisting of revenues, costs of resources, and costs of stations. The revenue per component o_j and the recovered quantities q_j determine the overall revenue per cycle. The costs for resources arise by their allocation among stations and workplaces y_{kre}, the respective cost rates c_r^{R} of resources r per time unit, and the cycle time ct. The costs of stations are based on the number of opened stations u, the cost rate of stations c^{MF} per time unit, and the cycle time ct.

$$\max Z = P^{\text{total}} = \sum_{j \in J} o_j \cdot q_j - ct \cdot \sum_{k \in K} \sum_{r \in R} \sum_{e \in E} c_r^{\text{R}} \cdot y_{kre} - ct \cdot c^{\text{MF}} \cdot u \quad (1)$$

Overall, there are four categories of constraints. In the first category, the general allocation of tasks and resources is described. To this end, constraints ensure that each task is assigned upmost once and to a resource type capable of conducting the task. The number of opened stations is determined by coupling variables u and x_{ikre}. By coupling variables x_{ikre} and y_{kre} the assignment of required resources is ensured. Further constraints ensure that upmost one resource is assigned to each workplace. Finally, the number of recovered components of each type is evaluated. In the second category, compliance of the allocation with the precedence relations is ensured between the stations. To this end, AND-predecessors of a task are forced to an upstream or the same station. Furthermore, it is enforced that at least one among several OR-predecessors is assigned upstream or at the same station as the considered task. For the OR-successors of a task, constraints ensure their allocation to a downstream or the same station. Also, they enforce upmost one of the OR-successor to be assigned at all. The third category comprises the scheduling constraints. One set of constraints ensures that the cycle time is not exceeded. Further constraints enforce tasks assigned to an identic station to be conducted after each other and not in parallel if these tasks are subject to precedence relations or assigned to the same workplace. Finally, the domains of the decision variables are declared in the fourth group.

The resulting model classifies as a mixed-integer linear programming model. It was implemented in Java and solved using the Gurobi 9.1.2 Java API on a standard machine with i7-1180G7 @ 1.30 GHz CPU and 16 GB RAM. For the illustrative example reported in the next section, the models comprise of around 6,200 linear constraints and 1,200 variables. The models are solved optimally in less than 3 s.

4 Illustrative Example

In this section, we modify the illustrative example of the disassembly of PCs. We adopt the precedence relations, components, processing times, and the cycle time as reported by [7] and [1]. As these contributions refer to manual disassembly only, further assumptions have to be made. We assume worker costs of 0.67 EUR/min corresponding to the German wage level in the industrial sector [4]. Based on an investment estimation conducted in [9], we derive costs of 0.26 EUR/min per cobot and 0.09 EUR/min per station. The revenues for the components are based on careful market estimation. We assume the cobots to comprise limited capabilities, i.e., they can only perform Tasks 1, 2, 3, and 4 autonomously, while the human workers can conduct all tasks. Additionally, we assume the cobots to require 1.5 times the workers' processing time. The number of workplaces per station is restricted to $|E| = 2$. The precedence graph of the product, the recoverable components, and their revenues are illustrated in Fig. 1.

We compute optimal line configurations for the manual and partially automated cases to evaluate the economic and ecological potential of the deployment of cobots. In the manual case, only human workers are available and can be deployed among the stations. In the partially automated case, both human workers and cobots are available. The aggregate results are given in Table 1. Accordingly, a higher profit can be achieved in the partially automated case

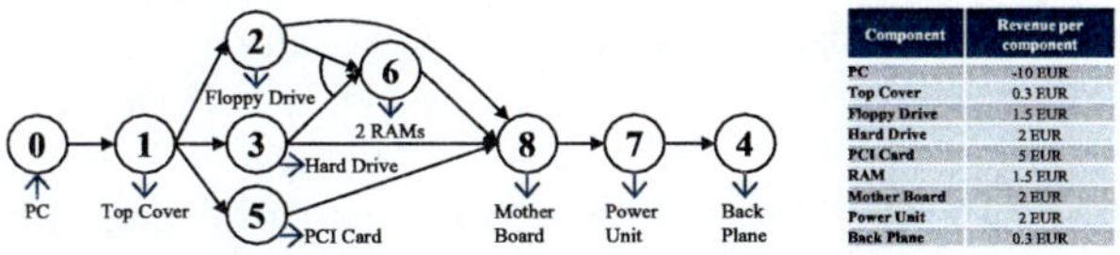

Fig. 1. Precedence graph of the illustrative example, recoverable components, and their revenue

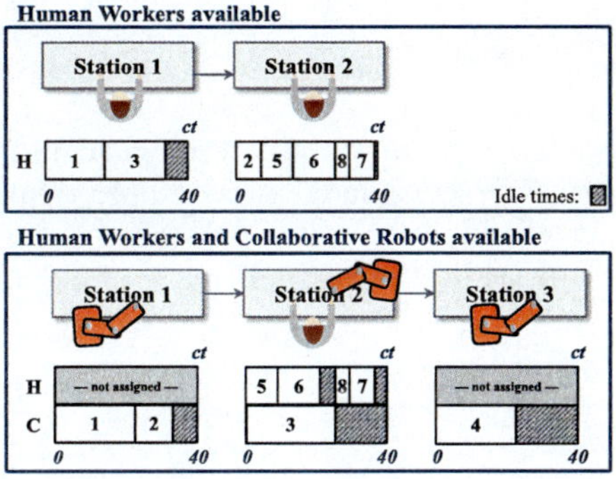

Fig. 2. Illustration of the optimal line configurations with and without the availability of cobots

(4.95 EUR per cycle) than in the manual case (4.79 EUR per cycle). Two workers (# W) are utilized in the manual case at two stations (# S). If cobots (# C) are available, three stations with one worker and three cobots are used. We consider the disassembly depth as a provisional estimation of the ecological quality of the disassembly activities. In the partially automated case, the disassembly depth increases, leading to an economically and ecologically advantageous line configuration compared to the corresponding manual disassembly line.

The resulting line configurations are illustrated in Fig. 2. In the manual case, one worker is assigned to each of the two stations. Consequently, the disassembly tasks are conducted sequentially. However, the Back Plane is not recovered from the PCs since Task 4 is not completed. Opening another station and allocating an additional worker to conduct this task is economically disadvantageous and therefore omitted. Consequently, the disassembly depth remains at 89%. In the partially automated case, cobots perform each of the tasks they are capable of

Table 1. Results of the illustrative example

Resources	Profit	Revenue	Costs	Dis. depth	# W	# C	# S
Workers	4.79 EUR	5.80 EUR	1.01 EUR	89%	2	–	2
Workers & cobots	4.95 EUR	6.10 EUR	1.15 EUR	100%	1	3	3

(Tasks 1–4), and a disassembly depth of 100% is achieved. At Station 2, worker and cobot simultaneously work on the same workpiece. Please note that Tasks 3 and 6 can feasibly be conducted in parallel as Tasks 2 and 3 are OR-Predecessors of Task 6. Therefore, only one the two tasks has to be finished before Task 6 can start. Task 0 serves as a dummy task to assure the purchase of the PCs before recovering their components (not visualized in Fig. 2). In the considered illustrative example cobots serve as an efficient means of partial automation, enhancing economic and ecological indicators of disassembly lines.

5 Outlook

In future, we will develop a more suitable ecological indicator of disassembly activities than the currently considered disassembly depth. Also, more recent industrial examples in combination with a detailed estimation of the market value of components will facilitate the application of our approach.

References

1. Altekin, F.T.: Profit oriented disassembly line balancing. Dissertation (2005)
2. Altekin, F.T., Kandiller, L., Ozdemirel, N.E.: Profit-oriented disassembly-line balancing. Int. J. Prod. Res. **46**(10), 2675–2693 (2008)
3. Boysen, N., Fliedner, M., Scholl, A.: A classification of assembly line balancing problems. Eur. J. Oper. Res. **183**, 674–693 (2007)
4. Eurostat: Hourly labour costs (2021). https://bit.ly/2zge5ky
5. Fang, Y., Liu, Q., Li, M., Laili, Y., Pham, D.T.: Evolutionary many-objective optimization for mixed-model disassembly line balancing with multi-robotic workstations. Eur. J. Oper. Res. **276**, 160–174 (2019)
6. Güngör, A., Gupta, S.M.: A solution approach to the disassembly line balancing problem in the presence of task failures. Int. J. Prod. Res. **39**(7), 1427–1467 (2001)
7. Güngör, A., Gupta, S.M.: Disassembly line in product recovery. Int. J. Prod. Res. **40**(11), 2569–2589 (2002)
8. Özceylan, E., Kalayci, C.B., Güngör, A., Gupta, S.M.: Disassembly line balancing problem: a review of the state of the art and future directions. Int. J. Prod. Res. **57**(15–16), 4805–4827 (2019)
9. Weckenborg, C.: Modellbasierte Gestaltung von Fließproduktionssystemen im Spannungsfeld von Ergonomie und Ökonomie. Springer, Wiesbaden (2021). https://doi.org/10.1007/978-3-658-32888-7
10. Weckenborg, C., Kieckhäfer, K., Müller, C., Grunewald, M., Spengler, T.S.: Balancing of assembly lines with collaborative robots. Bus. Res. **13**(1), 93–132 (2019). https://doi.org/10.1007/s40685-019-0101-y
11. Xu, W., Cui, J., Liu, B., Liu, J., Yao, B., Zhou, Z.: Human-robot collaborative disassembly line balancing considering the safe strategy in remanufacturing. J. Clean. Prod. **324**, 129158 (2021)

Systems Modeling and Simulation

Maintenance of a System Subject to Multiple Degradation Processes with Cox Arrivals

Lucía Bautista[1](✉), Inma Torres Castro[2], and Luis Landesa[3]

[1] Department of Mathematics, School of Technology, Universidad de Extremadura, 10001 Cáceres, Spain
luciabb@unex.es

[2] Department of Mathematics, Faculty of Sport Sciences, Universidad de Extremadura, 10001 Cáceres, Spain
inmatorres@unex.es

[3] Department of Computers and Communications Technology, School of Technology, Universidad de Extremadura, 10001 Cáceres, Spain
llandesa@unex.es

Abstract. A system subject to multiple deterioration is studied. This deterioration is represented through the arrival process and the growth process. The degradation processes arrive at the system with stochastic intensity following a Cox process and they grow according to a homogeneous gamma process. Analytic formulas are developed for the expected intensity, the expected number of arrivals and the survival function.

Keywords: Condition-based maintenance · Cox process · Gamma deterioration · Survival function

1 Introduction

Condition based-maintenance (CBM) is one of the most widely used maintenance strategies. It has many advantages in terms of lowering costs and avoiding unnecessary maintenance tasks compared to maintenance based on the age or use of the system. CBM is based on the system state, which is generally analysed by direct observation or by periodic inspections of the system [5].

A system subject to multiple degradation processes is studied. Two stochastic processes are involved: the arrival or initiation of the degradation processes and the degradation growth process. We suppose that the starting times of the degradation processes follow a Cox process with stochastic intensity $\lambda^*(t)$. The degradation paths of these processes grow according to a homogeneous gamma process. The gamma distribution is a good and suitable probability distribution for modelling continuous and non-decreasing deterioration. This combined model of stochastic Cox and gamma processes is quite realistic for the modelling of real phenomena such as crime, stock market movements, earthquakes or even in epidemics.

N. Trautmann and M. Gnägi (Eds.): OR 2021, LNOR, pp. 397–402, 2022.
https://doi.org/10.1007/978-3-031-08623-6_58

2 Degradation Process

A subsequent external process with arrival times $T_1, T_2, \ldots T_n$ is necessary to be defined for the shot-Cox noise process. Each arrival time has influence on the arrival rate of new degradation processes. For example, supposing that $t > T_1$, The contribution of T_1 to the process for the arrival of new degradation processes at time $T_1 + t$ is

$$h(t) = e^{-\delta(t-T_1)}, \quad \text{with} \quad \delta > 0. \tag{1}$$

The same happens for the times $T_2, \ldots, T_n$. This degradation model is useful to represent processes involving random shocks that continuously deteriorate the system. For example, one can think of a pipe or sheet metal. The effects of corrosion create small pits or cracks randomly distributed over the surface. These cracks become larger and larger, a process that is modelled by gamma decay. In addition, further deterioration of the system leads to the appearance of new cracks, hence the Cox model is appropriate.

Let $T_1, T_2, \ldots, T_n$ be the arrival times of the subsequent Poisson process. Then, the stochastic intensity of the shot-Cox noise process has the following form:

$$\lambda^*(t) = \lambda_0 + \sum_{i=1}^{N(t)} \exp\left(-\delta(t - T_i)\right). \tag{2}$$

Since T_i, for $i = 1, 2, \ldots$ follow an homogeneous Poisson process, conditioning on $\{N(t) = n\}$, the sequence $(T_1, T_2, \ldots, T_n)$ has the same distribution as sequence of uniform variables of size n on the interval $(0, t)$. Hence, the intensity's expectation is easily obtained:

$$\mathbb{E}[\lambda^*(t)] = \sum_{n=0}^{\infty} \mathbb{E}[\lambda^*(t)|N(t) = n]P(N(t) = n) = \lambda_0 + \frac{\mu}{\delta}(1 - \exp(-\delta t)). \tag{3}$$

We recall a general result in stochastic processes, useful for calculating the expected number of the process' arrivals.

Theorem 1. *Let $\lambda^*(t)$ be the stochastic intensity function of a counting process $\{N^*(t), t \geq 0\}$, that fulfils $N^*(0) = 0$. Then,*

$$\mathbb{E}[N^*(t)] = \int_0^t \mathbb{E}[\lambda^*(s)]\, ds. \tag{4}$$

2.1 Description of the Model

The degradation processes initiate following a Cox process. The deterioration level grow follow a gamma process with shape parameter α and scale parameter β. The gamma process is a good choice for modelling the deterioration due it is a continuous and non-decreasing stochastic process with independent increments.

The density function of the gamma distribution with parameters $\alpha > 0$ and $\beta > 0$ is:

$$f_{\alpha,\beta} = \frac{\beta^\alpha}{\Gamma(\alpha)} x^{\alpha-1} \exp(-\beta x), \quad x > 0. \tag{5}$$

In order to control the degradation level of the system, two thresholds are included in the model: the preventive threshold, denoted by M, and the corrective threshold, denoted by L. The system is considered to be failed if some of the degradation processes exceed the corrective threshold.

The deterioration level at time t is denoted by $X(t)$. Suppose that a degradation process starts at time 0 and it grow according to an homogeneous gamma process with parameters α and β.

The first time at which a degradation process exceeds L is defined as the random variable σ_L:

$$\sigma_L = \{\inf t > 0 : X(t) \geq L\}, \tag{6}$$

which follows the probability distribution

$$F_{\sigma_L}(t) = P(X(t) \geq L) = \int_L^\infty f_{\alpha t,\beta}(x) = \frac{\Gamma(\alpha t, \beta L)}{\Gamma(\alpha t)}, \quad t \geq 0. \tag{7}$$

Deterioration processes arrive at the system at times $S_1, S_2, \ldots$, following a shot-Cox noise process. The degradation of each process grow independently following a homogeneous gamma process with parameters α and β. The degradation level of the k-th process at time t is defined as:

$$X_k(t) = X^{(k)}(t - S_k), \quad t \geq S_k, \tag{8}$$

where $X^{(k)}, k \in \mathbb{N}$ are independent and identically distributed homogeneous gamma processes with shape and scale parameters α and β.

The instant of time when the k-th process exceeds the corrective threshold is the random variable W_k, defined as

$$W_k = S_k + \sigma_L \tag{9}$$

With that, the number of degradation processes whose deterioration level exceeds the threshold L at time t is given by

$$N_L(t) = \sum_{k=1}^{\infty} \mathbf{1}_{\{W_k \leq t\}}. \tag{10}$$

As Cox processes are generalizations of Poisson processes, most of the results for Poisson processes can be applied to Cox processes, for example, the conservation of basic operations in point processes [3].

Definition 1. Let $\{N^*(t), t \geq 0\}$ be a counting process with occurrence times $S_1, S_2, \ldots$ and let $\{D_i\}_i$ be, for $i = 1, 2, \ldots$ a sequence of non-negative, independent and identically distributed random variables.

The point process $\{N(t), t \geq 0\}$ with occurrence times $\{S_i + D_i\}$, for $i = 1, 2, \ldots$ is called the *displaced process*.

A well-known result for point processes [4] is the following.

Lemma 1. If $\{N^*(t), t \geq 0\}$ is a Poisson process with intensity $\lambda^*(t)$, then the displaced process $\{N(t), t \geq 0\}$ with occurrence times $\{T_i+D_i\}_i$ is also a Poisson process with intensity

$$\lambda(t) = \int_0^t \lambda^*(t) f(t-u)\, du, \quad t \geq 0, \tag{11}$$

where f is the density function of the variables D_i.

This result can be generalize to Cox processes with the following Lemma.

Lemma 2. If $\{N^*(t), t \geq 0\}$ is a Cox process with stochastic intensity $\lambda^*(t)$, then the displaced process $\{N(t), t \geq 0\}$ with occurrence times given by $\{T_i+D_i\}$ is also a Cox process with stochastic intensity

$$\lambda(t) = \int_0^t \lambda^*(u) f(t-u)\, du, \quad t \geq 0, \tag{12}$$

where f is the density function of the variables D_i.

Using the previous results, we have

Lemma 3. Let $S_1, S_2, \ldots$ be the occurrence times of the shot-Cox noise process with intensity given by Eq. 2. The displaced process $\{N_L(t), t \geq 0\}$ with occurrence times $\{S_i + \sigma_L\}$ is also a Cox process with stochastic intensity

$$\lambda_L(t) = \int_0^t \lambda^*(u) f_{\sigma_L}(t-u)\, du, \quad t \geq 0, \tag{13}$$

where f_{σ_L} is the density function of the variable σ_L, obtained from Eq. 7.

3 Time to the System Failure

We consider that the system has failed when, at least, one of the degradation processes reaches its corresponding corrective threshold. We assume for the sake of simplicity that $L_i = L$ for all $i = 1, \ldots, n$.

Let $W_{[1]}$ be the instant of time at which, for the first time, the degradation level of a process reaches the corrective threshold L:

$$W_{[1]} = \min_{i=1,2,\ldots} \{W_i\} \tag{14}$$

Proposition 1. The survival function of the random variable $W_{[1]}$ is given by:

$$\bar{F}_{W_{[1]}}(t) = \exp\left(-\lambda_0 \int_0^t F_{\sigma_L}(u)du - \mu \int_0^t \left(1 - \exp\left(-\int_0^u e^{-\delta w} F_{\sigma_L}(u-w)dw\right)\right) du\right), \tag{15}$$

where F_{σ_L} is given by Eq. (7).

Proof. Since $\{N_L(t), t \geq 0\}$ is a Cox process with intensity $\lambda_L(t)$. Then,

$$\bar{F}_{W_{[1]}}(t) = P(N_L(t) = 0) = \mathbb{E}\left[\exp\left\{-\int_0^t \lambda^*(u) F_{\sigma_L}(t-u)\, du\right\}\right] \quad (16)$$

Conditioning on $N(t) = n$, we can develop:

$$\begin{aligned}
\bar{F}_{W_{[1]}}(t) &= \sum_{n=0}^{\infty} P(W_{[1]} > t | N(t) = n) P(N(t = n)) \\
&= C_1(t) \exp(-\mu t) + \sum_{n=1}^{\infty} P(W_{[1]} > t) P(N(t = n)) \\
&= C_1(t) \exp(-\mu t) + C_1(t) \exp(-\mu t) \sum_{n=1}^{\infty} \frac{A(t)^n}{t^n} \frac{(\mu t)^n}{n!}, \quad (17)
\end{aligned}$$

where $A(t)$ and $C_1(t)$ are given by

$$A(t) = \int_0^t \exp\left(-\int_0^x e^{-\delta w} F_{\sigma_L}(x-w)\, dw\right) dx. \quad (18)$$

$$C_1(t) = \exp\left(-\lambda_0 \int_0^t F_{\sigma_L}(u)\, du\right), \quad t \geq 0. \quad (19)$$

Substituting the terms of Eq. (18) and Eq. (19) in Eq. (17), the result holds.

4 A Numerical Example

Considering a system subject to multiple degradation processes that start according to the following stochastic intensity:

$$\lambda(t) = 1 + \sum_{i=1}^{N(t)} e^{-0.5(t-T_i)}, \quad t \geq 0, \quad (20)$$

where $N(t)$ is a Poisson process with rate $\mu = 2$ processes per unit time.

The processes' degradation grows according to a gamma process with shape parameter $\alpha = 1.1$ and scale parameter $\beta = 2.5$. The failure threshold is $L = 10$.

A fail is supposed to occur when the degradation level of a process exceeds the failure threshold. The system state is periodically inspected and each maintenance action has an associated cost.

A grid of size 10 is used to find numerically the optimal values for the time between inspections and the preventive threshold that minimize the expected cost rate [1]. The search intervals [0, 10] for the preventive threshold M_{opt} and [0, 25] for the time between inspections T_{opt} are considered. Through Monte-Carlo simulation, with 10,000 iterations, the optimal values obtained are:

$$T_{opt} = 8.53 \qquad M_{opt} = 2.85$$

Figure 1 represents a realization of the intensity function of the shot-Cox noise process, using a thinning algorithm. Note that the graph shows that the arrival of new processes to the system causes the intensity for the following processes to increase.

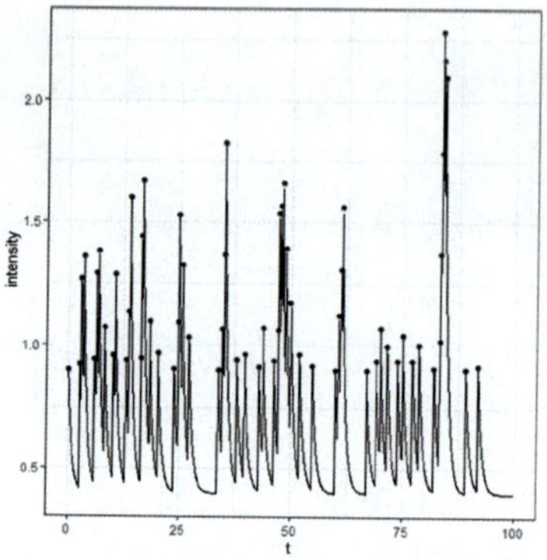

Fig. 1. Intensity function of the shot-Cox noise process.

5 Conclusions

A combined model has been applied to a system subject to multiple degradation processes. The novelty of this work is the use of a stochastic intensity instead of a deterministic one, modelled by a Cox type process introduced in [2].

Regarding to the numerical example, the optimal values for the time between inspections T and the preventive threshold M for this stochastic degradation are obtained. The realization of the intensity of the process is shown graphically. The corrective threshold usually remains fixed to define the system failure.

References

1. Caballé, N., Castro, I.T., Pérez, C.J., Lanza-Gutiérrez, J.: A condition-based maintenance of a dependent degradation-threshold-shock model in a system with multiple degradation processes. Reliab. Eng. Syst. Saf. **134**, 89–109 (2015)
2. Cha, J.H., Finkelstein, M.: On a new shot noise process and the induced survival model. Methodol. Comput. Appl. Probab. **20**, 897–917 (2018). https://doi.org/10.1007/s11009-017-9550-y
3. Serfozo, R.F.: Point processes. In: Heyman, D.P., Sobel, M.J. (eds.) Handbooks in Operations Research and Management Science, Volume 2: Stochastic Models, pp. 1–93. North Holland, Amsterdam (1990)
4. Tijms, H.C.: A First Course in Stochastic Models. Wiley, West Sussex (2013)
5. Wu, S., Castro, I.T.: Maintenance policy for a system with a weighted linear combination of degradation processes. Eur. J. Oper. Res. **280**(1), 124–133 (2020)

Optimization of the Patient Flow in a Forensic Psychiatric Hospital with Discrete Event Simulation

Samuel Kolb[1(✉)], Harold Tiemessen[1], Peter Wermuth[2], and Jörg Forrer[1]

[1] Eastern Switzerland University of Applied Sciences, St. Gallen, Switzerland
{samuel.kolb,harold.tiemessen,jorg.forrer}@ost.ch
[2] Psychiatrische Dienste Aargau, Brugg, Switzerland
peter.wermuth@pdag.ch

Abstract. We consider a forensic psychiatric hospital with multiple wards and forensic commitment patients as well as emergency patients arriving. Forensic commitment patients are awaiting their treatment outside the hospital, and are called in when a treatment place becomes available. Typically, forensic psychiatric hospitals have relatively few treatment places leading to long waiting lists and significant waiting times. Emergency patient arrivals arise due to crisis interventions and occur at random points in time. Forensic psychiatric hospitals are faced with the challenge of matching treatment programmes and security precautions to patients' psychosocial abilities and risk potential while maximizing patient flow in order to cover costs. We develop a discrete event simulation model and identify effective patient allocation strategies to optimize patient flow, considering treatment adequacy, occupancy of all wards, average waiting times inside and outside the hospital and rejection rates of patient arrivals. Based on real-life data we evaluate various future scenarios and provide valuable decision support.

Keywords: Decision support systems · Discrete event simulation · Queuing theory · Forensic psychiatry

1 Introduction

Forensic psychiatry is a subdiscipline of psychiatry and is related to criminology, as it considers legal questions that arise in relation to mentally ill people. This includes consultation with judges or public prosecutors e.g., on culpability or criminal prognosis. A forensic psychiatric hospital offers treatment to diagnosed mentally ill individuals who have come into conflict with the law. Since the committed crime has been in direct relation to symptoms of a severe mental illness, the risk of future delinquency will be reduced if the patient receives adequate therapy. Since the treatment conditions in prison settings are in general not suitable for severe mentally disturbed inmates, there is a general consent that the capacities in specialized forensic clinical facilities in Switzerland must be increased.

N. Trautmann and M. Gnägi (Eds.): OR 2021, LNOR, pp. 403–408, 2022.
https://doi.org/10.1007/978-3-031-08623-6_59

Forensic psychiatric hospitals face the challenge of finding an optimal trade-off between quality of treatment and safety (captured in key performance indicators such as average size of treatment groups, easing level homogeneity within treatment groups, average waiting time until admission, and rejection rate) versus pure economic figures like occupancy rate and total cost.

A very well-known concept to increase occupancy, reduce waiting times and rejection rates and thus reduce cost is to pool (i.e., aggregate) expensive or scarce resources [1]. Pooling is especially beneficial for queuing systems with high variability in arrivals and service times [2]. This implies that forensic psychiatric hospitals can decrease waiting times and rejection rates by reducing the number of wards. In this context, pooling also has a clear disadvantage: fewer wards lead to bigger and more heterogeneous patient groups at each ward. This negatively affects the security and the effectiveness of treatment [3].

Modeling and analysis of patient flow in healthcare systems has been studied extensively. Most studies have applied Markov Decision Processes (MDP) or Discrete Event Simulation (DES) [4]. In contrast to MDP, DES is better suited to handle complex business rules and does not suffer from the so-called curse of dimensionality (= exploding computation times) and is therefore popular in real-life applications. Literature on patient flow in forensic psychiatric hospitals is very scarce. In [5] the authors present a generic DES model for a prison to forecast the composition of the prison population.

Our main contribution consists of three parts:

- We develop a DES model to support patient flow planning in forensic psychiatric hospitals. This is a new application of Operations Research in healthcare. We discuss similarities and fundamental differences between patient flow in forensic psychiatric hospitals and patient flow in the hospital settings.
- We apply the concept of hold back levels from the field of inventory control to design smart patient allocation policies that allow to balance key performance indicators across various subpopulations of forensic patients.
- We present an overview of open logistic challenges in forensic psychiatric hospitals with big opportunities for Operations Research to provide valuable decision support.

2 Problem Description and Model Formulation

A forensic psychiatric hospital accepts patients sentenced to a forensic commitment (forensic commitment patients) and crisis interventions from prisons (emergency patients). Forensic commitment patients are placed on the waiting list after their registration and are admitted to the forensic psychiatric hospital as soon as a place on a compatible ward becomes available.

Depending on the psychosocial skills and risk potential, forensic commitment patients are assigned to easing levels. Each entering forensic commitment patient starts at level 0, which corresponds to a maximum restriction of freedom.

During treatment, most patients make progress and thus climb the levels step by step until they are finally allowed to be released at level 10. However, it also happens that due to bad behavior (drugs, violence, etc.) patients are downgraded. To treat patients adequately, forensic psychiatric hospitals consist of several wards, each exclusively dedicated to treat patients in a preferably small range of predefined easing levels.

The second stream of patients consists of emergency patients. They are either admitted immediately (and stay for a couple of days and then return to prison after stabilization) or they are rejected if no place is available. There is no waiting list and there are no easing levels for emergency patients.

Typically, requests for treatment places exceed available capacity, resulting in long waiting times for forensic commitment patients and significant rejection rates for emergency patients. In some countries (e.g. Switzerland), forensic psychiatric hospitals have contractual agreements with neighboring local governments (Switzerland: "cantons") that guarantee them a minimum number of treatment places (contingents).

This large number of subpopulations of patients, combined with a high occupancy rate make patient flow management extremely challenging. To reduce complexity, forensic psychiatric hospitals often strictly divide total capacity among all subpopulations in an ad-hoc manner.

Model Formulation

Complex business logic together with high randomness in patient arrival process, sojourn times at easing levels and easing level transitions call for a DES approach. The resulting simulator is used to evaluate scenarios defined by the hospital board. The results were obtained with Simio 12 from Simio LLC. Figure 1 shows a flow diagram for emergency and forensic commitment patients.

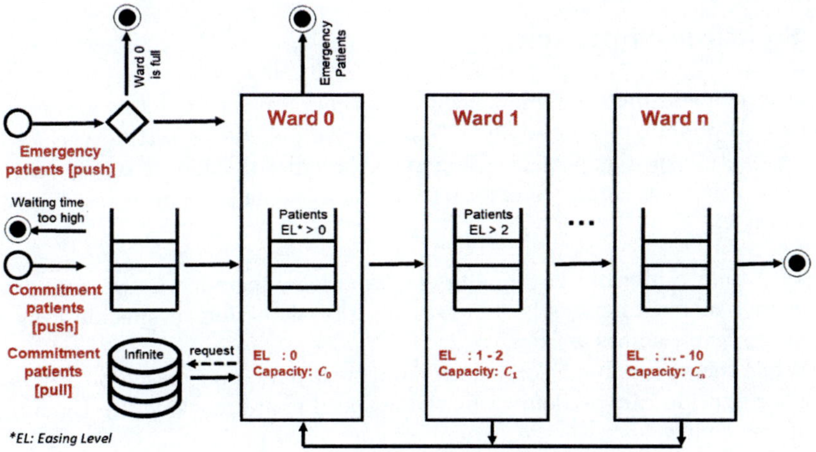

Fig. 1. Flow diagram of patient flow (own illustration)

Requests for emergency patients arrive via the push principle and the hospital must decide whether to accept or reject a patient from a particular canton.

Requests for a free place for forensic commitment patients, on the other hand, are made according to the pull principle and the hospital must decide which patient to call in when a place on a treatment ward becomes available.

Each ward has a waiting list for patients, who have reached an easing level that would allow them to progress to the next ward but cannot be transferred right now due to congestion. Those patients will be transferred as soon as a free place becomes available. During this waiting time, treatment and easing levels development continue, preventing system congestion from leading to increased hospital sojourn times.

The phenomenon that forensic commitment patients waiting outside the hospital for admission usually are removed from the waiting list after 6–18 months for various reasons has been modelled by introducing a randomly generated maximum waiting time. Since no reliable historical information is currently available, we rely on experts to estimate the maximum waiting time distribution. Arrival rates, sojourn times at easing levels and easing level transition probabilities for all subpopulations are estimated based on combination of historical data and expert knowledge.

Although there are obvious similarities between patient flow in a forensic psychiatric hospital and patient flow in a hospital, we also identify some fundamental differences. First, the waiting list with forensic commitment patients in the forensic psychiatric hospital is never empty and occupancy rate is close to 100%. Second, the number of subpopulations of patients with own contractual service agreements (not only emergency and forensic commitments, but also 12 cantons) and the number of easing level transitions is larger than in a regular hospital.

3 Solution Approach

In the developed simulation model, the hospital board is able to test and evaluate different (i) pooling strategies and (ii) admission policies by setting various control parameters such as patient interarrival time distribution, maximum waiting time on waiting list, contingents for patients of a particular canton and treatment time samples.

(i) To evaluate alternative pooling strategies, we include the policy variable quality of treatment, which represents the allocation of specific ranges of easing levels across wards.
(ii) While the admission of forensic commitment patients is based on a dynamic prioritization rule predefined by the hospital board, considering the number of free contingents, the admission of emergency patients is controlled using canton-specific holdback levels. The holdback level of a canton specifies how many emergency places must be free to allow a patient from that particular canton to enter. Via holdback levels we can deal with maximum rejection rate targets that differ per canton and optimize the occupancy rate of the places reserved for emergency patients.

The idea to use holdback levels for emergency patients originates from spare parts logistics where holdback levels are used to differentiate between request from various service contracts [6]. For each service contract i a holdback level h_i is specified, and requests underlying service contract i are only fulfilled if the currently available stock exceeds h_i. For the most expensive service contract (usually the one with the fastest response time) $h_i = 0$ ("send if available"). For service contracts with relatively slow response times, holdback levels are set higher such that parts are only sent if there is enough stock to fulfill future part requests underlying expensive service contracts. This mechanism is also very well suited for this problem setting to guarantee that each canton get its medically desired/contractual agreed service.

4 Numerical Experiments

Strategic Planning
To identify the optimal future operating concept, the board of the forensic psychiatric hospital developed several scenarios that differ in the pooling strategy. All these scenarios have been simulated and evaluated based on waiting time, rejection rate, sojourn times, annual admissions, and occupancy rate. By means of a bottleneck analysis, we identified the most important levers for an optimal patient flow and derived a new pooling scenario that slightly differs from the ones proposed by the board. In this new scenario we assign different easing levels to the wards. Thereby, rejection rates are halved and waiting times are reduced by a quarter.

Operational Planning
Via testing of different holdback levels for emergency patients, we could derive a strategy that maximizes the number of admitted emergency patients from neighboring cantons without significantly compromising the ability to admit patients from the home canton and without exceeding the target average occupancy rate more than 10%.

5 Conclusion

Thanks to the application of DES, different admission strategies and pooling strategies could be evaluated in terms of key performance indicators.

Overall, DES has proven to be a very suitable approach to model a complex system such as a forensic psychiatric hospital, to validate it together with the client and to derive optimal future scenarios.

With the help of the simulation model, we were able to provide the forensic psychiatric hospital with a decision-making basis for the development of the future operating concept. Thereby, we combine the strength of human experts with the power of computers to make the best decisions in complex situations.

Using our decision support system, we were able to derive a better performing pooling scenario than those proposed by the experts with respect to the trade-off between treatment quality and patient flow. Our proposed pooling scenario will be integrated into the operating concept and implemented in the near future. After 1–2 years, an evaluation will take place to review the improvements in patient flow.

Future Work

- Extend DSS with optimization capabilities. User defines target values for the key performance indicators and an optimizer automatically calculates optimal capacities and optimal allocation policy.
- Monitor forensic psychiatric hospital when new facilities open and validate simulation model.
- Recommend European forensic psychiatric hospitals what data and information must be tracked to allow for better parameter estimates and thus more precise simulation models.
- Extend model to capture networks of collaborating forensic psychiatric hospitals.

References

1. Cattani, K., Schmidt, G.M.: The pooling principle. INFORMS Trans. Educ. **5**(2), 17–24 (2005)
2. Anupindi, R., Chopra, S., Deshmukh, S.D., Van Mieghem, J.A., Zemel, E.: Managing Business Process Flows, vol. 400, pp. 102–120. Prentice Hall, Upper Saddle River (1999)
3. Müller, J., et al.: Standards für die Behandlung im Maßregelvollzug nach §§ 63 und 64 StGB: Interdisziplinäre Task-Force der DGPPN. Nervenarzt **88**(1), 1–29 (2017)
4. van de Vrugt, N.M., Schneider, A.J., Zonderland, M.E., Stanford, D.A., Boucherie, R.J.: Operations research for occupancy modeling at hospital wards and its integration into practice. In: Kahraman, C., Topcu, Y.İ (eds.) Operations Research Applications in Health Care Management. ISORMS, vol. 262, pp. 101–137. Springer, Cham (2018). https://doi.org/10.1007/978-3-319-65455-3_5
5. Bartos, B., McCleary, R.: Generic discrete-event simulation model of a prison, pp. 1–10 (2015)
6. Tiemessen, H., Fleischmann, M., Van Houtum, G., Van Nunen, J., Pratsini, E.: Dynamic demand fulfillment in spare parts networks with multiple customer classes. Eur. J. Oper. Res. **228**(2), 367–380 (2013)

Routing in Reinforcement Learning Markov Chains

Maximilian Moll(✉) and Dominic Weller

Universität der Bundeswehr München, Werner-Heisenberg-Weg 39, 85577 Neubiberg, Germany
{maximilian.moll,dominic.weller}@unibw.de

Abstract. With computers beating human players in challenging games like Chess, Go, and StarCraft, Reinforcement Learning has gained much attention recently. The growing field of this data-driven approach to control theory has produced various promising algorithms that combine simulation for data generation, optimization, and often bootstrapping. However, underneath each of those lies the assumption that the problem can be cast as a Markov Decision Process, which extends the usual Markov Chain by assigning controls and resulting rewards to each potential transition. This assumption implies that the underlying Markov Chain and the reward, the data equivalent of an inverse cost function, form a weighted network. Consequently, the optimization problem in Reinforcement Learning can be translated to a routing problem in such possibly immense and largely unknown networks. This paper analyzes this novel interpretation and provides some first approaches to its solution.

Keywords: Reinforcement Learning · Routing

1 Introduction

After the big wave of research into deep learning, interest started to turn more and more towards reinforcement learning (RL). The latter is next to supervised and unsupervised learning one of the three areas of machine learning and studies primarily dynamic problems. Many of its applications and success stories are focussed on solving games, like those on the atari consoles [6], Go [9,10], and StarCraft [12]. However, also more serious applications to OR problems are being studied. Some examples are: optimization of System Dynamics models [7], Combinatorial Optimization [5], traveling salesman problem [4], and many more. In this paper, we propose to turn this idea on its head and introduce a novel approach to solving RL problems based on routing algorithms taken from the toolkit of OR.

2 Background and Related Literature

In RL the agent usually navigates through a state space $\mathcal{S} \subseteq \mathbb{R}^n$ in discrete time steps $t = 1, \ldots, T$ by choosing an action $a_t \in \mathcal{A} \subseteq \mathbb{R}^m$ in a given state $s_t \in \mathcal{S}$.

N. Trautmann and M. Gnägi (Eds.): OR 2021, LNOR, pp. 409–414, 2022.
https://doi.org/10.1007/978-3-031-08623-6_60

For this paper n is assumed to be arbitrary and $m = 1$. A given task consists of a transition distribution

$$\mathbb{P}(s_{t+1}, r_t | s_t, a_t),$$

which gives the probability for reaching the next state s_{t+1} and a reward signal r_t. The goal is then to find a policy $\pi\colon \mathcal{S} \to \mathcal{A}$, which selects an action in each state, that maximizes $\mathbb{E}\left[\sum_{t=1}^{T-1} \gamma^t r_t\right]$ with $\gamma \in (0, 1)$. It should be noted that the transition distribution is assumed to be unknown. Furthermore, it satisfies the Markov property, i.e., the next state and reward only depends on the current state and the selected action. For this reason, this setting is known as Markov decision process.

Approaches to RL can be split into two possible directions: model-based RL, which tries to learn the distribution in some way and then perform planning based on this information, and model-free RL, which tries to get around doing this explicitly. Instead these methods usually try to find some form of approximation to the expected sum of future rewards from a given state on. Finally, there is the class of policy gradient methods, which parametrizes the policy function π directly and thus tries to choose better and better actions by adjusting them. For more details on RL, the interested reader is referred to [11]. The approach presented in the next section could be classified as model-based, as it learns model-dynamics. However, it works quite differently from most such approaches.

3 Routing in Markov Chains

Since the transition distribution satisfies the Markov Property, the RL problem can also be viewed as moving through the underlying Markov chain. To turn the search for the largest return into a shortest path problem, the inverse of the reward between states is interpreted as distance between nodes. However, usually the Markov chains are very large and largely not known.

This transformation is quite immediate as described above in the case of deterministic transitions and discrete state and action spaces. For the continuous setting, one option would be to discretize it first, which is a common approach in RL [11]. In stochastic environments, the Markov chain has to be modified first, depending on the exact nature of stochasticity. One example of such randomness is that the probability for the next state depends on the action and previous state, while the reward only depends on the latter deterministically. In this case, each node in the chain needs to be split in two: The first one was an edge leading outwards for each possible action, the end-nodes of which represent the selected action. From these points on there can then be edges leading to nodes representing various next states, which induces a stochastic jump, which depends overall then on both the previous state and the action.

Thus, the problem of finding optimal behavior can be reduced to that of routing in a large, partially known network. This problem, however, has been studied in the literature and one possible solution is the PHA* algorithm [3], which can be found in Algorithm 1. The key idea, given a current node n, is to split the total path cost, $f(n)$, into costs up to n, $g(n)$, and an estimate for costs from n to the goal, $h(n)$. Finding a good heuristic for this estimation is one of the core challenges and needs to be somewhat tailored to the application.

Algorithm 1. PHA*

procedure PHA*
  $openList = Array[rootNode]$
  $closedList = Array[]$
  while $goalNode$ not expanded **do**
    $bestNode$ = node with lowest *f-value* from $openList$
    remove $bestNode$ from $openList$
    add $bestNode$ to $closedList$
    travel to $bestNode$ ▷ with usage of navigation algorithm
    expand $bestNode$ ▷ explore therefore the children
    $children$ = explored nodes with $bestNode$ as parent
    for c in $children$ **do**
      $f(c) = g(c) + h(c)$
      if c not in $closedList$ **or** $f(c_{closed\,list}) > f(c)$ **then**
        add c to $openList$

4 Experiments

In this section, some first experiments are being reported based on the mountain car task [8] in the OpenAi Gym environment [1]. The goal is to drive a car up a hill, see Fig. 1. However, the car is not strong enough to reach the hill top directly. So, it first needs to drive back up the other slope to gain momentum. The action space is here the discrete decision, which direction to drive in. It should be noted, that the state space is continuous and was thus discretized using 3 (7×7) tile encodings [11]. As mentioned in the last section, a key aspect for the performance of PHA* is the heuristic chosen for $h(n)$.

Here, results on four different choices will be presented: euclidean distance p (position only), euclidean distance p+v (position and velocity), Q-learning, constant heuristic. Using the euclidean distance as heuristic was suggested in literature [2,3] and makes sense, given the physical exploration nature of PHA*. The choice to include the velocity is justified, since building up the right speed is integral to solving the task. The idea behind the Q-learning heuristic is to have a comparison with more traditional RL approaches. It uses an estimate of the expected sum of future rewards. It should be noted that for this first analysis, these values are pre-computed. Finally, a constant heuristic with $h(n) = 1$ was chosen for comparison. The results can be found in Table 1.

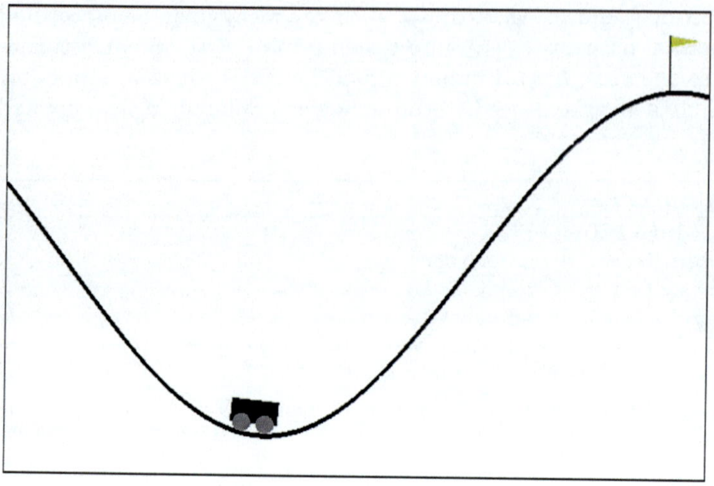

Fig. 1. The mountain car task.

Path length denotes the number of nodes from root goal node, while Euclidean path length sums the euclidean distances of all nodes along the path. Nodes visited (whether total or different) are nodes that have been physically visited during the algorithm, while nodes explored are those, that have been explored, but not necessarily visited. The number of existing states is calculated based on the tiling size and episodes states how many training epochs were used to train the Q-values if needed.

It can be seen, that the final path length agrees between all of them. However, the heuristic using the Euclidean distance of both position and velocity visits much fewer nodes. On the other hand, the simple heuristic needs roughly three times as many visited nodes as the average of the other three. This is not surprising, as other heuristics encode varying degrees of knowledge about the task. In the case of the first two it is the knowledge of the final position (and velocity). In the other case, assuming known Q-values is equivalent to having a solution of the task. It is interesting to observe, that this still leads to more total nodes visited. This is unexpected, since the Q-values already encode so much information and warrants further investigation in the future.

Table 1. Results for experiments

Feature	Euclidean distance p (position)	Euclidean distance p+v (position, velocity)	Q-Values	Simple
Path length	24	24	24	24
Euclidean path length	1.4241	1.4558	1.4558	1.4558
Total nodes visited	1010	762	1634	3560
Total nodes explored	390	345	357	711
Different nodes visited	131	116	120	238
Different nodes explored	149	132	141	252
Number of existing states	441	441	441	441
Episodes	–	–	5000	–

5 Conclusion and Outlook

In this paper some first explorations of a novel approach to RL based on routing in the underlying, largely unknown Markov chain were presented. Initial findings demonstrate that further exploration of this idea would be beneficial. Even to the extent presented in this paper, the approach needs to be tested on a variety of problems. Next, detailed studies of runtime and comparison with existing methods need to be conducted. Particular attention should be payed to handling different styles of tasks, such as various forms of stochasticity. Finally, there needs to be a theoretical analysis of required run time, proportion of the network to be explored and beneficial properties of possible reward structures.

In addition, various exploration approaches should be studied. Since the efficiency of the algorithm depends partially on how quickly certain parts of the graph are being explored, this could lead to significant performance increase. This could include simple random walks as well as more complicated procedures tailored to the RL problem. This could mean for example to retrace certain high-return paths more often and to explore their surrounding.

References

1. Brockman, G., et al.: Openai gym. arXiv preprint arXiv:1606.01540 (2016)
2. Cui, X., Shi, H.: A*-based pathfinding in modern computer games. Int. J. Comput. Sci. Netw. Secur. **11**(1), 125–130 (2011)
3. Felner, A., Stern, R., Ben-Yair, A., Kraus, S., Netanyahu, N.: PHA*: finding the shortest path with A* in an unknown physical environment. J. Artif. Intell. Res. **21**, 631–670 (2004)
4. Hu, Y., Yao, Y., Lee, W.S.: A reinforcement learning approach for optimizing multiple traveling salesman problems over graphs. Knowl. Based Syst. **204**, 106244 (2020)
5. Mazyavkina, N., Sviridov, S., Ivanov, S., Burnaev, E.: Reinforcement learning for combinatorial optimization: a survey. Comput. Oper. Res. **134**, 105400 (2021)
6. Mnih, V., et al.: Playing atari with deep reinforcement learning. arXiv preprint arXiv:1312.5602 (2013)

7. Moll, M.: Towards extending algorithmic strategy planning in system dynamics modeling. In: 2017 IEEE International Conference on Industrial Engineering and Engineering Management (IEEM), pp. 1047–1051. IEEE (2017)
8. Moore, A.W.: Efficient memory-based learning for robot control (1990)
9. Silver, D., et al.: A general reinforcement learning algorithm that masters chess, shogi, and Go through self-play. Science **362**(6419), 1140–1144 (2018)
10. Silver, D., et al.: Mastering the game of go without human knowledge. Nature **550**(7676), 354–359 (2017)
11. Sutton, R.S., Barto, A.G.: Reinforcement Learning: An Introduction. MIT Press, Cambridge (2018)
12. Vinyals, O., et al.: Grandmaster level in StarCraft II using multi-agent reinforcement learning. Nature **575**(7782), 350–354 (2019)

Resource Optimization in Mass Casualty Management: A Comparison of Methods

Marian Sorin Nistor(✉), Maximilian Moll, Truong Son Pham, Stefan Wolfgang Pickl, and Dieter Budde

Universität der Bundeswehr München, 85577 Neubiberg, Germany
Sorin.Nistor@unibw.de

Abstract. This paper studies and compares various optimization approaches ranging from classical optimization to machine learning to respond swiftly and optimally in casualty incidents. Key points of interest in the comparison are the solution quality and the speed of finding it. In multiple-casualty scenarios knowing both is essential to choosing the correct method. A set of 960 synthetic MCI scenarios of different settings are being considered here to give an indication of scalability. For these scenarios, the aim is to optimize the number of victims receiving specialized treatments at the nearest available hospital.

Keywords: Disaster and crisis management · Mass casualty incidents · Optimization · Casualty processing schedule

1 Introduction

The Emergency Medical Services resources, such as vehicles and personnel, are often challenged in mass casualty incidents (MCI). In any MCI response, the process of transporting injured persons to a hospital is the essential component. First, when people are trapped (i.e., trapped in debris or damaged buildings) and in an unstable condition, stabilizing treatment is required before they are released from the trap. Second, the injured are collected and taken to a nearby secure area called the triage area or casualty clearing station (CCS) [1]. In the CCS, casualties are color-coded to place them in the proper categories: red for immediate treatment needed, yellow for required urgent treatment, green can be delayed, and black for dead or missing. Medical treatment (usually first-aid) is often provided at CCS to stabilize injured persons during transport to hospitals. In this paper, we explore three approaches: a mixed-integer linear programming (MILP) formulation, an Iterated Greedy heuristic (IG), and a Genetic Algorithm (GA) to mitigate the mortality rate in synthetic MCI scenarios. This is performed by optimizing the number of casualties receiving specialized care at the nearest available hospital with the required profile, e.g., trauma, pneumonology, or surgery. Each casualty is assumed to be processed by the three operations rescue, first-aid and transportation, before receiving a specialized treatment at

N. Trautmann and M. Gnägi (Eds.): OR 2021, LNOR, pp. 415–420, 2022.
https://doi.org/10.1007/978-3-031-08623-6_61

hospitals. Forecasts and pre-allocation of resources from multiple hospitals in near regions (i.e. less than 120 min for driving) are expected for a large number of casualty requests. A large number of settings with regards to the number of casualties and responding teams is expected to reveal the different strengths and weaknesses of the investigated optimization algorithms.

2 Related Work

A casualty processing schedule problem (CPSP) is typically modeled as a Flexible Job-Shop Schedule Problem (FJSSP), and is optimized by meta-heuristic algorithms, such as GA and IG, to minimize makespan and mortality. In this section, we will briefly discuss some recent studies on applying optimization techniques for CPSP [2] and FJSSP [3–8].

Xiang et al. [2] proposed a triage scheduling optimization approach for MCI response. In this approach, a Markov Chain model was employed to represent the health levels and their stochastic transitions to estimate the probability of death for casualties during the processing time. The authors modeled casualties, the three processing tasks and medical groups as jobs, operations and machines in FJSSP, respectively. A GA was then utilized to find an optimal schedule for the CPSP minimizing mortality. The experimental results suggested that when minimizing the mortality rate, we also achieve a good solution on makespan and vice versa. Lately, Viana et al. [5] introduced the development of GAs for solving JSSP. Traditional GAs can easily fall into local optimal. The authors developed new crossover operators based on local search schemes for the standard GA. The evaluation of 58 instances of literature showed that the developed GA performed effectively on given JSSP scenarios. A comprehensive review on applying GAs to JSSP can be found in [6].

Aqel et at. [7] proposed a modification of the Iterated Greedy (MIG) algorithm in order to create a simple heuristic method for FJSSP. When applying IG to FJSSP, the algorithm consists of two iterative phases: destruct some parts of a current solution and then reconstruct these parts by using greedy techniques, typically the NEH heuristics introduced in [9]. The authors made an MIG by using dispatching rules (DRs) for the constructing phase instead of NEH. Their evaluation showed that MIG could be able to find global optima in most cases. Alternatively, mixed-integer linear programming and constraint programming were applied to solving FJSSP concerning minimization of the total processing time [8]. Based on numerical experiments, the authors suggested that the approaches can deal with small-sized FJSSP and medium-sized ones when considering some constraints.

3 Optimization Approaches for CPSP

3.1 A MILP Formulation

One approach considered for the problem described above is a MILP formulation:

$$\min \sum_{i=1}^{n} z_i \tag{1}$$

$$s_{i2} + d_{i2} \leq W_i + M z_i \qquad \forall i \in [n] \tag{2}$$

$$\sum_{k=1}^{K} x_{ij}^k = 1 \qquad \forall i \in [n], j = 0, 1, 2 \tag{3}$$

$$s_{i(j+1)} \geq s_{ij} + d_{ij} \qquad \forall i \in [n], j = \{0, 1, 2\} \tag{4}$$

$$y_{ijlm}^k \geq x_{ij}^k + x_{lm}^k - 1 \qquad \forall i, l \in [n], j, m = \{0, 1, 2\}, k \in [K] \tag{5}$$

$$a_{ijlm} + b_{ijlm} \leq 1 \qquad \forall i, l \in [n], j, m = \{0, 1, 2\} \tag{6}$$

$$M a_{ijlm}^k + s_{ij} + M(1 - y_{ijlm}^k) \geq s_{lm} + d_{lm}, \qquad \forall i, l \in [n], j, m = \{0, 1, 2\}, k \in [K] \tag{7}$$

$$M b_{ijlm}^k + s_{lm} + M(1 - y_{ijlm}^k) \geq s_{ij} + d_{ij}, \qquad \forall i, l \in [n], j, m = \{0, 1, 2\}, k \in [K] \tag{8}$$

$$s_{ij} \geq 0 \qquad i \in [n], j = 0, 1, 2 \tag{9}$$

$$z_i \in \{0, 1\} \qquad \forall i \in [n] \tag{10}$$

$$x_{ij}^k \in \{0, 1\} \qquad \forall i \in [n], j = 0, 1, 2, k \in [K] \tag{11}$$

$$a_{ijlm}, b_{ijlm}, y_{ijlm}^k \in \{0, 1\} \qquad \forall i, l \in [n], j, m = 0, 1, 2, k \in [K] \tag{12}$$

where $s_{ij} \geq 0$ is the starting time of operation j on casualty i, $z_i \in \{0, 1\}$ indicates the death of casualty i and $x_{ij}^k \in \{0, 1\}$ assigns team k to operation j on casualty i. $a_{ijlm}, b_{ijlm}, y_{ijlm}^k \in \{0, 1\}$ are auxiliary variables, M is a large number and $[n] = \{1, 2, \ldots, n\}$. The objective function (1) simply counts the number of deaths and thus minimizes it, as (2) forces $z_i = 1$ if the last operation cannot be completed before the waiting time. (3) assures that each operation for every casualty is assigned to one team, while (4) makes sure that the next operation on a casualty can only begin after completion of the previous one. Finally, equations (5)–(8) ensure the same thing for operations of the same team. To this end, (5) forces $y_{ijlm}^k = 1$ if both operation j on casualty i and operation m on casualty l are being performed by team k. If this is not the case, the $M(1 - y_{ijlm}^k)$ term will make both Eqs. 7, (8) trivially true. These two equations deal with performing operation j on casualty i first or second respectively. Finally, (6) in conjunction with the Ma_{ijlm}, Mb_{ijlm} terms makes sure, that only one of those cases is non-trivial. It should be noted, that a dependency on the team is not necessary here.

3.2 Genetic Algorithm Approach

GAs start with an initial population of individuals (solutions) in the first generation. At each iteration, a number of individuals is selected (parent) for creating new solutions (children) by performing GA operators. "Good" children will be selected to produce a new population in the next generation. The evolutionary process is controlled by a fitness function until it reaches the optimal target.

In CPSP, GA individuals are designed to represent the casualty processing schedules. A typical GA approach for CPSP can be found in [2].

3.3 Iterated Greedy Approach

We employ the modification of an iterated greedy heuristic (MIG) introduced for FJSSP in [7] for our CPSP scenarios. In this work, a sub set of DRs for selecting responding teams presented in [7] is utilized such as the 2nd, 4th, 5th and 7th rules. For re-sequencing operations, it is simple to insert an operation at the end of the sequence on the selected responding teams instead of using DRs. At each iteration, a candidate can be accepted as a new solution if its performance is not worse than the current one with a ξ parameter. ξ is set to 1 in the case of minimizing mortality.

4 Experiments

We generated 960 synthetic MCI scenarios for our experiments, details for which can be found in Table 1. The processing time of the operations can be varied on different casualties in pre-determined ranges of $[5, 30]$, $[5, 30]$ and $[5, 120]$ in minutes for *rescue*, *first-aid* and *transport*, respectively. We define four different waiting intervals (120, 480, 600, and 720 in minutes) associated with their triage levels that casualties can survive without any treatments.

Table 1. A synthetic dataset of 960 MCI scenarios

Cases of casualty	10 11 12 13 14 15	16 17 18 19 20 25	Total scenarios
Cases of team	2, 3	2, 3, 4, 5	960
No of scenarios	40 for each case	20 for each case	

We evaluate the GA, IG and MILP models on the 960 MCI scenarios. The performance of these models are measured on a metric of three features such as makespan, the number of deaths and running time. When solving the MILP models using CPLEX on the 80 scenarios of 10 casualties with 2 and 3 responding teams, the average, maximum and standard deviation of running time are 11.6, 874.3 and 97.1 min respectively. However, for a larger scenarios, 11 casualties with 2 teams, we get 90.0, 590.8 and 168.6 min on average, maximum and standard deviation of running time. In the two cases, This average and standard deviation of running time are significantly high, particularly the maximum values are much higher than the maximum waiting intervals. Thus, the MILP formulation is more likely to be unusable in its basic form for CPSP, and its performance will not be presented in Sect. 5. However, it has the benefit of demonstrating the optimal solution and some further techniques would have to be investigate to reduce run-time. Due to the space limitation, we report only the performance (average of deaths) of the best approach using GA model in Table 2. The overall results and comparisons are illustrated in Fig. 1.

5 Results and Discussion

Table 2. The average of deaths for 36 settings of MCI scenarios using GA

No of casualty	10	11	12	13	14	15	16	17	18	19	20	25
2 teams	2.15	2.68	4.38	5.45	6.50	7.73	8.50	10.05	11.25	12.40	13.85	19.25
3 teams	1.28	0.98	1.18	1.40	2.88	3.65	3.85	5.15	5.80	7.55	8.65	14.50
4 teams	–	–	–	–	–	–	2.10	1.90	3.05	3.90	4.20	9.85
5 teams	–	–	–	–	–	–	1.35	1.05	1.85	1.70	2.50	6.40

The Table 2 illustrates the average death counts as an aspect of GA performance on the 960 scenarios. Overall, the average of deaths is gradually increasing as the number of casualties increases. However, the mortality rate will fall down as the number of responding teams increases. Both of these are to be expected in the context of this application. To compare the performance between GA and IG on the 960 MCI scenarios, visualizations are being used for better comprehension. Figure 1 shows different performance aspects of GA and IG on the given scenarios. Firstly, two essential aspects of the MCI problem, makespan and the number of deaths, are observed in Sub-figs. 1(a) and 1(b). The Sub-fig. 1(a) shows a strong positive correlation between makespan and casualties for both methods (GA and IG). The makespan produced from GA is slightly higher than that of IG. On the contrary, the average number of deaths is plotted against the responding team resource parameters as in Sub-fig. 1(b). The sub-fig illustrates that the average number of deaths decreases sharply as the number of teams increases. On this aspect, GA outperforms IG in all scenarios. In sub-fig. 1(c), the average running time is then visualized against the MCI size (casualty size * team size). The average running time of GA slightly increases as the MCI size increases and is much higher than that of IG. Fortunately, GA is still considered a promising optimization method for CPSP because its running time (around 300 s) can be acceptable. At the same time, its resulting schedule causes a lower mortality rate than that of IG.

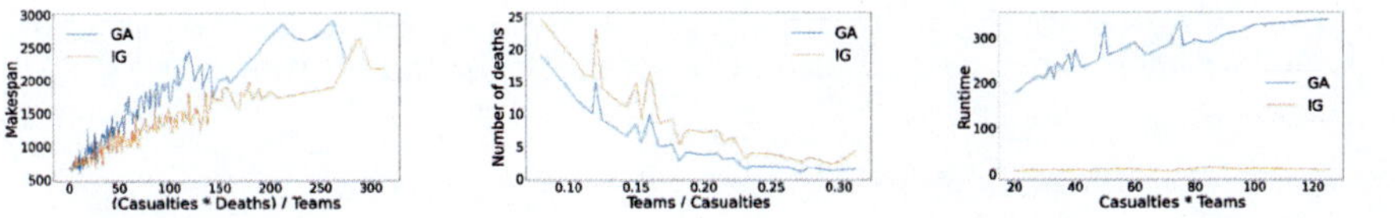

(a) Makespan (s) (b) Death vs team/casualty (c) Runtime (s) vs MCI size

Fig. 1. Different aspects of the performance of GA and IG on the 960 MCI scenarios

6 Conclusion and Future Work

This paper presents a comparison of different optimization algorithms on the problem of casualty processing schedules. In other words, three different performance aspects of GA, IG and CPLEX are evaluated on the 960 synthetic MCI scenarios. The experimental results suggest that GA can be considered a good method for CPSP because its solution can yield the lowest mortality rate with an acceptable running time.

Acknowledgements. This research was sponsored by the NATO Science for Peace and Security Programme under grant SPS MYP G5700.

References

1. Carr, B.G., Caplan, J.M., Pryor, J.P., Branas, C.C.: A meta-analysis of prehospital care times for trauma. Prehospital Emerg. Care **10**(2), 198–206 (2006)
2. Chu, X., Zhong, Q.Y., Khokhar, S.G.: Triage scheduling optimization for mass casualty and disaster response. APJOR **32**(06), 1550041 (2015)
3. Ji, S., Zheng, Y., Wang, Z., Li, T.: A deep reinforcement learning-enabled dynamic redeployment system for mobile ambulances. In: Proceedings of the ACM on Interactive, Mobile, Wearable and Ubiquitous Technologies, vol. 3, issue number 1, pp. 1–20 (2019)
4. Wilson, D.T., Hawe, G.I., Coates, G., Crouch, R.S.: A multi-objective combinatorial model of casualty processing in major incident response. EJOR **230**(3), 643–655 (2013)
5. Viana, M.S., Morandin Junior, O., Contreras, R.C.: A modified genetic algorithm with local search strategies and multi-crossover operator for job shop scheduling problem. Sensors **20**(18), 5440 (2020)
6. Amjad, M.K., et al.: Recent research trends in genetic algorithm based flexible job shop scheduling problems. Math. Probl. Eng. **2018**, 1–32 (2018). https://downloads.hindawi.com/journals/mpe/2018/9270802.pdf
7. Al Aqel, G., Li, X., Gao, L.: A modified iterated greedy algorithm for flexible job shop scheduling problem. CJME **32**(1), 1–11 (2019). https://doi.org/10.1186/s10033-019-0337-7
8. Lunardi, W.T., Birgin, E.G., Laborie, P., Ronconi, D.P., Voos, H.: Mixed integer linear programming and constraint programming models for the online printing shop scheduling problem. Comput. Oper. Res. **123**, 105020 (2020)
9. Nawaz, M., Enscore, E.E., Jr., Ham, I.: A heuristic algorithm for the m-machine, n-job flow-shop sequencing problem. Omega **11**(1), 91–95 (1983)

Influence of Forecast Error and Forecast Bias on Safety Stock on a MRP System with Rolling Horizon Forecast Updates

Wolfgang Seiringer[1(✉)], Fabian Brockmann[1], Klaus Altendorfer[1], and Thomas Felberbauer[2]

[1] Department for Production and Operations Management, University of Applied Sciences Upper Austria, 4400 Steyr, Austria
{Wolfgang.Seiringer,Fabian.Brockmann,Klaus.Altendorfer}@fh-steyr.at

[2] Department of Media and Digital Technologies, University of Applied Sciences St. Pölten, 3100 St. Pölten, Austria
Thomas.Felberbauer@fhstp.ac.at

Abstract. This paper investigates the effects of rolling horizon forecast updates on a production system relying on material requirements planning (MRP). The underlying demand model is the MMFE (martingale model of forecast evolution) model extended by forecast biases revealed certain periods before delivery, i.e. information quality is not strictly increasing as assumed in MMFE. Simulation is applied to model the MRP planning method and the shop floor behavior of a two stage production system including a two level bill-of-materials with 8 finished goods and 4 semi-finished materials. Several scenarios on the demand model parameterization are tested and a finite solution space for the MRP planning parameter safety stock is enumerated to minimize overall costs. In this numerical study, preliminary results to identify the influence of forecast uncertainty on MRP planning parameter safety stock are identified when rolling horizon forecast updates occur.

Keywords: Forecast errors · Production planning · Production order accuracy · Forecast evolution · Simulations

1 Introduction

The MRP (material requirements planning) method is often applied in practice for production planning, therefore, studying different effects on the optimal planning parameters of this method is a relevant field of research. Standard MRP is a deterministic planning approach assuming that demand and shop floor behavior have no stochastic effects. In practice, it is usually applied in a rolling horizon manner to react on changes on the shop floor level and customer demand [1]. For each MRP planning run, demand information is updated which leads to stochastic effects in gross requirements. In general it can be observed, that customer demand

N. Trautmann and M. Gnägi (Eds.): OR 2021, LNOR, pp. 421–426, 2022.
https://doi.org/10.1007/978-3-031-08623-6_62

information quality decreases for due dates further in the future. Customer forecasts can be biased, meaning an overbooking peak or underbooking trough several periods before delivery, whereby forecast values then gradually (or steeply) move back to the really needed amounts [2]. The martingale model of forecast evolution (MMFE) is a known modeling technique for evolving demand forecasts and is applicable in industry and investigated in inventory theory [3–5]. In the MMFE model, forecasts appear a certain time in advance and are gradually updated in a rolling horizon manner until due date. MRP parameters are planned lead time, lot-sizing rules and safety stock, all of which can be applied to counteract uncertainties [6]. Safety stock and safety lead time (i.e. considering buffer lead times) are effective ways to protect against stochastic demand. Some authors emphasize that lot-sizes and safety stocks should always be computed together to minimize the costs [7]. Other researchers even claim that it is only useful to consider all parameters at the same time [8]. Previous studies have shown the effects of demand uncertainty on unit costs, but with greater focus on lot sizing rules [9]. Zhao et al. [10] evaluated alternative methods of establishing the safety stock for the MPS under demand uncertainties by using the measures of historical forecast accuracy but assume an independent and identical distributed customer demand. Enns [11] discussed the use of planned lead times and safety stocks to mitigate forecast bias and demand uncertainty for a batch production system using MRP.

Multiple studies suggest that there is no analytical method to directly determine the safety stocks in an MRP environment with demand uncertainties [7], therefore, simulation is applied in this study to mimic the MRP planned production system with stochastic shop floor behavior and different stochastic demand model parameterizations whereby inventory and backorder costs are evaluated. Simulation is appropriate since the relation between MRP planning, stochastic demand forecast updates, and shop floor uncertainties cannot be treated in an analytical way. This paper presents preliminary results of optimal safety stock related to demand forecasts. Forecasts are received directly from the customers on a rolling horizon basis and evolve from a long-term forecast to the due date. The effect of different forecast uncertainty levels on the optimal safety stock in a rolling horizon forecast update system with and without forecast bias is studied. After the introduction, the applied demand forecast model is described. Next the production system used for the simulation study is introduced, followed by a description of the simulation experiments, selected results and a conclusion.

2 Demand and Forecast Model Description

To model the forecast behavior of the customer we use the following model. The demand model reflects the customer behavior of changing the amounts with an upcoming due date. This introduces uncertainty into the production system of the supplier. The forecast vector $F_{i,j}$ defines the forecasts for all finish goods (FG) at due date i for j periods before delivery. There also exists a long-term forecast vector μ for all FG. H periods before the due date the customer starts updating the forecast periodically based on a rolling horizon. The forecast updates are then modeled as the follows:

$$F_{i,j} = \mu \, for \, j \in \{H+1, \ldots, \infty\} \, F_{i,j} = F_{i,j-1} + \epsilon_{i,j} \, for \, j \in \{0, \ldots, H\} \quad (1)$$

whereby $\epsilon_{i,j}$ is the forecast update vector for due date i observed j periods before delivery (the updating period is period i-j). In the standard MMFE modelling this update vectors are identically disturbed multivariate normal random vectors with mean 0 (see [4] for more details). To create different customer behavior, the calculation of the update vector is varied. If the customer has unsystematic behavior, the update vector is calculated similar to the standard MMFE. In detail we simplify the stochastic updates to:

$$\epsilon_{i,j} = S_j(\mu, 0, \alpha) \sim N(0, \alpha a_j \mu) \quad (2)$$

whereby $S_j(\mu,\ 0,\ \alpha)$ is a vector of normal distributed random variables with mean 0 and standard variation $\alpha\mu$. Note that if a_j is constant for all j, all forecast updates have the same variance, independent of the periods before delivery j, α defines the level of uncertainty varied in the simulation study. If the customer behavior is systematic, we assume that forecast is several periods before delivery, on average, too high or too low. In this paper the forecast bias changes for periods before delivery j. To realize this, we define the update vector as follows:

$$\epsilon_{i,j} = S_j(\mu, \gamma, \alpha) \sim N(\gamma c_j \mu, \alpha a_j \mu) \quad (3)$$

whereby γ is a scaling factor varied in the simulation study and c_j is a shaping factor. This allows us to create different biased customer behaviors like over- and underbooking in different magnitudes with the same shape for each finish good. Note that a biased forecast is also stochastic, the uncertainty depends on αa_j, and that the biased information update does not necessarily increase information quality. In conclusion: α describes the level of uncertainty, γ the level of bias, a_j the shaping of uncertainty, and c_j the shaping of the bias.

3 Production System Simulation Model

The simulation setting consists of a two-stage production system including a two-level bill-of-materials (BOM) with eight FG and four semi-finished goods, whereby every semi-finished good is converted into two different FGs. The FGs are produced on two different machines with the same processing time 0.002933 periods/piece and setup times of 0.00033 periods/piece for all materials, both times are not deterministic during simulation runs. The long-term forecast vector μ is defined as μ = (200, 400,..., 1600). Each semi-finished good consumes the same raw material which is always available. The production system is continuously available. To evaluate a production system which is under stress due to high customer demand variability meaning irregular order times a low planned capacity utilization is assumed. This allows the production system to mitigate uncertainties with safety stock to hold service level. Therefore the system is designed for a planned capacity utilization of 83.55%, including 8.75% for set-ups. As a result, in 74.8% of the available time materials are produced. Fixed order Period (FOP) with a value of 3 is the selected lot sizing policy. The MRP run is calculated once a period and the forecasts are as well updated once a period. We define the following parameters for the simulation run: The safety stock vector ST for all FG is defined as:

$$ST = x * \mu,\ with\ x \in [0, 0.1, \ldots, 2] \tag{4}$$

We apply the same safety stock related to the long-term forecast for all FGs. For semi-finished goods no safety stock is applied. The planned lead timed is defined with three periods. The used simulation framework implements a discrete event simulation model and uses a customer order agent to mimic the periodic order behavior of updating the customer demand. Within the simulation framework, the standard MRP logic with netting, lot-sizing, offsetting and BOM explosion is applied [12]. The run-time for the iterations of the simulation experiments are set to 1800 periods and 30 replications are used to observe the stochastic in the production system. WIP costs of 0.5 CU/period, FGI costs of 1 CU/period and two levels of backorder costs, i.e. b = 99 and b = 198 CU/period, are applied.

4 Numerical Study

A set of scenarios is defined to answer the research questions and derive some managerial insights. The numerical study is conducted using the previous described production system and customer behavior with varying lognormally distributed interarrival times, that lead on average to 0.85 orders per product and day. The customer behavior is unsystematic in scenarios A1 and A2, therefore, α varies from 0 to 2 with step-size 0.25. The customer behavior is systematic in scenarios B1 and B2, therefore, γ varies from 0 to 2 with step-size 0.25. The following detailed specifications are applied. A1: Basic Scenario $a_j = 0.1$ for all j; $\gamma = 0$; $H = 10$. A2: Scenario with shorter forecast horizon $a_j = 0.1$ for all j; $\gamma = 0$; $H = 5$. B1: Biased forecast with overbooking $\alpha = 1$; $a_j = 0.1$ for all j; $c_j = [-0.05, -0.05, -0.1, 0.1, 0.05, 0.05]$ for $j = [3, 4, 5, 6, 7, 8]$; $H = 10$. B2: Biased forecast with underbooking $\alpha = 1$; $a_j = 0.1$ for all j; $c_j = [0.05, 0.05, 0.1, -0.1, -0.05, -0.05]$ for $j = [3, 4, 5, 6, 7, 8]$; $H = 10$. The obtained simulation results for the unsystematic customer behavior, i.e. A1 and A2, in Fig. 1 show a rising trend for safety stock factor (ssf) for increasing alpha value for all backorder cost rates and scenarios. Consequently more forecast introduced uncertainty in the customer demand requires a higher safety stock to fulfill customer demand and hold service level. Detailed results (omitted here due to space reasons) show that with an increasing α value, the minimal overall costs (inventory + backorder) represented by the ssf also tend to increase. The results of the biased scenarios, see Fig. 2, show that for B1, i.e. overbooking, the optimal safety stock is rather low with no clear trend related to the level of bias. However, for B2, i.e. underbooking, the ssf is rapidly increasing with respect to level of bias. This shows that overbooking implies a certain FGI buffer and, therefore, only few safety stock is necessary and underbooking needs to be hedged by a higher safety stocks. This finding is in line with [11], however, it extends his study since here also rolling horizon forecast updates are investigated.

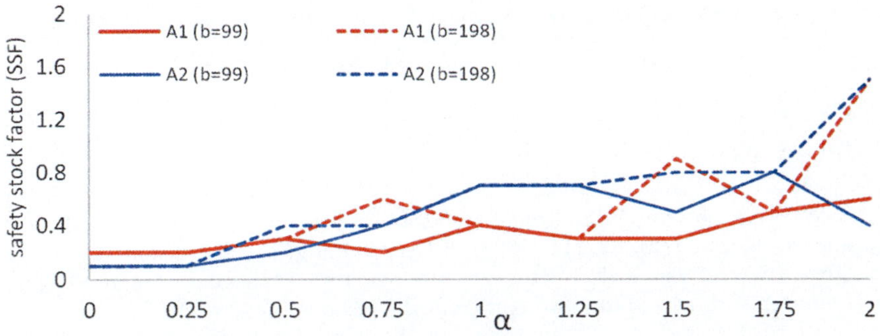

Fig. 1. Optimal safety stock factor unsystematic scenarios A1 and A2.

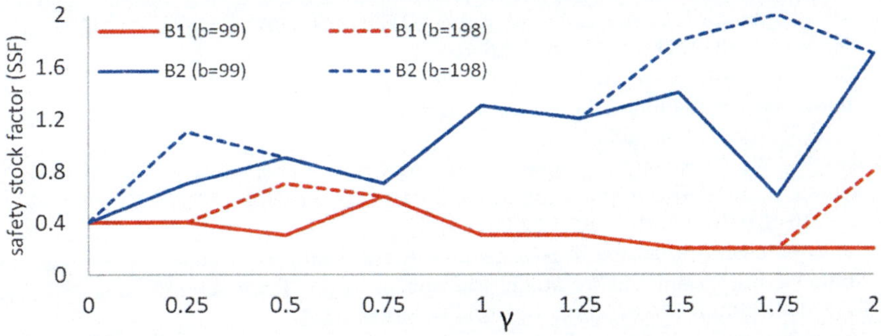

Fig. 2. Optimal safety stock factor for biased scenarios B1 and B2.

5 Conclusion

In this paper the optimal safety stock of finished goods in an MRP planned production system under different rolling horizon forecast update settings was investigated and first preliminary results are shown. In detail, it is assumed that customers update their long-term forecast ten or five periods before delivery with unbiased or biased forecast errors. The MRP planning is conducted each period applying the updated demand values. With a simple enumeration scheme, the safety stock to minimize inventory+backorder costs is identified. The numerical results show that for unbiased forecast updates, a higher uncertainty leads to higher optimal safety stocks and higher overall costs. For biased updates, overbooking already implies a certain buffer and only a low safety stock is necessary, while underbooking needs to be hedged by higher safety stocks. In our study several constraining assumptions were made that have to be relaxed in further research. For example, further studies will investigate the interrelation of the MRP parameters (lead time, lot-size and safety stock) in a broader solution space.

Acknowledgement. The work described in this paper was done within the Produktion der Zukunft Project (InnoFIT, #867471), funded by the Austrian Research Promotion Agency (FFG).

References

1. Tsay, A.A., Lovejoy, W.S.: Quantity flexibility contracts and supply chain performance. MSOM **1**(2), 89–111 (1999). https://doi.org/10.1287/msom.1.2.89
2. Silver, E.A., Pyke, D.F., Peterson, R.: Inventory Management and Production Planning and Scheduling, 3rd edn. Wiley, Chichester (1998)
3. Heath, D.C., Jackson, P.L.: Modeling the evolution of demand forecasts ITH application to safety stock analysis in production/distribution systems. IIE Trans. **26**(3), 17–30 (1994). https://doi.org/10.1080/07408179408966604
4. Norouzi, A., Uzsoy, R.: Modeling the evolution of dependency between demands, with application to inventory planning. IIE Trans. **46**(1), 55–66 (2014). https://doi.org/10.1080/0740817X.2013.803637
5. Chen, L., Lee, H.L.: Information sharing and order variability control under a generalized demand model. Manage. Sci. **55**(5), 781–797 (2009). https://doi.org/10.1287/mnsc.1080.0983
6. Dolgui, A., Prodhon, C.: Supply planning under uncertainties in MRP environments: a state of the art. Ann. Rev. Control **31**(2), 269–279 (2007). https://doi.org/10.1016/j.arcontrol.2007.02.007
7. Thevenin, S., Adulyasak, Y., Cordeau, J.-F.: Material requirements planning under demand uncertainty using stochastic optimization. Prod. Oper. Manage. **30**(2), 475–493 (2021). https://doi.org/10.1111/poms.13277
8. Altendorfer, K.: Effect of limited capacity on optimal planning parameters for a multi-item production system with setup times and advance demand information. Int. J. Prod. Res. **57**(6), 1892–1913 (2019). https://doi.org/10.1080/00207543.2018.1511925
9. Fildes, R., Kingsman, B.: Incorporating demand uncertainty and forecast error in supply chain planning models. J. Oper. Res. Soc. **62**(3), 483–500 (2011). https://doi.org/10.1057/jors.2010.40
10. Zhao, X., Lai, F., Lee, T.S.: Evaluation of safety stock methods in multilevel material requirements planning (MRP) systems. Prod. Plann. Control **12**(8), 794–803 (2001). https://doi.org/10.1080/095372800110052511
11. Enns, S.T.: MRP performance effects due to lot size and planned lead time settings. Int. J. Prod. Res. **39**(3), 461–480 (2001). https://doi.org/10.1080/00207540010002810
12. Hopp, W.J., Spearman, M.L.: Factory Physics, 3rd edn. Waveland Press, Long Grove (2011)

Combining Causal Loop Diagrams, Behavior-Over-Time Graphs, and Narratives to Structure and Explore Complex Decision-Making Situations

Adrian Stämpfli(✉)

Institute of Modeling and Simulation, Eastern Switzerland University of Applied Sciences, 9000 St. Gallen, Switzerland
adrian.staempfli@ost.ch

Abstract. Structuring and exploring complex problems is still one of the most significant challenges in strategic decision-making and management. On the one hand, we strive to add as much rigor as possible to the analyses made, for example, through simulation models or data analyses. On the other hand, we need to stay connected with all kinds of stakeholders - an essential precondition for implementation.

Following up on our earlier research, we present an approach that combines narratives, Causal Loop Diagrams, and Behavior-Over-Time Graphs to illustrate the structure, dynamic patterns and quantitative scale of the problem under study step-by-step, allowing exploration and reflection by a broad audience.

Keywords: Strategic decision-making · Causal Loop Diagrams · Behavior-Over-Time Graphs · System Dynamics · Domain-Specific Languages · R

1 Learning in and About Complex Problems

In a world of growing complexity, the understanding of complex systems is increasingly relevant. Since its beginning [1], one of the key motivations of System Dynamics was to enhance learning in and about complex systems. Scholars have repeatedly stated that flawed mental models, misunderstanding of feedback and the failure to take an endogenous perspective are the main reasons for the misperception of complex systems [2,3]. Therefore, several tools and processes aiming at making (flawed) mental models explicit - so they can be altered towards a better understanding of the complex systems under study - have been promoted. Among them are concepts of participatory modeling (Group Model Building [3] and Community Based System Dynamics [4]). Within those concepts (and generally within all kinds of projects including some participatory model building), some tools are of undisputed value: Causal Loop Diagrams (CLDs) and Behavior-Over-Time Graphs (BOTGs).

N. Trautmann and M. Gnägi (Eds.): OR 2021, LNOR, pp. 427–432, 2022.
https://doi.org/10.1007/978-3-031-08623-6_63

CLDs are a flexible and valuable tool for diagramming the feedback structure of systems. In strategic decision-making and management, CLDs are used to structure and explore complex problems, to foster learning, as a basis for simulation models, and to communicate simulation results. Often we combine CLDs with BOTGs as an initial step to understanding the dynamic patterns and the quantitative scale of the problem under study [4].[1] BOTGs are especially helpful in capturing dynamic, quantitative hypotheses about the problem at hand. As BOTGs illustrate the dynamic and quantitative scale of a variable, feedback loop, or a complete CLD, they help foster thinking about the structure-behavior relationships relevant to the problem under study.

2 How Not to Lose Decision Makers

While those tools prove valuable to engage with stakeholders closely involved in the modeling process [3,4], there is still the question of how we can transfer the possible learnings beyond those core teams. Eric Wolstenholme argues that those having the power to implement are not necessarily the same people that dig their minds deep into mathematical simulation models. He, therefore, argues that System Dynamics should not underestimate the power of qualitative results that might better fit decision-makers mental models than simulation results [5]. He further argues that although this situation might suggest that the quantitative people (analysts) should help the non-quantitative, it is well established that we cannot transfer insights quickly. The implication, therefore, is that we must develop methods to involve everyone in the modeling process to learn these insights for themselves [5]. One way to achieve this is through the usage of visual boundary objects [6]. Visual boundary objects are visual representations that capture the structure and behavior of the system under study. They represent mental models and help participants to stay in touch with each other and the modelers [4,6]. They further help to think differently about projects and serve as what Donald Schön calls generative metaphors [7]. For a visual representation to become a boundary object, the modeling process must find the evolving balance between participants' mental models, models under construction, models already finished, and good System Dynamics practice [4].

3 Combining Causal Loop Diagrams, Behavior-Over-Time Graphs, and Narratives

Following up on our earlier research [8], we thus present an approach that combines Causal Loop Diagrams (CLDs), Behavior-Over-Time Graphs (BOTGs), and narratives to structure and explore complex problems by generating visual representations that are accessible to a broad audience and thus have the potential to become boundary objects. The approach is made accessible through a

[1] Detailed instructions about how we can elicit BOTGs and CLDs in workshop settings can be found on Scriptapedia; https://en.wikibooks.org/wiki/Scriptapedia.

Domain-Specific Language (DSL) implemented in R.[2] The DSL allows generating visual representations of CLDs enriched with BOTGs and textual descriptions. We use a combination of CLDs, BOTGs, and textual descriptions to generate narratives. Narratives describe scenarios of how the system under study may behave under certain circumstances. Those narratives illuminate the structure, dynamic patterns, and quantitative scale of the problem under study step-by-step and thus accessible to a broad audience - including senior decision-makers not deeply involved in the modeling process.

4 Building Narratives Using the DSL

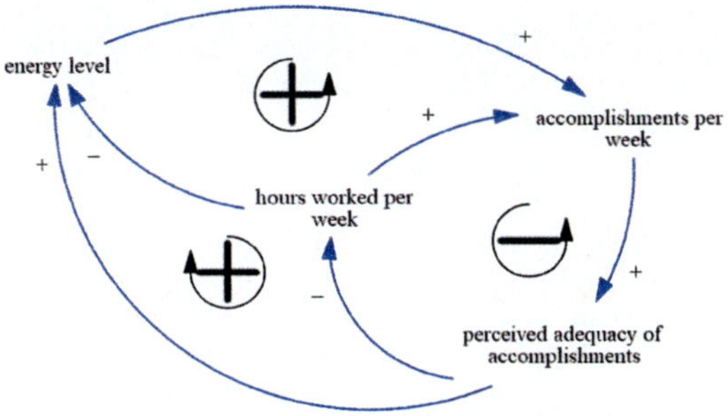

Fig. 1. A slightly simplified version of the worker burnout model [9]

In the current debate on the future and development of labor, maintaining workers' mental health is one of the most significant challenges. Nevertheless, there is still little known about the dynamic interplay between occupational workloads, life situations, and individual coping strategies [10]. In a series of ongoing research projects, we make that complex interplay accessible using the DSL presented here.[3] The target audience is diverse and includes affected individuals and professionals working in the field: Social workers, human resource personnel, psychologists, psychiatrists. The targeted use-cases are twofold: The narratives can be used during one-to-one counseling, and they can be used for educational purposes in classrooms or presentations.

Although this is not the place to discuss a complete example, we nevertheless want to show usage of the DSL by an example close to the domain. The example we are going to discuss in the following is a classic model of the System Dynamics community: Jack B. Homers' worker burnout model [9]. The model

[2] We described some underlying ideas in [8]. We host the code open-source on GitHub. https://github.com/ims-fhs/cld.

[3] Among them are i) a project funded by the Swiss National Science Foundation about psychosocial risks at work; ii) a project funded by Innosuisse - Swiss Innovation Agency aiming at improving ambulant mobile psychotherapy; and iii) a project about common work-life-balance conflicts funded by the Swiss Federal Office for Gender Equality.

explores the dynamics of worker burnout, a process in which a hard-working individual becomes increasingly exhausted, frustrated, and unproductive. The slightly simplified version shown in Fig. 1 consists of one balancing loop and two reinforcing mechanisms mediated through the variable energy level.

Start a Narrative. Assuming the `cld` has been imported to R, we can start a narrative using the following DSL statement:[4]

```
cld %>%
  link('perceived adequacy') %>%
  describe(type = "text", "You (or your boss) are
    unhappy with your accomplishments.") %>%
  plot()
```

The first `link` statement highlights the variable *perceived adequacy of accomplishments*. The `describe` statement adds a textual description and thus adds some context. The plot generated by the DSL is shown in Fig. 2.

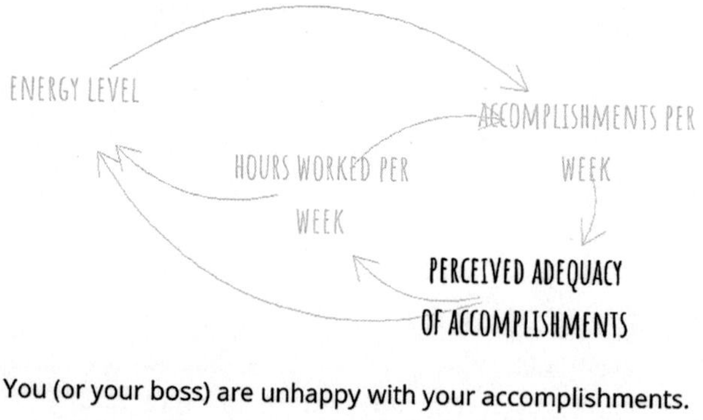

Fig. 2. A first step in explaining the worker burnout model.

Continue the Narrative and Add a Behavior-Over-Time Graph. In a second step we continue the narrative by including the reaction on the initial situation:

```
cld %>%
  link('perceived adequacy' %->% 'hours worked') %>%
  describe(type = "text", "As a reaction you start to work more
    hours per week.") %>%
  describe(type = "ref_mode", 0/.5 %)% .3/.7) %>%
  plot()
```

[4] See [8] for details about the DSL grammar and import options.

Here, the first `link` statement highlights the causal chain. The `describe` statements add a textual description and a BOTG.[5] (Fig. 3)

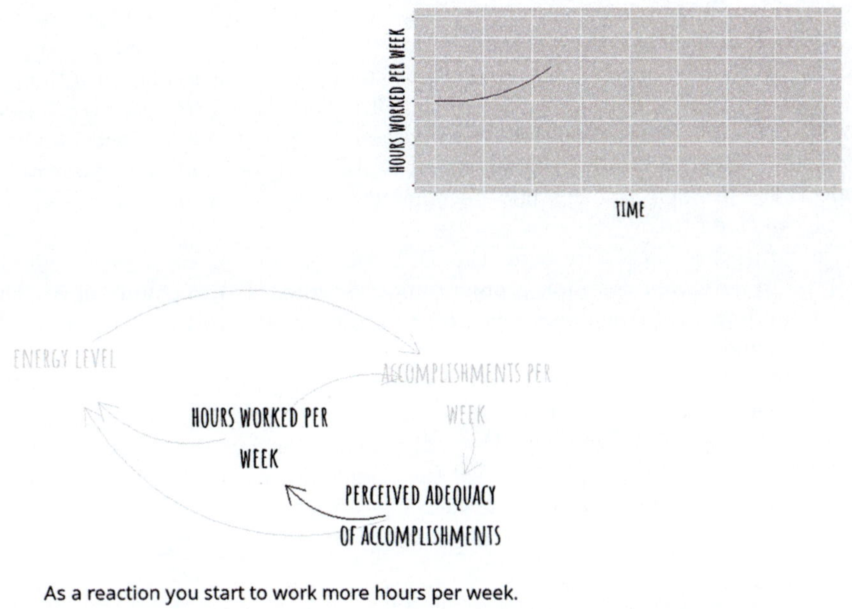

Fig. 3. A second step in explaining the worker burnout model.

A Full Narrative. Adding more steps allows us to discuss the structure, dynamic patterns, and quantitative scale of the CLD discussed in one or many what-if scenarios.[6]

5 Conclusions

Even the most valuable learnings (from modeling projects) do not solve any problems if they are not implemented. Implementation, however, might only happen when stakeholders, from the ones deeply involved in the projects - often not the senior decision-makers - to those only punctually involved - often including the senior decision-makers - build trust in models, simulation results, and identify themselves with those project results. Since qualitative results might better fit decision-makers mental models than simulation results [5], we promote

[5] Defined curve segments are: straight lines (`%-%`), upward (`%(%`) and downward (`%)%`) bent segments, and s-curves (`%s%`). Details in code documentation.

[6] An example of a complete narrative is accessible at https://fhsg.shinyapps.io/burnout/.

narratives to communicate with stakeholders of the non-quantitative kind. Our approach combines the universality of the natural language with more formal CLDs and BOTGs.

This allows communicating systemic complexity through three connected layers: (i) The CLD - visible in all graphics - is at the center. The CLD captures the causal structure of the problem under study. It allows explaining the problematic systemic behavior that needs to be changed in a particular situation; (ii) Through highlighting some aspects of the CLD and adding textual descriptions, the CLD becomes understandable and accessible for a broad audience; and (iii) By adding BOTGs to the visuals, the possible behavior and the quantitative scale of the problem under study become accessible - even for stakeholders not deeply involved.

In numerous client projects, the DSL turned out to be a very valuable tool: (i) to develop a common problem understanding; (ii) to communicate that understanding to stakeholders beyond the project team; (iii) to foster strategic decision-making.

Future research is needed to (i) explore further possibilities enhancing the DSL expressiveness; (ii) evaluate more formally what kind of learnings people draw from CLDs visualized with the DSL; and (iii) how and to what extent the DSL helps to bridge the gap between the people deeply involved in the project and the ones only punctually involved.

References

1. Forrester, J.W.: Industrial dynamics (1961)
2. Sterman, J.D.: Learning in and about complex systems. Syst. Dyn. Rev. **10**(2–3), 291–330 (1994)
3. Vennix, J.A.M.: Group model-building: tackling messy problems. Syst. Dyn. Rev. **15**(4), 379–401 (1999)
4. Hovmand, P.S.: Community based system dynamics (2014)
5. Wolstenholme, E.F.: Qualitative vs quantitative modelling: the evolving balance. J. Oper. Res. Soc. **50**(4), 422–428 (1999)
6. Black, L.J., Andersen, D.F.: Using visual representations as boundary objects to resolve conflict in collaborative model-building approaches. Syst. Res. Behav. Sci. **29**, 194–208 (2012)
7. Schon, D.A.: Generative metaphor and social policy. In: Metaphor and Thought (1979)
8. Stämpfli, A.: A domain-specific language to process causal loop diagrams with R. In: Neufeld, J.S., Buscher, U., Lasch, R., Möst, D., Schönberger, J. (eds.) Operations Research Proceedings 2019. ORP, pp. 651–657. Springer, Cham (2020). https://doi.org/10.1007/978-3-030-48439-2_79
9. Homer, J.B.: Worker burnout: a dynamic model with implications for prevention and control. Syst. Dyn. Rev. **1**(1), 42–62 (1985)
10. Paulus, S.: Gefährdungsbeurteilungen von psychosozialen Risiken in der Arbeitswelt. Zum Stand der Forschung. Zeitschrift für Arbeitswissenschaft **73**(2), 141–152 (2018)

Author Index

N. Trautmann and M. Gnägi (Eds.): OR 2021, LNOR, pp. 433–435, 2022.
https://doi.org/10.1007/978-3-031-08623-6